UNITEXT

# La Matematica per il 3+2

Volume 177

The **UNITEXT - La Matematica per il 3+2** series is designed for undergraduate and graduate academic courses, and also includes books addressed to PhD students in mathematics, presented at a sufficiently general and advanced level so that the student or scholar interested in a more specific theme would get the necessary background to explore it.

Originally released in Italian, the series now publishes textbooks in English addressed to students in mathematics worldwide.

Some of the most successful books in the series have evolved through several editions, adapting to the evolution of teaching curricula.

Submissions must include at least 3 sample chapters, a table of contents, and a preface outlining the aims and scope of the book, how the book fits in with the current literature, and which courses the book is suitable for.

For any further information, please contact the Editor at Springer: Francesca Bonadei, francesca.bonadei@springer.com.

**THE SERIES IS INDEXED IN SCOPUS.**

Paolo Biscari · Tommaso Ruggeri ·
Giuseppe Saccomandi · Maurizio Vianello

# Rational Mechanics

Paolo Biscari
Department of Physics
Politecnico di Milano
Milan, Italy

Tommaso Ruggeri
Department of Mathematics
Università di Bologna
Bologna, Italy

Giuseppe Saccomandi
Department of Engineering
Università degli Studi di Perugia
Perugia, Italy

Maurizio Vianello
(Retired), Department of Mathematics
Politecnico di Milan
Milan, Italy

*Translated by*
Simon G. Chiossi
Universidade Federal Fluminense (UFF)
Niterói, Rio de Janeiro, Brazil

ISSN 2038-5714 ISSN 2532-3318 (electronic)
UNITEXT
ISSN 2038-5722 ISSN 2038-5757 (electronic)
La Matematica per il 3+2
ISBN 978-3-032-07461-4 ISBN 978-3-032-07462-1 (eBook)
https://doi.org/10.1007/978-3-032-07462-1

Translation from the Italian language edition: "Meccanica Razionale" by Paolo Biscari et al., © The Editor(s) (if applicable) and The Author(s), under exclusive license to Springer-Verlag Italia S.r.l., part of Springer Nature 2022. Published by Springer Milan. All Rights Reserved.



Cover illustration: "Gyroscope Arnaldo Pomodoro". *Courtesy* Fondazione Arnaldo Pomodoro

This Springer imprint is published by the registered company Springer Nature Switzerland AG
The registered company address is: Gewerbestrasse 11, 6330 Cham, Switzerland

# Preface

For several years, we have published and refined four editions of an Italian textbook on Rational Mechanics and we now believe that an English version could be valuable for at least two reasons. First, many Italian universities now offer courses taught in English, attended by international students with diverse academic backgrounds, who may require suitable material to fully grasp the subject. Second, while English literature includes many excellent textbooks on "Classical Mechanics", "Theoretical Mechanics", "Analytical Mechanics", "Engineering Mechanics" (and the like), none, we dare say, fully capture the essence of a comprehensive course on Rational Mechanics as it has been traditionally taught in Italian universities for over a century.

This book presents topics that are undoubtedly classical and fundamental to any study of mechanics. As a glance at the table of contents will confirm, it covers essential subjects such as the kinematics of points and rigid bodies, Newton's laws, mass' properties and distribution, balance equations for linear and angular momentum, kinetic and potential energy, conservation principles, statics and dynamics of systems, along with many applications.

However, certain distinctive aspects, both in the style of presentation and the selection of topics, make a textbook on Rational Mechanics unique. This specificity arises from the cultural and didactic context in which the subject is traditionally placed within a student's academic curriculum.

Rational Mechanics is a mathematical discipline structured as rigorously as possible, following a deductive approach with well-defined assumptions, definitions, and proofs. It is typically introduced after at least one course in Mathematical Analysis and one in Physics, where students first encounter the fundamentals of mechanics. Conceptually, it serves as a foundational framework from which students can explore various disciplines, bridging the gap between pure mathematics, applied sciences, and engineering.

For many students, Rational Mechanics represents their first exposure to complex mathematical modeling of real systems. They learn not only its intrinsic importance but also how it serves as a gateway to various fields and applications. The course covers fundamental aspects of classical Newtonian mechanics, including the dynamics of particles, rigid bodies, and systems. Additionally, it introduces Analytical Mechanics (d'Alembert's Principle, Lagrange equations, state space, stability, nonholonomic constraints), some elements of Applied Mechanics (e.g., systems with friction, balancing of rotating masses), and topics leading to Engineering and Continuum Mechanics (e.g., internal stresses, equilibrium of one-dimensional continua such as strings, cables, rods, bars, trusses, and structures). We believe this multifaceted aspect of Rational Mechanics makes it particularly valuable and somewhat unique, with contents which in other cultural traditions are spread out over different subjects and books.

A note on our work as authors: despite our diverse backgrounds and the distinct educational contexts in which we operate, we believe that bringing together differing perspectives fosters a more engaging and less conventional approach, adding elements of originality. We hope this expectation has been at least partially met. We also warmly invite all readers to share their comments and feedback with us freely.

Finally, we would like to express our deepest gratitude to the late Maestro Arnaldo Pomodoro, who generously granted us permission to feature an image of his Gyroscope on the cover of this book. We see this gesture as a meaningful symbol of the connection between Art and Science.

Milan, Italy — Paolo Biscari
Bologna, Italy — Tommaso Ruggeri
Perugia, Italy — Giuseppe Saccomandi
Milan, Italy — Maurizio Vianello
July 2025

**Competing Interests** The authors have no competing interests to declare that are relevant to the content of this manuscript.

# Contents

# Chapter 1
# Point Kinematics

The description of motion is subordinated to the choice of a *reference frame*, which possesses a *time axis* on which an origin has been fixed, and a *frame system* made of three orthonormal vectors $\{\mathbf{i}, \mathbf{j}, \mathbf{k}\}$ and an origin $O$. There are obviously infinitely many bases one can employ in a reference frame, because in three-dimensional Euclidean space there exist infinitely many orthonormal bases.

The *motion* of a point $P$ in a time interval $[t_1, t_2]$ is given by a function that sends each instant $t \in [t_1, t_2]$ to a corresponding position $P(t)$ in space. Once the origin $O$ has been fixed, the position of $P$ is determined by the vector $\mathbf{r}(t) = OP(t)$, called *position vector* (see Fig. 1.1). The components of $OP$ with respect to the orthonormal basis $\{\mathbf{i}, \mathbf{j}, \mathbf{k}\}$ are the *Cartesian coordinates* of $P$ in the chosen frame system. For this reason one usually indicates the components of $OP$ by $(x, y, z)$, which are also functions of time, and one writes $OP(t) = x(t)\mathbf{i} + y(t)\mathbf{j} + z(t)\mathbf{k}$.

The set of points in space occupied by $P$ is called *trajectory* or *orbit* of the motion. If we assume the point's motion is described by a sufficiently regular function $P(t)$ we can suppose the trajectory is a *regular* curve, at least piecewise (see Appendix A.2).

The *velocity* of point $P$ is by definition the derivative of the position vector $OP(t)$, and is indicated by $\mathbf{v}(t)$ for short, or more explicitly by $\mathbf{v}_P(t)$. We shall follow the convention whereby differentiation with respect to *time* is denoted by a dot above the function, so that

$$\mathbf{v}_P(t) = \dot{P}(t) = \frac{dP}{dt} = \dot{x}(t)\mathbf{i} + \dot{y}(t)\mathbf{j} + \dot{z}(t)\mathbf{k}\,. \tag{1.1}$$

In the same way, we define the *acceleration* $\mathbf{a}$ to be the second derivative of $OP(t)$

$$\mathbf{a}_P(t) = \ddot{P}(t) = \frac{d^2P}{dt^2} = \ddot{x}(t)\mathbf{i} + \ddot{y}(t)\mathbf{j} + \ddot{z}(t)\mathbf{k}\,.$$



P. Biscari et al., *Rational Mechanics*, UNITEXT 177,
https://doi.org/10.1007/978-3-032-07462-1_1

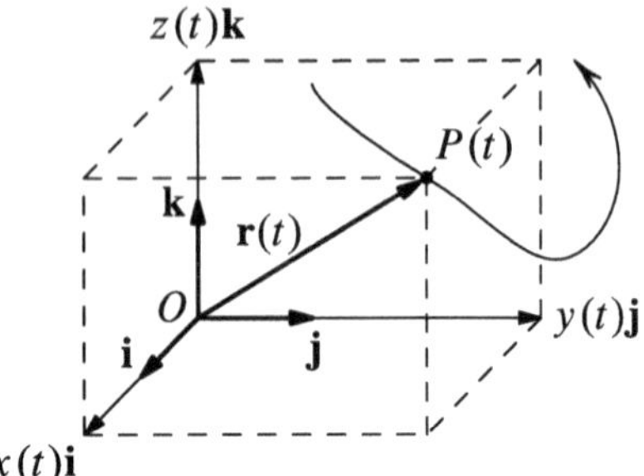


**Fig. 1.1** Motion and trajectory of point $P$

The derivative of the vector $AB$ joining two points in motion equals the difference of the points' velocities, and the second derivative is equal to the difference of the accelerations. In fact, as $OB = OA + AB$, differentiating repeatedly in time gives

$$AB = OB - OA \quad \Longrightarrow \quad \frac{d(AB)}{dt} = \mathbf{v}_B - \mathbf{v}_A \quad \Longrightarrow \quad \frac{d^2(AB)}{dt^2} = \mathbf{a}_B - \mathbf{a}_A \,. \tag{1.2}$$

The *infinitesimal* (or *elementary*) displacement of point $P$ in an infinitesimal time interval $dt$ is indicated by $dP$ and corresponds to the differential of the map $P(t)$, that is

$$dP = \mathbf{v}dt = \big(\dot{x}(t)\mathbf{i} + \dot{y}(t)\mathbf{j} + \dot{z}(t)\mathbf{k}\big)\, dt \,.$$

The *modulus* of such infinitesimal displacement is identified (by definition) with the length of the infinitesimal arc travelled by the point, and is denoted by

$$ds = |dP| = |\mathbf{v}(t)|dt = \sqrt{\dot{x}^2 + \dot{y}^2 + \dot{z}^2}\, dt \,.$$

The quantity

$$s(t) = \int_{t_1}^{t} |\mathbf{v}(\tau)|\, d\tau \,, \tag{1.3}$$

obtained integrating, gives rise to the *arclength time law*, which expresses the length of the finite part of trajectory travelled between time $t_1$ and the generic instant $t$. This length $s$ is called *arclength*. Differentiating (1.3) we find $\dot{s} = |\mathbf{v}(t)| \geq 0$. This implies that over time intervals that only contain isolated zeroes of the map $\dot{s}(t) = |\mathbf{v}(t)|$, the function $s(t)$ will be strictly monotone, and therefore *invertible* (see Fig. 1.2). Hence the instant $t$ itself is completely determined by the arclength travelled by the point along the trajectory by means of the function $t(s)$.

We can also express the point's position as the arclength $s$ varies, instead of using time $t$: for this it suffices to insert in the expression of the motion $P(t)$ the function $t(s)$ to obtain $\hat{P}(s) = P(t(s))$. The mapping $\hat{P}(s)$ expresses the position as function of the arclength $s$ computed starting from the position corresponding to time $t_1$, and hence describes the *trajectory*, though without any information on how the point

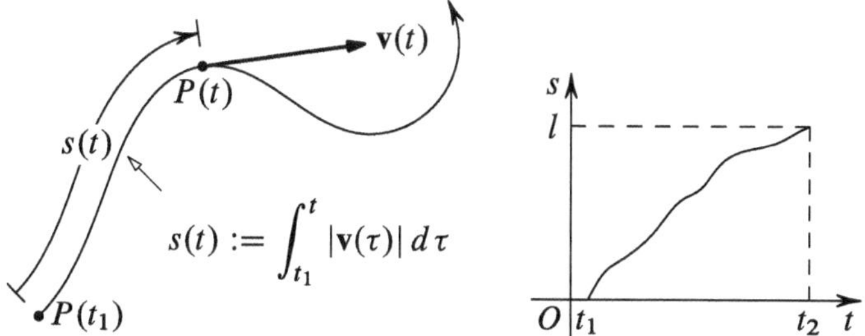


**Fig. 1.2** Arclength time law and velocity

moves along it. To fully know the motion, therefore, it is not enough to assign the trajectory $\hat{P}(s)$; we also need the arclength time law $s(t)$.

To sum up, we have two equivalent ways to describe the motion of a point: the position as function of time, or the pair $\{\hat{P}(s), s(t)\}$ given by trajectory and arclength time law. As we have seen, from the knowledge of the motion $P(t)$ we can deduce both the trajectory $\hat{P}(s)$ and the arclength time law $s(t)$. Conversely if we know trajectory and arclength time law we can recover the motion $P(t)$:

$$\text{motion } P(t) \Longleftrightarrow \begin{cases} \hat{P}(s) \text{ trajectory} \\ s(t) \text{ arclength time law} \end{cases}$$

Note that the function $\hat{P}(s)$ only describes a curve in space, the motion's trajectory, with no reference whatsoever to time. The usefulness of describing motion with $\{\hat{P}(s), s(t)\}$ therefore consists in the possibility of making a clear distinction between the geometric aspect, namely the trajectory, and the properly kinematic side, i.e. the time law.

## 1.1 Frenet Frame Along a Trajectory

It is useful to be able to decompose the components of the velocity vector $\mathbf{v}$ and the acceleration vector $\mathbf{a}$ along tangential and normal directions to the trajectory. As we will see right now, while the velocity only has a tangential component the acceleration possesses both a tangential component and one along the trajectory's normal direction.

Before expanding further on these statements it is necessary to briefly review some facts on the geometry of curves.

The curve corresponding to the trajectory of a point is described by the function $\hat{P}(s)$. From the theory of curves we know that the derivative of $\hat{P}(s)$ with respect to $s$ is the unit tangent vector to the curve (see (A.16)):

$$\mathbf{t}(s) = \frac{d\hat{P}}{ds}\,.$$

Moreover, differentiating the relation $\mathbf{t} \cdot \mathbf{t} = 1$ in $s$ we immediately deduce

$$\frac{d\mathbf{t}}{ds} \cdot \mathbf{t} = 0$$

and so the vector $d\mathbf{t}/ds$ is always perpendicular to the trajectory (see property (A.20) in the Appendix).

The *modulus* of $d\mathbf{t}/ds$ can be given a precise geometric meaning, in that it retains an indication of the speed with which the trajectory changes as the arclength varies. This observation explains the name *curvature* given to the quantity

$$c = \left|\frac{d\mathbf{t}}{ds}\right|\,.$$

We call $\rho = 1/c$ *curvature radius*, at the points where $c \neq 0$ (and when the trajectory is a circle $\rho$ is equal to the radius itself).

The unit vector corresponding to $d\mathbf{t}/ds$, defined whenever the curvature $c$ is non-zero, is the trajectory's *(principal) unit normal* and is indicated by $\mathbf{n}$ (the term "principal" specifies that this particular normal vector, among the infinitely many choices of unit normals, is obtained differentiating $\mathbf{t}(s)$). Hence

$$\frac{d\mathbf{t}}{ds} = c\mathbf{n} \tag{1.4}$$

(observe that at the points on the trajectory where $c = 0$ the unit normal is not defined).

Having introduced $\mathbf{t}$ and $\mathbf{n}$, which are orthogonal, it is only natural to define a third unit vector, called *binormal* and denoted by $\mathbf{b}$, as the their cross product

$$\mathbf{b} = \mathbf{t} \times \mathbf{n}\,.$$

In this way we obtain a right-handed orthonormal basis $(\mathbf{t}, \mathbf{n}, \mathbf{b})$, called *Frenet frame* at the point in question, which is a function of the point's position along the trajectory and thus a function of the arclength $s$.

## 1.2 Intrinsic Components of the Velocity and the Acceleration

The velocity and acceleration vectors can be decomposed both along the fixed reference basis $\{\mathbf{i}, \mathbf{j}, \mathbf{k}\}$ and, in the light of the previous section's notions, along the Frenet basis $(\mathbf{t}, \mathbf{n}, \mathbf{b})$. We shall now show how to express the components of $\mathbf{v}$ and $\mathbf{a}$ with respect to the latter basis of vectors.

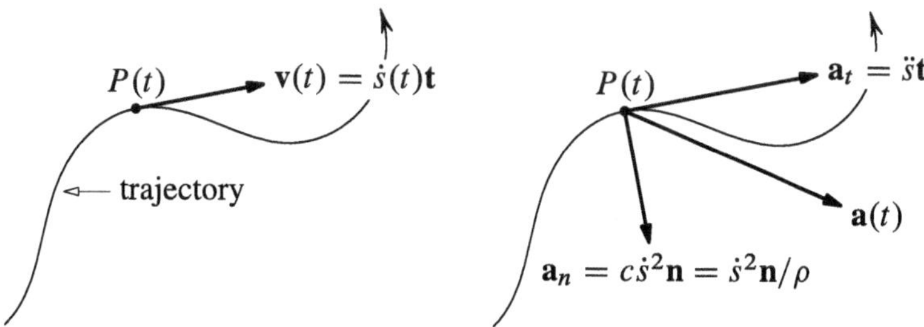


**Fig. 1.3** Components of **v** and **a** with respect to the Frenet frame

**Proposition 1.1** (Velocity and acceleration in the Frenet frame) *Let $s(t)$ be the arclength time law describing a point's motion and $\{\mathbf{t}, \mathbf{n}, \mathbf{b}\}$ the Frenet frame associated with its trajectory. Then*

$$\mathbf{v} = \dot{s}\mathbf{t} \qquad \textit{and} \qquad \mathbf{a} = \ddot{s}\mathbf{t} + c\dot{s}^2\mathbf{n}\,, \tag{1.5}$$

*where $c = 1/\rho$ is the trajectory's curvature.*

***Proof*** To prove the first relationship in (1.5) we invoke the chain rule (differentiation rule of composite functions) applied to $\hat{P}(s(t))$:

$$\mathbf{v} = \frac{dP}{dt} = \frac{d\hat{P}}{ds}\frac{ds}{dt} = \dot{s}\mathbf{t}\,.$$

The situation is more complicated as regards the acceleration, in which case we will show first that in general there is not only a component tangential to the trajectory but also one component along the unit normal. Indeed, let us differentiate the velocity expressed by $(1.5)_1$. Recalling (1.4) together with the formula $c = 1/\rho$ relating curvature and curvature radius, we obtain

$$\mathbf{a} = \frac{d\mathbf{v}}{dt} = \frac{d(\dot{s}\mathbf{t})}{dt} = \ddot{s}\mathbf{t} + \dot{s}\frac{d\mathbf{t}}{dt} = \ddot{s}\mathbf{t} + \dot{s}\frac{d\mathbf{t}}{ds}\frac{ds}{dt} = \ddot{s}\mathbf{t} + c\dot{s}^2\mathbf{n} = \ddot{s}\,\mathbf{t} + \frac{\dot{s}^2}{\rho}\mathbf{n}\,.$$

□

Equations (1.5) are particularly significant. The first one says that the velocity is at each instant tangential to the trajectory, and its component along **t** is $\dot{s}$. The second relation shows instead that the acceleration is not always tangential: apart from the tangential component $\ddot{s}$ it also has a component along the unit normal, equal to $c\dot{s}^2$, as depicted in Fig. 1.3. Only at the positions where the trajectory has zero curvature, or at the instants for which $\dot{s} = 0$, the acceleration reduces to purely tangential.

Notice at last that both velocity and acceleration have zero component along the unit binormal **b**.

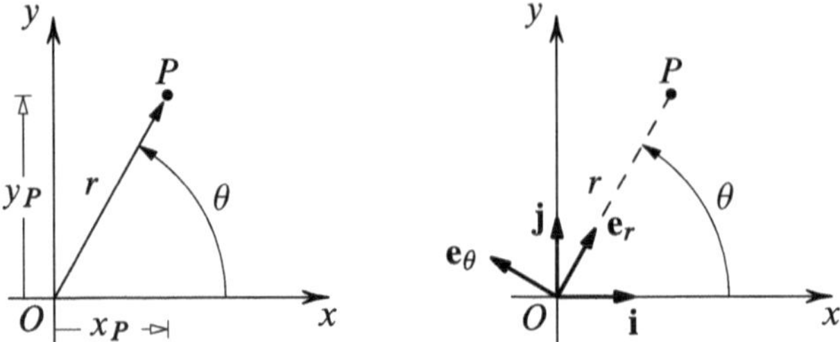


**Fig. 1.4** Polar coordinates and the variable unit vectors $\mathbf{e}_r$ and $\mathbf{e}_\theta$

## 1.3 Polar Coordinates in Planar Motions

The motion of a point is called *planar* when the whole trajectory is a planar curve. If so we can identify the point's position not only with the two Cartesian coordinates corresponding to the components of the position vector on the given plane, but also by means of polar coordinates $(r, \theta)$. This is very convenient in certain applications.

Let us fix a polar axis, which for simplicity we assume is the $x$-axis, oriented as the unit vector $\mathbf{i}$. We introduce the coordinates $r$ and $\theta$ such that $OP = r\cos\theta\,\mathbf{i} + r\sin\theta\,\mathbf{j}$, with $r = |OP| \geq 0$ and $\tan\theta = y_P/x_P$. The motion of the point can then be given using the functions $r(t)$ and $\theta(t)$ (clearly there are geometrical constraints to remember, because for instance the origin is a singular point, given that $\theta$ is not defined there).

It is useful to find the form taken by the velocity and the acceleration in this picture, and in particular to calculate their components along directions that vary with the motion. To that end we introduce unit vectors $\mathbf{e}_r$ and $\mathbf{e}_\theta$, defined as

$$\mathbf{e}_r = \cos\theta\,\mathbf{i} + \sin\theta\,\mathbf{j} \qquad \mathbf{e}_\theta = -\sin\theta\,\mathbf{i} + \cos\theta\,\mathbf{j}\,. \tag{1.6}$$

The unit vector $\mathbf{e}_r$ indicates the so-called *radial* direction, since it is oriented from the origin to the position of $P(t)$, while $\mathbf{e}_\theta$ corresponds to the *transversal* direction, obtained by rotating the former by a counter-clockwise angle $\pi/2$ (see Fig. 1.4).

**Proposition 1.2** (Velocity and acceleration in polar coordinates) *In a planar motion described in polar coordinates the following relations hold:*

$$\mathbf{v} = \dot{r}\,\mathbf{e}_r + r\dot{\theta}\,\mathbf{e}_\theta \qquad \mathbf{a} = \left(\ddot{r} - r\dot{\theta}^2\right)\mathbf{e}_r + \left(2\dot{r}\dot{\theta} + r\ddot{\theta}\right)\mathbf{e}_\theta\,. \tag{1.7}$$

***Proof*** The position vector of point $P$ with respect to the origin $O$ turns into the simple expression $OP = r\,\mathbf{e}_r$. To obtain the velocity $\mathbf{v}$ it is then sufficient to compute its derivative

$$\mathbf{v} = \frac{d(OP)}{dt} = \dot{r}\,\mathbf{e}_r + r\,\dot{\mathbf{e}}_r\,.$$

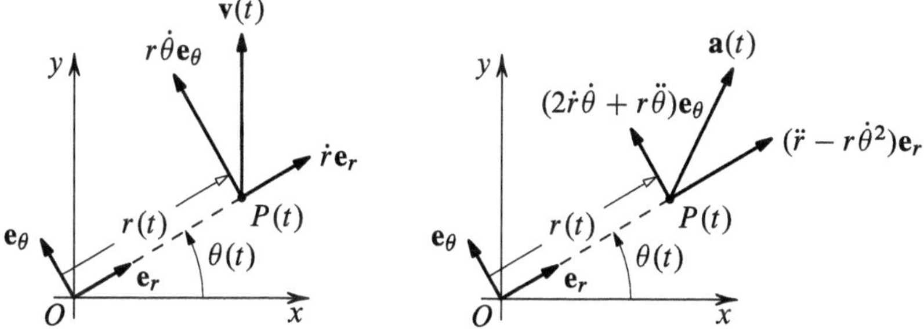


**Fig. 1.5** Radial and transversal components of $\mathbf{v}$ and $\mathbf{a}$

From definition (1.6) we deduce $\dot{\mathbf{e}}_r = (-\sin\theta\mathbf{i} + \cos\theta\mathbf{j})\dot{\theta} = \dot{\theta}\mathbf{e}_\theta$, so in the end $\mathbf{v} = \dot{r}\,\mathbf{e}_r + r\dot{\theta}\,\mathbf{e}_\theta$. We conclude that the velocity's radial component is $\dot{r}$, whereas the transversal component is $r\dot{\theta}$, as shown in Fig. 1.5.

As for the acceleration, we just have to differentiate in time expression $(1.7)_1$ for the velocity, to find

$$\mathbf{a} = \ddot{r}\mathbf{e}_r + \dot{r}\dot{\mathbf{e}}_r + \dot{r}\dot{\theta}\mathbf{e}_\theta + r\ddot{\theta}\mathbf{e}_\theta + r\dot{\theta}\dot{\mathbf{e}}_\theta\,. \tag{1.8}$$

In analogy to what we saw above, the second formula in (1.6) produces $\dot{\mathbf{e}}_\theta = -\dot{\theta}\mathbf{e}_r$, so (1.8) reduces to $(1.7)_2$, as depicted in Fig. 1.5. □

It is convenient and natural to indicate the radial and transversal components of velocity and acceleration of a point in planar motion using the subscripts $r$ and $\theta$ respectively. The results found above can then be recast as

$$v_r = \dot{r}\,, \qquad v_\theta = r\dot{\theta}\,, \qquad a_r = \ddot{r} - r\dot{\theta}^2\,, \qquad a_\theta = 2\dot{r}\dot{\theta} + r\ddot{\theta}\,.$$

In view of the applications it is further useful to observe that the *radial* and *transversal* components of velocity and acceleration are quite different quantities from the same vectors' components along the *tangential* and *normal* directions.

We finally stress that the acceleration's radial and transversal components do not coincide with the time derivatives of the corresponding velocity components. This property appears every time $\mathbf{v}$ and $\mathbf{a}$ are written with respect to a moving basis (such as the Frenet frame for example, see (1.5)).

# Chapter 2
# Rigid Body Kinematics

The motion of a system is called *rigid* if the distances between all pairs of points do not change in time. Similarly, we say a body $\mathcal{B}$ is rigid when its only possible motions are rigid ones.

It is important to introduce the concept of *comoving point*. Any rigid body $\mathcal{B}$ occupies at each instant a bounded region in the space in which it is contained. Notwithstanding, we may imagine of extending its motion to points that do not belong to it. We can indeed conceive that every point $P$, even if external to the rigid body and not actually belonging to it, is so to speak "dragged along" by the motion of $\mathcal{B}$, in such a way that the distance of $P$ to any point of the body keeps constant in time. In this way we can easily define a whole Euclidean space made of points comoving with the body during its motion. Such a *comoving space* moves together with the rigid body and shares with it all of the motion's features.

The notion of comoving space can be visualized more easily in the special but significant case of a rigid body made of a finite thin plate moving on a given plane. If we suppose the plate is glued to a glass pane, ideally infinite, which therefore moves with the plate, we have a concrete representation of the notion of comoving space and comoving point. We therefore see that from a kinematical point of view the motion of a rigid system can be identified with that of an entire Euclidean space.

We call *comoving* any reference frame dragged along with the motion of the comoving space. With respect to such a frame the rigid body's points do not change position, and in particular neither do their coordinates. Evidently every rigid motion admits infinitely many comoving reference frames, because we can pick in infinitely many ways both the comoving point used as origin and the basis of comoving unit vectors chosen as orthonormal basis.

The first issues we shall address when analyzing the motion of a rigid body are the following. How can we determine the position of a rigid body? Which and how many are the parameters we need? As it is clearly *not* enough to know a point's position, we will need to provide more information regarding the body's orientation in the space surrounding the given point. We will later prove that it suffices to ask



P. Biscari et al., *Rational Mechanics*, UNITEXT 177,
https://doi.org/10.1007/978-3-032-07462-1_2

that the distances between $P$ and *three* non-collinear points of $\mathcal{B}$ are constant in order to identify the motion completely.

## 2.1 Characterization of a Rigid Motion

Let us show how during a rigid motion a right-handed (or left-handed) orthonormal comoving basis stays orthonormal and does not change orientation.

Consider three comoving points $Q$, $A$, $B$ and note that

$$AB = QB - QA$$

so

$$|AB|^2 = |QB|^2 + |QA|^2 - 2\,QB \cdot QA\,. \tag{2.1}$$

During a rigid motion the distance of pairs of comoving points does not change, so from (2.1) we deduce that the dot product of two comoving vectors, and therefore the angle between them, stays the same.

It follows from this observation that any orthonormal comoving basis $\mathbf{e}_i$ $(i = 1, 2, 3)$ stays orthonormal during the motion because their unit lengths do not vary, nor do the right angles between them. Therefore at each instant we necessarily have

$$\mathbf{e}_1(t) \times \mathbf{e}_2(t) \cdot \mathbf{e}_3(t) = \pm 1 \tag{2.2}$$

with $+$ (or $-$) sign on the right when the basis is right-handed (respectively left-handed). Since we only consider regular motions, which depend continuously on time, the mixed product in (2.2) must stay constant during motion, and the basis will stay right- or left-handed, depending on how it was at the start.

We next examine how to determine unambiguously the position and so the motion of every point $P$ of $\mathcal{B}$.

**Proposition 2.1** *The motion of a rigid body is known when we know, as functions of time, the position of a point $Q$ and the orientation of an orthonormal comoving basis $\{\mathbf{e}_1, \mathbf{e}_2, \mathbf{e}_3\}$.*

***Proof*** Let $(y_1, y_2, y_3)$ denote the components of $QP$ along the given basis. It suffices to note that these components will stay constant in time, since

$$y_k = QP \cdot \mathbf{e}_k$$

and the dot product of comoving vectors is constant. But

$$QP(t) = y_1\mathbf{e}_1(t) + y_2\mathbf{e}_2(t) + y_3\mathbf{e}_3(t)\,, \tag{2.3}$$

from which we deduce that knowing $Q(t)$ and $\mathbf{e}_k(t)$ allows to determine the motion of a generic comoving point $P$ if we know its (constant) coordinates in the comoving frame with origin at $Q$. □

The most natural way to assign the vectors $\mathbf{e}_k$ at a given instant is simply to write them as linear combinations of the fixed unit vectors $\mathbf{i}_h$, using the $3 \times 3 = 9$ Cartesian components $R_{hk}$

$$\mathbf{e}_k(t) = \sum_{h=1}^{3} R_{hk}(t)\,\mathbf{i}_h \qquad (R_{hk} \equiv \mathbf{i}_h \cdot \mathbf{e}_k)\,. \tag{2.4}$$

Each coefficient $R_{hk}$ is the dot product of two unit vectors, and therefore equals the cosine of the angle they form. We may thus interpret these numbers as *direction cosines* of the comoving basis $\mathbf{e}_k$ with respect to the fixed basis $\mathbf{i}_h$.

The nine numbers $R_{hk}$ *cannot* be assigned randomly if we expect, as it should be, that the moving unit vectors $\mathbf{e}_k$ form an orthonormal system with the same orientation as the fixed basis. What condition must they satisfy then? Precisely the ones that say the comoving basis is orthonormal and further guarantee it is oriented coherently with the fixed basis:

$$\mathbf{e}_j \cdot \mathbf{e}_k = \delta_{jk} = \begin{cases} 1 & \text{if } j = k \\ 0 & \text{if } j \neq k \end{cases} \qquad \mathbf{e}_1 \times \mathbf{e}_2 \cdot \mathbf{e}_3 = \mathbf{i}_1 \times \mathbf{i}_2 \cdot \mathbf{i}_3\,. \tag{2.5}$$

Substituting (2.4) in these relations we obtain

$$\sum_{h=1}^{3} R_{hj} R_{hk} = \delta_{jk} = \begin{cases} 1 & \text{if } j = k \\ 0 & \text{if } j \neq k \end{cases} \qquad \det[R_{hk}] = 1\,. \tag{2.6}$$

To summarize, if the coefficients $R_{hk}$ satisfy all of the above relationships we can be sure that the vectors obtained from (2.4) form at each instant an orthonormal basis (oriented as the fixed basis) which, together with the function $Q(t)$, defines a rigid motion.

We can look at conditions (2.6) from another viewpoint. Let us define a linear transformation (or operator) $\mathbf{R}$ whose matrix with respect to the orthonormal basis $\mathbf{i}_h$ reads

$$\begin{bmatrix} R_{11} & R_{12} & R_{13} \\ R_{21} & R_{22} & R_{23} \\ R_{31} & R_{32} & R_{33} \end{bmatrix}$$

that is, $R_{hk} = \mathbf{i}_h \cdot \mathbf{R}\mathbf{i}_k$. It is easy to see that the conditions imposed on the components $R_{hk}$ are equivalent to saying $\mathbf{R}$ is an orthogonal mapping. More precisely, it is a *rotation* (see Appendix A.3), in other words a linear mapping such that $\mathbf{R}^T\mathbf{R} = \mathbf{R}\mathbf{R}^T = \mathbf{I}$, $\det \mathbf{R} = 1$. As shown in Fig. 2.1, this rotation sends each unit vector of the fixed basis to the corresponding unit vector in the comoving basis:

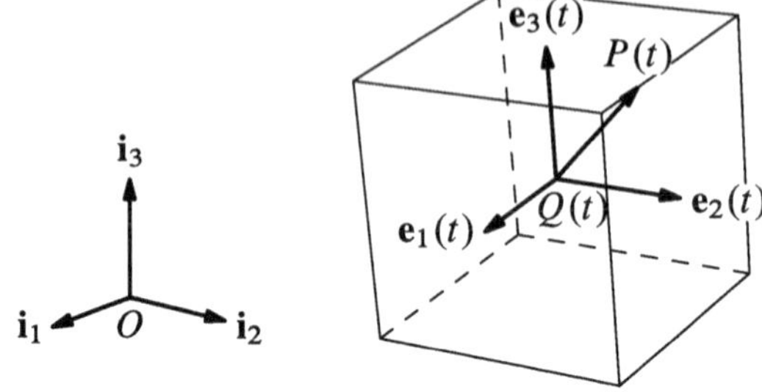


**Fig. 2.1** In a rigid motion an orthonormal basis moving with the body is related to the fixed basis by a rotation: $\mathbf{e}_h(t) = \mathbf{R}(t)\mathbf{i}_h$

$$\mathbf{e}_k = \mathbf{R}\mathbf{i}_k \qquad (\mathbf{i}_h \cdot \mathbf{R}\mathbf{i}_k = \mathbf{i}_h \cdot \mathbf{e}_k = R_{hk}), \tag{2.7}$$

while the inverse rotation, which coincides with the transpose $\mathbf{R}^T$, maps the $\mathbf{e}_k$ to the $\mathbf{i}_h$: $\mathbf{i}_h = \mathbf{R}^T\mathbf{e}_h$.

In the light of (2.3) and (2.7) we may express the vector $QP$ using $\mathbf{R}$, which during the motion will depend on time,

$$QP(t) = y_1\mathbf{e}_1 + y_2\mathbf{e}_2 + y_3\mathbf{e}_3 = \mathbf{R}(t)(y_1\mathbf{i}_1 + y_2\mathbf{i}_2 + y_3\mathbf{i}_3) \tag{2.8}$$

and therefore

$$OP(t) = OQ(t) + QP(t) = OQ(t) + \mathbf{R}(t)(y_1\mathbf{i}_1 + y_2\mathbf{i}_2 + y_3\mathbf{i}_3).$$

Now let us call $x_k(P)$ and $x_k(Q)$ the Cartesian coordinates of $P$ and $Q$ in the fixed orthonormal frame $\mathbf{i}_k$ with origin $O$ (time dependency is omitted to simplify the notation). Recalling relations (2.7) we deduce that each coordinate $x_k(P) = \mathbf{i}_k \cdot OP(t)$ can be written as

$$\begin{aligned} x_k(P) &= \mathbf{i}_k \cdot OQ(t) + \mathbf{i}_k \cdot \mathbf{R}(t)(y_1\mathbf{i}_1 + y_2\mathbf{i}_2 + y_3\mathbf{i}_3) \\ &= x_k(Q) + R_{k1}y_1 + R_{k2}y_2 + R_{k3}y_3 \end{aligned}$$

so in conclusion

$$x_k(P) = x_k(Q) + \sum_{h=1}^{3} R_{kh}y_h \qquad (k = 1, 2, 3). \tag{2.9}$$

This last relation is called *Cartesian equation of a rigid motion* and allows to determine the motion of the generic comoving point $P$ once we know its coordinates $y_h$ in the comoving frame. This is possible for all points provided we know the motion of a point $Q$ and the components $R_{kh}$ of the rotation $\mathbf{R}$ instant by instant.

## 2.2 Euler Angles

Conditions (2.6) give 6 independent relations on the components of $\mathbf{R}$ and so suggest the 9 direction cosines $R_{hk}$ can be expressed in terms of $9 - 6 = 3$ free parameters. Keeping (2.9) in account we conclude that to know the motion of a rigid body we must assign at each instant the 3 coordinates of any of its points $Q$ and 3 independent parameters that give the rotation matrix $\mathbf{R}$. For this reason we say a rigid body without constraints has 6 *degrees of freedom* (for the precise definition and discussion on degrees of freedom see Sect. 4.8).

The choice and the construction of these 3 parameters that allow to determine the rotation matrix is a delicate issue. There exist several possibilities, so here we shall only indicate two alternative constructions based on a common idea: we introduce two angles to determine the position in space of one of the comoving frame's axes, and use a third angle to assign the rotation around that axis in order to determine the position of the other two moving axes.

We remark from the start that any construction we choose to prescribe the direction cosines in terms of three arbitrary parameters has drawbacks: more precisely, it can be proved that there always exists orientations of the comoving basis for which the local one-to-one correspondence with the three chosen parameters fails. This means that all parametrizations we may define are by nature only "partial", in the sense that they do not consider certain orientations for the body's comoving basis. This fact is however not that bad, since in many applications there are *a priori* reasons to exclude these "singular" configurations from the set of allowed configurations for the rigid body's motion we wish to describe.

A classical method for introducing three free and independent parameters allowing to detect the comoving frame's position with respect to the position of the fixed one relies on the so-called *Euler angles*. In all cases where $\mathbf{i}_3 \times \mathbf{e}_3 \neq \mathbf{0}$ it is possible to define the unit vector

$$\mathbf{n} = \text{unit}(\mathbf{i}_3 \times \mathbf{e}_3) = \frac{\mathbf{i}_3 \times \mathbf{e}_3}{|\mathbf{i}_3 \times \mathbf{e}_3|} \tag{2.10}$$

namely the unit vector of the intersection line between the planes perpendicular to the directions $\mathbf{e}_3$ and $\mathbf{i}_3$, which goes by the name of *line of nodes* (see Fig. 2.2). The line of nodes enables us to define the following three angles:

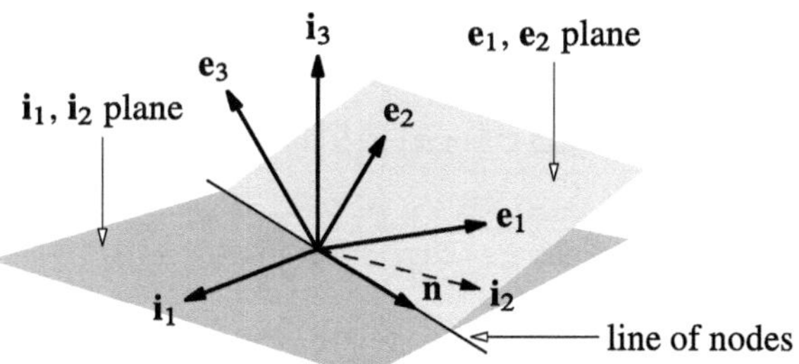


**Fig. 2.2** Line of nodes

1. *Precession angle* $\psi$: the angle by which $\mathbf{i}_1$ rotates counter-clockwise, on the orthogonal plane to $\mathbf{i}_3$, onto the line of nodes $\mathbf{n}$;
2. *Nutation angle* $\theta$: angle by which $\mathbf{i}_3$ rotates onto $\mathbf{e}_3$ on the plane orthogonal to $\mathbf{n}$;
3. *Intrinsic rotation angle* $\phi$: angle by which $\mathbf{n}$ rotates counter-clockwise, on the orthogonal plane to to $\mathbf{e}_3$, onto the line of nodes $\mathbf{e}_1$.

Note that (2.10) only makes sense for $0 < \theta < \pi$. The *precession* angle $\psi$ and the *intrinsic rotation* angle $\phi$ are understood instead as rotational angles, and can assume any value.

It is possible to show that there is a local one-to-one correspondence between the values of the Euler angles and the orientations a rigid body admits. Given in fact an absolute basis $\mathbf{i}_h$ and a basis $\mathbf{e}_h$ comoving with the rigid body, for each orientation such that $\mathbf{e}_3$ is not parallel to $\mathbf{i}_3$ (so $0 < \theta < \pi$) one can find unique values for the three Euler angles. (The fact that $\mathbf{e}_3$ cannot be parallel to $\mathbf{i}_3$ shows that the correspondence is only locally bijective. When the two unit vectors are parallel the line of nodes cannot be defined, and one parameter is enough to describe the mutual orientation of the bases, as we shall see for the special but remarkable case of planar motions.)

Let us now show how given specific Euler angles we can transform the fixed basis $\mathbf{i}_h$ into the comoving basis $\mathbf{e}_h$ by three successive rotations. In this way we will also manage to express the components of the $\mathbf{e}_h$, i.e. the numbers $R_{hk}$, as functions of the angles $(\psi, \theta, \phi)$.

The first operation, shown in Fig. 2.3, consists in rotating the unit vectors $\mathbf{i}_h$ by an angle $\psi$ about the axis $\mathbf{i}_3$, and obtain a basis $\tilde{\mathbf{e}}_h$ such that

$$\tilde{\mathbf{e}}_1 = \cos\psi\,\mathbf{i}_1 + \sin\psi\,\mathbf{i}_2\,, \qquad \tilde{\mathbf{e}}_2 = -\sin\psi\,\mathbf{i}_1 + \cos\psi\,\mathbf{i}_2\,, \qquad \tilde{\mathbf{e}}_3 = \mathbf{i}_3\,.$$

The second rotation, depicted in Fig. 2.4, corresponds to the angle $\theta$ and revolves around the unit vector $\tilde{\mathbf{e}}_1 = \mathbf{n}$. It maps the basis $\tilde{\mathbf{e}}_h$ to the basis $\hat{\mathbf{e}}_h$ whose unit vectors are defined by

$$\hat{\mathbf{e}}_1 = \tilde{\mathbf{e}}_1\,, \qquad \hat{\mathbf{e}}_2 = \cos\theta\,\tilde{\mathbf{e}}_2 + \sin\theta\,\tilde{\mathbf{e}}_3\,, \qquad \hat{\mathbf{e}}_3 = -\sin\theta\,\tilde{\mathbf{e}}_2 + \cos\theta\,\tilde{\mathbf{e}}_3\,.$$

We complete the transformation with a rotation by $\phi$ about the unit vector $\hat{\mathbf{e}}_3$, thus obtaining the comoving basis $\mathbf{e}_h$ from $\hat{\mathbf{e}}_h$, as in Fig. 2.5. We have

$$\mathbf{e}_1 = \cos\phi\,\hat{\mathbf{e}}_1 + \sin\phi\,\hat{\mathbf{e}}_2\,, \qquad \mathbf{e}_2 = -\sin\phi\,\hat{\mathbf{e}}_1 + \cos\phi\,\hat{\mathbf{e}}_2\,, \qquad \mathbf{e}_3 = \hat{\mathbf{e}}_3\,.$$

After a few substitutions, all conceptually simple but rather exacting, we arrive at

$$\begin{aligned}
\mathbf{e}_1 &= (\cos\psi\cos\phi - \sin\psi\cos\theta\sin\phi)\mathbf{i}_1 + (\sin\psi\cos\phi + \cos\psi\cos\theta\sin\phi)\mathbf{i}_2 \\
&\qquad + \sin\theta\sin\phi\,\mathbf{i}_3 \\
\mathbf{e}_2 &= (-\sin\phi\cos\psi - \cos\phi\sin\psi\cos\theta)\mathbf{i}_1 - (\sin\phi\sin\psi - \cos\phi\cos\psi\cos\theta)\mathbf{i}_2 \\
&\qquad + \cos\phi\sin\theta\,\mathbf{i}_3 \\
\mathbf{e}_3 &= \sin\theta\sin\psi\,\mathbf{i}_1 - \sin\theta\cos\psi\,\mathbf{i}_2 + \cos\theta\,\mathbf{i}_3.
\end{aligned} \tag{2.11}$$

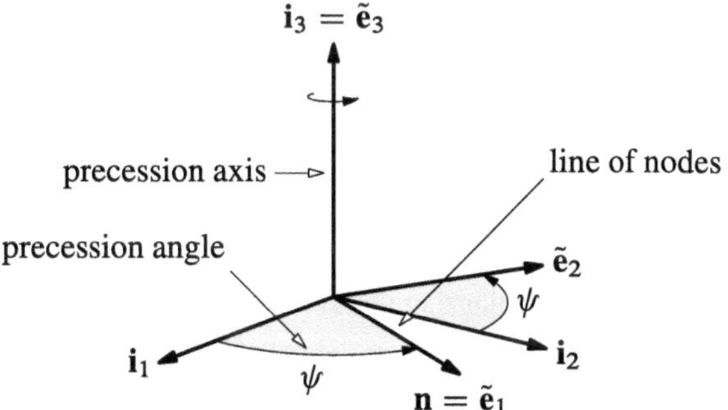


**Fig. 2.3** Euler angles: first rotation (precession)

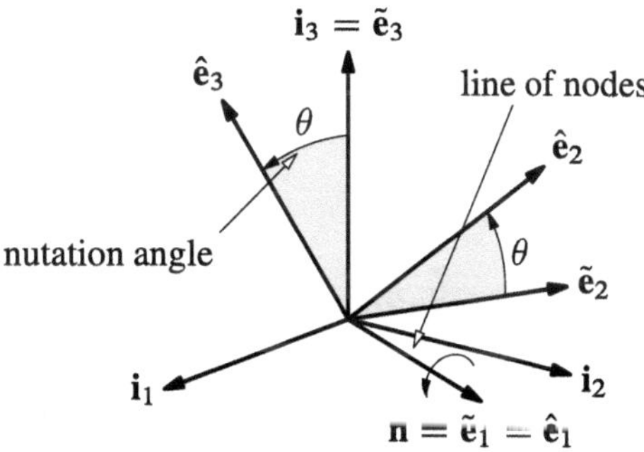


**Fig. 2.4** Euler angles: second rotation (nutation)

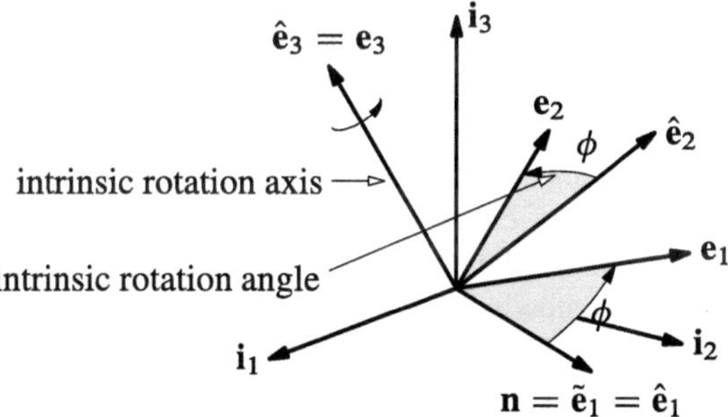


**Fig. 2.5** Euler Angles: third rotation (intrinsic rotation)

Taking (2.4) into account, the above allow to write each component $R_{hk}$ of the rotation matrix $\mathbf{R}$ in terms of the angles $(\psi, \theta, \phi)$.

Using the inverse rotation, coinciding with the transpose $\mathbf{R}^T$, we can also write the equations expressing the fixed unit vectors in terms of the comoving ones:

$$\mathbf{i}_1 = (\cos\psi\cos\phi - \sin\psi\cos\theta\sin\phi)\mathbf{e}_1 - (\sin\phi\cos\psi + \cos\phi\sin\psi\cos\theta)\mathbf{e}_2 + \sin\theta\sin\psi\,\mathbf{e}_3$$
$$\mathbf{i}_2 = (\sin\psi\cos\phi + \cos\psi\cos\theta\sin\phi)\mathbf{e}_1 - (\sin\phi\sin\psi - \cos\phi\cos\psi\cos\theta)\mathbf{e}_2 - \sin\theta\cos\psi\,\mathbf{e}_3$$
$$\mathbf{i}_3 = \sin\theta\sin\phi\,\mathbf{e}_1 + \cos\phi\sin\theta\,\mathbf{e}_2 + \cos\theta\,\mathbf{e}_3 \tag{2.12}$$

Notice in particular that the unit vector of the line of nodes can be written as

$$\mathbf{n} = \tilde{\mathbf{e}}_1 = \hat{\mathbf{e}}_1 = \cos\psi\,\mathbf{i}_1 + \sin\psi\,\mathbf{i}_2 = \cos\phi\,\mathbf{e}_1 - \sin\phi\,\mathbf{e}_2\,.$$

Therefore we conclude that, at least during a time interval where $\mathbf{i}_3 \times \mathbf{e}_3 \neq \mathbf{0}$, the motion can be prescribed by the quantities $\{x_Q(t), y_Q(t), z_Q(t); \psi(t), \theta(t), \phi(t)\}$, the six independent parameters necessary to know the position of a rigid body in space.

### 2.2.1 Cardano Angles

An alternative to the description that uses the Euler angles relies on the so-called *Cardano angles*. In engineering contexts the latter are often employed to identify the so-called roll, pitch and yaw of an aircraft or ship as seen by a fixed observer.

After we introduce a comoving orthonormal basis $\mathbf{e}_i$ we focus on the unit vector $\mathbf{e}_1$ and note that we need *two* angles to describe its position with respect to the fixed basis: the angle $\beta$ it forms with the plane spanned by $\mathbf{i}_1, \mathbf{i}_2$, and the angle $\alpha$ between its projection of that plane and $\mathbf{i}_1$ (in practice these are the "latitude" and "longitude" of the "tip" of $\mathbf{e}_1$ on a unit sphere). There remains to assign the rotation about $\mathbf{e}_1$ in order to place the other unit vectors $\mathbf{e}_2, \mathbf{e}_3$. This can be done by a third angle $\gamma$. All in all, the knowledge of $\{\alpha, \beta, \gamma\}$ allows to reconstruct the comoving basis' position, as shown in Fig. 2.6.

To better visualize these angles we may imagine the axis $\mathbf{e}_1$ points in the direction of motion (parallel to the aircraft's nose-tail axis), and that the $\mathbf{i}_3$ axis (fixed) is vertical. In this case the pitch $\beta$ represents the inclination above/below the horizontal

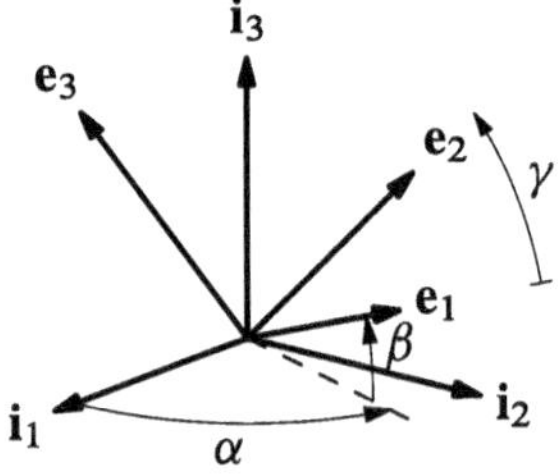


**Fig. 2.6** Cardano angles

plane, the yaw $\alpha$ characterizes the direction of motion (north/south/east/west), while the roll $\gamma$ controls the aircraft's tilt about its axis.

In a completely similar manner to what we have seen for the Euler angles, we can write the 9 coefficients $R_{hk}$ in terms of the Cardano angles. The functions $R_{hk}(\alpha, \beta, \gamma)$ thus obtained (which we shall not write explicitly) satisfy the six orthonormality conditions (2.6) for *every* value of the angles, which then take the role of *essential* and *independent* parameters. That means they satisfy two requests: (1) typically we cannot take fewer; (2) for any value given to two of them, there are infinitely many admissible values for the third.

Observe that the configurations $\beta = \pm\pi/2$ are singular, meaning that the other rotation angles are undefined in this case (this is an unavoidable problem, irrespective of the type of angles chosen). There is in fact no set of 3 angles capable of setting up a regular bijection to comoving bases. In more precise terms, one cannot define a regular parametrization of the entire collection (manifold) of rotations.

We conclude by remarking that the difference between Euler angles and Cardano angles disappears if we rename the comoving unit vectors in Fig. 2.6 from $\mathbf{e}_1, \mathbf{e}_2, \mathbf{e}_3$ to $\mathbf{e}_3, \mathbf{e}_1, \mathbf{e}_2$ (in this order). Comparing the pictures describing Euler angles we will have $\alpha = \pi + \psi$, $\beta = \pi/2 - \theta$, $\gamma = \phi$. Note though that if we keep the names of the comoving vectors as in Fig. 2.6, the relationship between Euler and Cardano angles in much more complicated. Put differently: these sets of rotation angles differ whenever we pin down the single unit vectors in the bases.

### 2.2.2 Rotations Around a Fixed Axis

Definition (2.10) for the line of nodes does not require that $\mathbf{e}_3 \times \mathbf{i}_3 \neq \mathbf{0}$. Though if the rotation occurs around an axis with unit vector $\mathbf{k}$, it is often convenient to take the latter as unit vector common to both the fixed and the rotated bases, by setting $\mathbf{e}_3 = \mathbf{i}_3 = \mathbf{k}$.

Under these assumptions, and despite (2.10) now loses its meaning, relations (2.11) simplify because we can now take $\theta = 0$. Using standard trigonometric formulas we thus obtain

$$\begin{cases} \mathbf{e}_1 = \cos(\psi + \phi)\mathbf{i}_1 + \sin(\psi + \phi)\mathbf{i}_2 \\ \mathbf{e}_2 = -\sin(\psi + \phi)\mathbf{i}_1 + \cos(\psi + \phi)\mathbf{i}_2 \\ \mathbf{e}_3 = \mathbf{i}_3 \end{cases} \tag{2.13}$$

It is clear that in this case the angle $\psi + \phi$ alone is enough to describe the comoving basis' configuration with respect to the fixed one, although there is no way to distinguish $\psi$ from $\phi$ as the line of nodes is undefined. The simplest choice consists in identifying the overall angle with $\psi$ and just call it *rotation angle*. The situation is easy to visualise via Fig. 2.3, where clearly the basis $\tilde{\mathbf{e}}_k$ coincides with the comoving basis $\mathbf{e}_k$ from the start.

## 2.3 Angular Velocity

During a rigid motion every point in the system has its own velocity and acceleration, which in general are different from point to point and vary instant by instant. However, as we can easily imagine, these quantities are not distributed randomly across the points. To understand this we should think of the ends of a rigid rod and note that, for example, the velocities will never be aligned with the rod and be *opposite*: such a situation would correspond to the two ends moving towards each other (or moving away), and that is impossible due to the rod's rigidity. So now we seek to understand, apart from this simple example, which properties characterize the possible distributions of velocities and accelerations in rigid motions (or in rigid bodies). We begin our study by introducing the concept of angular velocity, a vectorial quantity of fundamental importance for this subject.

### 2.3.1 Poisson Formulas

Differentiating in time the orthonormality conditions (2.5) for a basis $\mathbf{e}_h(t)$ in motion we obtain

$$\dot{\mathbf{e}}_j \cdot \mathbf{e}_k + \mathbf{e}_j \cdot \dot{\mathbf{e}}_k = 0 \quad \Longrightarrow \quad \dot{\mathbf{e}}_j \cdot \mathbf{e}_k = -\mathbf{e}_j \cdot \dot{\mathbf{e}}_k \qquad j, k = 1, 2, 3. \tag{2.14}$$

In particular, choosing $j = k$, Eq. (2.14) imply that $\dot{\mathbf{e}}_k \cdot \mathbf{e}_k = 0$ for any $k = 1, 2, 3$.

**Theorem 2.2** (Poisson) *Let the system $\mathcal{B}$ move rigidly and call $\mathbf{e}_1, \mathbf{e}_2, \mathbf{e}_3$ the orthonormal vectors of a comoving basis. There exists a* unique *vector $\boldsymbol{\omega}$ such that*

$$\dot{\mathbf{e}}_1 = \boldsymbol{\omega} \times \mathbf{e}_1\,, \qquad \dot{\mathbf{e}}_2 = \boldsymbol{\omega} \times \mathbf{e}_2\,, \qquad \dot{\mathbf{e}}_3 = \boldsymbol{\omega} \times \mathbf{e}_3\,. \tag{2.15}$$

*This vector* does not *depend on the chosen comoving basis and is called* angular velocity *of $\mathcal{B}$. The relations* (2.15) *are known as* Poisson formulas.

*The angular velocity vector can be written*

$$\boldsymbol{\omega} = (\dot{\mathbf{e}}_2 \cdot \mathbf{e}_3)\,\mathbf{e}_1 + (\dot{\mathbf{e}}_3 \cdot \mathbf{e}_1)\,\mathbf{e}_2 + (\dot{\mathbf{e}}_1 \cdot \mathbf{e}_2)\,\mathbf{e}_3 = \frac{1}{2}\sum_{h=1}^{3} \mathbf{e}_h \times \dot{\mathbf{e}}_h\,. \tag{2.16}$$

*Finally, given any vector* $\mathbf{w}$ *comoving with the rigid body, we have*

$$\dot{\mathbf{w}} = \boldsymbol{\omega} \times \mathbf{w}\,. \tag{2.17}$$

***Proof*** The argument consists of two parts regarding the properties of $\boldsymbol{\omega}$: (1) existence and uniqueness; (2) independence from the chosen comoving basis. Without loss of generality we assume the chosen basis is *right-handed*; if not, it suffices to modify little details to adapt the proof.

Take the dot product of the first Poisson formula (2.15) with $\mathbf{e}_2$, of the second one with $\mathbf{e}_3$ and of the third with $\mathbf{e}_1$:

$$\dot{\mathbf{e}}_1 \cdot \mathbf{e}_2 = \boldsymbol{\omega} \times \mathbf{e}_1 \cdot \mathbf{e}_2 \,, \qquad \dot{\mathbf{e}}_2 \cdot \mathbf{e}_3 = \boldsymbol{\omega} \times \mathbf{e}_2 \cdot \mathbf{e}_3 \,, \qquad \dot{\mathbf{e}}_3 \cdot \mathbf{e}_1 = \boldsymbol{\omega} \times \mathbf{e}_3 \cdot \mathbf{e}_1 \,.$$

Recall that in a mixed product we are allowed to swap the dot and cross product, so that

$$\dot{\mathbf{e}}_1 \cdot \mathbf{e}_2 = \boldsymbol{\omega} \cdot \mathbf{e}_1 \times \mathbf{e}_2 \,, \qquad \dot{\mathbf{e}}_2 \cdot \mathbf{e}_3 = \boldsymbol{\omega} \cdot \mathbf{e}_2 \times \mathbf{e}_3 \,, \qquad \dot{\mathbf{e}}_3 \cdot \mathbf{e}_1 = \boldsymbol{\omega} \cdot \mathbf{e}_3 \times \mathbf{e}_1 \,. \tag{2.18}$$

Since in a right-handed basis we have

$$\mathbf{e}_1 \times \mathbf{e}_2 = \mathbf{e}_3 \,, \qquad \mathbf{e}_2 \times \mathbf{e}_3 = \mathbf{e}_1 \,, \qquad \mathbf{e}_3 \times \mathbf{e}_1 = \mathbf{e}_2 \,,$$

equations (2.18) become

$$\dot{\mathbf{e}}_1 \cdot \mathbf{e}_2 = \boldsymbol{\omega} \cdot \mathbf{e}_3 = \omega_3 \,, \qquad \dot{\mathbf{e}}_2 \cdot \mathbf{e}_3 = \boldsymbol{\omega} \cdot \mathbf{e}_1 = \omega_1 \,, \qquad \dot{\mathbf{e}}_3 \cdot \mathbf{e}_1 = \boldsymbol{\omega} \cdot \mathbf{e}_2 = \omega_2 \,,$$

and therefore we necessarily have

$$\boldsymbol{\omega} = (\dot{\mathbf{e}}_2 \cdot \mathbf{e}_3)\, \mathbf{e}_1 + (\dot{\mathbf{e}}_3 \cdot \mathbf{e}_1)\, \mathbf{e}_2 + (\dot{\mathbf{e}}_1 \cdot \mathbf{e}_2)\, \mathbf{e}_3 \,. \tag{2.19}$$

In this way we have deduced that for the chosen comoving basis $\boldsymbol{\omega}$ is unique and must be as in (2.19). We still need to check that this vector truly satisfies (2.15). Let us do the computation explicitly in the case $h = 1$. Take the cross product of expression (2.19) with $\mathbf{e}_1$

$$\boldsymbol{\omega} \times \mathbf{e}_1 = (\dot{\mathbf{e}}_2 \cdot \mathbf{e}_3)\, \mathbf{e}_1 \times \mathbf{e}_1 + (\dot{\mathbf{e}}_3 \cdot \mathbf{e}_1)\, \mathbf{e}_2 \times \mathbf{e}_1 + (\dot{\mathbf{e}}_1 \cdot \mathbf{e}_2)\, \mathbf{e}_3 \times \mathbf{e}_1 \,.$$

Recalling that $\mathbf{e}_1 \times \mathbf{e}_1 = \mathbf{0}$, $\mathbf{e}_2 \times \mathbf{e}_1 = -\mathbf{e}_3$ and $\mathbf{e}_3 \times \mathbf{e}_1 = \mathbf{e}_2$, we find

$$\boldsymbol{\omega} \times \mathbf{e}_1 = -\,(\dot{\mathbf{e}}_3 \cdot \mathbf{e}_1)\, \mathbf{e}_3 + (\dot{\mathbf{e}}_1 \cdot \mathbf{e}_2)\, \mathbf{e}_2 \,.$$

Using (2.14) gives

$$\boldsymbol{\omega} \times \mathbf{e}_1 = (\dot{\mathbf{e}}_1 \cdot \mathbf{e}_1)\, \mathbf{e}_1 + (\dot{\mathbf{e}}_1 \cdot \mathbf{e}_2)\, \mathbf{e}_2 + (\dot{\mathbf{e}}_1 \cdot \mathbf{e}_3)\, \mathbf{e}_3 \tag{2.20}$$

(where on the right we have added the first term, which is zero). From the identity

$$\dot{\mathbf{e}}_1 = (\dot{\mathbf{e}}_1 \cdot \mathbf{e}_1)\, \mathbf{e}_1 + (\dot{\mathbf{e}}_1 \cdot \mathbf{e}_2)\, \mathbf{e}_2 + (\dot{\mathbf{e}}_1 \cdot \mathbf{e}_3)\, \mathbf{e}_3 \,,$$

(true since $(\dot{\mathbf{e}}_1 \cdot \mathbf{e}_j)$ coincides with the $j$th component of $\dot{\mathbf{e}}_1$) the right-hand side of (2.20) is precisely $\dot{\mathbf{e}}_1$. This concludes the check of the Poisson formula (2.15) for $h = 1$. The proof that $\boldsymbol{\omega}$ satisfies the remaining two as well is completely similar.

There remains to show the $\boldsymbol{\omega}$ is *independent* of the basis. Until now we have seen that to every comoving basis there corresponds a unique vector satisfying the Poisson formulas in that basis, but we still do not known whether this vector "works" for *all* bases. So, we need to prove that the *same* $\boldsymbol{\omega}$ defined by the first relation in (2.16) also satisfies the Poisson formulas for another orthonormal comoving basis $\{\mathbf{e}'_1, \mathbf{e}'_2, \mathbf{e}'_3\}$. To that end consider an arbitrary vector $\mathbf{w}$ comoving with the rigid system, and call $(w_1, w_2, w_3)$ its components with respect to the basis $\{\mathbf{e}_1, \mathbf{e}_2, \mathbf{e}_3\}$, so that $\mathbf{w} = w_1\mathbf{e}_1 + w_2\mathbf{e}_2 + w_3\mathbf{e}_3$. Since the vector $\mathbf{w}$ comoves with the system in motion, the components $w_h$ $(h = 1, 2, 3)$ are *constant* in time, so

$$\frac{d\mathbf{w}}{dt} = w_1\dot{\mathbf{e}}_1 + w_2\dot{\mathbf{e}}_2 + w_3\dot{\mathbf{e}}_3 \,. \tag{2.21}$$

Using the Poisson formulas we discover

$$w_1\dot{\mathbf{e}}_1 + w_2\dot{\mathbf{e}}_2 + w_3\dot{\mathbf{e}}_3 = \boldsymbol{\omega} \times (w_1\mathbf{e}_1 + w_2\mathbf{e}_2 + w_3\mathbf{e}_3) = \boldsymbol{\omega} \times \mathbf{w}$$

and comparing with (2.21) we deduce the important relationship (2.17), which must hold for *every* vector comoving with the rigid body. Choosing in particular $\mathbf{w}$ equal to $\mathbf{e}'_1$, $\mathbf{e}'_2$ and then $\mathbf{e}'_3$ (the second basis' unit vectors) we have

$$\frac{d\mathbf{e}'_h}{dt} = \boldsymbol{\omega} \times \mathbf{e}'_h \qquad h = 1, 2, 3.$$

We have then proved that the same vector $\boldsymbol{\omega}$ satisfies the Poisson formulas not only for the comoving basis $\{\mathbf{e}_h\}$ but also for $\{\mathbf{e}'_h\}$. Hence $\boldsymbol{\omega}$ does not depend on the basis we choose to define it.

Eventually, let us prove the second expression in (2.16) for the angular velocity vector. The Poisson formulas force $\sum_h \mathbf{e}_h \times \dot{\mathbf{e}}_h = \sum_h \mathbf{e}_h \times (\boldsymbol{\omega} \times \mathbf{e}_h)$. From this, and using identity (A.8), we obtain

$$\mathbf{e}_h \times (\boldsymbol{\omega} \times \mathbf{e}_h) = (\mathbf{e}_h \cdot \mathbf{e}_h)\boldsymbol{\omega} - (\boldsymbol{\omega} \cdot \mathbf{e}_h)\mathbf{e}_h = \boldsymbol{\omega} - \omega_h\mathbf{e}_h \,.$$

Therefore $\sum_h \mathbf{e}_h \times \dot{\mathbf{e}}_h = \sum_h [\boldsymbol{\omega} - \omega_h\mathbf{e}_h] = 3\boldsymbol{\omega} - \boldsymbol{\omega} = 2\boldsymbol{\omega}$, and so

$$\boldsymbol{\omega} = \frac{1}{2}\sum_{h=1}^{3} \mathbf{e}_h \times \dot{\mathbf{e}}_h \,.$$

□

## 2.4 General Properties of Rigid Motions

Equation (2.17), which we found in the previous section, allows to characterize the velocity distributions that can occur during a rigid motion.

**Theorem 2.3** (Velocity distribution law) *The motion of a system is rigid if and only if its velocity distribution satisfies at all times the relation*

$$\mathbf{v}_P(t) = \mathbf{v}_Q(t) + \boldsymbol{\omega}(t) \times QP \quad \textit{for every} \quad P, Q. \tag{2.22}$$

Equation (2.22) is called *velocity distribution law* and the vector $QP$ connects the positions in space of the particles $Q$ and $P$ at the instant considered. It gives a relationship between the velocities, and as we will show later it completely characterizes every rigid motion: for any pair of comoving points $P$ and $Q$ we can compute the velocity of one ($P$, say) if we know the velocity of the other ($Q$) and the angular velocity $\boldsymbol{\omega}$.

***Proof*** To show the necessity of (2.22) it suffices to use (2.17) with $\mathbf{w} = QP$, where $Q$ and $P$ comove with the moving system, as is the vector $\mathbf{w}$ between them. Using (1.2) we then obtain

$$\underbrace{\frac{d(QP)}{dt}}_{\mathbf{v}_P - \mathbf{v}_Q} = \boldsymbol{\omega} \times QP, \tag{2.23}$$

which gives (2.22).

To prove the sufficient part, there remains to show that this relation, at each instant, warrants the motion's rigidity. Let $P$ and $Q$ be points of a system in motion, for which we assume (2.22) holds at *each* instant. Differentiate the vector $QP$ squared:

$$\frac{d(QP)^2}{dt} = 2\frac{d(QP)}{dt} \cdot QP = 2\big(\mathbf{v}_P - \mathbf{v}_Q\big) \cdot QP = 2\big(\boldsymbol{\omega} \times QP\big) \cdot QP = 0.$$

In the last passage, we invoked the property whereby the mixed product vanishes whenever two of the factors are equal or parallel. Hence the distances between the points remain constant, so the corresponding motion is rigid. □

Differentiating the velocity distribution law with respect to time one can prove that also the possible accelerations of the points of a rigid body obey precise requirements.

**Proposition 2.4** (Acceleration distribution law) *The distribution of accelerations in a rigid body satisfies the equation*

$$\mathbf{a}_P(t) = \mathbf{a}_Q(t) + \dot{\boldsymbol{\omega}} \times QP + \boldsymbol{\omega} \times (\boldsymbol{\omega} \times QP) \quad \textit{for every} \quad P, Q, \tag{2.24}$$

*called the* acceleration distribution law.

***Proof*** To obtain (2.24) we must differentiate in time (2.22). Because of (2.23) we find

$$\mathbf{a}_P = \mathbf{a}_Q + \dot{\boldsymbol{\omega}} \times QP + \boldsymbol{\omega} \times (\mathbf{v}_P - \mathbf{v}_Q) = \mathbf{a}_Q + \dot{\boldsymbol{\omega}} \times QP + \boldsymbol{\omega} \times (\boldsymbol{\omega} \times QP),$$

equivalent to (2.24). □

Once the acceleration of any comoving point $Q$ is given, together with the angular velocity $\boldsymbol{\omega}$ and its derivative $\dot{\boldsymbol{\omega}}$, we can find the acceleration of *any* other point $P$ of the system.

**Remark 2.5** (Elementary rigid displacement) The elementary displacement of a moving point is $dP = \mathbf{v}_P\,dt$, which we may interpret as the main part of the finite displacement $\Delta P$ as $\Delta t$ tends to zero. By (2.22), the displacement $dP = \mathbf{v}_P\,dt$ is related to the displacement $dQ = \mathbf{v}_Q\,dt$ by

$$dP = dQ + \boldsymbol{\omega}\,dt \times QP. \tag{2.25}$$

The quantity $\boldsymbol{\omega}\,dt$ has a certain importance, and is often referred to as the *infinitesimal rotation vector* $\boldsymbol{\varepsilon}$ relative to the instant and the rigid motion to which $\boldsymbol{\omega}(t)$ refers. With this notation the elementary rigid displacement (2.25) reads

$$dP = dQ + \boldsymbol{\varepsilon} \times QP \qquad (\boldsymbol{\varepsilon} = \boldsymbol{\omega}\,dt), \tag{2.26}$$

an expression we will use again later. □

## 2.5 Special Rigid Motions

Under certain conditions a rigid body might undergo special motion during a time interval $t \in [t_1, t_2]$. In this section we analyze a few of these particular situations. Before we proceed to the description, we can impose without loss of generality that the fixed and comoving reference frames coincide at time $t = 0$, that is

$$Q(0) = O, \qquad \mathbf{e}_k(0) = \mathbf{i}_k \quad \forall k = 1, 2, 3.$$

### 2.5.1 *Translations*

**Definition 2.6** A rigid motion is said to be translational if *every* comoving straight line does not change orientation with respect to the fixed observer.

**Proposition 2.7** *A rigid motion is translational if and only if* $\boldsymbol{\omega}(t) \equiv \mathbf{0}$ *for every* $t$.

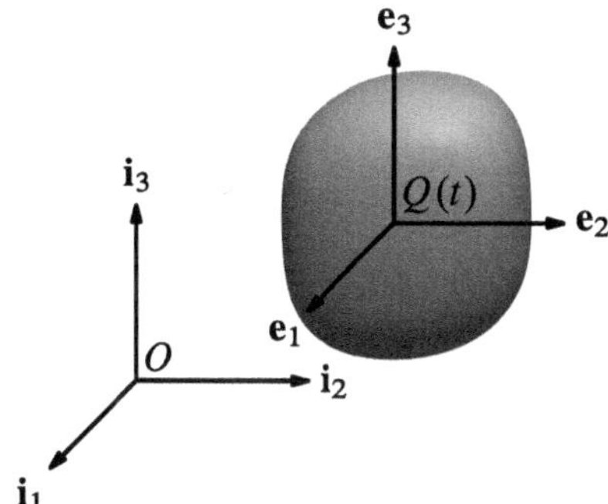


**Fig. 2.7** Translational motion

***Proof*** By definition in a translational motion every comoving straight line has constant orientation. In particular, the comoving basis will have constant unit vectors, from which we deduce $\dot{\mathbf{e}}_k = \mathbf{0}$, $\forall k = 1, 2, 3$ and, by (2.16), $\boldsymbol{\omega} = \mathbf{0}$.

Conversely, if $\boldsymbol{\omega} = \mathbf{0}$ Eq. (2.17) implies that every comoving unit vector stays constant, so the motion is translational. □

The choice of axes made for $t = 0$ implies that in a translational motion $\mathbf{e}_k = \mathbf{i}_k$ for any $t$ (see Fig. 2.7). The rotation matrix transforming the fixed basis into the comoving one is therefore $\mathbf{R} = \mathbf{I}$, so the Cartesian equation of rigid motion (2.9) now becomes:

$$x_k(P) = x_k(Q) + y_k\,, \qquad k = 1, 2, 3.$$

During a translational motion the position of any point $P$ is known once we know the motion of one of the points $Q$ of the rigid body, because the comoving unit vector $QP$ is constant at all times. This implies we only need *three* parameters (the variable coordinates of $Q$) to determine the position of every point in the system. The degrees of freedom are now 3. Finally, Eqs. (2.22), (2.24) and (2.26) provide

$$\mathbf{v}_P = \mathbf{v}_Q\,, \qquad \mathbf{a}_P = \mathbf{a}_Q\,, \qquad dP = dQ \qquad \forall P, Q.$$

In a translational motion, at each instant $t \in [t_1, t_2]$ all points have the same velocity, the same acceleration and the same elementary displacements. For this reason during a translational motion we can speak of velocity of the body without distinguishing between points.

In the particular case where the velocity also has constant *direction* the translational motion is called *linear*, since the single points' trajectories are straight lines, and it is called *uniform linear motion* when the velocity is further constant in time.

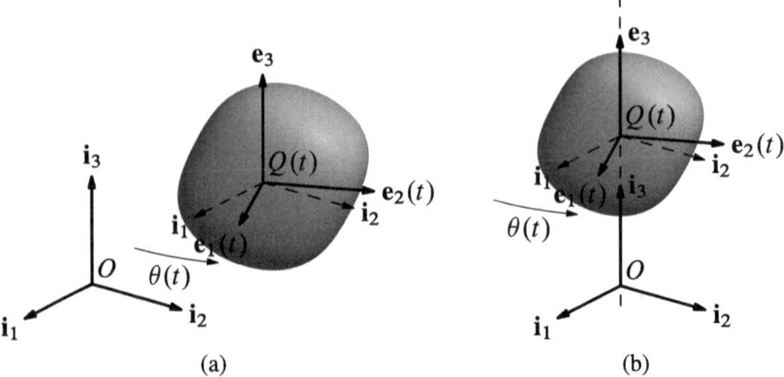


**Fig. 2.8** Rototranslational and helicoidal rigid motion

## 2.5.2 Roto-Translations

**Definition 2.8** A rigid motion is roto-translational if there exists *one* orientation comoving with the body (or a privileged direction) that stays constant with respect to the fixed observer.

**Proposition 2.9** *A rigid motion is roto-translational if and only if the direction of* $\boldsymbol{\omega}$ *is constant, in which case this direction is the one that keeps its orientation fixed.*

***Proof*** Assume there is a constant direction, and choose $\mathbf{e}_3 = \mathbf{i}_3$, parallel to it. From Poisson's formula (2.15) we have $\dot{\mathbf{e}}_3 = \boldsymbol{\omega} \times \mathbf{e}_3 \equiv \mathbf{0}$, which implies $\boldsymbol{\omega}$ is parallel to $\mathbf{e}_3$.

Conversely, the same Poisson formula shows that if $\boldsymbol{\omega}$ is always parallel to $\mathbf{e}_3$, the latter is constant and the motion is roto-translational. □

Roto-translational motions clearly include translational ones. In fact the latter are obtained in the special case where the angular velocity is zero, and not just one but all comoving unit vectors are constant.

In roto-translational motions we have a similar situation to what we studied in Sect. 2.2.2. Referring to Fig. 2.8a, let $\theta$ be the angle that $\mathbf{e}_1$ forms with $\mathbf{i}_1$. Equations (2.13) imply

$$\mathbf{e}_1 = \cos\theta \mathbf{i}_1 + \sin\theta \mathbf{i}_2 , \qquad \mathbf{e}_2 = -\sin\theta \mathbf{i}_1 + \cos\theta \mathbf{i}_2 , \qquad \mathbf{e}_3 = \mathbf{i}_3 , \tag{2.27}$$

and the rotation matrix is

$$\mathbf{R} \equiv \begin{bmatrix} \cos\theta & -\sin\theta & 0 \\ \sin\theta & \cos\theta & 0 \\ 0 & 0 & 1 \end{bmatrix} . \tag{2.28}$$

The Cartesian equation of motion (2.9) with $\mathbf{R}$ given by (2.28) shows that in a roto-translational motion we typically need *four* parameters to identify the position of every point in the system. Given in fact the three coordinates of a point $Q$ and the angle $\theta$ that determined the comoving unit vectors, we can recover the position of any other point $P$. Moreover, differentiating (2.27) in time gives

$$\dot{\mathbf{e}}_1 = \dot{\theta}\mathbf{e}_2\,, \qquad \dot{\mathbf{e}}_2 = -\dot{\theta}\mathbf{e}_1\,, \qquad \dot{\mathbf{e}}_3 = 0\,,$$

which by (2.16) implies

$$\boldsymbol{\omega} = \dot{\theta}\mathbf{i}_3\,. \tag{2.29}$$

**Remark 2.10** It is important to stress that the angle employed to describe the comoving basis' rotation with respect to the fixed basis might differ from the angle indicated in Fig. 2.8a, and that different choices of the rotation angle might result in small modifications in (2.29). For instance, if we decided to use the angle $\psi(t) = \pi/2 - \theta(t)$ between the comoving vector $\mathbf{e}_1(t)$ and $\mathbf{i}_2$ (instead of $\mathbf{i}_1$), the angular velocity's expression would rather be $\boldsymbol{\omega} = -\dot{\psi}(t)\mathbf{i}_3$. Informally speaking, to compute the angular velocity of a roto-translational rigid motion it suffices to differentiate the rotation angle, but we must pay attention to certain so-to-speak "practical" details.

- The name *rotation angle* should only be reserved for an angle between a *fixed* direction for the observer and a direction *comoving* with the system in motion. This means the moving direction should be thought of as "drawn" onto the moving body.
- It is often convenient to think of the rotation angle as oriented from the fixed direction towards the moving direction. A practical rule of thumb to fix the sign is to check if the result agrees or not (using the right-hand rule) with $\mathbf{i}_3$.

□

From (2.22), (2.24) and (2.26) we obtain the relationship between velocities, accelerations and elementary displacements for points of a rigid body in roto-translational motion:

$$\begin{aligned} \mathbf{v}_P &= \mathbf{v}_Q + \dot{\theta}\,\mathbf{i}_3 \times QP\,, \\ dP &= dQ + \mathrm{d}\theta\,\mathbf{i}_3 \times QP\,, \\ \mathbf{a}_P &= \mathbf{a}_Q + \ddot{\theta}\,\mathbf{i}_3 \times QP - \dot{\theta}^2 P^*P\,, \end{aligned} \tag{2.30}$$

where $P^*$ is the projection of $P$ onto the line through $Q$ and parallel to the angular velocity. Given $P$ and $Q$ so that the segment $PQ$ is parallel to the privileged direction (that is, the direction of $\mathbf{i}_3$ and therefore of $\boldsymbol{\omega}$), Eqs. (2.30) imply

$$\mathbf{v}_P = \mathbf{v}_Q\,, \qquad \mathbf{a}_P = \mathbf{a}_Q\,, \qquad dP = dQ\,, \qquad \forall P, Q\,:\, PQ \parallel \mathbf{i}_3\,. \tag{2.31}$$

Then (2.30) can be better understood once we note that

$$\mathbf{i}_3 \times QP = \mathbf{i}_3 \times \big(\underbrace{QP^*}_{\parallel \mathbf{i}_3} + P^*P\big) = \mathbf{i}_3 \times P^*P\,.$$

Therefore in (2.30) both the terms proportional to $\mathbf{i}_3 \times QP$ and the term proportional to $P^*P$ are contained in the plane orthogonal to $\mathbf{i}_3$. Their directions are respectively tangential and radial, with respect to the axis parallel to $\mathbf{i}_3$ and passing through $P^*$. At last, their modulus is proportional to the distance $|P^*P|$.

**Remark 2.11** Expression (2.29) for the angular velocity affects the infinitesimal rotation vector $\boldsymbol{\varepsilon}$ introduced in (2.26). Indeed,

$$\boldsymbol{\varepsilon} = \boldsymbol{\omega}\, dt = \dot{\psi}\, \mathbf{k}\, dt = d\psi\, \mathbf{k}\,, \tag{2.32}$$

so in this particular case $\boldsymbol{\varepsilon}$ is the differential of the function $\psi(t)\mathbf{k}$.

In roto-translational motions the infinitesimal rotation vector is therefore the differential of a function of the body's rotation angle (and this justifies in part the name given to $\boldsymbol{\varepsilon}$ itself). That said, as we shall see more comfortably in Sect. 3.5, Chap. 3, this property is no longer valid for several generic rigid motions. More precisely: in general there is *no* vector $\boldsymbol{\gamma}$ depending on the Euler angles such that $\boldsymbol{\varepsilon} = d\boldsymbol{\gamma}$. □

**Helicoidal motion**

**Definition 2.12** A roto-translational motion is helicoidal whenever there is a straight line, parallel to the privileged direction, whose points have parallel velocity to the line.

Based on $(2.31)_1$, it is enough for one point $Q$ of the rigid body to have velocity parallel to $\boldsymbol{\omega}$ for all points on the line through $Q$ parallel to the angular velocity to enjoy the same property (see Fig. 2.8b).

If the velocity of $Q$ is always parallel to $\mathbf{i}_3$, the coordinates $x_{Q1}$ and $x_{Q2}$ will be constant, so *two* parameters $\{x_{Q3}, \theta\}$ will suffice to identify the configuration of a rigid body in helicoidal motion.

Observe that irrespective of $\boldsymbol{\omega} \neq \mathbf{0}$, the line of points with velocity parallel to $\mathbf{i}_3$ is unique. In fact, if we still call $Q$ a point on said line, $(2.30)_1$ shows that the velocity of any other point $P$ has a component orthogonal to $\mathbf{i}_3$ unless $QP \parallel \mathbf{i}_3$.

**Rotational motion**

**Definition 2.13** A roto-translational motion is called rotational if there is a straight line (the *axis of rotation*) that lies parallel to the privileged direction and whose points have zero velocity.

Rotational motions are special helicoidal motions in which the $x_{Q3}$ coordinate of the aforementioned point is constant. So there remains only one parameter, represented by $\theta(t)$ (see Fig. 2.9a). Moreover, still assuming $Q$ lies on the axis of rotation, Eqs. (2.30) simplify to:

$$\mathbf{v}_P = \dot{\theta}\, \mathbf{i}_3 \times P^*P\,, \qquad \mathbf{a}_P = \ddot{\theta}\, \mathbf{i}_3 \times P^*P - \dot{\theta}^2 P^*P\,, \qquad dP = d\theta\, \mathbf{i}_3 \times P^*P\,,$$

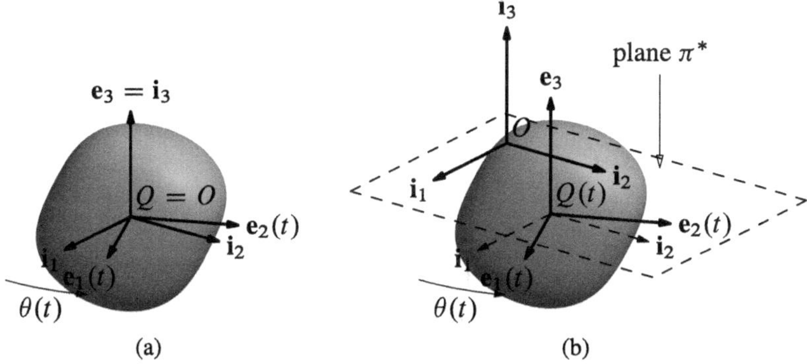


**Fig. 2.9** Rotational and planar rigid motion

where $P^*$ is not the projection of $P$ onto the axis of rotation. Hence we can say that the velocity, the acceleration and the elementary displacements are orthogonal to the axis of rotation, and their modulus is proportional to the distance to the axis. Furthermore, while velocity and elementary displacements only have a tangential component (along the line that goes to the axis of rotation), accelerations contain both a tangential term and a radial term (often called centripetal acceleration).

**Planar motion**

**Definition 2.14** A rigid motion is planar if there is a plane $\pi$, comoving with the body, that always stays parallel to, and at constant distance from, a fixed plane $\pi^*$.

Rigid planar motions are roto-translational, since the condition $\pi \parallel \pi^*$ is completely equivalent to $\mathbf{e}_3 \parallel \mathbf{i}_3$, as long as we take the two unit vectors to be respectively perpendicular to $\pi$, $\pi^*$. Moreover, that $\pi$ stays at a constant distance from $\pi^*$ implies the coordinates of all of its points along the axis $\mathbf{i}_3 \perp \pi^*$ are constant. Planar motions are therefore determined by the *three* parameters $\{x_{Q1}, x_{Q2}, \theta\}$ (see Fig. 2.9b).

During a planar motion the position of an arbitrary $P$ of the rigid body is known when we know the position of its projection $P^*$ on the fixed plane $\pi^*$, because (2.31) implies the two points always move with the same velocity and so the projection point evolves like the original point. For this reason the study or description of a planar rigid motion can be reduced to the study of a planar figure, obtained projecting the rigid body onto the plane $\pi^*$. Consequently, and without any loss of generality, one usually speaks of *planar systems*, by considering only points lying on $\pi^*$.

**Polar motion**

**Definition 2.15** A rigid motion is polar if one of the points comoving with the rigid body stays fixed.

In a polar motion the coordinates of some point $Q$ in the rigid body are constant. In view of the discussion of Sect. 2.2 therefore, *three* parameters are necessary, for

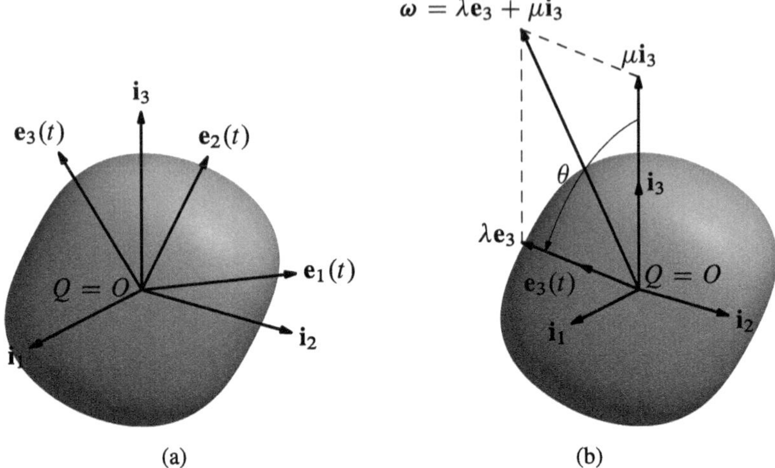


**Fig. 2.10** Polar and precessional rigid motion

example the Euler angles $\{\theta, \varphi, \psi\}$, to identify the configuration of the whole rigid body (see Fig. 2.10a).

Calling $Q$ the fixed point of a polar motion, the velocity distribution with respect to it takes the form $\mathbf{v}_P = \omega \times QP$. From this relation it follows immediately that all points on the line through the fixed point $Q$ and parallel to $\omega$ have zero velocity. It is important though to note that this line, called *instantaneous axis of rotation*, changes direction with $\omega$ in space. In other words the fact that the direction of $\omega$ can vary during a polar motion implies the latter will not in general be rotational, and that the instantaneous axis of rotation is therefore not the axis of a rotation.

### Precessional motion

**Definition 2.16** A polar motion is said to be precessional (or a precession) if there exist two straight lines passing through the fixed point $Q \equiv O$: one comoving with the rigid body, with unit direction $\mathbf{e}_3$ (axis of proper rotation), and another fixed one, of unit direction $\mathbf{i}_3$ (precession axis), such that their angle $\theta$ is constant during motion.

**Proposition 2.17** *A polar motion is a precession if and only if, at each instant, $\omega$ lies on the plane spanned by a fixed unit vector $\mathbf{i}_3$ and a comoving unit vector $\mathbf{e}_3$. Equivalently,*

$$\omega = \lambda \mathbf{e}_3 + \mu \mathbf{i}_3 , \tag{2.33}$$

*where $\lambda$ and $\mu$ are functions of time in general.*

***Proof*** Suppose a polar motion is precessional. That the angle formed by $\mathbf{e}_3$ and $\mathbf{i}_3$ is constant implies (see Fig. 2.10b)

$$\mathbf{e}_3 \cdot \mathbf{i}_3 = \cos\theta = \text{constant}. \tag{2.34}$$

Differentiating (2.34) in time and remembering that the comoving vector $\mathbf{e}_3$ satisfies Poisson's formula $\dot{\mathbf{e}}_3 = \boldsymbol{\omega} \times \mathbf{e}_3$, we obtain $\boldsymbol{\omega} \times \mathbf{e}_3 \cdot \mathbf{i}_3 = 0$. From that, $\boldsymbol{\omega}$, $\mathbf{e}_3$ and $\mathbf{i}_3$ belong on the same plane, so there exist $\lambda$ and $\mu$ such that $\boldsymbol{\omega} = \lambda \mathbf{e}_3 + \mu \mathbf{i}_3$.

Conversely, if $\boldsymbol{\omega}$ can be decomposed as in (2.33), the angle between $\mathbf{e}_3$ and $\mathbf{i}_3$ is constant since

$$\frac{d}{dt}(\mathbf{e}_3 \cdot \mathbf{i}_3) = \dot{\mathbf{e}}_3 \cdot \mathbf{i}_3 = (\boldsymbol{\omega} \times \mathbf{e}_3) \cdot \mathbf{i}_3 = 0\,. \qquad \square$$

The components $\lambda(t)$ and $\mu(t)$ characterizing $\boldsymbol{\omega}$ in a precession are respectively called *proper angular velocity* and *precessional angular velocity*. When the components $\lambda(t)$ and $\mu(t)$ are constant the precession is said to be *regular*.

**Remark 2.18** An important observation is that the classification of motion completed above is not exhaustive, in the sense that there is no guarantee an arbitrary rigid motion will fall into one of the previous categories. A generic rigid motion, in fact, is characterized by the arbitrary motion of a point $Q(t)$ (giving rise to a polar motion in the special case of $Q$ fixed), together with a similarly arbitrary angular velocity $\boldsymbol{\omega}$ (producing a roto-translation only when at least the direction of $\boldsymbol{\omega}$ does not change). □

## 2.6 Angular Velocity and Rotations

While in a rigid roto-translation the angular velocity can be expressed in terms of the derivative of the angle of rotation, as we saw in (2.29), when we deal with a generic rigid motion the picture is more complicated, because the body's rotation depends on *three* parameters, like the Euler angles $\{\psi, \theta, \phi\}$, in terms of which we can write the components $R_{hk} = \mathbf{i}_h \cdot \mathbf{e}_k$ of the comoving basis' unit vectors $\{\mathbf{e}_k\}$.

By substituting (2.11), which give the vectors $\{\mathbf{e}_1, \mathbf{e}_2, \mathbf{e}_3\}$ in function of the Euler angles, in definition (2.16) for the angular velocity vector, allows to find, after computing all derivatives, the components of $\boldsymbol{\omega}$ in terms of $\{\psi, \theta, \phi\}$ and their time derivatives $\{\dot\psi, \dot\theta, \dot\phi\}$. This lengthy calculation, that we will not do explicitly but nonetheless suggest that readers carry out, will produce the desired expressions.

That said, there is a more convenient alternative to write the angular velocity of a rigid motion as function of the rotation angles and their derivatives. It consists in introducing suitable moving observers and decomposing the spatial problem into a "sequence" of planar problems. The idea, at present necessarily vague, will be clarified in the chapter devoted to relative kinematics, to which we refer for details.

Still, it is important to say straightaway that in the general case the infinitesimal rotation vector $\boldsymbol{\varepsilon}$ will not be expressible by means of the differentials of the Euler angles, in contrast to what we found in formula (2.32).

## 2.7 Velocity Distribution in a Rigid Motion

### 2.7.1 Eulerian and Lagrangian Description of Motions

The kinematics of an extended body (not necessarily rigid) can be studied from two distinct perspectives.

- The *Lagrangian* or *global* point of view consists in following each point of the body by its evolution in time, and is the perspective we have adopted thus far to examine certain special rigid motions. In this context the symbols $\mathbf{v}_P(t_1)$, $\mathbf{v}_P(t_2)$ indicate the velocities of the same particle $P$ at the instants $t_1, t_2$.
- On the other hand, the *Eulerian* or *local* point of view consists in fixing a control region and considering the body's motion at the instant the body occupies the region. Hence we focus on the distribution of velocities at a generic instant, and we associate with each point $P$ in *space* the vector $\mathbf{v}(P)$ corresponding to the velocity of the point of the system that at said instant passes through $P$, supposing there is one at all. At the given instant this particular velocity vector field is called *velocity distribution*. Evidently the velocity distribution can vary instant by instant, if only because the particles that pass through the control region's points vary.

Let us emphasize the difference in notation and the distinct meaning of the symbols. From the Lagrangian viewpoint $\mathbf{v}_P(t)$ is the velocity of particle $P$ in time. From the Eulerian point of view, on the contrary, time is fixed and the velocity is a function of the position. This justifies the notation $\mathbf{v}(P)$ or $\mathbf{v}(Q)$, where now $P$ and $Q$ do not indicate the particles $P$ and $Q$ but rather the points in the control space.

**Example 2.19** Let us consider two special motions. In the first, non-rigid motion a collection of points is in uniform linear motion, but the points move at different speed; in the second, uniform and rotational, a rigid cylinder rotates with constant angular velocity around its axis.

In the uniform linear motion each point has constant velocity, so in the *Lagrangian* picture we would see $\mathbf{v}_P(t)$ (the velocity of particle $P$) staying constant in time. In an *Eulerian* picture on the other hand we could easily have $\mathbf{v}(P)$ (the velocity of particle *when passing through* $P$) varying over time, since at distinct instants points with distinct velocities can transit through $P$.

The above uniform rotational motion is from the second point of view dual to the first: in the Lagrangian picture the velocities $\mathbf{v}_P(t)$ vary in time since every point changes velocity during the rotational motion. On the contrary, in the Eulerian picture the velocities $\mathbf{v}(P)$ would be constant in time because each particle, when passing through $P$, would always have the same velocity. □

Going back to rigid motions, the problem we want to address is the following: how does one characterize rigid velocity distributions, i.e. velocity distributions a system in rigid motion can admit?

Before we answer that, it proves useful, as we did for motions, to classify certain velocity distributions. It will be crucial to remember that, while for motions the classification was based on comparing the body's configurations at different instants, to classify velocity distributions we must take into account that time is fixed. In order to study the velocity distribution in a rigid motion, we focus on the control space $\mathcal{C}$ at any specific time $t$. From the properties of velocities in rigid motions (see Sect. 2.5) the following classification becomes natural.

**Definition 2.20** A velocity distribution is called *translational* when all points have the same velocity: $\mathbf{v}(P) = \mathbf{v}(Q)$ for any $P, Q \in \mathcal{C}$.

**Definition 2.21** A velocity distribution is *roto-translational* if it is rigid and there is a direction in the control space (privileged direction) such that every line parallel to it is a locus of points with identical velocity. Calling $\mathbf{u}$ the privileged unit vector we therefore have: $\mathbf{v}(P) = \mathbf{v}(Q)$ for any $P, Q \in \mathcal{C}$ such that $PQ \parallel \mathbf{u}$.

**Definition 2.22** A roto-translational velocity distribution is *helicoidal* if there is a line $r$ parallel to the privileged direction that is a locus of points with velocity parallel to the line itself: $\mathbf{v}(P) = \mathbf{v}(Q) = \lambda\mathbf{u}$ for every $P, Q \in r \cap \mathcal{C}$.

**Definition 2.23** A roto-translational velocity distribution is *rotational* if there is a line $r$ parallel to the privileged direction that is a locus of points with velocity zero: $\mathbf{v}(P) = \mathbf{v}(Q) = \mathbf{0}$ for any $P, Q \in r \cap \mathcal{C}$.

The velocity distribution law in a rigid motion (2.22) requires that

$$\mathbf{v}(P) = \mathbf{v}(Q) + \boldsymbol{\omega} \times QP. \tag{2.35}$$

More precisely, there exists a vector $\boldsymbol{\omega}$ such that (2.35) is satisfied by the velocities $\mathbf{v}(P)$ and $\mathbf{v}(Q)$ of the particles occupying positions $P$ and $Q$, as they vary, and with $\boldsymbol{\omega}$ independent of them. Equation (2.35) clearly shows that *the most general rigid velocity distribution is roto-translational*, since any line parallel to $\boldsymbol{\omega}$ is a locus of points of identical velocity, and the privileged direction is then precisely the one given by the angular velocity.

The properties of rigid motions described in Sect. 2.4 allow to infer rather easily that these motions are characterized by having a roto-translational velocity distribution at each instant.

**Proposition 2.24** *A system's motion is rigid if and only if at each instant its velocity distribution is roto-translational, as defined in* (2.35).

Due to this fact (2.35) is also known as a *rigid velocity distribution*.

In the sequel we shall discuss some properties of roto-translational velocity distributions that are paramount in the applications. We will write $\mathcal{B}$ for a system that at a given instant has roto-translational velocity distribution, and hence is described by (2.35). This relation may be read as a law that gives the velocity of a generic point $P$ when one knows the velocity of another point $Q$ and the angular velocity $\boldsymbol{\omega}$. These two vector quantities correspond to six scalar quantities, not accidentally the same number of independent parameters that characterize the position of a rigid body in space.

It is also essential to understand that the points $P$, $Q$ in (2.35) are not at all special: they are *arbitrary* points belonging in the region occupied by $\mathcal{B}$ at the given instant, so the *same* roto-translational velocity distribution can be *described* in several ways. If for instance $H$ is another point in the system we can write the velocity of $P$ as

$$\mathbf{v}(P) = \mathbf{v}(H) + \boldsymbol{\omega} \times HP.$$

Here one usually says the velocity distribution is given *with reference* or *with respect to H* (or $Q$, in case of (2.35)). Note that the angular velocity vector $\boldsymbol{\omega}$ is clearly still the same: in fact it depends on the velocity distribution overall, and not on the particular reference point used to describe it.

A special case is that in which $\boldsymbol{\omega} = \mathbf{0}$; this condition in fact forces $\mathbf{v}(P) = \mathbf{v}(Q)$, which in turn means all points have equal velocity and therefore the rigid velocity distribution is *translational*.

Suppose now $\mathcal{B}$ has *non-translational* velocity distribution (so we assume (2.35) with $\boldsymbol{\omega} \neq \mathbf{0}$), and let us prove a few properties.

**Proposition 2.25** *The quantity $I = \mathbf{v}(P) \cdot \boldsymbol{\omega}$ is independent of the point $P$ used to compute it, and is called* kinematic scalar invariant.

***Proof*** Take the dot product of (2.35) with $\boldsymbol{\omega}$:

$$\mathbf{v}(P) \cdot \boldsymbol{\omega} = \mathbf{v}(Q) \cdot \boldsymbol{\omega} + \underbrace{\boldsymbol{\omega} \times QP \cdot \boldsymbol{\omega}}_{=0} .$$

The mixed product in the last tem vanishes because there are two equal vectors, so we have $\mathbf{v}(P) \cdot \boldsymbol{\omega} = \mathbf{v}(Q) \cdot \boldsymbol{\omega}$. Hence $I$ is an invariant, i.e. it does not depend on the point used to compute it. □

The above property of the kinematic invariant expresses the fact that as the point varies *the velocity component parallel to $\boldsymbol{\omega}$ does not change, while only the orthogonal component varies.*

**Proposition 2.26** *The components of the velocity of two points along their joining line are equal.*

***Proof*** Assuming $P \neq Q$, take the dot product of (2.35) with $QP$. We obtain $\mathbf{v}(P) \cdot QP = \mathbf{v}(Q) \cdot QP$, where we have used the usual properties of the mixed product. Defining the unit vector $\mathbf{e} = QP/|QP|$, parallel to the segment joining the points, we then have $\mathbf{v}(P) \cdot \mathbf{e} = \mathbf{v}(Q) \cdot \mathbf{e}$. □

**Proposition 2.27** *Points on a line parallel to $\boldsymbol{\omega}$ have equal velocities. This common component goes by the name of* drift velocity *of the line.*

***Proof*** Suppose $P$ and $Q$ lie on a line parallel to $\boldsymbol{\omega}$. Equation (2.35) then gives $\mathbf{v}(P) = \mathbf{v}(Q)$ because $\boldsymbol{\omega} \times QP = \mathbf{0}$ as $QP$ is parallel to $\boldsymbol{\omega}$. Therefore $QP \parallel \boldsymbol{\omega}$ implies $\mathbf{v}(P) = \mathbf{v}(Q)$. □

## 2.8 Mozzi's Theorem

Consider a generic roto-translational velocity distribution (2.35) for which $\boldsymbol{\omega} \neq \mathbf{0}$ (hence a non-translational distribution), and let $I = \mathbf{v}(Q) \cdot \boldsymbol{\omega}$ be the kinematic scalar invariant. We want to prove there exists a line, called *Mozzi axis*, made of the points $P$ with the same velocity, *parallel* to $\boldsymbol{\omega}$.

**Theorem 2.28** (Mozzi) *For a rigid velocity distribution with angular velocity* $\boldsymbol{\omega} \neq \mathbf{0}$ *there exist a line, called* Mozzi axis, *whose points have velocity* parallel *to the angular velocity and of smallest modulus among the velocities of any other comoving point. The common velocity to all axis points is the* translational velocity *of the velocity distribution.*

*In the special case* $I = 0$ *the axis consists of points with* zero *velocity, and is called* instantaneous axis of rotation.

*The vanishing of the kinematic scalar invariant* $I$ *is therefore equivalent to asking that a non-translational rigid velocity distribution is rotational.*

***Proof*** The velocity $\mathbf{v}(O)$ at a generic point $O$ will typically have a component parallel to $\boldsymbol{\omega}$ and one perpendicular to it.

As the velocity does not vary along any line parallel to $\boldsymbol{\omega}$, as shown in Fig. 2.11 left, to find the Mozzi axis it will suffice to determine its intersection $H$ with the plane perpendicular to $\boldsymbol{\omega}$ through $O$, see Fig. 2.11 right. The vector $OH$ we seek must therefore determine a point $H$ such that $OH \cdot \boldsymbol{\omega} = 0$ and

$$\mathbf{v}(H) \times \boldsymbol{\omega} = \mathbf{0}$$

(for $\mathbf{v}(H)$ and $\boldsymbol{\omega}$ to be parallel). From the formula for rigid velocity distributions we then deduce

$$(\mathbf{v}(O) + \boldsymbol{\omega} \times OH) \times \boldsymbol{\omega} = \mathbf{0},$$

and by identity (A.9) this amounts to

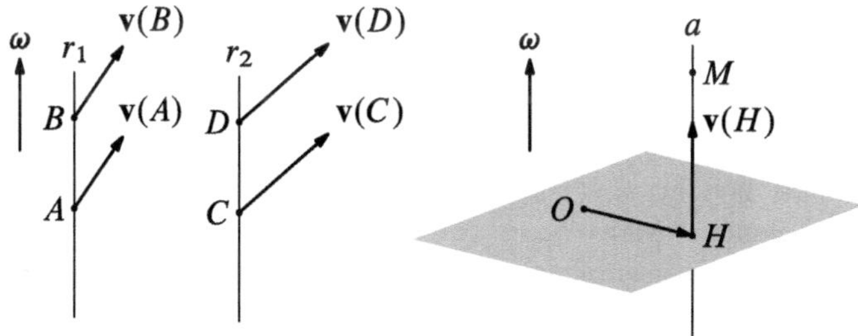


**Fig. 2.11** Points on each line parallel to $\boldsymbol{\omega}$ have the same velocity (left). The velocity of all points on Mozzi's axis $a$ is parallel to $\boldsymbol{\omega}$ (right)

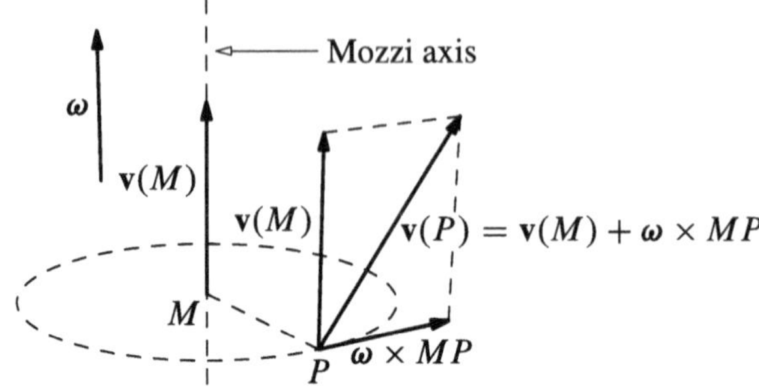


**Fig. 2.12** Mozzi axis: $\mathbf{v}(M) \parallel \omega$, $|\mathbf{v}(P)| > |\mathbf{v}(M)|$

$$\mathbf{v}(O) \times \omega + \omega^2 OH - \underbrace{(OH \cdot \omega)}_{=0}\omega = \mathbf{0}$$

where we used the fact that $OH$ and $\omega$ are orthogonal. Hence we must have

$$OH = \frac{\omega \times \mathbf{v}_O}{\omega^2}\,.$$

The vector $OM$ joining $O$ to a generic point $M$ on the line through $H$ and parallel to $\omega$ (called $a$ in Fig. 2.11 right) decomposes as $OM = OH + HM$, where $HM = \lambda\omega$ for some value of the parameter $\lambda$. Consequently

$$OM(\lambda) = \frac{\omega \times \mathbf{v}(O)}{\omega^2} + \lambda\omega\,. \tag{2.36}$$

Line (2.36), called *Mozzi axis*, is then made of points with velocity *parallel* to $\omega$.

For any other point $P$ in space

$$\mathbf{v}(P) = \mathbf{v}(M) + \omega \times MP$$

where $\mathbf{v}(M)$ is parallel to $\omega$ because $M$ lies on Mozzi's axis, while the second term is orthogonal to it, as we see in Fig. 2.12. The sum of orthogonal vectors fulfils Pythagoras' theorem, so

$$|\mathbf{v}(P)|^2 = |\mathbf{v}(M)|^2 + |\omega \times MP|^2 \geq |\mathbf{v}(M)|^2$$

and therefore the modulus of $\mathbf{v}(P)$ is larger than the modulus of $\mathbf{v}(M)$ (with equality only when $P$ belongs to Mozzi's axis as well).

The Mozzi axis consists of points with velocity parallel to the axis itself. By the definition of helicoidal velocity distribution contained in Proposition 2.22 we conclude that *every* rigid velocity distribution can be written as *helicoidal* velocity distribution.

Finally, in case $I = 0$, the Mozzi axis has a further feature. As $I = \omega \cdot \mathbf{v}(O)$ is independent of $O$, its vanishing implies the velocity of every point is *orthogonal* to

$\boldsymbol{\omega}$. But since along the Mozzi axis (2.36) the velocity is instead *parallel* to $\boldsymbol{\omega}$, we must conclude that when $I = 0$ the Mozzi axis is made of points with *zero velocity*. In this case the rigid velocity distribution is rotational and line (2.36) is called *instantaneous axis of rotation*. □

Here is an immediate consequence of Mozzi's theorem. Suppose $I = 0$ with $\boldsymbol{\omega} \neq \mathbf{0}$. If, in (2.35), instead of choosing a generic $Q$ we pick a point $C$ on the instantaneous axis of rotation, for which then $\mathbf{v}(C) = \mathbf{0}$, we can write

$$\mathbf{v}(P) = \boldsymbol{\omega} \times CP\,. \tag{2.37}$$

Put in other terms: a rigid velocity distribution becomes purely rotational whenever there is at least one point (and so an entire line, the instantaneous axis of rotation) with zero speed.

It is not hard to deduce that during a rotational or polar motion the system's velocity distribution is at each instant rotational. Note though that while in a rotational motion the instantaneous axis of rotation comoves with the system and is made of fixed points around which it rotates, for a polar motion the instantaneous axis of rotation passes though the fixed point, but in general it varies instant by instant in keeping parallel to $\boldsymbol{\omega}$, without comoving with the system.

In this regard it is anyhow relevant to recall the distinction between *motions*, which occur during a time interval, and *velocity distributions*, associated with a given instant, so that we avoid any confusion when classifying the two. Thus, for instance, bodies in translational motion have a translational velocity distribution, whereas bodies in polar or rotational motion most certainly have rotational velocity distribution. At any rate, from the fact that the velocity distribution is at any instant rotational, it *does not* follow that the motion is rotational or polar, in general.

### 2.8.1 *Instantaneous Center of Rotation*

A remarkable consequence of the characterization of planar rigid motions is that for each such motion the kinematic scalar invariant is always null, as $I$ is the dot product of $\boldsymbol{\omega}$, perpendicular to the direction plane, and $\mathbf{v}(Q)$, parallel to it. This remark leads us to an important conclusion.

**Proposition 2.29** (Euler) *The velocity distribution of a system in planar rigid motion is translational or rotational.*

Thus, if $\boldsymbol{\omega} \neq \mathbf{0}$ we know there is an instantaneous axis of rotation parallel to the angular velocity, and so perpendicular to the plane. This implies the axis is uniquely determined by its intersection with the plane. This intersection is the *instantaneous center of rotation*. Finding this point is simplified by a consideration, also known as Chasles' theorem:

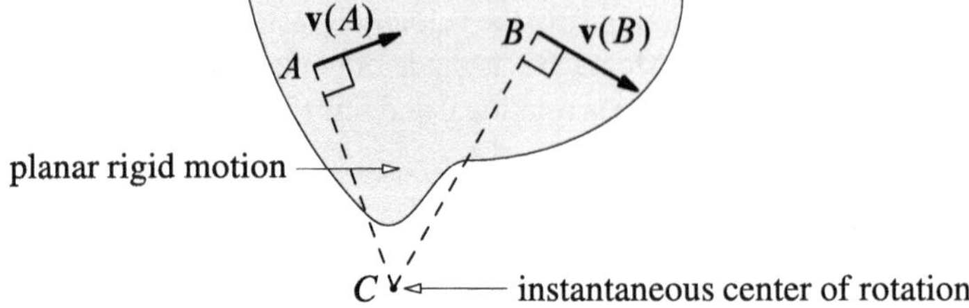


**Fig. 2.13** Chasles' theorem

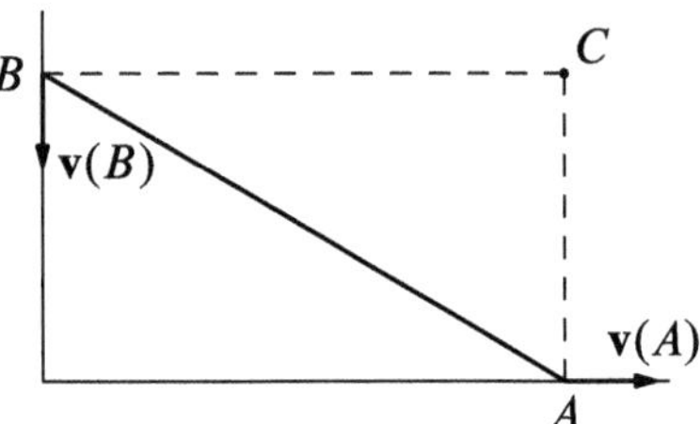


**Fig. 2.14** The position of the instantaneous center of rotation $C$ of a rod $AB$ with $A$ and $B$ constrained to move on two orthogonal fixed guides

**Theorem 2.30** (Chasles) *In a planar rigid motion, the normals to the velocities* $\mathbf{v}(A)$ *and* $\mathbf{v}(B)$ *of two points pass through a unique point C, the instantaneous center of rotation, when the velocity distribution is rotational. The normals are instead parallel when the velocity distribution is translational.*

***Proof*** The existence of a unique intersection of the normals to $\mathbf{v}(A)$ and $\mathbf{v}(B)$ implies the two velocities *are not* parallel, and hence that the velocity distribution is *rotational*, by Proposition 2.29. Calling $C$ the instantaneous center of rotation, from (2.37) it follows that $\mathbf{v}(A) = \boldsymbol{\omega} \times CA$ and $\mathbf{v}(B) = \boldsymbol{\omega} \times CB$.

By definition of cross product the above conditions imply $CA$ and $CB$ are respectively orthogonal to $\mathbf{v}(A)$ and $\mathbf{v}(B)$. Then $C$ lies both on the perpendicular to $\mathbf{v}(A)$ through $A$, and on the perpendicular to $\mathbf{v}(B)$ through $B$ (see Fig. 2.13), so it must be their common point. □

Chasles' theorem, albeit completely intuitive, is pivotal in the applications, where it is employed over and over.

**Example 2.31** Consider a rigid rod $AB$ with ends constrained to two perpendicular guides (we are anticipating the notion of "constraint" that will be studied carefully in Chap. 4, but this should not be a problem). Clearly the endpoints' velocities will be directed as in Fig. 2.14, so from Chasles's Theorem 2.30 we infer the position of the instantaneous center of rotation $C$ of $AB$. This means that $C$, assumed comoving with the rod, has zero velocity at the instant of concern. □

## 2.9 Acceleration Distribution in a Rigid Motion

Let us revert to the acceleration distribution law (2.24). We associate with every point in space $P$, occupied at a generic instant by a particle of a moving system, the acceleration $\mathbf{a}$ of the point. In this way we obtain the so-called *acceleration distribution*. In this case too, it is easy to deduce its structure for systems in rigid motion.

**Proposition 2.32** *During a rigid motion the acceleration distribution satisfies*

$$\mathbf{a}(P) = \mathbf{a}(Q) + \dot{\boldsymbol{\omega}} \times QP + \boldsymbol{\omega} \times (\boldsymbol{\omega} \times QP) . \tag{2.38}$$

Comparing with (2.35), we note that while to know the velocity distribution we need *two* vectors, $\mathbf{v}(Q)$ and $\boldsymbol{\omega}$, for the acceleration distribution we need *three*: $\mathbf{a}(Q)$, $\boldsymbol{\omega}$ and $\dot{\boldsymbol{\omega}}$.

For later use it is also useful to expand the triple product on the right as

$$\boldsymbol{\omega} \times (\boldsymbol{\omega} \times QP) = (\boldsymbol{\omega} \cdot QP)\boldsymbol{\omega} - \omega^2 QP$$

and note that if $Q$ is chosen to coincide with the orthogonal projection $H$ of $P$ onto a line parallel to $\boldsymbol{\omega}$ (say, the instantaneous axis of rotation, or more generally the Mozzi axis) we have

$$\mathbf{a}(P) = \mathbf{a}(H) + \dot{\boldsymbol{\omega}} \times HP - \omega^2 HP \tag{2.39}$$

since $\boldsymbol{\omega} \cdot HP = 0$, obviously.

Relation (2.39) finds its natural area of application in planar motions, as shown in Fig. 2.15.

In the special case where $\boldsymbol{\omega}$ is constant and $H$ is a fixed point (uniform rotational motion around the fixed line through $H$ parallel to $\boldsymbol{\omega}$) we conclude that $\mathbf{a}(P) = -\omega^2 HP$. In other words the acceleration of $P$ points towards the instantaneous axis of rotation (centripetal acceleration) and its modulus equals the angular velocity

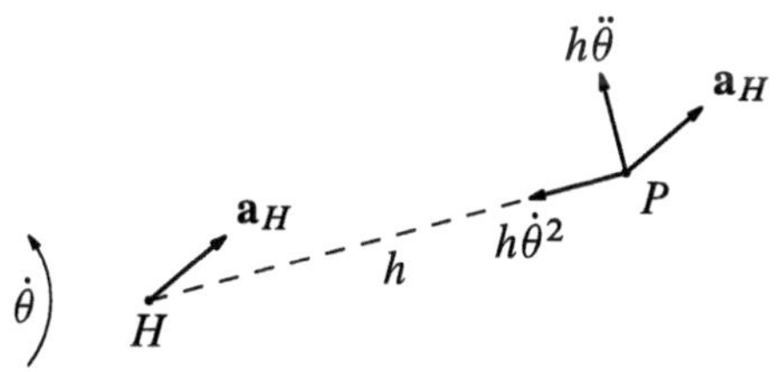


**Fig. 2.15** In a planar motion the acceleration of $P$ is the sum of the acceleration of $H$ with a vector of component $h\ddot{\theta}$, perpendicular to $HP$, and one of component $h\dot{\theta}^2$, oriented towards $H$ (in the picture the angular velocity is oriented counter-clockwise)

squared times the distance of $P$ to the axis. This can be seen again in Fig. 2.15 by letting $\mathbf{a}_H = \mathbf{0}$ and $\ddot{\theta} = 0$.

**Remark 2.33** (*Rivals' Theorem*) Relation (2.38), which expresses the acceleration distribution law, is also known as *Rivals' Theorem*, particularly in some texts on Applied Mechanics. □

## 2.10 Angular Velocity and Vorticity Tensor

We will show, for completeness, how it is possible to recover the velocity distribution's expression by other means, starting from formula (2.8), and without invoking Poisson's Theorem 2.2.

It is an alternative to the description we have adopted, but still interesting since it allows to understand the matter from a different perspective.

The understanding of this section depends in an essential manner on concepts present in the Appendix, especially section Appendix A.3 about the properties of rotations and skew-symmetric tensors.

In pursuing this approach we will introduce a tensor $\mathbf{W}$, built from the rotation $\mathbf{R}(t)$ and its derivative $\dot{\mathbf{R}}(t)$, that corresponds one-to-one to the angular velocity vector $\boldsymbol{\omega}$ in a sense we shall explain later.

To simplify notations, let us put

$$\mathbf{p} = y_1\mathbf{i}_1 + y_2\mathbf{i}_2 + y_3\mathbf{i}_3$$

in (2.8) so that we may recast the latter as

$$QP(t) = \mathbf{R}(t)\mathbf{p} \tag{2.40}$$

where, as we know, $\mathbf{R}(t)$ is the time-dependent rotation of a comoving basis with respect to a fixed basis. We differentiate this formula at a generic instant $t$ to find

$$\mathbf{v}_P - \mathbf{v}_Q = \dot{\mathbf{R}}\mathbf{p}\,, \tag{2.41}$$

since $\mathbf{p}$ is a constant vector. Recall that rotations are characterized by $\mathbf{R}^T\mathbf{R} = \mathbf{R}\mathbf{R}^T = \mathbf{I}$ and $\det \mathbf{R} = 1$. Hence, multiplying (2.40) on the left by $\mathbf{R}^T$ produces $\mathbf{R}^T QP = \mathbf{R}^T\mathbf{R}\mathbf{p}$, equivalent to $\mathbf{p} = \mathbf{R}^T QP$. Now substitute this expression of $\mathbf{p}$ into (2.41) to find

$$\mathbf{v}_P - \mathbf{v}_Q = \dot{\mathbf{R}}\mathbf{R}^T QP.$$

After putting

$$\mathbf{W} = \dot{\mathbf{R}}\mathbf{R}^T \tag{2.42}$$

the above reads

$$\mathbf{v}_P = \mathbf{v}_Q + \mathbf{W}QP\,. \tag{2.43}$$

The tensor $\mathbf{W}$, as we shall see straightaway, corresponds bijectively to the angular velocity, for which reason we may call it *spin tensor* or *vorticity tensor* (the latter is typical of Continuum Mechanics).

At this stage it becomes significant to remark that $\mathbf{W} = \dot{\mathbf{R}}\mathbf{R}^T$ is a *skew-symmetric* linear transformation. In fact, if we differentiate in time the identity $\mathbf{R}(t)\mathbf{R}^T(t) = \mathbf{I}$ we get

$$\dot{\mathbf{R}}\mathbf{R}^T + \mathbf{R}\dot{\mathbf{R}}^T = \mathbf{0}\,. \tag{2.44}$$

As $\mathbf{R}\dot{\mathbf{R}}^T = (\dot{\mathbf{R}}\mathbf{R}^T)^T$, Eq. (2.44) reads

$$\dot{\mathbf{R}}\mathbf{R}^T + (\dot{\mathbf{R}}\mathbf{R}^T)^T = \mathbf{0}$$

and by definition (2.42) this is equivalent to $\mathbf{W} + \mathbf{W}^T = \mathbf{0}$. Hence $\mathbf{W}$ is skew-symmetric, as per (A.25).

In the Appendix (see (A.32)) we show that any skew-symmetric transformation $\mathbf{W}$ acting on three-dimensional space corresponds to a unique vector, here denoted precisely $\boldsymbol{\omega}$, called *axial vector* associated with $\mathbf{W}$, such that $\mathbf{W}\mathbf{a} = \boldsymbol{\omega} \times \mathbf{a}$ for any vector $\mathbf{a}$ of Euclidean 3-space. By this, the angular velocity is the axial vector of the vorticity tensor $\mathbf{W} = \dot{\mathbf{R}}\mathbf{R}^T$.

Due to the above property Eq. (2.43) can be rewritten in the form

$$\mathbf{v}_P = \mathbf{v}_Q + \boldsymbol{\omega} \times QP$$

which is precisely formula (2.22) arising from Poisson's theorem. The conclusion is that $\boldsymbol{\omega}$ can be seen as the axial vector associated with the skew-symmetric transformation $\mathbf{W}$ obtained differentiating in time the rotation $\mathbf{R}$ and then right-multiplying by $\mathbf{R}^T$.

# Chapter 3
# Relative Motions

The velocities and the accelerations of the points of a system, just like the angular velocity of a rigid body, are not absolute quantities but they depend on the *reference frame* describing the motion. The aim of this chapter is to deduce the laws that govern how these three vectorial quantities change as the reference frame varies.

Before we proceed let us recall that a reference frame is schematically represented by an orthonormal basis and an origin. We suppose that the distance of two points in space and the time interval separating two events are *invariant* quantities, that is, independent of the reference frame. These two axioms characterize Classical Mechanics and distinguish it from Relativistic Mechanics, in which distances between points and time intervals do not have an absolute nature.

From a kinematical perspective there are privileged frames of reference, in contrast to what happens in dynamics (where, we shall see, *inertial* reference frames play a special role). Each reference frame has the right to be considered *fixed* and view the others as *moving* with respect to itself. In deducing the transformation rules of kinematic quantities it would therefore be a good idea to distinguish the two reference frames at play, just by using neutral labels such as "one" and "two". However, it turns out to be more convenient to distinguish them by calling them *fixed* and *moving*, even if that is formally incorrect. Let us begin by introducing the terminology and notation we shall use in this chapter. The *fixed* reference frame will be denoted by an orthonormal basis $\{\mathbf{i}_1, \mathbf{i}_2, \mathbf{i}_3\}$ with origin $\bar{O}$. The basis corresponding to the *moving* frame of reference is denoted $\{\mathbf{e}_1, \mathbf{e}_2, \mathbf{e}_3\}$ and origin $O$ (see Fig. 3.1). The kinematic quantities associated with the fixed reference frame will be labelled with the subscript "a" (standing for absolute), while for those associated with the moving frame we shall use "r" (for relative). As mentioned above, we should not be misled by the word "absolute", which is not intended to give the "fixed" reference frame any special feature at all, except that it coincides with the point of view of one of the observers we identify with. Finally, it is useful to emphasize that the orthonormal bases used to represent reference frames are themselves comoving with rigid systems, which

 
P. Biscari et al., *Rational Mechanics*, UNITEXT 177,
https://doi.org/10.1007/978-3-032-07462-1_3

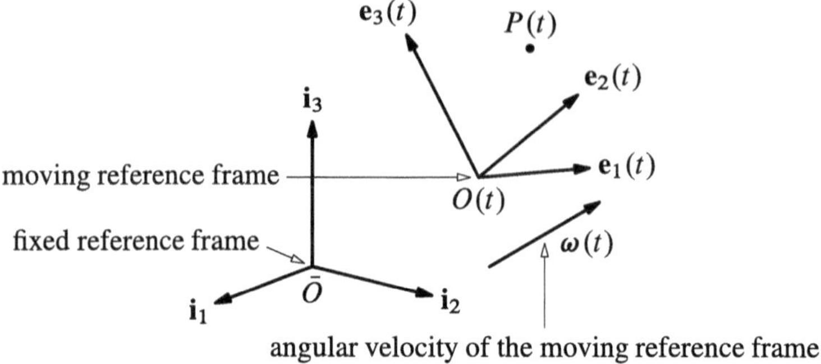


**Fig. 3.1** A fixed reference frame and a moving one

possess angular velocities relative to one another. So it will make sense to speak of the angular velocity of the moving frame with respect to the fixed frame, for instance.

## 3.1 Time Derivatives of a Vector

A generic vector $\mathbf{u}(t)$ depending on time has in general different derivatives with respect to distinct reference frames. Think for example of the unit vectors of a moving basis: they vary in the fixed reference frame, but are constant (and have thus zero time derivative) in the moving frame. This section's goal is that of finding a general law relating the time derivative of $\mathbf{u}(t)$ as seen in the fixed frame, which we denote

$$\dot{\mathbf{u}} = \frac{d_a \mathbf{u}}{dt},$$

and the time derivative of the same vector $\mathbf{u}(t)$ in the moving frame, which moves with angular velocity $\boldsymbol{\omega}$ with respect to the fixed frame. We indicate the latter by

$$\mathbf{u}' = \frac{d_r \mathbf{u}}{dt}$$

**Theorem 3.1** *The time derivative* $\dot{\mathbf{u}}$ *computed in the fixed reference frame is related to the time derivative* $\mathbf{u}'$ *computed in the moving reference frame by the relationship*

$$\dot{\mathbf{u}} = \mathbf{u}' + \boldsymbol{\omega} \times \mathbf{u}\,. \tag{3.1}$$

***Proof*** Using the components of $\mathbf{u}$ in the moving basis $\{\mathbf{e}_h\}$ we write

$$\mathbf{u} = u_1\mathbf{e}_1 + u_2\mathbf{e}_2 + u_3\mathbf{e}_3$$

where both the components and the unit vectors are functions of time. Differentiating in time (in the fixed frame) and using the Poisson formulas we obtain

$$\dot{\mathbf{u}} = \dot{u}_1\mathbf{e}_1 + \dot{u}_2\mathbf{e}_2 + \dot{u}_3\mathbf{e}_3 + \boldsymbol{\omega} \times u_1\mathbf{e}_1 + \boldsymbol{\omega} \times u_2\mathbf{e}_2 + \boldsymbol{\omega} \times u_3\mathbf{e}_3 \,. \tag{3.2}$$

Now, the time derivative of $\mathbf{u}$ in the moving frame is precisely equal to $\dot{u}_1\mathbf{e}_1 + \dot{u}_2\mathbf{e}_2 + \dot{u}_3\mathbf{e}_3$, since with respect to it the basis' unit vectors $\{\mathbf{e}_h\}$ are *fixed*. Hence

$$\mathbf{u}' = \dot{u}_1\mathbf{e}_1 + \dot{u}_2\mathbf{e}_2 + \dot{u}_3\mathbf{e}_3$$

and putting together the cross products in the right-hand side of (3.2) we find

$$\dot{\mathbf{u}} = \mathbf{u}' + \boldsymbol{\omega} \times \mathbf{u} \,.$$

□

This theorem has a simple but remarkable consequence, whose proof is trivial: taking $\mathbf{u} = \boldsymbol{\omega}$ in (3.1) has the effect that $\boldsymbol{\omega} \times \boldsymbol{\omega} = \mathbf{0}$.

**Corollary 3.2** *The time derivative of the angular velocity of the moving frame is the same in the fixed frame and in the moving frame*

$$\dot{\boldsymbol{\omega}} = \boldsymbol{\omega}' = \dot{\omega}_1\mathbf{e}_1 + \dot{\omega}_2\mathbf{e}_2 + \dot{\omega}_3\mathbf{e}_3 \,. \tag{3.3}$$

## 3.2 Composition of Velocities

Consider a point $P$ during a motion described in two frames. The velocity $\mathbf{v}_\mathrm{a}$ of $P$ measured in the fixed frame is clearly different (in general) from the velocity $\mathbf{v}_\mathrm{r}$ measured in the moving frame. We set out to determine the quantity we need to add to the latter to obtain the former. The answer is contained in a theorem due to Galilei, also known as *velocity composition law*. In the statement $\mathbf{v}_O$ and $\boldsymbol{\omega}$ are the origin's velocity and the angular velocity of the moving frame with respect to the fixed frame, respectively.

**Theorem 3.3** (Galilei) *The absolute velocity* $\mathbf{v}_\mathrm{a}$ *of a point* $P$ *is related to the relative velocity* $\mathbf{v}_\mathrm{r}$ *by*

$$\mathbf{v}_\mathrm{a} = \mathbf{v}_\mathrm{r} + \mathbf{v}_\tau \tag{3.4}$$

*where*

$$\mathbf{v}_\tau = \mathbf{v}_O + \boldsymbol{\omega} \times OP \tag{3.5}$$

*is called* drag velocity.

***Proof*** The position vector joining the origin $\bar{O}$ of the fixed frame to $P$ can be decomposed as the sum $\bar{O}P = \bar{O}O + OP$. Differentiating in time we find

$$\underbrace{(\bar{O}P)^{\boldsymbol{\cdot}}}_{\mathbf{v}_\mathrm{a}} = \underbrace{(\bar{O}O)^{\boldsymbol{\cdot}}}_{\mathbf{v}_O} + (OP)^{\boldsymbol{\cdot}}$$

and using (3.1) for the vector $\mathbf{u} = OP$, we obtain

$$\mathbf{v}_\mathrm{a} = \mathbf{v}_O + (OP)' + \boldsymbol{\omega} \times OP\,.$$

But $(OP)'$ is precisely the time derivative of the position vector of $P$ in the moving frame computed in the same reference frame, so it is by definition the relative velocity $\mathbf{v}_\mathrm{r}$. Hence the last relation may be recast as

$$\mathbf{v}_\mathrm{a} = \mathbf{v}_\mathrm{r} + \mathbf{v}_O + \boldsymbol{\omega} \times OP \tag{3.6}$$

which coincides with (3.4) once we define the drag velocity as in (3.5). □

Let us point out a few ideas present in this theorem.

- The time derivative of the proof is computed with respect to the fixed reference frame, in which the unit vectors $\{\mathbf{e}_h\}$ and the origin $O$ are moving.
- We should understand the physical meaning of the *drag velocity*. Its expression is similar to that of a roto-translational velocity distribution. If we imagine $P$ is *comoving* with the moving frame, and hence static with respect to the basis it represents it, its velocity is exactly $\mathbf{v}_O + \boldsymbol{\omega} \times OP$. Put differently: the drag velocity is the velocity the point *would have* if it were comoving with the moving frame. We may also say that $\mathbf{v}_\tau$ is the velocity that $P$ would have if it was simply *dragged* along by the moving frame. This observation justifies the choice of name.
- In case the moving frame has *zero* angular velocity, and so *translational* velocity distribution with respect to the fixed frame, we immediately deduce $\mathbf{v}_\tau = \mathbf{v}_O$. This means the drag velocity reduces to the velocity of the moving frame's origin.
- Two frames of reference observe the same velocity if and only if $\mathbf{v}_\tau = \mathbf{0}$. In other words, the velocity measurements coincide even if the origins and the bases are different, as long as neither the origin nor the basis used by the second frame move with respect to the first frame.

## 3.3 Composition of Accelerations

Employing the same conventions and notation as above we can deduce the relationship between the absolute acceleration $\mathbf{a}_\mathrm{a}$ and the relative acceleration $\mathbf{a}_\mathrm{r}$. The following theorem, due to Coriolis, establishes the so-called *composition law of the accelerations*. The quantities $\mathbf{a}_O$ and $\boldsymbol{\omega}$ obviously are the origin's acceleration and the angular velocity of the moving frame, while $\dot{\boldsymbol{\omega}}$ is the derivative of this angular velocity at the instant considered.

**Theorem 3.4** (Coriolis) *The absolute acceleration* $\mathbf{a}_\mathrm{a}$ *and the relative acceleration* $\mathbf{a}_\mathrm{r}$ *of a point* $P$ *are related by*

$$\mathbf{a}_\mathrm{a} = \mathbf{a}_\mathrm{r} + \mathbf{a}_\tau + \mathbf{a}_\mathrm{c} \tag{3.7}$$

*where* $\mathbf{a}_\tau$ *and* $\mathbf{a}_c$ *are defined as*

$$\mathbf{a}_\tau = \mathbf{a}_O + \dot{\boldsymbol{\omega}} \times OP + \boldsymbol{\omega} \times (\boldsymbol{\omega} \times OP) \qquad \mathbf{a}_c = 2\boldsymbol{\omega} \times \mathbf{v}_r \tag{3.8}$$

*and are respectively called* drag acceleration *and* Coriolis acceleration.

***Proof*** Differentiate (3.6) with respect to time to find

$$\mathbf{a}_a = (\mathbf{v}_r)^{\cdot} + \mathbf{a}_O + \dot{\boldsymbol{\omega}} \times OP + \boldsymbol{\omega} \times (OP)^{\cdot}.$$

Now use (3.1), first putting $\mathbf{u} = OP$ and then $\mathbf{u} = \mathbf{v}_r$, so that

$$\mathbf{a}_a = (\mathbf{v}_r)' + \boldsymbol{\omega} \times \mathbf{v}_r + \mathbf{a}_O + \dot{\boldsymbol{\omega}} \times OP + \boldsymbol{\omega} \times (OP)' + \boldsymbol{\omega} \times (\boldsymbol{\omega} \times OP).$$

At this point there only remains to note that $\mathbf{v}_r' = \mathbf{a}_r$ and $(OP)' = \mathbf{v}_r$ to conclude

$$\mathbf{a}_a = \mathbf{a}_r + \mathbf{a}_O + \dot{\boldsymbol{\omega}} \times OP + \boldsymbol{\omega} \times (\boldsymbol{\omega} \times OP) + 2\boldsymbol{\omega} \times \mathbf{v}_r.$$

If the drag and Coriolis accelerations are defined with (3.8), the above expression returns (3.7). □

For the Coriolis theorem as well a few comments are in order.

- A glance at (2.38) explains the name *drag* acceleration for $\mathbf{a}_\tau$, as defined in the first of (3.8). It is in fact the acceleration $P$ *would have* if it were rigidly attached to the moving frame and hence dragged by the latter's motion with respect to the fixed frame.
- The Coriolis acceleration, which might be called *complementary* acceleration, radically changes the composition law for the accelerations compared to velocity composition. The Coriolis acceleration vanishes when $\boldsymbol{\omega} = \mathbf{0}$, $\mathbf{v}_r = \mathbf{0}$ or more generally whenever $\boldsymbol{\omega} \times \mathbf{v}_r = \mathbf{0}$.
- In the very simple case in which the moving frame has constant angular velocity equal to zero ($\boldsymbol{\omega} = \dot{\boldsymbol{\omega}} = \mathbf{0}$), we deduce $\mathbf{a}_a = \mathbf{a}_r + \mathbf{a}_O$: the absolute acceleration is the sum of the relative acceleration plus the acceleration of moving frame's origin.
- A necessary and sufficient condition for two frames of reference to measure the same accelerations is the vanishing of $\mathbf{a}_O$, $\boldsymbol{\omega}$ and $\dot{\boldsymbol{\omega}}$. This means the two frames agree on the accelerations' measurements even if the origin used by the second frame is in motion with respect to the first frame, so long as this origin is in rectilinear translation motion. The relationships between the velocities and accelerations as measured in two such frames are called *Galilean transformations*: $\mathbf{v}_a = \mathbf{v}_r + \mathbf{v}_\tau$, $\mathbf{a}_a = \mathbf{a}_r$.

## 3.4 Composition of Angular Velocities

The angular velocity of a rigid body itself depends on the reference frame that describes the motion. Next we will find the relation between the angular velocity measured in the fixed frame, written $\boldsymbol{\omega}_{\rm a}$, and the angular velocity measured in the moving frame, written $\boldsymbol{\omega}_{\rm r}$. (We shall still call $\boldsymbol{\omega}$ the angular velocity of the second frame as seen by the first.)

**Theorem 3.5** *The angular velocity $\boldsymbol{\omega}_{\rm a}$ of a rigid body in a fixed frame is equal to the sum of its angular velocity $\boldsymbol{\omega}_{\rm r}$ in the moving frame and the angular velocity $\boldsymbol{\omega}$ of this frame with respect to the fixed one:*

$$\boldsymbol{\omega}_{\rm a} = \boldsymbol{\omega}_{\rm r} + \boldsymbol{\omega}\,. \tag{3.9}$$

*(Note that $\boldsymbol{\omega}$, the angular velocity of the moving frame, here plays the role of angular drag velocity, because it is the velocity the rigid body would have if it was comoving with the moving frame).*

***Proof*** Consider a vector $\mathbf{w}$ comoving with the rigid body in motion (see Fig. 3.2). Due to (2.17) the time derivative of $\mathbf{w}$ in the fixed frame is

$$\dot{\mathbf{w}} = \boldsymbol{\omega}_{\rm a} \times \mathbf{w} \tag{3.10}$$

where $\boldsymbol{\omega}_{\rm a}$ is the body's angular velocity seen in this frame. On the other hand the derivative of $\mathbf{w}$ for the moving frame equals

$$\mathbf{w}' = \boldsymbol{\omega}_{\rm r} \times \mathbf{w} \tag{3.11}$$

where now $\boldsymbol{\omega}_{\rm r}$ is the angular velocity of the body as seen by the moving frame. Relation (3.1) applied to $\mathbf{w}$ implies

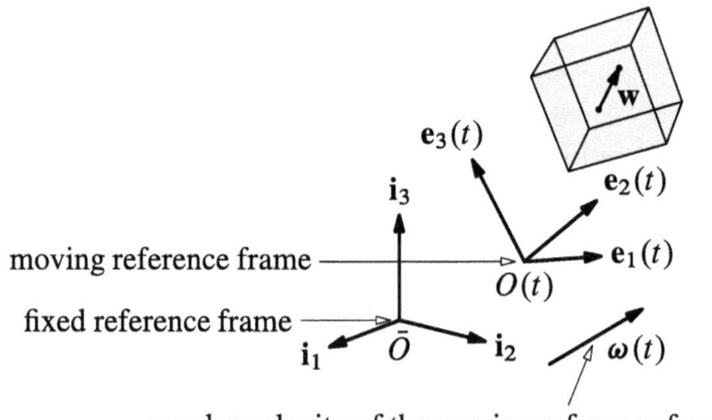


**Fig. 3.2** A vector **w** comoving with a rigid body

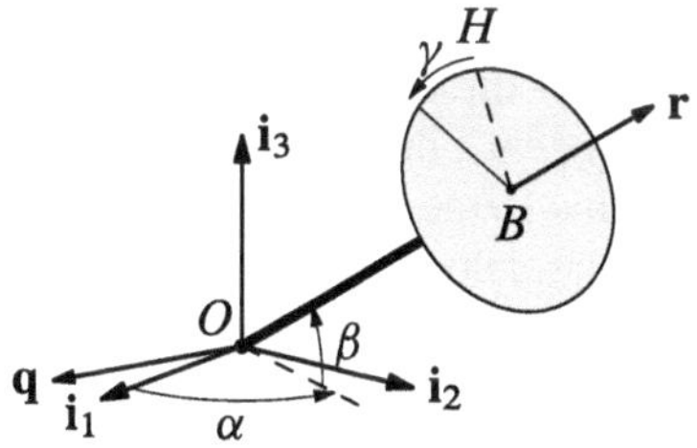


**Fig. 3.3** Deduction of $\boldsymbol{\omega}$ through the Cardano angles

$$\dot{\mathbf{w}} = \mathbf{w}' + \boldsymbol{\omega} \times \mathbf{w}$$

where $\boldsymbol{\omega}$ is the angular velocity of the moving frame with respect to the fixed one. Substituting (3.10) and (3.11) in this relation gives

$$\boldsymbol{\omega}_\mathrm{a} \times \mathbf{w} = \boldsymbol{\omega}_\mathrm{r} \times \mathbf{w} + \boldsymbol{\omega} \times \mathbf{w}$$

which we can rewrite as

$$[\boldsymbol{\omega}_\mathrm{a} - \boldsymbol{\omega}_\mathrm{r} - \boldsymbol{\omega}] \times \mathbf{w} = \mathbf{0} .$$

As $\mathbf{w}$ is arbitrary, we conclude $\boldsymbol{\omega}_\mathrm{a} = \boldsymbol{\omega}_\mathrm{r} + \boldsymbol{\omega}$. □.

The composition law for angular velocities (3.9) can be conveniently applied when one needs to find the angular velocity of a rigid body moving in space. In cases like this it is often useful to introduce one or more moving frames, each in planar motion with respect to the previous one, in order to simplify the process of recovering the (relative) angular velocity by means of the derivatives of the rotation angle. The required absolute angular velocity is then obtained adding up the quantities thus found.

**Example 3.6** Consider the system in Fig. 3.3, which consists of a rod with an end pinned to a *fixed* point $O$ and a comoving disc that is orthogonal to the rod, with center $B$ at the free end of the rod. To describe the system's configuration we only need the three angles in the picture: $\alpha$, $\beta$ and $\gamma$. Note that the angle $\gamma$ is formed by a direction $BH$ (dashed), that lies on the vertical plane containing the rod and the fixed unit vector $\mathbf{i}_3$, and another direction comoving with the disc (a radius). (It is not hard to recognize in this example the use of the Cardano angles, introduced in Sect. 2.2.1.)

We want to express the components of the angular velocity $\boldsymbol{\omega}$ *of the body* with respect to the fixed basis $\{\mathbf{i}_1, \mathbf{i}_2, \mathbf{i}_3\}$, in terms of the given angles and their derivatives.

The easiest way is to introduce a first moving frame that rotates about the unit vector $\mathbf{i}_3$, together with the vertical plane containing the rod, by an angle $\alpha$. It then has angular velocity $\boldsymbol{\omega}_1 = \dot{\alpha}\mathbf{i}_3$ for the fixed frame. Consider next a unit vector $\mathbf{q}$ orthogonal to the aforementioned plane and oriented coherently with $\beta$. We can choose a second frame with angular velocity $\boldsymbol{\omega}_2$, relative to the previous frame, equal to $\dot{\beta}\mathbf{q}$: it is merely the frame characterized by a basis with two unit vectors parallel to $\mathbf{q}$ and the rod, and the third vector perpendicular to these. This frame sees the rod's direction $OB$ as *not moving* with respect to itself, and from its point of view the system's motion is planar because it reduces to a rotation about $OB$. Notice that for this frame the direction $BH$ defined above (see Fig. 3.3) is *fixed*. The system's angular velocity $\omega_3$ relatively to the last reference frame is therefore equal to $\dot{\gamma}\mathbf{r}$, where $\mathbf{r}$ is a unit vector parallel to the rod. Using the composition law for the three relative angular velocities we conclude

$$\omega = \omega_1 + \omega_2 + \omega_3 = \dot{\alpha}\,\mathbf{i}_3 + \dot{\beta}\,\mathbf{q} + \dot{\gamma}\,\mathbf{r}\,. \tag{3.12}$$

By elementary trigonometry we find

$$\mathbf{q} = \sin\alpha\,\mathbf{i}_1 - \cos\alpha\,\mathbf{i}_2 \quad \mathbf{r} = \cos\beta\cos\alpha\,\mathbf{i}_1 + \cos\beta\sin\alpha\,\mathbf{i}_2 + \sin\beta\,\mathbf{i}_3$$

and substituting into (3.12) produces

$$\omega = (\dot{\beta}\sin\alpha + \dot{\gamma}\cos\beta\cos\alpha)\mathbf{i}_1 + (\dot{\gamma}\cos\beta\sin\alpha - \dot{\beta}\cos\alpha)\mathbf{i}_2 + (\dot{\alpha} + \dot{\gamma}\sin\beta)\mathbf{i}_3$$

as we wanted. □

This discussion, in a special case, solves the problem we posed in Sect. 2.6: write the components of the angular velocity vector in terms of rotation angles and their derivatives.

## 3.5 Angular Velocity and Euler Angles

It is also possible, and important, to express the angular velocity using the Euler angles and their first derivatives. To carry out the computations we refer to the notation and terminology adopted in Sect. 2.2, where the angles were defined.

Let us introduce three moving frames: the first, whose position is described by the basis $\{\tilde{\mathbf{e}}_i\}$, rotates by $\psi$ with respect to the fixed frame, with axis oriented as the unit vector $\mathbf{i}_3 = \tilde{\mathbf{e}}_3$; the second frame, denoted by the basis $\{\hat{\mathbf{e}}_i\}$, rotates by $\theta$ with respect to the first frame around the unit vector $\hat{\mathbf{e}}_1 = \tilde{\mathbf{e}}_1$; finally, the third frame, comoving with the rigid body and the basis $\{\mathbf{e}_i\}$, rotates by $\phi$ with respect to the second frame about the unit vector $\hat{\mathbf{e}}_3 = \mathbf{e}_3$.

With this construction we can immediately say that the relative angular velocities of each frame with respect to the preceding one are, respectively: $\dot{\psi}\,\mathbf{i}_3$, $\dot{\theta}\,\tilde{\mathbf{e}}_1$, $\dot{\phi}\,\hat{\mathbf{e}}_3$. Applying the composition law of angular velocities we obtain a first expression for $\omega$, the rigid body's angular velocity with respect to the fixed frame:

$$\omega = \dot{\psi}\,\mathbf{i}_3 + \dot{\theta}\,\tilde{\mathbf{e}}_1 + \dot{\phi}\,\hat{\mathbf{e}}_3\,.$$

Now we use the formulas for the components of each basis of unit vectors in terms of the preceding basis, and find that

$$\boldsymbol{\omega} = \dot{\psi}\,\mathbf{i}_3 + \dot{\theta}(\cos\psi\,\mathbf{i}_1 + \sin\psi\,\mathbf{i}_2) + \dot{\phi}(-\sin\theta\,\tilde{\mathbf{e}}_2 + \cos\theta\,\tilde{\mathbf{e}}_3)/,$$

Keeping into account

$$\tilde{\mathbf{e}}_2 = -\sin\psi\,\mathbf{i}_1 + \cos\psi\,\mathbf{i}_2 \qquad \tilde{\mathbf{e}}_3 = \mathbf{i}_3$$

we therefore obtain

$$\boldsymbol{\omega} = \dot{\psi}\,\mathbf{i}_3 + \dot{\theta}(\cos\psi\,\mathbf{i}_1 + \sin\psi\,\mathbf{i}_2) + \dot{\phi}\big(-\sin\theta(-\sin\psi\,\mathbf{i}_1 + \cos\psi\,\mathbf{i}_2) + \cos\theta\,\mathbf{i}_3\big)$$

which simplifies to

$$\boldsymbol{\omega} = (\dot{\theta}\cos\psi + \dot{\phi}\sin\theta\sin\psi)\mathbf{i}_1 + (\dot{\theta}\sin\psi - \dot{\phi}\sin\theta\cos\psi)\mathbf{i}_2 + (\dot{\psi} + \dot{\phi}\cos\theta)\mathbf{i}_3\,. \tag{3.13}$$

The above then gives the angular velocity's components in the fixed basis as functions of the Euler angles and their first derivatives.

Using formulas (2.12), a relatively demanding computation allows to express the angular velocity's components in the comoving basis $\{\mathbf{e}_i\}$. The result is

$$\boldsymbol{\omega} = \big(\dot{\theta}\cos\phi + \dot{\psi}\sin\theta\sin\phi\big)\mathbf{e}_1 + \big(\dot{\psi}\sin\theta\cos\phi - \dot{\theta}\sin\phi\big)\mathbf{e}_2 + \big(\dot{\psi}\cos\theta + \dot{\phi}\big)\mathbf{e}_3\,, \tag{3.14}$$

which we will need later on.

**Example 3.7** (*Precessional motions*) We introduced in Sect. 2.5.2 precessions, which are particular polar motions for which a fixed axis $p$ forms a constant angle with a comoving axis $r$. We showed that a precession is characterized by having angular velocity coplanar to $p$ and $r$, and hence is decomposable into

$$\boldsymbol{\omega} = \boldsymbol{\omega}_{\text{pre}} + \boldsymbol{\omega}_{\text{rot}} \quad (\boldsymbol{\omega}_{\text{pre}} \parallel p,\ \boldsymbol{\omega}_{\text{rot}} \parallel r) \tag{3.15}$$

where $\boldsymbol{\omega}_{\text{pre}}$, parallel to the fixed axis $p$, is called precessional angular velocity, while $\boldsymbol{\omega}_{\text{rot}}$, parallel to the comoving axis $r$, is the proper angular velocity. We can recover this result by introducing a frame of reference that rotates around the fixed axis $p$ and comoves with $r$. This frame will simply see a system in rotational motion with angular velocity parallel to $r$, say $\boldsymbol{\omega}_{\text{rot}}$. Furthermore, since this moving frame rotates around $p$ with its own angular velocity $\boldsymbol{\omega}_{\text{pre}}$, from the composition law of angular velocities (3.9) we infer decomposition (3.15).

The characteristics of a precessional motion for the system in Fig. 3.3, which was used to explain how to compute the angular velocity, can be deduced by assuming that the angle $\beta$ is constant, as is its complementary angle $\theta$ (see Fig. 3.4a). In this way the axis aligned as $\mathbf{i}_3$ will play the role of precession axis, while the axis parallel to the rod and the moving vector $\mathbf{r}$ will be the axis of proper rotation. The precessional angular

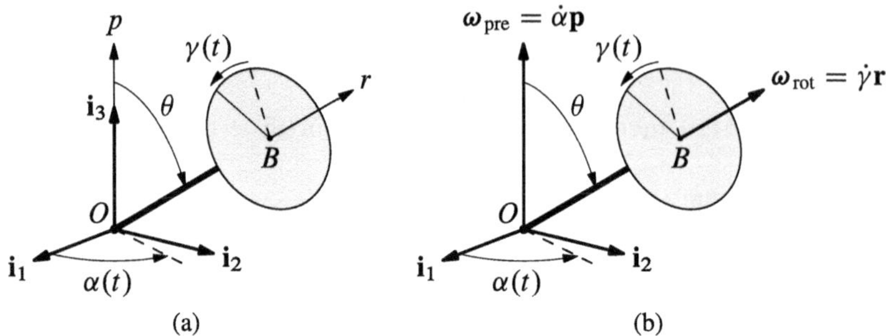


**Fig. 3.4** Angular velocity of a precessional motion: $\omega = \omega_{\text{pre}} + \omega_{\text{rot}}$. (a) precessional motion ($\theta =$ cost.); (b) angular velocity: $\omega = \dot{\alpha}\mathbf{p} + \dot{\gamma}\mathbf{r}$

velocity then equals $\omega_{\text{pre}} = \dot{\alpha}\,\mathbf{p}$, where $\mathbf{p} = \mathbf{i}_3$, and the proper angular velocity is just $\omega_{\text{rot}} = \dot{\gamma}\,\mathbf{r}$. The total angular velocity is therefore the sum of the two, and is coplanar to the axes $p$ and $r$ (see Fig. 3.4b).

We remind that a precession is called regular if the components $\dot{\alpha}$, $\dot{\gamma}$ are constants. □

**Remark 3.8** Contrary to what we have did in Chap. 2, almost trivially, with (2.32) in the case of planar motions, now we can no longer write the infinitesimal rotation vector $\varepsilon$ as the differential of a function of the Euler angles, in general. In other words there exists no vector $\gamma(\psi, \theta, \phi)$ such that $\omega = \dot{\gamma}$. In fact (3.13) implies that

$$\begin{aligned}\varepsilon = \omega\, dt = (\cos\psi d\theta + \sin\theta\sin\psi d\phi)\mathbf{i}_1 + (\sin\psi d\theta - \sin\theta\cos\psi d\phi)\mathbf{i}_2 \\ +(d\psi + \cos\theta d\phi)\mathbf{i}_3\end{aligned} \tag{3.16}$$

but one can prove that the expression on the right is not the differential of any function of the Euler angles. Put equivalently, it is a *differential* 1-*form* though *not* an exact one. Indeed if there were a vector $\gamma(\psi, \theta, \phi)$ such that $\varepsilon = d\gamma$, the components $(\varepsilon_1, \varepsilon_2, \varepsilon_3)$, written in (3.16) with the Euler angles's differentials, would have to satisfy

$$\begin{aligned}\cos\psi d\theta + \sin\theta\sin\psi d\phi &= \frac{\partial\gamma_1}{\partial\psi}d\psi + \frac{\partial\gamma_1}{\partial\theta}d\theta + \frac{\partial\gamma_1}{\partial\phi}d\phi \\ \sin\psi d\theta - \sin\theta\cos\psi d\phi &= \frac{\partial\gamma_2}{\partial\psi}d\psi + \frac{\partial\gamma_2}{\partial\theta}d\theta + \frac{\partial\gamma_2}{\partial\phi}d\phi \\ d\psi + \cos\theta d\phi &= \frac{\partial\gamma_3}{\partial\psi}d\psi + \frac{\partial\gamma_3}{\partial\theta}d\theta + \frac{\partial\gamma_3}{\partial\phi}d\phi\end{aligned}$$

But from the first relation we see that

$$\frac{\partial \gamma_1}{\partial \psi} = 0 \qquad \frac{\partial \gamma_1}{\partial \theta} = \cos \psi \qquad \frac{\partial \gamma_1}{\partial \phi} = \sin \theta \sin \psi \,,$$

differentiating which would produce the incompatible expressions

$$\frac{\partial^2 \gamma_1}{\partial \theta \, \partial \psi} = 0 \qquad \frac{\partial^2 \gamma_1}{\partial \psi \, \partial \theta} = - \sin \psi \,.$$

□

# Chapter 4
# Constraints

The systems we wish to describe are made of particles and rigid bodies that are not totally free to move, but are instead subject to *constraints*, that is, *a priori* restrictions on the motions allowed. Constraints are in general made of pins and similar devices, although we should make clear that we are not mainly interested in the way they are physically constructed, but only in their mathematical modelling, and especially the consequences they cause on the system's kinematics.

It is important to remember that the following instance is a constraint: imposing that two surfaces roll on one another without slipping, or more generally that they have contact points (that said, we will explain that these restrictions are of a different nature from those that merely constrain the position of the points).

This chapter is structured so to present the main concepts via a series of meaningful examples. The discussion will be summarized in Sect. 4.11, where the formal definitions will become easier to understand.

## 4.1 Examples of Constrained Systems

For each example of constrained mechanical system we will proceed as if we were "extracting" from a toolbox the pieces that form the system (rods, points etc.), initially free on the plane and in space, and build the system by successively introducing the constraints, which we will describe mathematically. Beware that these examples are only meant to introduce gradually and concretely more general ideas and concepts, so that the corresponding definitions will later seem sufficiently motivated.

 
P. Biscari et al., *Rational Mechanics*, UNITEXT 177,
https://doi.org/10.1007/978-3-032-07462-1_4

### 4.1.1 Point on a Fixed Circular Guide

Consider a point $P$ constrained to move on a circle of radius $R$ and center $O$. To describe this and other similar conditions it is customary to think of a small ring threaded on a guide (fixed or moving), a picture that well represents the physical reality we wish to model, as shown in Fig. 4.1. The ring should clearly be thought of as being dimensionless, while the guide is described by a regular curve.

After we choose a suitable coordinate system, the restriction that the constraint imposes on the position vector $OP = x\mathbf{i} + y\mathbf{j} + z\mathbf{k}$ is described by the relations

$$z = 0 \qquad x^2 + y^2 = R^2 .$$

Here it is convenient to use polar coordinates, and use the angle $\theta$ at the center as variable of the position vector $OP$:

$$OP(\theta) = R\cos\theta\mathbf{i} + R\sin\theta\mathbf{j} . \tag{4.1}$$

In absence of constraints the description of the position of $P$ in space requires three parameters, its three Cartesian coordinates. But the above constraint reduces this number to one: the angle at the center, or any quantity equivalent to it. Notice that knowledge of the function $\theta(t)$ and its derivatives allows us to write the constrained ring's velocity

$$\mathbf{v}_P(t) = \frac{\partial P}{\partial\theta}\dot{\theta} = -R\dot{\theta}\sin\theta\mathbf{i} + R\dot{\theta}\cos\theta\mathbf{j} \tag{4.2}$$

and acceleration

$$\mathbf{a}_P(t) = (-R\ddot{\theta}\sin\theta - R\dot{\theta}^2\cos\theta)\mathbf{i} + (R\ddot{\theta}\cos\theta - R\dot{\theta}^2\sin\theta)\mathbf{j} \tag{4.3}$$

at each instant.

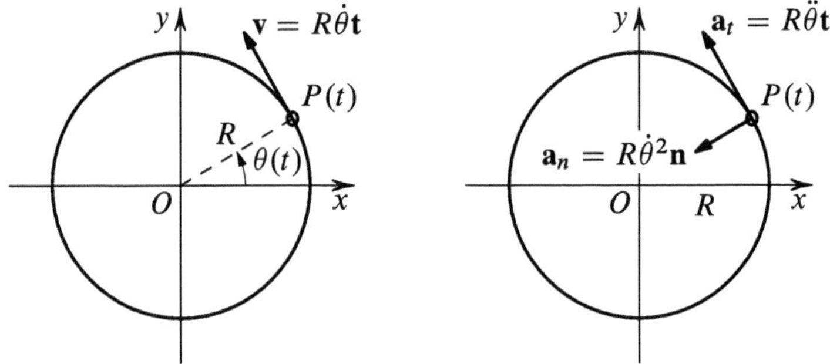


**Fig. 4.1** A point constrained to move on a fixed circular guide

Calling $\mathbf{t} = -\sin\theta\mathbf{i} + \cos\theta\mathbf{j}$ and $\mathbf{n} = -\cos\theta\mathbf{i} - \sin\theta\mathbf{j}$ the unit tangent and unit normal to the circle, from formulas (4.2) and (4.3) we obtain

$$\mathbf{v} = R\dot{\theta}\mathbf{t} \qquad \mathbf{a} = R\ddot{\theta}\mathbf{t} + R\dot{\theta}^2\mathbf{n}\,.$$

We can then express the point's position on the circle by means of the parameter $\theta$, which we view as *free* because it can be changed arbitrarily and still produce configurations that respect the constraints. Moreover, the velocity is given as a function of $\theta$ and $\dot{\theta}$.

The angle $\theta$ thus plays the role of a parameter that allows to describe the motion of the system's points, here clearly only one, and by differentiating therefore gives the velocity and the acceleration.

We anticipate the fact that the parameters (one or more) playing such a role are typically called *generalized* or *Lagrangian coordinates* and denoted by $q$ (in particular, here $q = \theta$), labelled by indices when there is more than one. In the present example we needed exactly one generalized coordinate to know the position and the motion of $P$, so

$$P(q) \qquad \mathbf{v}_P = \frac{\partial P}{\partial q}\dot{q}$$

are the natural generalizations of (4.1) and (4.2) for a generic systems with a single generalized coordinate $q$.

### 4.1.2 Slider-Crank

*Pins* are among the most common constraints in planar systems. They are usually employed when we want to prevent a point from moving or we want it to coincide with another point, still leaving the body free to rotate around it. In the first case we speak of *fixed* pins, whose representation is usually similar to Fig. 4.2, left. As is immediate to understand, the presence of such a constraint on the plane lowers by 2 the number of parameters necessary to describe the system's configuration, because in this case the coordinates of the point coinciding with the pin are prescribed a priori.

A rigid body thus constrained must have rotational velocity distribution around the fixed pin, which acts as instantaneous center of rotation. The velocities are therefore always orthogonal to the lines joining the points to the pin, and proportional to the distance (as shown in Fig. 4.2).

Naturally we can also consider parts of a system attached by moving pins, which then act as internal constraints. Using this type of constraint and a roller we now build a simple but interesting mechanical system called slider-crank, which transforms a rotational motion into a linear motion and conversely.

Suppose we have two rigid rods $HK$ and $AB$ of respective lengths $2l$ and $l$, which a priori move on a plane, so that their configurations are determined by the position

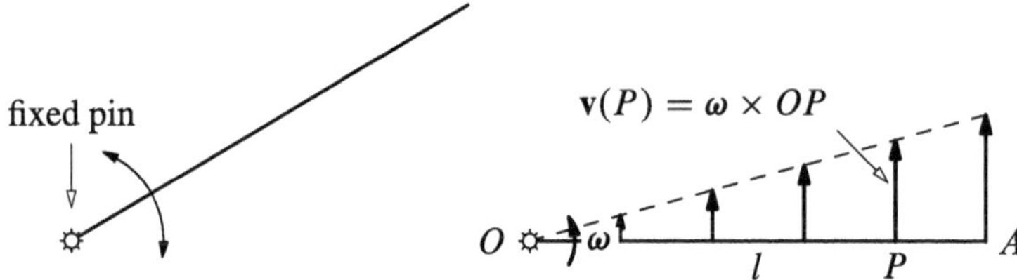


**Fig. 4.2** A rod constrained by a fixed pin and a compatible velocity distribution

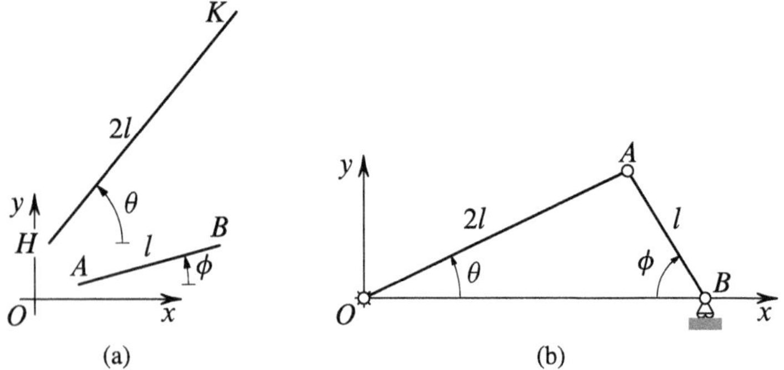


**Fig. 4.3** A slider-crank linkage: (a) rods free to move on a plane; (b) rods constrained by two pins and a roller

of the ends $H$ and $A$ and by the angles of rotation $\theta$ and $\phi$, as in Fig. 4.3a. We then have 6 parameters, at the moment all free and independent:

$$x_H, \quad y_H, \quad x_A, \quad y_A, \quad \theta, \quad \phi. \tag{4.4}$$

The first step to build the system of Fig. 4.3b is to use a *fixed* pin to make the end $H$ of the first rod coincide with a point on the plane, which we choose as the origin $O$ of a system of Cartesian coordinates with $x$-axis directed as the unit vector $\mathbf{i}$ parallel to the guide on which the second rod's end $B$ will glide.

Next we connect with a second pin the ends $K$ and $A$ and then impose that $B$ moves along the fixed axis. Each constraint can be described by a mathematical condition: we have

$$H = O, \quad K = A, \quad y_B = 0. \tag{4.5}$$

We can classify the first and last constraint as *external*, since they constrain points in the system to points outside of the system. On the other hand the pin forcing $K = A$ should be considered an *internal* constraint because it imposes a restriction on two points belonging to the system.

Equations (4.5) may be recast as

$$x_H = y_H = 0, \quad x_K = x_A, \quad y_K = y_A, \quad y_B = 0.$$

Now the question is: what restrictions do these five relations impose on the parameters (4.4) chosen to describe the system? Evidently $x_H$ and $y_H$ are now fixed and equal to zero, and the given constraints reduce to the 5 conditions

$$x_H = 0, \quad y_H = 0, \quad x_A - 2l\cos\theta = 0, \quad y_A - 2l\sin\theta = 0, \quad y_A - l\sin\phi = 0, \tag{4.6}$$

(note that for simplicity the angle $\phi$ is oriented as in Fig. 4.3b and the last constraint forces $y_B = 0$, as required by the presence of the roller).

Relations (4.6) must be read as a system of five equations in the six parameters (4.4). We may summarize them as

$$\lambda_i(x_H, y_H, x_A, y_A, \theta, \phi) = 0, \quad i = 1, \ldots, 5.$$

The existence of five, in general nonlinear relationships, among the six parameters suggests we try to express 5 of them as regular functions of the remaining one, or even all of them in function of a further sixth one to be introduced. Mathematically, this boils down to an application of the theory of implicit functions, in particular of the Implicit function theorem. In the present context it seems though more instructive to adopt a more direct mechanical approach, which sacrifices the maximum possible generality in favour of a more concrete understanding of the problems involved.

A glance at the system in Fig. 4.3b leads us to think, correctly, that a single parameter (in particular, $\theta$ or $\phi$) is necessary and sufficient to construct a system of generalized coordinates. This statement immediately raises two objections which we must address.

Suppose we have fixed in Fig. 4.3b the angle $\theta$ and note that its value does not determine a *unique* compatible configuration. In fact, Fig. 4.4 shows a different configuration that is compatible with the *same* value of $\theta$ as in Fig. 4.3b. So one could ask if it is correct to say that $\theta$ by itself forms a system of generalized coordinates (i.e., a collection of essential and independent parameters) for the system, given that it does not determine the system's configuration uniquely.

The answer is still yes, notwithstanding the highlighted ambiguity. Although the knowledge of $\theta$ does not determine *uniquely* the value of $\phi$, in fact, it nevertheless restricts its range to a *discrete* set of values, as shown in the pictures. This is enough to recognize $\{\theta\}$ as a set of generalized coordinates. If instead the knowledge of $\theta$ had allowed $\phi$ to take values in an *interval* then $\theta$ alone would not have constitute a system of generalized coordinates.

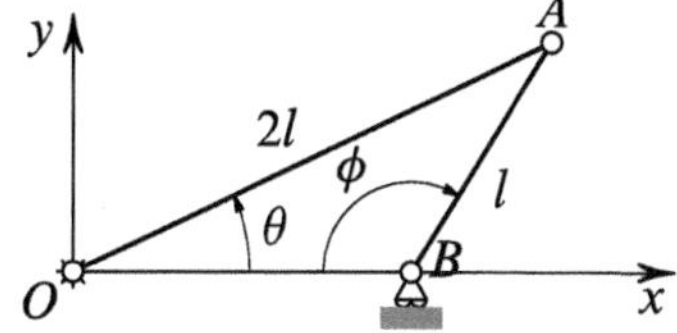


**Fig. 4.4** The two rods of Fig. 4.3b in a different configuration but with the same value of $\theta$

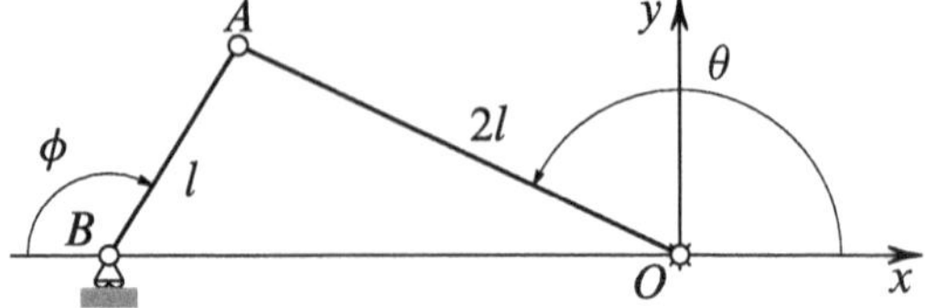

**Fig. 4.5** The two rods of Fig. 4.3b in a different configuration, for which the value of $\phi$ is nevertheless the same

Put equivalently, fixing the generalized coordinates must prevent the system's motion; but we do not need the compatible configuration to be unique.

Let us now consider the angle $\phi$ and for example note that the value shown in Fig. 4.3b is compatible with the configuration of Fig. 4.5 as well. In this case though we must say that the latter configuration cannot be reached in any way starting from the former (that is, using an exquisitely mathematical notion, the space of configurations of the system is assumed to be connected). What we see in Fig. 4.5 is indeed a *different mechanical system*, albeit made of the same constituent elements. We can therefore say that the coordinates $\phi$ determines the system's configuration *uniquely*.

In the sequel we study in more detail this simple but instructive system.

It is quite easy to express the coordinates of $A$ and $B$ via the angles $\theta$ and $\phi$

$$A = (2l\cos\theta, 2l\sin\theta) \qquad B = (2l\cos\theta + l\cos\phi, 2l\sin\theta - l\sin\phi)\,,$$

but the roller's presence at $B$ forces $y_B = 0$, in other words

$$f(\theta, \phi) = 2\sin\theta - \sin\phi = 0\,. \tag{4.7}$$

It is precisely this relation that shows the link between $\theta$ and $\phi$ and allows to recover $\phi = \tilde{\phi}(\theta)$ or $\theta = \tilde{\theta}(\phi)$, depending on whether we choose $\theta$ or $\phi$ as generalized coordinate.

Let us ask the following: is there a conceptual difference, apart from practical ones such as the simplification of calculations, between the coordinates $\theta$ and $\phi$? Equation (4.7) can be solved for one or the other angle. We immediately discover a few important peculiarities, found either analytically or by direct inspection of the system's geometry: while $\phi$ can take any value (modulo $2\pi$), the angle $\theta$ is forced to stay between $\theta_{\min} = -\pi/6$ and $\theta_{\max} = \pi/6$ (the situation for $\theta_{\max}$ is shown in Fig. 4.10a). Moreover for each value of $\theta$ in the interval $-\pi/6 \le \theta \le \pi/6$ there exist *two* possible configurations, corresponding to $\phi = \arcsin(2\sin\theta)$, $\phi = \pi - \arcsin(2\sin\theta)$ (this is 100% evident when directly analyzing the system, but it can as well be deduced mathematically).

From these considerations it follows immediately that $\theta$ can be employed to describe separately the set of configurations where $B$ stays at the right of $A$ ($-\pi/2 < \phi < \pi/2$), or those where $B$ stays at the left of $A$ ($\pi/2 < \phi < 3\pi/2$). On the other hand $\phi$ can parametrize *globally* the set of possible configurations. The latter is not the general situation though: for complex mechanical systems there exist no generalized coordinates that "work" globally. Rather, we must make do with local parametrizations, for specific subsets of possible configurations. All of this has a mathematical interpretation: in a nutshell we are describing a (hyper)surface (the set of possible configurations for the mechanical system) by different coordinate systems, a problem that can be discussed and properly understood with the tools of Differential Geometry.

There remains to discuss and clarify the situation corresponding to the extremal values $\theta_{\max}$ and $\theta_{\min}$. In each case the system's position is *uniquely determined* by the value of $\theta$, and therefore we might think that this angle is a generalized coordinate also for special configurations such as the above. However, solving (4.7) for $\phi = \tilde{\phi}(\theta)$ gives

$$\tilde{\phi}(\theta) = \arcsin(2\sin\theta)\,,$$

which is *not* differentiable for $\sin\theta = 1/2$ ($\theta_{\max} = \pi/6$). (Even if we had not been able to make $\tilde{\phi}(\theta)$ explicit from (4.7) because $f_\phi = -\cos\phi$ vanishes at $\phi = \pi/2$, i.e. for $\theta = \pi/6$, we could still have reached the same conclusion by relying on the Implicit function theorem and related results). The lack of differentiability at $\pi/6$ prevents $\theta$ from being a generalized coordinate for the corresponding configuration: we would not be able to write the points' *velocities* for the motions in this configuration.

The conclusion is that the constraints introduced to define and build the slider-crank system in Fig. 4.3b reduce the number of parameters necessary to determine the system's configuration from six to one: it suffices to assign $\phi$ in order to know the value of every other parameter. In this way a generic motion is given by an arbitrary function $\phi(t)$: the positions of *all* points can be deduced from the value of $\phi$, which can vary freely. Hence we say that $\phi$ plays the part of a *generalized coordinate*: to each value we freely give it there corresponds a well-defined configuration of the system.

The angle $\theta$, varying between $-\pi/6$ and $\pi/6$, can instead be used as generalized coordinate only on two disjoint subsets of configurations, with the extreme values excluded. It is then manifest, in this case, why choosing $\phi$ is more convenient than $\theta$.

### 4.1.3 Rod Constrained at One End

We shall call *roller* a particularly useful kinematical restriction. The representation via pictures can take several forms, all of which are meant to convey the idea that while a point of a rigid body is constrained to move along a given guide, the body is free to rotate around it on the plane of motion.

Looking at Fig. 4.6 left, where the possible movements are indicated by arrows, we see that in practice this constraint limits the range of motions of the system, reducing by one the number of parameters needed to describe the motion.

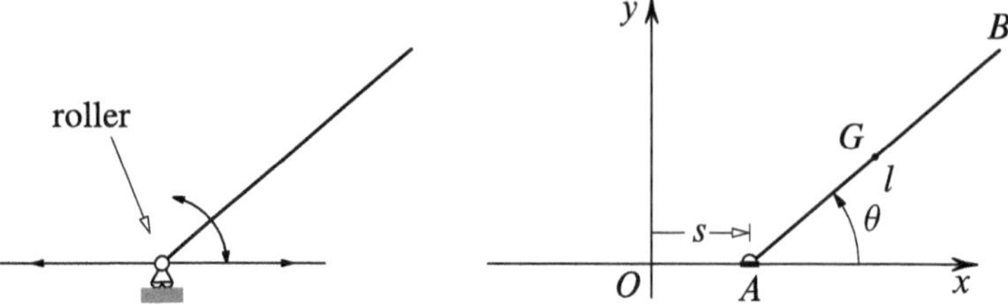


**Fig. 4.6** A rod with one end constrained by a roller and generalized coordinates $(s, \theta)$

Consider now a rod $AB$ of length $l$ and let us force the end $A$ to glide along a fixed guide, as in Fig. 4.6 right (note that here the roller is represented in a different way from how it appears in the left figure). Without the roller the system's configurations would be detected by *three* independent parameters, because in general we need three parameters to know the position of a rigid body on the plane (see Sect. 2.5.2). The presence of the constraint reduces this number to *two*. Indeed, the abscissa $s$ of $A$ with respect to a fixed point $O$ and the angle $\theta$ between $AB$ and the guide clearly determine uniquely the position, since $y_A = 0$ describes the restriction imposed by the roller's presence.

Next we show how to express the position of an arbitrary point $P$ of the rod in function of the parameters $s$ and $\theta$ more concretely. For example, for $A$ and the midpoint $G$, we have

$$OA(s, \theta) = s\,\mathbf{i}, \qquad OG(s, \theta) = \left(s + \frac{l}{2}\cos\theta\right)\mathbf{i} + \frac{l}{2}\sin\theta\,\mathbf{j}.$$

We will start using from now on a terminology that will be made precise in the sequel, and say $s$ and $\theta$ are *generalized coordinates* for the system. These are *essential* parameters, because it is not possible to drop either, and *independent*, because if one is fixed the other can still vary to produce infinitely many configurations for the system.

At this point we can give a more general and formal definition:

- *generalized coordinates*: a collection of $N - 1$ *essential* and *independent* parameters $q_k$ that allow to describe the system's configurations.

The velocities can be found by differentiating the previous relations, and computed explicitly once the functions $s(t)$ and $\theta(t)$ that determine the motion are known. In fact

$$\mathbf{v}_A = \frac{dA}{dt} = \frac{\partial A}{\partial s}\dot{s} + \frac{\partial A}{\partial \theta}\dot{\theta} = \dot{s}\mathbf{i}$$

$$\mathbf{v}_G = \frac{dG}{dt} = \frac{\partial G}{\partial s}\dot{s} + \frac{\partial G}{\partial \theta}\dot{\theta} = \left(\dot{s} - \frac{l}{2}\sin\theta\dot{\theta}\right)\mathbf{i} + \frac{l}{2}\cos\theta\dot{\theta}\mathbf{j}$$

Clearly the same considerations apply to a generic point $P$ whose position is given by a function $P(s, \theta)$. The velocity is found differentiating in time the composite map $P(t) = P\big(s(t), \theta(t)\big)$:

$$\mathbf{v}_P = \frac{dP}{dt} = \frac{\partial P}{\partial s}\dot{s} + \frac{\partial P}{\partial \theta}\dot{\theta}\,.$$

Calling $q_1 = s$ and $q_2 = \theta$ the chosen parameters, the relations can be written in the more compact form

$$\mathbf{v}_P = \frac{\partial P}{\partial q_1}\dot{q}_1 + \frac{\partial P}{\partial q_2}\dot{q}_2 = \sum_{k=1}^{2}\frac{\partial P}{\partial q_k}\dot{q}_k\,. \tag{4.8}$$

The *elementary displacement* the point undergoes in the infinitesimal time interval during which the generalized coordinates have changed by $dq_k$ is obviously

$$dP = \frac{\partial P}{\partial q_1}dq_1 + \frac{\partial P}{\partial q_2}dq_2 = \sum_{k=1}^{2}\frac{\partial P}{\partial q_k}dq_k\,. \tag{4.9}$$

Let us now suppose the system is in a *given* position, compatible with the constraint at some *given* instant. Speaking informally, we might say that we are looking at a "snapshot" of the system and we are ignoring the "history" of its motion in the preceding and successive instants. Let us think now of a velocity distribution compatible with the constraints at the instant and position considered. We will call such a velocity distribution *virtual*. The *virtual velocities* of the single points are therefore not found by substituting in (4.8) the $\dot{q}_k$ relative to a real motion $q_k(t)$. Rather, they arise by substituting the $\dot{q}_k$ with some coefficients, suggestively denoted $\delta q_k/\delta t$, that are *not* time derivatives, but parameters of suitable physical dimensions, the same as $\dot{q}_k$. Traditionally virtual velocities are marked by a dash, to distinguish them conceptually and graphically from effective velocities. With this convention we have

$$\mathbf{v}'_P = \frac{\partial P}{\partial q_1}\frac{\delta q_1}{\delta t} + \frac{\partial P}{\partial q_2}\frac{\delta q_2}{\delta t} = \sum_{k=1}^{2}\frac{\partial P}{\partial q_k}\frac{\delta q_k}{\delta t}\,. \tag{4.10}$$

To simplify the notation and also highlight the nature of the $\delta q_k/\delta t$ as *parameters* and *not* derivatives one can set $\nu_k = \delta q_k/\delta t$, so to recast (4.10) as

$$\mathbf{v}'_P = \frac{\partial P}{\partial q_1}\nu_1 + \frac{\partial P}{\partial q_2}\nu_2 = \sum_{k=1}^{2}\frac{\partial P}{\partial q_k}\nu_k\,. \tag{4.11}$$

It is important to perceive the subtle theoretical distinction between expression (4.8) for the effective velocity and (4.10) for the virtual velocity. In the former we have chosen a motion described by functions $q_k(t)$ and we have found the velocity of $P$ associated to it. In the latter, instead, the formula intends to give the collection (the totality) of velocities of $P$ that are compatible with the constraints at the instant of concern. What is true is that by substituting, in (4.10), to the arbitrary quantities $\delta q_k/\delta t$ the $\dot{q}_k$ coming from a given motion, we do find the corresponding effective velocity of the point. This can be summarized, in *this particular example*, by saying the effective velocity is among the virtual velocities. As we will see later, this fact is *not* true in general, even though here it seems almost obvious.

It is further possible to define the corresponding *virtual displacements* by replacing in (4.9) the effective infinitesimal variations $dq_k$, subordinated to a motion $q_k(t)$, with the virtual variations $\delta q_k$, which might not correspond to the infinitesimal displacements of the generalized coordinates during the motion. In this way we clearly obtain a quantity of the dimensions of a length, i.e.

$$\delta P = \frac{\partial P}{\partial q_1}\delta q_1 + \frac{\partial P}{\partial q_2}\delta q_2 = \sum_{k=1}^{2} \frac{\partial P}{\partial q_k}\delta q_k \,. \tag{4.12}$$

In conclusion, a virtual displacement is therefore an arbitrary infinitesimal displacement that respects the constraints at the given position and instant. Note once more the use of the symbol $\delta P$, chosen to distinguish the displacement from the effective infinitesimal displacement $dP$, which instead occurs in correspondence to a given motion.

Here, too, one should pay attention to the conceptual difference between (4.9), which describes the effective infinitesimal displacement subordinated to a motion of the system, and (4.12), which instead furnishes the totality of possible infinitesimal displacements compatible with the constraints.

Distinguishing between velocities and virtual displacements is more of a formal than a substantial matter: qualitatively they are the same vectors, viewed as velocities or infinitesimal displacements depending on the case. The notion of virtual displacement traditionally crops up in problems of statics, where it is more "natural", whereas in dynamics virtual velocities are preferred.

At last, we underline how comparing (4.10), or the equivalent (4.11), and (4.12) leads us to say, informally but suggestively, that a virtual velocity is obtained by "dividing" the virtual displacement by an infinitesimal time.

We wrap up this example with the formal definition of virtual displacement and virtual velocity distribution:

- virtual velocity: a velocity compatible with the constraints, thought of as *fixed* at the considered instant;
- virtual displacement: an infinitesimal displacement compatible with the constraints, thought of as *fixed* at the instant considered.

In the example that we have discussed, pointing out that the constraints should be viewed as fixed seems completely superfluous, because they are fixed anyway. We will see in the next example how this becomes important, because in the following simple system the constraints are *moving*.

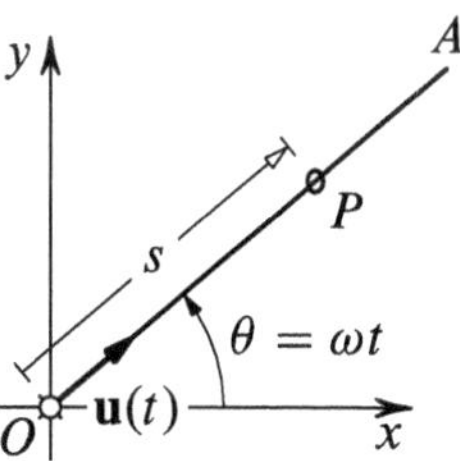


**Fig. 4.7** A variable constraint: a small ring inserted on a guide that is rotating with constant angular velocity

### *4.1.4 A Variable Constraint: Point on a Rotating Guide*

The example we wish to study is made of a rigid rod, pinned at its end $O$ to a fixed point, along which a small ring $P$ representing a particle is constrained to move freely. Suppose the rod is in uniform rotational motion with constant angular velocity $\omega$ around $O$, thus becoming a moving constraint that restricts the possible motions of the ring. In cases like this, when a rod or a one-dimensional body only acts as constraint for the system of interest (the ring), we call it more appropriately a *guide*.

The position of $P$ on the revolving guide is determined by the *generalized coordinate* $s$, abscissa of $P$ with respect to $O$ (Fig. 4.7). The constraint requires that the coordinates $x$ and $y$ of $P$ with respect to axes centered at $O$ are not arbitrary, as for a free point on the plane. They must satisfy

$$y\cos(\omega t) - x\sin(\omega t) = 0\,, \tag{4.13}$$

obtained under the assumption the guide is parallel to the $x$-axis at time zero. This is the representation of the constraint $OP \times \mathbf{u}(t) = \mathbf{0}$, where $\mathbf{u}(t) = (\cos\omega t, \sin\omega t)$ is the time-dependent unit vector giving the guide's direction.

The constraint thus introduced is interesting because it depends *explicitly* on time $t$. We will call *moving* (or *rheonomous*) any constraint of this sort. Let us state formally the distinction between fixed and moving constraints:

- *fixed* (or *scleronomous*) constraint: a priori restriction on the possible configurations or motions of a system, whose analytic expression *does not depend on time explicitly*;
- *moving* (or *rheonomous*) constraint: a priori restriction on the possible configurations or motions of a system, whose analytic expression *does depend explicitly on time*.

(That the constraint under exam is moving is then proved by expression (4.13), where time appears explicitly.)

The position of $P$ depends on the generalized coordinate $s$ and also explicitly on $t$. In fact *it is not* sufficient to know $s$ to determine the position of $P$, unless we have first established the guide's position, which in turn depends on the time instant.

So, we can write $OP(s,t)$ as

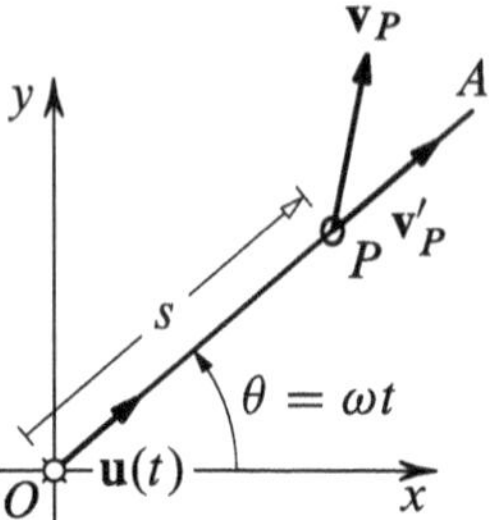


**Fig. 4.8** Virtual velocity $\mathbf{v}'_P$ and effective velocity $\mathbf{v}_P$. Each virtual velocity is tangent to the rotating guide, while a part of the effective velocity is necessarily perpendicular to it

$$OP(s,t) = s\mathbf{u}(t) = s\cos(\omega t)\mathbf{i} + s\sin(\omega t)\mathbf{j} \tag{4.14}$$

and then

$$\begin{aligned} \mathbf{v}_P &= \frac{dP}{dt} = \frac{\partial P}{\partial s}\dot{s} + \frac{\partial P}{\partial t} \\ &= \big(\dot{s}\cos(\omega t) - s\omega\sin(\omega t)\big)\mathbf{i} + \big(\dot{s}\sin(\omega t) + s\omega\cos(\omega t)\big)\mathbf{j} \\ &= \dot{s}\mathbf{u} + s\omega\big(-\sin(\omega t)\mathbf{i} + \cos(\omega t)\mathbf{j}\big) \\ &= \dot{s}\mathbf{u} + \frac{\partial P}{\partial t} \end{aligned} \tag{4.15}$$

where the vector

$$\frac{\partial P}{\partial t} = s\omega\big(-\sin(\omega t)\mathbf{i} + \cos(\omega t)\mathbf{j}\big)$$

is *orthogonal* to the guide.

What can we say about *virtual* velocities and displacements? Let us recall the definitions:

- virtual velocity: a velocity compatible with the constraints, thought of as *fixed* at the considered instant;
- virtual displacement: an infinitesimal displacement compatible with the constraints, thought of as *fixed* at the considered instant.

So let us imagine the system at some *given* position and instant, and consider all velocity distributions compatible with the setup. In this case we are dealing with a point, so the velocity distributions boil down to the velocity of $P$. But we must pay attention to the fact that time should be thought of as fixed and thus the constraint is "frozen", to use a whimsical but significant word.

As shown in Fig. 4.8, the ring's *virtual* velocities $\mathbf{v}'_P$ are all *tangent* to the guide, at the position in which the guide is at the given instant. The *effective* velocity $\mathbf{v}_P$, on the other hand, also shown in the picture, necessarily has an orthogonal component too, equal to the drag velocity of a reference frame rotating with the guide.

Now the subtle but important difference between the velocities of points in *effective* motion and *virtual* velocities, given by fixed position and constraints, becomes evident. It is immediate to see that in this case the effective velocity *is not* one of the infinitely many virtual velocities. The effective velocity has a component perpendicular to the guide, caused by its rotation.

All of this can be understood better if we write down the expression of the virtual velocity. Since the time $t$ is to be considered fixed, differentiating (4.14) only in $s$ gives

$$\mathbf{v}'_P = \frac{\partial P}{\partial s}\frac{\delta s}{\delta t} = (\cos(\omega t)\mathbf{i} + \sin(\omega t)\mathbf{j})\frac{\delta s}{\delta t} = \mathbf{u}(t)\nu_s \tag{4.16}$$

where $\mathbf{u}(t)$ is the unit vector along the rotating guide and $\nu_s = \delta s/\delta t$ *is not* a derivative but an arbitrary quantity. The vectors $\mathbf{v}'_P$ thus found are then *precisely* the tangent vectors to the guide (multiples of $\mathbf{u}$), coherently with the definition of virtual velocity as *velocity compatible with the constraints as if they were fixed at the time considered.*

Finally, the term $\partial P/\partial t$ on the right of (4.15) is precisely the one giving the component of the total effective velocity of $P$ due to the constraint's motion (the guide); as it should, it *does not* appear in (4.16).

The virtual displacements at time $t$ are therefore given by

$$\delta P = \frac{\partial P}{\partial s}\delta s = \mathbf{u}(t)\delta s$$

and are tangential (here too, $\delta s$ denotes an arbitrary infinitesimal variation of the coordinate $s$).

One last important consequence emerges from comparing (4.15) and (4.16), i.e. the effective velocities and the virtual velocities of $P$. It is indeed *impossible* to obtain the effective velocities from (4.16), even by replacing $\delta s/\delta t$ with $\dot{s}$, corresponding to a given motion $s(t)$. In fact, in (4.15) there is the additional term $\partial P/\partial t$, consequence of the constraint's motion.

This observation is general and can be formalized as follows:

- In systems with *moving* constraints the effective velocity *does not* appear among the virtual velocities.

This example should then be seen under two spotlights: (1) introduction of the notion of moving constraint; (2) clarification of the difference between *effettive velocity*, relative to possible motions, and *virtual velocities*, compatible with *fixed* or *frozen* constraints at the instant of concern.

Compare this to the discussion of the example in Fig. 4.6: there, emphasizing the constraints were frozen would have been irrelevant since they they already were *fixed*, i.e. independent of time.

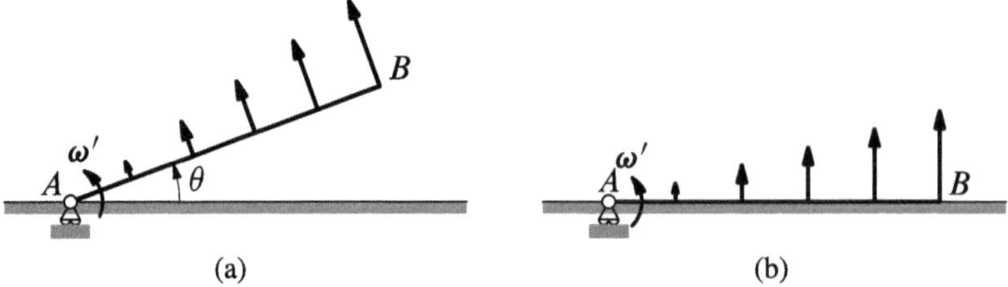


**Fig. 4.9** A system subject to a unilateral constraint: (a) reversible virtual velocity distribution ($0 < \theta < \pi$); (b) irreversible virtual velocity distribution ($\theta = 0$, boundary configuration)

### *4.1.5 Unilateral and Bilateral Constraints*

We now examine a system subject to a *unilateral* constraint. Consider again the rod of Sect. 4.1.3, as in Fig. 4.6. Suppose we impose that the end $B$ stays *above* the horizontal guide, because of some obstacle that prevents it from crossing over. This condition can be expressed by $AB \cdot \mathbf{j} \geq 0$, and once we have introduced generalized coordinates $s$ and $\theta$, it forces $\sin\theta \geq 0$ hence $0 \leq \theta \leq \pi$. We see that the further constraint only has an effect on the configurations corresponding to $\theta = 0$ and $\theta = \pi$, called *boundary configurations*. As for the motions where $\sin\theta$ is strictly positive the system is unchanged. It is important to observe that in these configurations every virtual velocity distribution is *reversible*, meaning that the opposite velocity distribution is also virtual (Fig. 4.9a).

Consider instead the configuration for $\theta = 0$. It is almost obvious that the virtual velocities must point upwards, as in Fig. 4.9b, since the constraint imposes so. We are looking at a special feature of this system: in certain configurations, the boundary ones, there exist virtual velocities whose opposites *are not* virtual.

Here are some useful definitions.

- A virtual velocity (or a displacement) is *reversible* if its *opposite* is virtual.
- A system with configurations admitting *non*-reversible virtual velocity (or displacements) is said to be subject to *unilateral* (one-sided) constraints. The configurations where this occurs are called *boundary* configurations. If instead in all configurations *every* virtual velocity (or displacement) is reversible the constraints are *bilateral* (two-sided).

In the light of these definitions we conclude that the constraint of our example is *unilateral*. Note that also in boundary configurations any virtual displacement can be written as

$$\delta P = \frac{\partial P}{\partial s}\delta s + \frac{\partial P}{\partial \theta}\delta\theta\,,$$

with $\theta = 0, \pi$, $\delta s$ arbitrary and $\delta\theta$ arbitrary but non-negative (when $\theta = 0$) or non-positive (when $\theta = \pi$), in order for the unilateral constraint to be respected. The dependence of the positions and velocities of the system's points upon $s$, $\theta$, $\dot{s}$ and $\dot{\theta}$

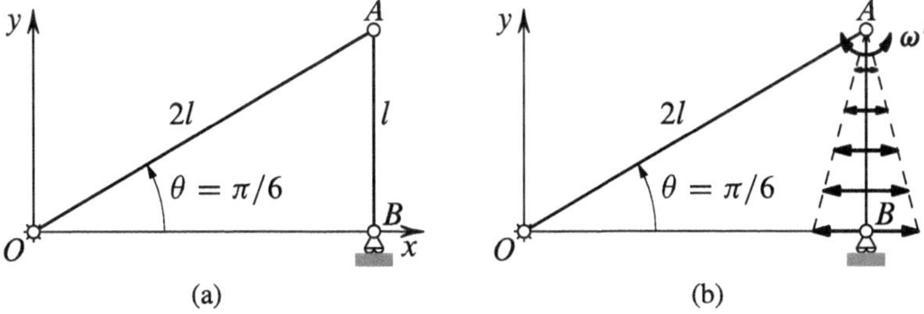


**Fig. 4.10** (a) The system with configuration $\theta_{max} = \pi/6$; (b) a virtual velocity distribution, which is rotational around $A$ for $AB$, and null for $OA$

are anyhow *regular* also for boundary configurations. For this reason $s$ and $\theta$ preserve their role of generalized coordinates in those cases too.

The definition of bilateral or unilateral constraint must necessarily rely on the reversibility feature of *virtual* velocity distributions, not the effective ones. To motivate this criterion it suffices to look at Fig. 4.8: the virtual velocities, tangent to the rotating guide, are reversible (and the constraint is thus bilateral, as to be expected), whereas the effective velocity has an orthogonal component, with orientation necessarily coherent with the rotation, so it can never be reversed.

The notion of unilateral constraint should not be associated automatically with the presence of generalized coordinates which can only range over bounded intervals, as this example might lead us to believe.

Let us return to the slider-crank of Fig. 4.3. As we saw, for configurations where $B$ is, say, on the right of $A$ we could use as generalized coordinate the angle $\theta$. It is very clear that the value $\theta_{max} = \pi/6$ is a maximum for $\theta$, so if we start from the configuration in Fig. 4.10a this angle can only *decrease*. This though *does not* mean the system is subject to a unilateral constraint. Consider in fact a generic virtual velocity distribution relative to this configuration: using Chasles's Theorem and easy geometrical arguments we deduce that any such virtual velocity distribution is necessarily *zero* for $OA$, while it is *rotational* around $A$ for the rod $AB$, as per Fig. 4.10b. Immediately, in these conditions every virtual velocity distribution is *reversible*, so the system is to all effects subject to *bilateral* constraints, as we should have expected.

## 4.2 Virtual Displacements and Virtual Velocity Distributions

The velocity distribution of a *rigid* system has the form

$$\mathbf{v}(P) = \mathbf{v}(Q) + \boldsymbol{\omega} \times QP,$$

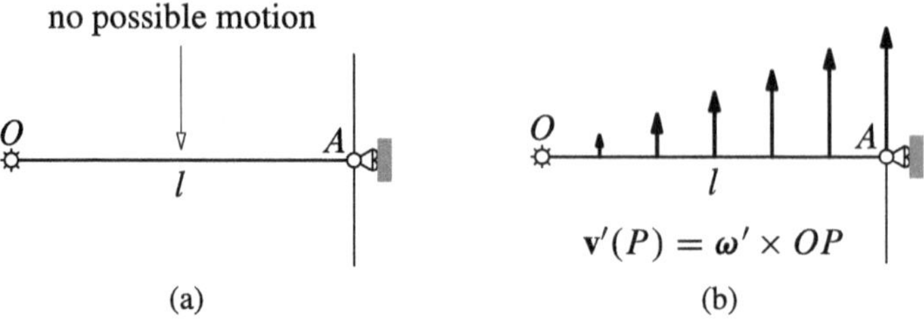


**Fig. 4.11** A labile system. (a) no motion possible, no generalized coordinates; (b) possible virtual velocity distribution

where $Q$ and $P$ are arbitrary points in the system or comoving with it, while $\omega$ is the angular velocity vector. To find the collection of *virtual* velocities compatible with the system's rigidity constraint we can use the same formula, in which we will insert *dashes* to remind us that the velocities are virtual, merely *compatible* with the constraints, and not subordinates to some effective motion. So we will put

$$\mathbf{v}'(P) = \mathbf{v}'(Q) + \omega' \times QP, \tag{4.17}$$

where, in absence of other constraints, $\mathbf{v}'(Q)$ and $\omega'$ are arbitrary vectors. When *additional* constraints are present, we must express the virtual velocity distribution by choosing $\mathbf{v}'(Q)$ and $\omega'$ so that all velocities coming from (4.17) are effectively *compatible*.

We should take into account that in general $\mathbf{v}'(Q)$ and $\omega'$ might not have anything to do with the effective velocity and angular velocity, as clearly shown by the system of Fig. 4.11a: a rod $OA$ of length $l$ pinned to a fixed point $O$ and constrained at $A$ by a roller moving on a fixed guide at a distance $l$ from $O$ and *perpendicular* to $OA$. Because of the constraints the rod *cannot move* in any way, yet it admits *infinitely many* rotational virtual velocity distributions around the axis perpendicular to the plane through $O$. Each one, as shown in Fig. 4.11b, is in fact compatible with the constraints: the velocity distribution

$$\mathbf{v}'(P) = \omega' \times OP \quad \omega' = \omega'\mathbf{k}$$

(where $\omega'$ is an arbitrary scalar quantity and $\mathbf{k}$ is the plane's unit normal) respects the rod's rigidity, the presence of a fixed pin at $O$ and also the roller, which forces $A$ to move along the guide and hence to have velocity *parallel* to it. Although this is a particular case, it is still quite significant: the system *has virtual* velocity distributions, but most certainly *does not have* any *effective* velocity distribution, as it admits *a unique* configuration. Systems of this kind are called *labile*, as we will see better in Sect. 4.4.

It is important to remark that if the guide was tilted, as in Fig. 4.12, there would exist no virtual velocity distribution, since the velocity prescribed at $A$ by any rotational velocity distribution around $O$ would no longer be tangent to the guide, making

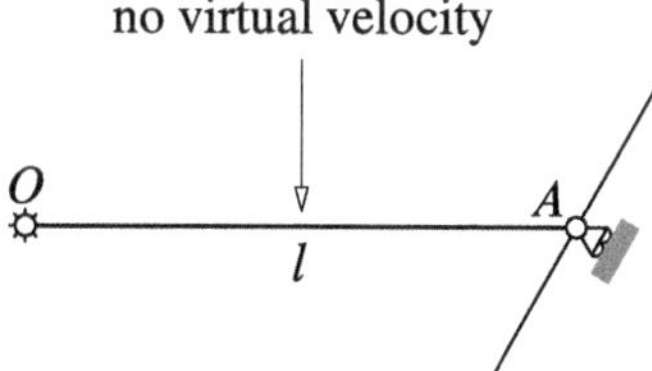


**Fig. 4.12** Isostatic system

it incompatible with the roller's presence. Systems that admit a unique configuration, are not subject to superfluous constraints and do not have virtual velocity distributions are labelled *isostatic* (cf. Sect. 4.4).

The above discussion should be enough to explain why we need to use different symbols for the velocities and angular velocities appearing in *effective* and *virtual* velocity distributions, these being *different* notions.

As we saw in Sect. 2.4, the elementary displacement of the points of a system in given rigid motion is expressible by (2.26)

$$dP = dQ + \varepsilon \times QP, \tag{4.18}$$

where $\varepsilon = \omega\, dt$ is the so-called infinitesimal rotation vector corresponding to the velocity distribution at the given instant. So it is natural to write the *virtual displacement* of the points of a *rigid system* as

$$\delta P = \delta Q + \varepsilon' \times QP, \tag{4.19}$$

in perfect analogy to what we did for the virtual velocity distribution (4.17). But it is important to understand the reason for using $\delta P$ instead of $dP$ and $\varepsilon'$ instead of $\varepsilon$, as is clear when comparing (4.18) to (4.19). We use the letter $\delta$ and a dash (′) to indicate *virtual* quantities, i.e. arbitrary but compatible with the constraints (imagined fixed at the considered instant), and *not* infinitesimal quantities associated with some effective displacement of the body or system. As anticipated, for a generic point $P$ we will agree to use $\delta P$ to denote a *virtual* displacement, and that $\varepsilon'$ is the *virtual* infinitesimal rotation vector of a rigid body.

## 4.3 Generalized Coordinates

We finally present a procedure to define a system of *generalized coordinates* for generic systems. The basic idea is that they should serve to *parametrize* the set of configurations in a sufficiently regular way: the generalized coordinates are thus essential, independent parameters varying in a region that describe the system's configurations, os a subset thereof, regularly.

A *system of generalized coordinates* for a collection of configurations $\mathcal{C}$ is then a correspondence associating, at each instant $t$ of a time interval, with every $N$-tuple of real numbers $\mathsf{q} = (q_1, q_2, \ldots, q_N)$ in a set $\mathcal{U}$ of $\mathbb{R}^N$ a configuration in $\mathcal{C}$. We ask that $\mathcal{U}$ is open and connected, possibly including a portion of its frontier, and that for any point $P$ the function

$$P(q_1, q_2, \ldots, q_N; t)\,, \tag{4.20}$$

that gives the point's *position* as the parameters $(q_1, q_2, \ldots, q_N)$ and time $t$ vary, is *regular* (at least twice differentiable with continuous second derivative). We suppose the coordinates $(q_1, q_2, \ldots, q_N)$ are, as said, *essential* and *independent*. This means: (1) none can be eliminated without affecting the unique determination of the system's position; (2) there is no finite relationship between them.

Any motion transiting through a configuration in the set $\mathcal{C}$ is then describable, at least for some time, by an $N$-tuple of functions $q_k(t)$. The corresponding motion of a generic point $P$ is

$$P(t) = P(q_1(t), q_2(t), \ldots, q_N(t), t) = P(\mathsf{q}(t), t)\,.$$

These hypotheses ensure that the generalized coordinates satisfy all of the reasonably necessary properties, which we will not attempt to justify formally. Also observe that:

- typically we will need several systems of generalized coordinates, since each one can only be used for a *proper* subset of system's configurations. It is good to remind though that in the simplest cases there are generalized coordinate systems that comprise all possible configurations, so this issue often is not present.
- even for the same set $\mathcal{C}$ of configurations there exist several systems of generalized coordinates, conceptually equivalent, but sometimes quite different from one another from a practical viewpoint.
- The *explicit* dependence of (4.20) on time corresponds to the presence of *moving* constraints. If *fixed*, that dependence disappears.

In the most common cases the construction of a system of generalized coordinates can be done intuitively, and does not require any mathematical technicality. It is only for the need of a general definition that we addressed the question more abstractly. The presentation we made of this notion *is not* of constructive nature: it does not tell *how* to build a system of generalized coordinates concretely, but it simply lists its core characteristics.

Once we are given the motion of a system by an $N$-tuple $q_k(t)$ it is easy to obtain the velocity of a generic point $P$ by differentiating (4.20). Recalling the chain rule we have

$$\mathbf{v}_P = \frac{dP}{dt} = \sum_{k=1}^{N} \frac{\partial P}{\partial q_k} \dot{q}_k + \frac{\partial P}{\partial t}\,, \tag{4.21}$$

which implies

$$dP = \mathbf{v}_P dt = \sum_{k=1}^{N} \frac{\partial P}{\partial q_k} dq_k + \frac{\partial P}{\partial t} dt\,. \tag{4.22}$$

It is important to stress that the elementary displacements and effective velocities are subordinated and referring to the choice of a motion $q_k(t)$, and they are derived from it by differentiating with respect to time, as in (4.21) and (4.22).

The expressions of velocity and virtual displacements for systems with $N$ generalized coordinates $(q_1, q_2, \dots, q_N)$ are instead:

$$\begin{aligned} \mathbf{v}'(P) &= \sum_{k=1}^{N} \frac{\partial P}{\partial q_k} \frac{\delta q_k}{\delta t} \\ \delta P &= \sum_{k=1}^{N} \frac{\partial P}{\partial q_k} \delta q_k \end{aligned} \tag{4.23}$$

(where the quantities $\delta q_k$ and $\delta q_k/\delta t$ should be understood as arbitrary coefficients).

Comparing (4.21) and (4.22) with (4.23) allows to confirm the observation contained in Sect. 4.1.4:

- for systems subject to moving constraints the elementary displacements $dP$ and the effective velocities $\mathbf{v}_P$ *are not* among the virtual displacements $\delta P$ and virtual velocities $\mathbf{v}'_P$.

Yet, in case the constraints are *fixed*, and so (4.20) does not contain time $t$ explicitly, then (4.21) and (4.22) reduce to

$$\mathbf{v}_P = \frac{dP}{dt} = \sum_{k=1}^{N} \frac{\partial P}{\partial q_k} \dot{q}_k \qquad dP = \mathbf{v}_P dt = \sum_{k=1}^{N} \frac{\partial P}{\partial q_k} dq_k\,.$$

The above follow from (4.23) when substituting $\dot{q}_k$ and $dq_k$, coming from an effective motion $q_k(t)$, in place of $\delta q_k/\delta t$ and $\delta q_k$.

In a nutshell:

- in systems subject to fixed constraints the elementary displacements $dP$ and the effective velocities $\mathbf{v}_P$ are special instances of the virtual displacements $\delta P$ and virtual velocities $\mathbf{v}'_P$.

## 4.4 Labile, Hyperstatic and Isostatic Systems

Consider the system seen in Sect. 4.2 and represented in Fig. 4.11: it evidently has one possible configuration, depicted, so it does not admit generalized coordinates. But it is equally true that any rotational velocity distribution around $O$ is *compatible*

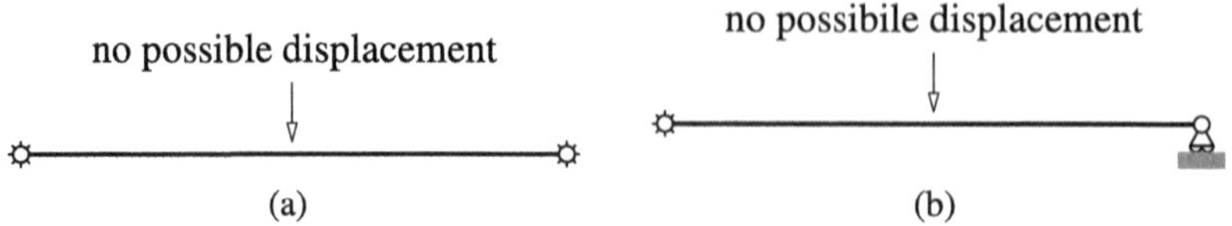


**Fig. 4.13** A hyperstatic system (a) transformed into an isostatic one (b) through the reduction of the constraint at the right end

with the constraints. In fact for any velocity distribution of this type the velocity of $A$ is tangent to the guide, but there is no way to deduce it via a virtual variation of the generalized coordinates (that here do not exist). So, the system admits (at least) one virtual velocity distribution that cannot be written in form (4.23). The presence of virtual velocities (or displacements) that cannot be formally expressed by varying the generalized coordinates is what characterizes so-called *labile systems*. The number of independent virtual displacements (or velocity distributions) is therefore *bigger* than the number of generalized coordinates.

Labile systems may also be called *singular* since mathematically they correspond to the vanishing of a Jacobian determinant coming from the constraints' equations. Without going into these details we limit ourselves to anticipate that labilily usually generates paradoxical situations and equilibrium equations that cannot be solved. So it is important to be able to recognize labile mechanical systems, not an easy task in general.

A similar problem arises when there are more constraints that necessary. As a very simple example take a rod pinned to *two* fixed points, see Fig. 4.13a.

Clearly such system does not allow for any motion. This would not be true if one pin was replaced by a roller, as in Fig. 4.13b. Here we have *hyperstatic* system, where the set of possible motions (at present, the empty set, trivially) does not change even if we get rid of some of the constraints, see Fig. 4.13. Systems with this property are normally called *hyperstatic*, and are characterized by superfluous constraints, meaning useless from the point of view of limiting motions.

As we will see better later, when we will be able to write down the equilibrium equations, they are typically characterized by "too many" unknowns compared to the available equations. To discuss them it is sometimes necessary to abandon rigid bodies and adopt more flexible models such as elastic bodies (cf. Sect. 9.7.2).

That said, hyperstatic systems are not irrelevant in the applications. The two fixed pins of Fig. 4.13a are certainly excessive if we aim at preventing a rigid rod from moving. But they become more understandable if we agree, as happens in reality, that apparently rigid bodies undergo small deformations. In this case it is evident that the rod in Fig. 4.13a is certainly better constrained than the one in Fig. 4.13b.

Let us finally remind that systems that cannot move (those with one admissible configuration only) and are neither labile nor hyperstatic are called *isostatic*. An example are the immobile rods in Figs. 4.12 and 4.13, which do not have virtual velocity distributions and for which we cannot eliminate or reduce any constraint without changing the set of possible motions. In other words: the constraints of an isostatic system are exactly those that are strictly necessary to prevent *finite* and *infinitesimal* displacements.

## 4.5 Holonomic Bilateral Constraints

Each one of the bilateral constraints examined is in the end translated into a finite relationship between the parameters $(q_1, q_2, \ldots, q_N)$ that give the system's position. Put alternatively, the motion's restriction produced by the constraint is equivalent to asking that the parameters $q_k$ are not free to take any value in a given region of $\mathbb{R}^N$, but must obey one or more relations of the type

$$f(q_1, q_2, \ldots, q_N; t) = 0\,. \tag{4.24}$$

Any bilateral constraint that takes this mathematical form is called *holonomic* (and may be moving or fixed, depending on the presence or absence of $t$ in its expression). Any equation like (4.24) usually reduces by one the number of generalized coordinates needed to describe the system's configurations.

The bilateral constraints seen thus far are all holonomic. But there exist situations that must be modelled with so-called *mobility constraints*, i.e. a priori restrictions on the *velocity distributions* of the system rather than the positions. These constraints *cannot* in general writable in holonomic form.

## 4.6 Pure Rolling Constraint and Contact Constraint

A constraint of great theoretical and applicative importance is the *pure rolling* (or *rolling without slipping*) between two bodies, which we discuss mathematically together with its consequences. The surfaces of the two bodies that roll on one another without slipping touch at a common contact point $P$, as in Fig. 4.14. This

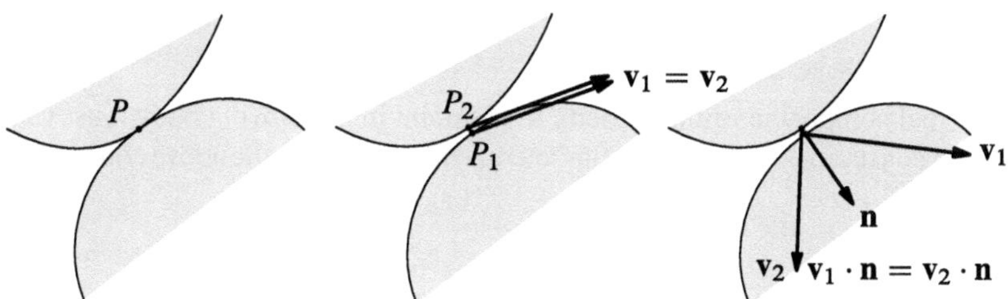


**Fig. 4.14** For *pure rolling constraints* the velocities of points $P_1$ and $P_2$ in contact at $P$ must be equal. For *bilateral contact constraints* only equality of the normal components is required

point corresponds at each instant to *two* particles $P_1$ and $P_2$ belonging to the bodies in motion. They are *physically distinct* even though *geometrically coinciding* (they represent different particles at each instant).

What is the restriction on the bodies' velocity distribution caused by this constraint? By definition, in agreement with the physical intuition, it imposes that at each instant the velocities of the points in contact are the same: $\mathbf{v}_1 = \mathbf{v}_2$. In fact, if the normal components to the surfaces' common tangent plane were different, the surfaces would *detach* (or even penetrate one another), while if the tangential components were different we would witness *slipping*. In this case, the difference of the velocities' tangential components in the contact points is called *slip velocity*.

If in particular the point $P_2$ lies on a fixed surface the rolling constraint with no slipping simplifies to $\mathbf{v}_1 = \mathbf{0}$. That is, the point of the body that is in motion has zero velocity.

If we only require instead that the surfaces stay in *contact*, allowing for one to slip on the other, we should only impose the velocity of $P_1$ and $P_2$ have *the same normal component* to the surfaces. More precisely, referring to Fig. 4.14,

$$\mathbf{v}_1 \cdot \mathbf{n} = \mathbf{v}_2 \cdot \mathbf{n}\,. \tag{4.25}$$

This is the mathematical form of a *bilateral contact constraint* with slipping, another frequent situation.

A weaker (unilateral) version of contact constraint exists which prevents one surface from ramming into the other, but still allows for *detachment*. In this case, see Fig. 4.14, the normal component of the velocities of $P_2$, oriented towards $P_1$, must not exceed that of $P_1$:

$$\mathbf{v}_2 \cdot \mathbf{n} < \mathbf{v}_1 \cdot \mathbf{n}\,. \tag{4.26}$$

A particularly simple case bilateral contact constraint is a point that stays on a fixed or moving surface, while we have a unilateral contact constraint when we permit the point to leave the surface but only move towards one of the spatial regions the surface itself separates.

If one of the surfaces is fixed (say, the one to which $P_2$ belongs) then conditions (4.25) and (4.26) respectively read

$$\mathbf{v}_1 \cdot \mathbf{n} = 0\,, \qquad \mathbf{v}_1 \cdot \mathbf{n} > 0\,.$$

The normal component of the velocity of the point in the moving body must vanish (bilateral constraint) or be positive (unilateral constraint, with the above conventions).

### 4.6.1 Disk Rolling Without Slipping

Let us study in detail the pure rolling constraint (no slipping) by applying it to the simplest case: the motion of a disk on a fixed straight guide.

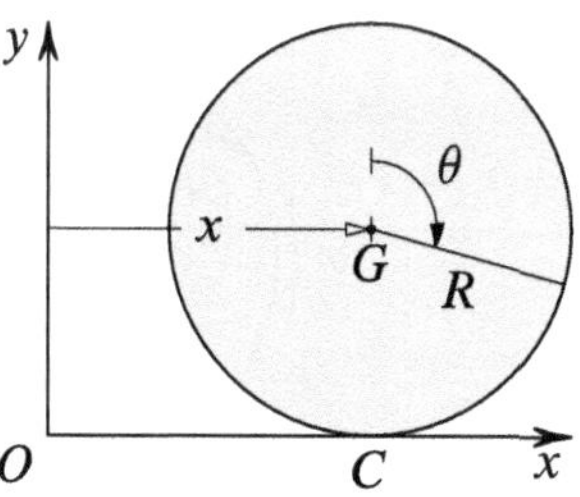


**Fig. 4.15** A disk rolling on a straight line

The position of a disk on the plane is given by the coordinates $x$ and $y$ of the center $G$ and a rotation angle $\theta$, as in Fig. 4.15. Clearly if we want the disk to touch the $x$-axis, which we assume is the fixed guide, we will need to impose $y = R$, a simple *holonomic* constraint of type (4.24). Only with that, though, the disk might still *roll and slip* and for its motion we would need two generalized coordinates: $x$ and $\theta$. To ensure rolling without slipping we impose the velocity of the contact point $C$ equals the guide's velocity, i.e. zero. (This is equivalent to demanding that the disk's velocity distribution is *rotational* around $C$, and $C$ becomes the instantaneous center of rotation.) Using the unit vectors $\mathbf{i}$ and $\mathbf{j}$ of the axes and $\mathbf{k} = \mathbf{i} \times \mathbf{j}$ we write $\mathbf{v}(G) = \dot{x}\mathbf{i}$, $\boldsymbol{\omega} = -\dot{\theta}\mathbf{k}$ (since the angle $\theta$ in Fig. 4.15 is measured clockwise) and $GC - -R\mathbf{j}$. Using the formula for the roto-translational velocity distribution, we discover

$$\mathbf{v}(C) = \mathbf{v}(G) + \boldsymbol{\omega} \times (GC) = \dot{x}\mathbf{i} + (-\dot{\theta}\mathbf{k}) \times (-R\mathbf{j}) = (\dot{x} - R\dot{\theta})\mathbf{i}\,.$$

The condition $\mathbf{v}_C = \mathbf{0}$ is then equivalent to

$$\dot{x} = R\dot{\theta}\,. \tag{4.27}$$

The above has a rather intuitive interpretation: it says the velocity of the center $G$ and the disk's angular velocity are proportional. At first sight the constraint thus obtained seems different from (4.24), as here we have a relation between the *derivatives* of the generalized coordinates and not the coordinates themselves. It is trivial to see however that (4.27) is equivalent to

$$x = R\theta + \text{constant} \tag{4.28}$$

where the arbitrary constant is determined by the value of $x$ when $\theta$ is zero. Evidently, the rolling without slipping in *this case* is equivalent to a holonomic constraint. In this way we can reduce the generalized coordinates from two to one, using (4.28) to express, say, $x$ in terms of $\theta$.

Due to the major impact this example has in the applications we should discuss the kinematics in more detail. The formula for the rotational velocity distribution assigns to each point $P$ of the disk a velocity $\mathbf{v}(P)$ of modulus proportional to the distance to the instantaneous center of rotation $C$, i.e. the contact point with the

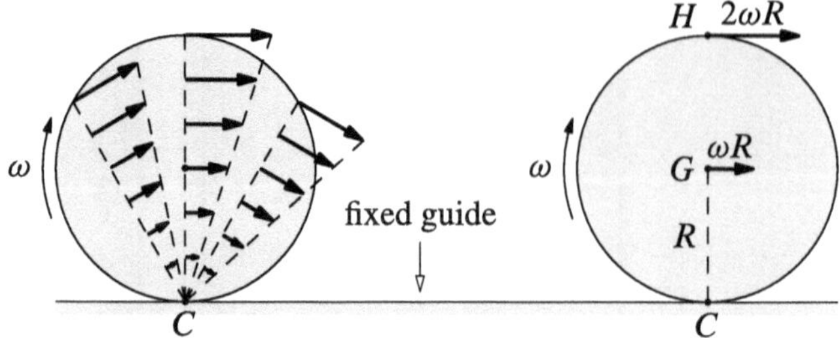


**Fig. 4.16** The velocity distribution for a disk which is rolling without slipping on a fixed guide

fixed guide, according to $\mathbf{v}(P) = \boldsymbol{\omega} \times CP$. The velocity is therefore *perpendicular* to $CP$, oriented coherently with $\boldsymbol{\omega}$, and of modulus proportional to the distance from $C$. The velocity distribution of a disk that rolls without slipping on fixed guide is described in Fig. 4.16 left.

In view of several applications, it is also important to clarify which are the velocities of the disk's center $G$ and the point $H$ antipodal to $C$. Setting to $R$ the radius' length, the situation is described by Fig. 4.16 right: the velocity of the center is parallel to the guide and of component equal to $\omega R$, while the point at $H$ has *twice* that velocity.

As shown, the condition of pure rolling translates into a constraint for the velocity distribution rather than on the system's position. In the special case discussed, the planar motion of a disk along a guide, we could check that this condition is equivalent to a finite relation between the disk's rotation and the abscissa of the center. This allows to conclude, in this case, that rolling without slipping is the same as a holonomic constraint, for it reduces the number of parameters necessary to describe the system's motions and restricts the set of possible configurations.

It is important to observe though that this conclusion is certainly *not* valid in general: for example, by studying a sphere rolling without slipping on a fixed plane we realize that the constraint is not equivalent to a number of finite relations in the system's coordinates; this system is not in the same category as those with holonomic constraint.

We postpone to a future section the study of these constraints, and present in the sequel a system with a contact constraint.

**Example 4.1** In Fig. 4.17 left we see a disk of radius $r$ and center $G$ constrained to roll without slipping on a fixed horizontal guide identified with the $x$-axis; it touches the hypotenuse of a plate shaped as a right triangle, which in turn is constrained to move vertically with one leg attached to the $y$-axis. We want to know what is the relationship between the angular velocity $\omega$ of the disk and the vertical velocity $\dot{y}$ of the plate.

The disk's velocity distribution has instantaneous center of rotation at the contact point with the $x$-axis and, as already said, in this case the velocity of the center $G$ equals $r\omega$, see the picture on the right.

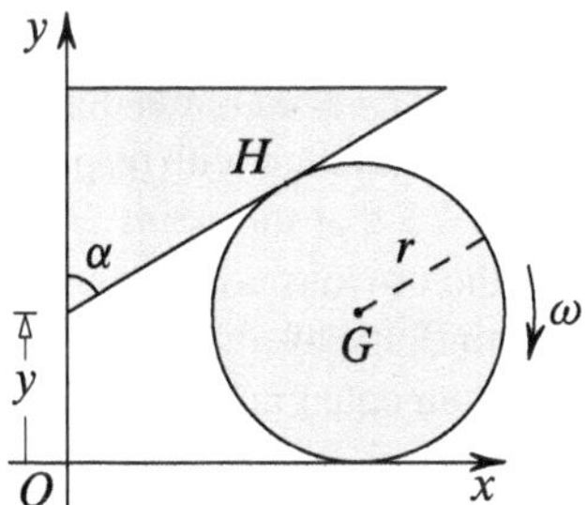

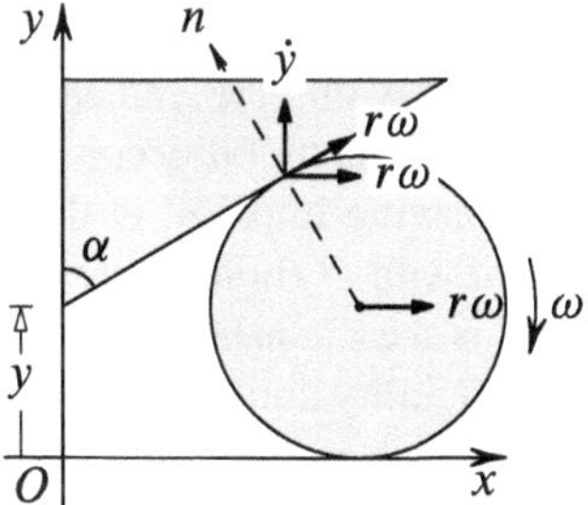


**Fig. 4.17** A rolling disk touching a triangular plate

The disk and the plate's long edge touch at $H$ where, conceptually, at each instant two particles coincide, one on the disk and one on the plate. These *do not* have the same velocity but the same velocity component along $n$, normal to the surfaces at the contact point.

Let us first consider the disk's point at $H$ and find its velocity using $\mathbf{v}(H) = \mathbf{v}(G) + \boldsymbol{\omega} \times GH$. We know $\mathbf{v}(G)$, which we shift to $H$, and add to it $\boldsymbol{\omega} \times GH$. The latter has modulus $r\omega$ and is tangent to the circle, so parallel to the hypotenuse, see Fig. 4.17 right.

The point of the plate at $H$, like all points of the plate by the way, has vertical velocity of component $\dot{y}$, since the plate has translational velocity distribution caused by the constraint it is subject to.

We impose that components along $n$ of the velocities of $H$ as plate point and disk point are equal (Fig. 4.17 right). Basic trigonometry guarantees that the former is $\dot{y} \sin\alpha$, where $\alpha$ is the angle between the vertical leg and the hypotenuse. Thinking $H$ as disk point, one part of its velocity has zero projection onto $n$, since parallel to the hypotenuse, while the other, horizontal, component projects onto $n$ to give $-r\omega\cos\alpha$ (the minus sign is because the projection is oppositely oriented to the axis). Hence, setting these two equal eventually yields

$$\dot{y} \sin\alpha = -r\omega\cos\alpha\,.$$

A naïve check on the above sign is useful: it is evident that to positive angular velocity $\omega$ there corresponds a disk moving to the right and a decreasing $y$, hence $\dot{y} < 0$. Vice versa, for $\omega < 0$ we have $\dot{y} > 0$. □

## 4.7 Mobility and Nonholonomic Constraints

There exist restrictions to a system's motion that must be modelled by constraints on the velocities rather than on the positions: the case of rolling without slipping and the contact between two bodies, seen earlier, are two significant and important examples.

In such cases one speaks of *mobility* constraints instead of *holonomic* constraints . Let us immediately observe, rather trivially, that from any holonomic constraint we can find an equivalent mobility constraint, just by differentiating with respect to time. For instance, having imposed in the system of Fig. 4.3 that the point $H$ coincides at each instant with $O$ (holonomic constraint) has the obvious consequence that the velocity of $H$ is at each instant null (mobility constraint). It may also happen that from one or more mobility constraints we can go back to an equal number of holonomic constraints *equivalent* to the former set. That means the restrictions imposed on the possible motions are exactly the same, as in the case just mentioned: forcing the velocity of $H$ to vanish identically (mobility constraint) is equivalent to imposing that $H$ is pinned to a fixed point (holonomic constraint).

We have also seen in Sect. 4.6.1 how the mobility constraint corresponding to a disk's rolling without slipping on a fixed guide can be reduced to a finite relation between $x$ and $\theta$, the parameters that determine its position when we allow slipping to occur. So we conclude that the mobility constraint introduced is here *holonomic*. This is not the general situation, as we will see in the sequel. Let us say for now that those mobility constraints that *are not* equivalent to holonomic constraints are called *nonholonomic*, or *pure* mobility constraints.

Consider a system with holonomic constraints and equipped with $N$ generalized coordinates $(q_1, \ldots, q_N)$, and suppose we subject it to an *additional* mobility constraint. Since by (4.21) we can write the velocity of a generic point in function of $q_k$, $\dot{q}_k$ and $t$, we can definitely translate the new constraint into a relationship, perhaps complicated, between those quantities. This relationship must be interpreted as a restriction on the $\dot{q}_k$ at a generic position and time instant. Now there are two possibilities: the restriction is equivalent to what we would find by differentiating with respect to time a suitable bunch of holonomic constraints, or not. In the first case, if convenient, we can simply replace the new constraint by equivalent holonomic constraints, thus lowering the number of generalized coordinates. In the second case we call *nonholonomic* the constraint in question (together with the mechanical system in which it appears), and we must keep it in its original form of a restriction on the possible velocity distributions. The generalized coordinates do not become less, but we reduce the number of independent $\dot{q}_k$, which are now related by the nonholonomic constraint.

In certain particularly simple cases we can justify intuitively that a constraint is nonholonomic, by using physical considerations to show it *does not* affect the set of possible configurations of the system, as a holonomic constraint would do. It only restricts the way in which the system can pass from a configuration to another. It is not hard to use this remark to understand why a ball rolling without slipping on a fixed plane has a nonholonomic character. It is quite intuitive in fact that the ball can, under suitable motions, attain any possible configuration: it can reach any point on the plane, for any possible orientation for the comoving basis that, together with the center's position, defines the ball's configuration. Now we wish to prove rigorously that a given constraint is nonholonomic. Instead of rolling without slipping in space we discuss a simpler instance.

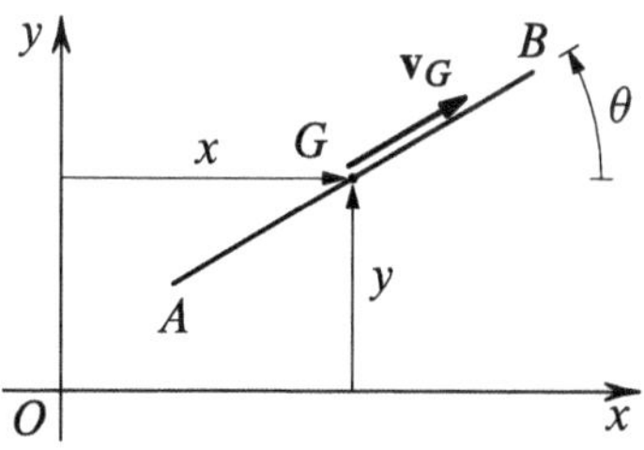


**Fig. 4.18** Nonholonomic constraint: a rod with velocity of point $G$ parallel to $AB$

**Example 4.2** (*Nonholonomic constraint*) A rod $AB$ of length $l$ is free to move on the plane under the sole constraint that its midpoint $G$ at each instant has velocity *parallel* to the rod (imagine that this situation is modelling the constraint of a bobsled's runner blades). Let us show it is indeed a *nonholonomic* situation.

The system's position is given by three generalized coordinates: the Cartesian coordinates $x$ and $y$ of $G$ and the angle of rotation $\theta$ the rod forms with the $x$-axis, see Fig. 4.18. Using the properties of the cross product, we express that the velocity of $G$ is parallel to $AB$ as $\mathbf{v}(G) \times (AB) = \mathbf{0}$, where $\mathbf{v}(G) = \dot{x}\mathbf{i} + \dot{y}\mathbf{j}$ and $AB = l\cos\theta\mathbf{i} + l\sin\theta\mathbf{j}$. The calculations show that

$$\dot{x}\sin\theta - \dot{y}\cos\theta = 0\,. \tag{4.29}$$

As this relation links the generalized coordinates $(x, y, \theta)$ and their derivatives $(\dot{x}, \dot{y}, \dot{\theta})$ the suspicion is that the constraint *might* be nonholonomic. To make sure we must prove that it is *not* equivalent to the derivative of a relation of the coordinates. In other words, we have to prove there *does not exist* any function $f(x, y, \theta)$ such that the triples $(\dot{x}, \dot{y}, \dot{\theta})$ satisfying

$$\frac{\partial f}{\partial x}\dot{x} + \frac{\partial f}{\partial y}\dot{y} + \frac{\partial f}{\partial \theta}\dot{\theta} = 0 \tag{4.30}$$

are *exactly* the triples satisfying the constraint (4.29). If one such function $f$ existed, we would deduce that the constraint given by (4.29) is holonomic, since equivalent to $f(x, y, \theta) =$ constant.

Observe that the constraint (4.29) does not restrict $\dot{\theta}$, which in any configuration can assume an arbitrary value. It does relate the quantities $\dot{x}$ and $\dot{y}$, which satisfy it if and only if we assign the relationships

$$\dot{x} = \lambda\cos\theta\,, \qquad \dot{y} = \lambda\sin\theta\,,$$

with $\lambda$ arbitrary.

All in all, the set of triples $(\dot{x}, \dot{y}, \dot{\theta})$ that are admissible under the nonholonomic constraint (4.29) can be described by two *arbitrary* parameters $\lambda$ and $\mu$, called *generalized velocities*, via

$$\dot{x} = \lambda \cos\theta\,, \quad \dot{y} = \lambda \sin\theta\,, \quad \dot{\theta} = \mu\,. \tag{4.31}$$

Choosing in particular $\lambda = 0$ we find $\dot{x} = \dot{y} = 0$, and if we leave $\dot{\theta}$ arbitrary, from (4.30) we deduce that in a generic configuration we must have $\partial f/\partial\theta = 0$. This means $f$ cannot depend on $\theta$ but only on $x$ and $y$.

Inserting in (4.30) the generic expressions of $\dot{x}$ and $\dot{y}$ from (4.31), and keeping in account the arbitrariness of $\lambda$, we discover

$$\frac{\partial f(x, y)}{\partial x} \cos\theta + \frac{\partial f(x, y)}{\partial y} \sin\theta = 0$$

where we have highlighted the possible dependence of $f$ on $x$ and $y$ only. Imposing that the equality holds for $\theta = 0$ and $\theta = \pi/2$ produces the conditions

$$\frac{\partial f}{\partial x} = \frac{\partial f}{\partial y} = 0\,,$$

which then imply $f$ is actually constant. This makes it impossible for (4.30) and (4.29) to be equivalent. We have thus reached a contradiction and finished the proof that (4.29) is nonholonomic.

Note that there still are 3 generalized coordinates $(x, y, \theta)$ even after the nonholonomic constraint (4.29) has been imposed. For this reason the set of possible configurations does not reduce by the presence of the constraint, and the rod of Fig. 4.18 can still be in any position of the plane. What does become smaller is instead the set of velocity distributions and virtual displacements: without the constraint there would be $\infty^3$ of such, depending on three independent parameters $(\delta x, \delta y, \delta\theta)$, while the nonholonomic constraint reduces the number to $\infty^2$ depending on two generalized velocities $\lambda$ and $\mu$, introduced in (4.31). The nonholonomic constraints therefore do not reduce the number of generalized coordinates, as the holonomic constraints do, but affect instead the ways in which the system can pass from one configuration to another. □

**Example 4.3** (*Nonholonomic constraint*) Consider a disk of radius $R$, ideally modelling a car's wheel, constrained to roll without slipping on a horizontal plane and staying in a vertical position with respect to it. In Fig. 4.19 we see that the parameters that a priori are necessary to express the disk's position are: (1) the coordinates $x$ and $y$ of the center $G$; (2) the angle $\theta$ between the disk's plane and the $x$-axis; (3) the angle $\phi$ the comoving radius forms with the vertical direction.

As highlighted in Fig. 4.19 right, the angular velocity $\boldsymbol{\omega}$ of the disk is the sum of a vertical part $\dot{\theta}\mathbf{k}$ and a horizontal part with component $\dot{\phi}$ in the direction perpendicular to the disk. Hence

$$\boldsymbol{\omega} = \dot{\phi} \sin\theta \mathbf{i} - \dot{\phi} \cos\theta \mathbf{j} + \dot{\theta}\mathbf{k}\,.$$

The constraint of rolling without slipping imposes that the velocity of the contact point $C$ with the plane vanishes, at all times. The formula for the roto-translational velocity distribution, after computing the cross product of $\boldsymbol{\omega}$ and $GC = -R\mathbf{k}$, gives

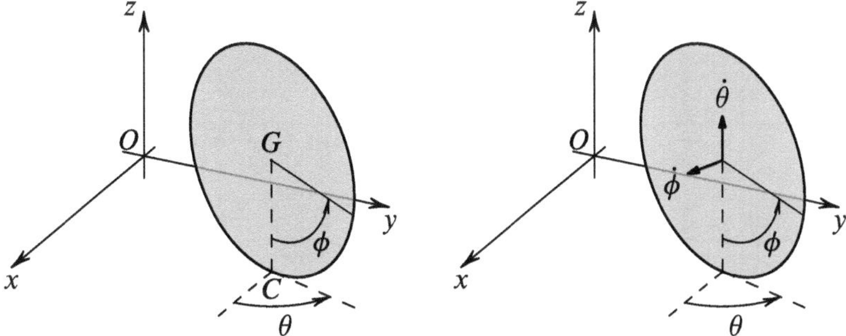


**Fig. 4.19** A disk which is rolling without slipping and is perpendicular to a fixed plane

$$
\begin{aligned}
\mathbf{v}(C) &= \mathbf{v}(G) + \boldsymbol{\omega} \times GC \\
\mathbf{v}(C) &= \dot{x}\mathbf{i} + \dot{y}\mathbf{j} + (\dot{\phi}\sin\theta\mathbf{i} - \dot{\phi}\cos\theta\mathbf{j} + \dot{\theta}\mathbf{k}) \times (-R\mathbf{k}) \\
&= (\dot{x} + R\dot{\phi}\cos\theta)\mathbf{i} + (\dot{y} + R\dot{\phi}\sin\theta)\mathbf{j}
\end{aligned}
$$

The vanishing of $\mathbf{v}(C)$ is then guaranteed by

$$\dot{x} + R\dot{\phi}\cos\theta = 0\,, \qquad \dot{y} + R\dot{\phi}\sin\theta = 0\,. \tag{4.32}$$

In analogy to the previous example the set of quadruples $(\dot{x}, \dot{y}, \dot{\theta}, \dot{\phi})$ subject to the linear restrictions (4.32) can be parametrized by two *arbitrary* quantities (called generalized velocities) $\lambda$ and $\mu$ as follows:

$$\dot{x} = -R\lambda\cos\theta\,, \qquad \dot{y} = -R\lambda\sin\theta\,, \qquad \dot{\phi} = \lambda \qquad \dot{\theta} = \mu\,. \tag{4.33}$$

Now suppose, by contradiction, that (4.32) were equivalent to the following conditions arising from two functions $f$ and $g$ in the coordinates $(x, y, \phi, \theta)$:

$$\frac{\partial f}{\partial x}\dot{x} + \frac{\partial f}{\partial y}\dot{y} + \frac{\partial f}{\partial \phi}\dot{\phi} + \frac{\partial f}{\partial \theta}\dot{\theta} = 0\,, \quad \frac{\partial g}{\partial x}\dot{x} + \frac{\partial g}{\partial y}\dot{y} + \frac{\partial g}{\partial \phi}\dot{\phi} + \frac{\partial g}{\partial \theta}\dot{\theta} = 0\,. \tag{4.34}$$

From (4.32) and (4.33) it follows, in a generic configuration, that the values of $\dot{\theta}$ and $\dot{\phi}$ can be assigned freely, as the latter coincide with the generalized velocities $\lambda$ and $\mu$, whereas $\dot{x}$ and $\dot{y}$ come from

$$\dot{x} = -R\dot{\phi}\cos\theta, \qquad \dot{y} = -R\dot{\phi}\sin\theta\,.$$

Consider the first equation in (4.34), relative to $f$, and substitute in it $\dot{\phi} = 0$, so also $\dot{x} = 0$, $\dot{y} = 0$ in view of (4.32). We find

$$\frac{\partial f}{\partial \theta}\dot{\theta} = 0\,,$$

showing that $f$ *cannot* depend on $\theta$ as $\dot{\theta}$ is arbitrary. In this way, due to (4.32) the first of (4.34) reads

$$\frac{\partial f}{\partial x}(-R\dot{\phi}\cos\theta) + \frac{\partial f}{\partial y}(-R\dot{\phi}\sin\theta) + \frac{\partial f}{\partial \phi}\dot{\phi} = 0\,.$$

Dividing by $\dot{\phi}$ gives

$$-R\frac{\partial f}{\partial x}\cos\theta - R\frac{\partial f}{\partial y}\sin\theta + \frac{\partial f}{\partial \phi} = 0\,.$$

But as the three partial derivatives do not depend on $\theta$, this restriction can hold for *any* $\theta$ just by putting

$$\frac{\partial f}{\partial x} = \frac{\partial f}{\partial y} = \frac{\partial f}{\partial \phi} = 0\,.$$

Hence $f$ is a constant, and cannot represent a constraint of the system. A similar reasoning applies to $g$. In conclusion conditions (4.32) express a nonholonomic, or pure mobility, constraint.

For this reason the vehicle, despite having velocity distribution constrained by the fact each of the four wheels rolls without slipping, can be parked in any position by steering suitably. In this case in fact the constraint does not translate into a finite relation between the generalized coordinates, but remains a linear condition between the first derivatives.

Here too we see that restrictions (4.32), while they *do not* set up any relation for the generalized coordinates $(x, y, \theta, \phi)$, they do impose a relation for the $\dot{q}_k$. In particular, they make sure the virtual variations of the $q_k$ are no longer independent, but satisfy

$$\delta x + R\delta\phi\cos\theta = 0, \qquad \delta y + R\delta\phi\sin\theta = 0\,. \tag{4.35}$$

Equations (4.33) allow to express the variations $\delta \mathsf{q} = (\delta x, \delta y, \delta\theta, \delta\phi)$ by means of two arbitrary parameters, which we called $\lambda$ and $\mu$:

$$\delta x = -R\lambda\cos\theta \qquad \delta y = -R\lambda\sin\theta \qquad \delta\phi = \lambda \qquad \delta\theta = \mu\,. \tag{4.36}$$

In view of further considerations, as a last observation let us say how, in this case, in the displacements and virtual velocities' expressions (4.23) we cannot think of $\delta q_k$ or $\nu_k = \delta q_k/\delta t$ as arbitrary independent parameters. They are related by (4.35) or rather, they are functions of two arbitrary quantities as in (4.36). □

**Example 4.4** (*Nonholonomic constraint*) A ball rolling without slipping on a fixed plane is an example of a system with nonholonomic constraint (Fig. 4.20).

Let $\mathbf{i}_h$ be the unit vectors of a fixed right-handed reference frame, for which the ball's center $G$ has Cartesian coordinates $(x, y, R)$, where $R$ is the radius. Utilizing formula (3.13) for the ball's angular velocity, and calling $C$ the contact point with the plane, the condition of rolling without slipping $\mathbf{v}_C = \mathbf{0}$ reads

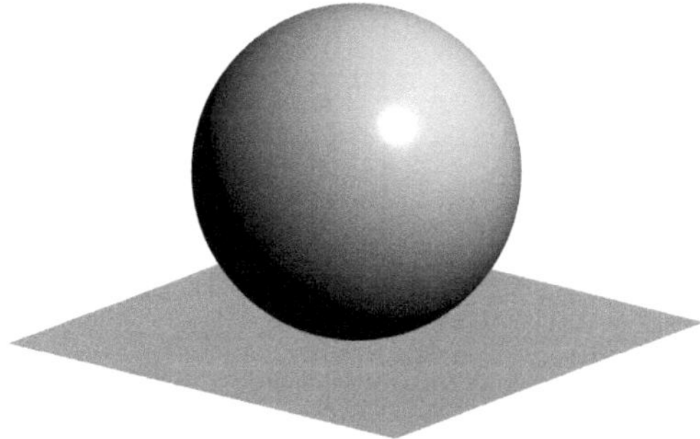
**Fig. 4.20** A sphere which is rolling without slipping on a plane is subject to a nonholonomic constraint

$$\mathbf{v}_G + \boldsymbol{\omega} \times GC = \mathbf{0}\,.$$

In Cartesian components,

$$\begin{cases} \dot{x} - R(\dot{\psi} \sin\theta \cos\phi - \dot{\theta} \sin\phi) = 0 \\ \dot{y} + R(\dot{\theta} \cos\psi + \dot{\psi} \sin\theta \sin\phi) = 0 \end{cases} \tag{4.37}$$

A completely similar analysis to the previous examples would allow to show that this system of two relations is not integrable, that is, it is not equivalent to finite relationships for the 5 generalized coordinates $(x, y, \theta, \phi, \psi)$. □

**Remark 4.5** We may prove that the constraint of rolling without slipping seen in Example 4.4 is nonholonomic in a less formal way. The experimental evidence would show that the ball can reach any configuration, corresponding to any value of the generalized coordinates $(x, y, \theta, \phi, \psi)$ without restrictions, provided we make suitable manoeuvers. This means the constraint *is not equivalent* to any finite relation between the generalized coordinates. □

### *4.7.1 Nonholonomic Linear Constraints*

The theory of nonholonomic constraints is rather complex, and here we shall merely discuss *linear* constraints. They are the most important ones in view of the applications. For a system with $n$ generalized coordinates $\mathsf{q} = (q_1, q_2, \ldots, q_N)$, a set of $m$ $(< n)$ nonholonomic linear constraints in the $\dot{\mathsf{q}}$ is given by

$$\sum_{k=1}^{N} b_{jk}(\mathsf{q}, t)\dot{q}_k = b_j(\mathsf{q}, t) \quad j = 1, \ldots, m\,. \tag{4.38}$$

To write all of this more compactly we conveniently define an $m \times N$ matrix $\mathsf{B}$ of elements $b_{jk}(\mathsf{q}, t)$ and a column vector $\mathsf{b}$ with $m$ components, and then write the set of constraints (4.38) as

$$
\begin{bmatrix} b_{11} & b_{12} & \cdots\cdots & b_{1N} \\ b_{21} & b_{22} & \cdots\cdots & b_{2N} \\ \vdots & \vdots & \vdots \quad \vdots & \vdots \\ b_{m1} & b_{m2} & \cdots\cdots & b_{mN} \end{bmatrix} \begin{bmatrix} \dot{q}_1 \\ \dot{q}_2 \\ \vdots \\ \vdots \\ \dot{q}_N \end{bmatrix} = \begin{bmatrix} b_1 \\ b_2 \\ \vdots \\ b_m \end{bmatrix} \tag{4.39}
$$

or

$$
\mathsf{B}\dot{\mathsf{q}} = \mathsf{b}\,. \tag{4.40}
$$

Easily, (4.29) can take such a form. Setting $\mathsf{q} = (x, y, \theta)$ with $m = 1$ we have

$$
\mathsf{B} = [\sin\theta\,, -\cos\theta\,, 0] \qquad \mathsf{b} = [0]\,.
$$

Similarly, (4.32) can be obtained from (4.39) by setting $\mathsf{q} = (x, y, \phi, \theta)$ and $m =$ 2:

$$
\mathsf{B} = \begin{bmatrix} 1 & 0 & (R\cos\theta) & 0 \\ 0 & 1 & (R\sin\theta) & 0 \end{bmatrix} \qquad \mathsf{b} = \begin{bmatrix} 0 \\ 0 \end{bmatrix}
$$

At last, also (4.37) for the ball rolling without slipping on a plane takes form (4.39) if

$$
\mathsf{B} = \begin{bmatrix} 1 & 0 & (R\sin\phi) & 0 & (-R\sin\theta\cos\phi) \\ 0 & 1 & (R\cos\psi) & 0 & (R\sin\theta\sin\phi) \end{bmatrix} \qquad \mathsf{b} = \begin{bmatrix} 0 \\ 0 \end{bmatrix}
$$

and $\mathsf{q} = (x, y, \theta, \phi, \psi)$.

The system of constraints (4.38) is *nonholonomic* when *there do not exist m* functions $f_j(q_1, q_2, \ldots, q_n, t)$ such that the system of linear equations in the $\dot{q}_k$ obtained by differentiating with respect to time

$$
\sum_k \frac{\partial f_j}{\partial q_k}\dot{q}_k + \frac{\partial f}{\partial t} = 0 \qquad j = 1, \ldots, m \tag{4.41}
$$

is *equivalent* to (4.38), in the sense that for any configuration $\mathsf{q}$ the values of the $\dot{q}_k$ satisfying (4.38) are *exactly those* satisfying (4.41).

This condition refers to the totality of (4.38), and not to each single restriction taken separately. Put differently, a collection of nonholonomic constraints taken one by one might produce a holonomic constraint when we consider the system they form. For example, take (4.29) (velocity of the midpoint *parallel* to the rod) together with the analogous nonholonomic constraint that the midpoint's velocity is *orthogonal* to the rod. The overall constraint would be equivalent to asking the midpoint always had zero velocity, which is the well-known holonomic constraint built with a pin.

In general one assumes the matrix $\mathsf{B}$ in (4.39) has *maximal* rank $m$ in any configuration $\mathsf{q}$. If so, the theory of linear systems warrants we can write the $\dot{q}_k$ in function of $s = N - m$ parameters, called *generalized velocities* and traditionally denoted by $e_i$:

$$\dot{q}_k = \sum_{i=1}^{s} g_{ki} e_i + g_k \qquad k = 1, \ldots, N \tag{4.42}$$

where each $g_{ki}$ and each $g_k$ is a function of $(\mathsf{q}, t)$.

As we have seen in Examples 4.2 and Sect. 4.1.2, in the simplest cases the generalized velocities might be some of the $\dot{q}_k$ themselves.

Observe at last that for *fixed* constraints, i.e. time-independent, one always considers the rest state $\dot{\mathsf{q}} = 0$ as admissible. Hence system (4.39) becomes homogeneous

$$\mathsf{B}\dot{\mathsf{q}} = 0$$

with the entries of $\mathsf{B}$ depending on the $q_k$ only. Equations (4.42) for the $\dot{q}_k$ in terms of the generalized velocities $e_i$ also simplify:

$$\dot{q}_k = \sum_{i=1}^{s} g_{ki} e_i \qquad k = 1, \ldots, N \tag{4.43}$$

with $g_{ki}$ not depending explicitly on time, but only on the $q_k$.

This is obviously the case in the aforementioned examples. Instead, for a nonhomogeneous moving constraint it would be enough to think of a disk rolling without slipping on a plane moving in a prescribed way.

### 4.7.2 Virtual Nonholonomic Displacements and Velocity Distributions

We have seen that in systems with $N$ generalized coordinates $q_k$ the displacement $\delta P$ and the virtual velocity $\mathbf{v}'_P$ of a generic point of the system are given by (4.23), where the quantities $\delta q_k$ and $\delta q_k/\delta t$ should be considered *arbitrary*. This is certainly correct only when there are no nonholonomic constraints, those of the type appearing in Examples 4.2 and (4.3), or more generally given by (4.38) or in matrix form (4.40).

Let us recast (4.38) trivially as

$$\sum_{k=1}^{N} b_{jk}(\mathsf{q}, t)\frac{dq_k}{dt} = b_j(\mathsf{q}, t) \quad j = 1, \ldots, m$$

and also in the equivalent differential form

$$\sum_{k=1}^{N} b_{jk}(\mathsf{q}, t)dq_k = b_j(\mathsf{q}, t)dt \quad j = 1, \ldots, m\,. \tag{4.44}$$

As both displacements and virtual velocities must obey the constraints and the finite relations and differentials that describe them when time si supposed *fixed*, from (4.44) we deduce

$$\sum_{k=1}^{N} b_{jk}(\mathsf{q}, t)\delta q_k = 0 \quad j = 1, \ldots, m$$

Hence the quantities $\nu_k = \delta q_k$ must satisfy

$$\sum_{k=1}^{N} b_{jk}(\mathsf{q}, t)\nu_k = 0 \tag{4.45}$$

or in matrix form,

$$\mathsf{B}\boldsymbol{\nu} = \mathbf{0} \tag{4.46}$$

where $\boldsymbol{\nu} = (\nu_1, \ldots, \nu_N)$.

Therefore in presence of *noholonomic* constraints the $\delta q_k$ or $\nu_k = \delta q_k/\delta t$ should *not* be considered independent but related by (4.45) or (4.46). For this reason to find the expressions of displacements and virtual velocities in function of independent arbitrary parameters we will need to use (4.43), which now reads

$$\nu_k = \sum_{i=1}^{s} g_{ki} e_i \qquad k = 1, \ldots, N\,,$$

and write

$$\mathbf{v}_P = \sum_{k=1}^{N} \frac{\partial P}{\partial q_k} \sum_{i=1}^{s} g_{ki} e_i = \sum_{k=1}^{N} \sum_{i=1}^{s} \frac{\partial P}{\partial q_k} g_{ki} e_i$$

or a similar expression for the virtual displacements $\delta P$. The generalized velocities $e_i$ are precisely the *arbitrary* and *independent* parameters whose variations give all virtual velocities.

Nonholonomic constraints *are not* relations for the generalized coordinates, which stay independent, but for their infinitesimal *variations* $\delta q_k$ at a given instant.

## 4.8 Degrees of Freedom

The number of generalized coordinates corresponds, as seen in Sect. 4.3, to the number of independent parameters necessary to describe the system's position.

In the absence of labile or nonholonomic constraints we call *degree of freedom* the number of generalized coordinates. In these hypotheses we will say a system with $N$ generalized coordinates possesses $N$ degrees of freedom, because it admits $\infty^N$ possible configurations and $\infty^N$ virtual displacements.

More in general, the term "degree of freedom" denotes the number of *independent parameters* that characterize the set of *virtual displacements*, which may be *different* from the number of generalized coordinates, as the ensuing examples will clarify.

All in all, a system has $N$ generalized coordinates and $G$ degrees of freedom if it admits $\infty^N$ possible configurations and $\infty^G$ virtual displacements.

- The system of Fig. 4.6 has two generalized coordinates $(s, \theta)$ and hence $\infty^2$ possible configurations. Similarly, two are the independent quantities $(\delta s, \delta\theta)$ that allow to prescribe a virtual displacement. The system has therefore $\infty^2$ virtual displacements, so we say it has two generalized coordinates and two degrees of freedom. The situation in which there are as many generalized coordinates as degrees of freedom is the most common. As already said, it is what happens in systems only subject to holonomic constraints and no lability.
- The system in Example 4.2, represented by Fig. 4.18, has three generalized coordinates $(x, y, \theta)$. Due to relation (4.29), their virtual variations are constrained by
$$\delta x \sin\theta - \delta y \cos\theta = 0\,.$$
For this reason the set of virtual displacements is not described by three independent variations $(\delta x, \delta y, \delta\theta)$ but only two arbitrary parameters. We conclude that this system possesses *three* generalized coordinates, so $\infty^3$ possible configurations, but *two* degrees of freedom, because it has $\infty^2$ virtual displacements.
- In Example 4.3, see Fig. 4.19, the system has four generalized coordinates $(x, y, \theta, \phi)$ whose virtual variations are related by
$$\delta x + R\delta\phi \cos\theta = 0\,, \qquad \delta y + R\delta\phi \sin\theta = 0\,.$$
Hence the set of virtual displacements depends on two independent parameters only.
The system has *four* generalized coordinates, and $\infty^4$ possible configurations, but only *two* degrees of freedom, because it admits $\infty^2$ virtual displacements.
- Figure 4.11 depicts a labile system *without* generalized coordinates that vary in an interval (there is only one possible configuration), but it has infinitely many virtual displacements depending on *one* arbitrary parameter. The system's $\infty^1$ virtual displacements are precisely the rotational infinitesimal displacements around $O$. We say that the system has *one* degree of freedom.
- Compare the previous example with the system in Fig. 4.12, which also has a unique configuration. In this case however no virtual displacement is permitted, so the system *does not have* degrees of freedom.

**Remark 4.6** The discussion of the notion of degree of freedom presented here is based on virtual displacements. As is easy to understand, it could as well be carried out using virtual velocity distributions instead, with no substantial difference. □

## 4.9 Configuration Space

Consider a holonomic mechanical system with time-independent constraints of time whose configuration is determined by the generalized coordinates $(q_1, \ldots, q_N)$. For such systems it is very convenient to introduce the concept of *space of configurations*. This is an $N$-dimensional space whose generic point has coordinates $\mathsf{q} = (q_1, \ldots, q_N)$. It is particularly handy in equilibrium problems.

In particular, the configuration space is useful to detect the presence of further unilateral constraints. In fact if the system only has bilateral constraints the entire configuration space is admissible. On the other hand for unilateral constraints the admissible configurations form subsets of the configuration space, which may be bounded or unbounded, as the following examples will show.

For the mechanical system described in Fig. 4.6 and discussed in Sect. 4.1.3 the configuration space has dimension 2 ($q_1 = s, q_2 = \theta$). If we add two walls restricting the system to the first quadrant (Fig. 4.21 left) the space of admissible configurations will be given by (Fig. 4.21 right):

$$\begin{array}{ll} 0 \le q_1 \le l & 0 \le q_2 \le \pi/2 + \arcsin(q_1/l) \\ q_1 > l & 0 \le q_2 \le \pi. \end{array} \tag{4.47}$$

where $l$ is the rod's length.

The ordinary positions are those where the inequalities in (4.47) are strict, while the boundary positions correspond to having at least one equality in (4.47), namely for points on the boundary of the grey area in Fig. 4.21 right.

Another interesting example of configuration space is that in Fig. 4.22, where two particles connected by an inextensible string of length $l$ are constrained to stay on a line, the $x$-axis.

In this case there are two further unilateral constraints: one forces the points to have distance at most equal to the string's length, and the other prevents $Q$ from passing to the left side of $P$. The configuration space has dimension 2 and the unilateral constraints read:

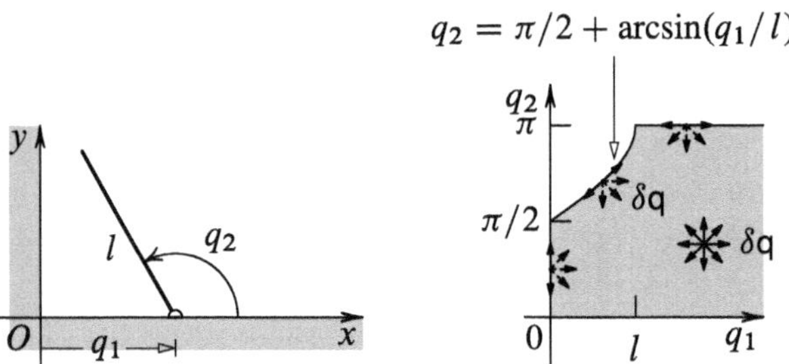


**Fig. 4.21** Physical space and configuration space for a rod subject to unilateral constraints. The virtual variations of the generalized coordinates are shown on the right, using a solid line for reversible virtual displacements and a dashed line for irreversible displacements

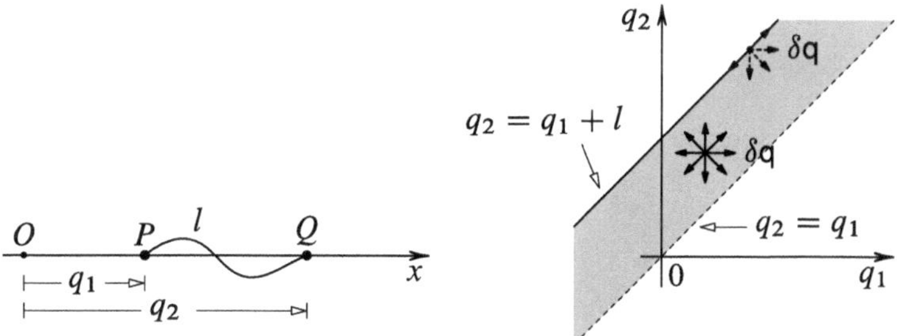


**Fig. 4.22** A description in the physical space and in the configuration space of a system subject to unilateral constraints: two points constrained to move on the $x$-axis and linked by an inextensible string

$$q_1 < q_2 \leq q_1 + l$$

so in this case the boundary positions are those on the line $q_2 = q_1 + l$ (Fig. 4.22).

The configuration space is also very useful to represent virtual displacements in a simple manner and recognize reversible and irreversible ones. Think of the examples in Figs. 4.21 and 4.22. If we virtually displace the physical system by varying the generalized coordinates from $q_1$ to $q_1 + \delta q_1$ and from $q_2$ to $q_2 + \delta q_2$, in the configuration space the point is displaced by $\delta \mathsf{q} \equiv (\delta q_1, \delta q_2)$, as in Fig. 4.22 right. In the picture the reversible virtual displacements are drawn with a continuous line and the irreversible ones with a dashed line. In agreement with the previous remarks the virtual displacements in ordinary positions are all reversible while in boundary positions there (usually) are both reversible and irreversible displacements.

## 4.10 Base and Roulette

In every *planar* motion of a rigid body it is possible to find a *fixed* curve, called *base*, and a *comoving* curve, called *roulette*, such that during motion the roulette rolls without slipping on the base.

Base and roulette are defined as the curves traced out by the position of the instantaneous center of rotation on the plane comoving with the fixed frame and on the plane comoving with the moving frame, respectively. Finding an explicit expression for the base and roulette of the plane motion of a rigid body is interesting only when the rotation angle $\theta$ can be utilized as only generalized coordinate (the system is also supposed subject to *fixed* constraints). Hence when the position of a point $A$ comoving with the body is expressible in terms of $\theta$ itself: $A(\theta)$.

The velocity of $A$, found by the formula for the roto-translational velocity distribution referred to the instantaneous center of rotation $C$, is $\mathbf{v}(A) = \boldsymbol{\omega} \times CA$. As the motion is planar, $\mathbf{v}(A) = \dot{\theta}\mathbf{k} \times CA$.

Let us introduce the coordinates of $A$ and $C$ with respect to a fixed reference frame with basis $\{\mathbf{i}, \mathbf{j}, \mathbf{k}\}$, here assumed *right-handed*. The above equation becomes

$$\begin{aligned}\dot{x}_A\mathbf{i} + \dot{y}_A\mathbf{j} &= \dot{\theta}\mathbf{k} \times [(x_A - x_C)\mathbf{i} + (y_A - y_C)\mathbf{j}] \\ &= \dot{\theta}[(x_A - x_C)\mathbf{k} \times \mathbf{i} + (y_A - y_C)\mathbf{k} \times \mathbf{j}] \\ &= \dot{\theta}[(x_A - x_C)\mathbf{j} - (y_A - y_C)\mathbf{i}]\end{aligned}$$

and then

$$\dot{x}_A = \dot{\theta}(y_C - y_A), \quad \dot{y}_A = \dot{\theta}(x_A - x_C). \tag{4.48}$$

As

$$\dot{x}_A = \frac{dx_A}{d\theta}\dot{\theta}, \quad \dot{y}_A = \frac{dy_A}{d\theta}\dot{\theta}, \tag{4.49}$$

we can compare (4.48) to (4.49) and divide by $\dot{\theta}$:

$$\frac{dx_A}{d\theta} = y_C - y_A, \quad \frac{dy_A}{d\theta} = x_A - x_C,$$

so eventually

$$x_C = x_A - \frac{dy_A}{d\theta}, \quad y_C = y_A + \frac{dx_A}{d\theta}. \tag{4.50}$$

Since $OC = x_C\mathbf{i} + y_C\mathbf{j}$, we can write it vectorially as

$$\begin{aligned}OC(\theta) &= \left(x_A(\theta) - \frac{dy_A}{d\theta}\right)\mathbf{i} + \left(y_A(\theta) + \frac{dx_A}{d\theta}\right)\mathbf{j} \\ &= x_A\mathbf{i} + y_A\mathbf{j} - \frac{dy_A}{d\theta}\mathbf{i} + \frac{dx_A}{d\theta}\mathbf{j} \\ &= OA(\theta) - \frac{dy_A}{d\theta}\mathbf{i} + \frac{dx_A}{d\theta}\mathbf{j}.\end{aligned} \tag{4.51}$$

Then (4.51) allows us to find, from a known $A(\theta)$, the position of $C$ for any value of the generalized coordinate. Equation (4.50), together with its vectorial form (4.51), should then be interpreted as the parametric equation of a plane curve, the locus of the instantaneous centers of rotation $C$ as seen by the fixed frame. This curve is known as the *base of the motion.*

Referring to Fig. 4.23, where the $x'$- and $y'$-axes are parallel to the unit vectors $\mathbf{i}'$ and $\mathbf{j}'$ comoving with the body, we have

$$\mathbf{i} = \cos\theta\mathbf{i}' - \sin\theta\mathbf{j}', \qquad \mathbf{j} = \sin\theta\mathbf{i}' + \cos\theta\mathbf{j}'.$$

Substituting the latter in (4.51) and doing the calculations gives

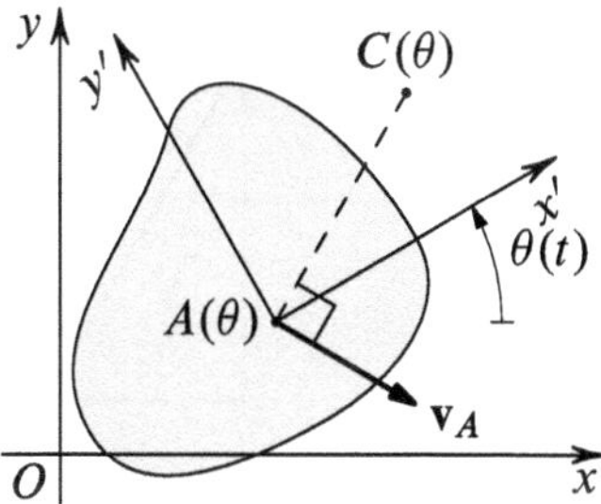


**Fig. 4.23** Analytical computation of base and roulette

$$OC(\theta) = OA(\theta) + \left(\frac{dx_A}{d\theta}\sin\theta - \frac{dy_A}{d\theta}\cos\theta\right)\mathbf{i}' + \left(\frac{dx_A}{d\theta}\cos\theta + \frac{dy_A}{d\theta}\sin\theta\right)\mathbf{j}'$$

and because $OC = OA + AC$,

$$AC(\theta) = \left(\frac{dx_A}{d\theta}\sin\theta - \frac{dy_A}{d\theta}\cos\theta\right)\mathbf{i}' + \left(\frac{dx_A}{d\theta}\cos\theta + \frac{dy_A}{d\theta}\sin\theta\right)\mathbf{j}'. \tag{4.52}$$

This corresponds to the *roulette*'s equation, in that it describes in terms of the generalized coordinate $\theta$ the locus of instantaneous centers of rotation, as seen in the comoving frame now, whose origin is $A$ and has comoving vectors $\{\mathbf{i}', \mathbf{j}'\}$. Hence (4.52) must be interpreted as the parametric form of a plane curve, known as roulette, that comoves with the body. Introducing the coordinates $(x'_C, y'_C)$ of $C$ with respect to the moving axes $x'$, $y'$ with origin $A$, (4.52) reads

$$x'_C = \frac{dx_A}{d\theta}\sin\theta - \frac{dy_A}{d\theta}\cos\theta\,, \quad y'_C = \frac{dx_A}{d\theta}\cos\theta + \frac{dy_A}{d\theta}\sin\theta\,.$$

**Example 4.7** A classical case of computation of base and roulette is a rod $AB$ of length $l$ with ends constrained to the two Cartesian axes, as in Fig. 4.24. Here we need not even use the previous equations to find the two curves. By Chasles' theorem the instantaneous center of rotation $C$ is, as shown, the intersection of the two orthogonal lines to the velocities of $A$ and $B$. The picture on the right shows that during the motion the rod changes position so also $C$ moves. The isolated dot indicates where $C$ was at a previous instant.

Figure 4.25 shows instead that the segment $OC$ has length equal to $l$ at all times, being the diagonal of the rectangle $OACB$. Therefore $C$ keeps at constant distance to $O$, and the fixed frame, comoving with these axes, will see $C$ trace a circle centered at $O$ with radius $l$, of which we drew only a portion. This is the *base* curve of the rod $AB$.

Regarding the roulette, it suffices to note that, as in Fig. 4.25 right, $C$ keeps at constant distance $l/2$ to the rod's midpoint, which is clearly fixed for the frame comoving with the rod. Hence the *roulette* is the circle of radius $l/2$ centered at the midpoint. The roulette rolls without slipping on the base curve, and we can reconstruct the trajectories of the body's points from the base and the roulette. In

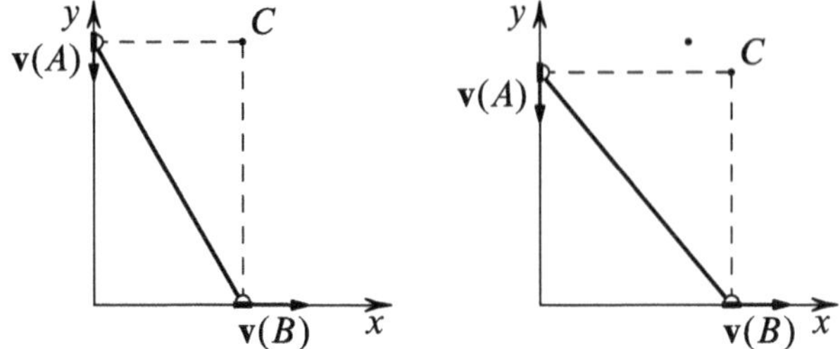


**Fig. 4.24** A rod with the two ends constrained to the $x$- and $y$-axis

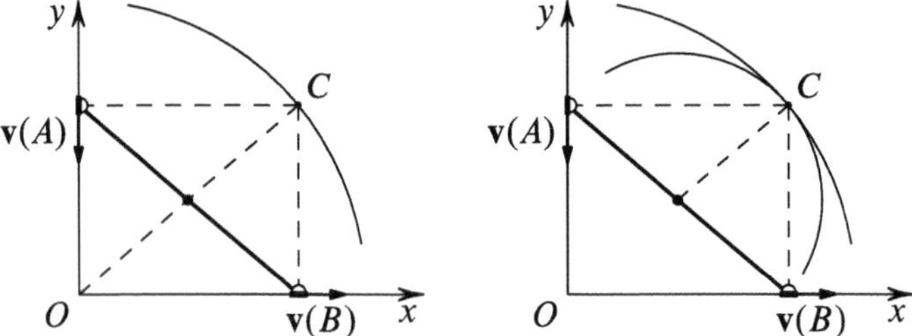


**Fig. 4.25** Base and roulette of a rod with the ends constrained to the $x$- and $y$-axis

fact, if we impose the rod comoves with the roulette whilst rolling along the base, the ends $A$ and $B$ automatically stay on the cartesian axes, irrespective of having imposed the constraints. □

## 4.11 Recap on Constraints

At the end of the discussion of constrained systems, their classification and complicated surrounding notions, we propose a summary for the reader's benefit. Each definition was introduced, discussed and illustrated in this chapter by the previous sections' examples. One cannot prescind from them, and they should be revisited for a deeper understanding.

- **Constraint**: an *a priori* restriction on the configurations or possible velocity distributions of a system.
- **Generalized** (or **Lagrangian**) **coordinates**: a set of $N$ *essential* and *independent* parameters $q_k$ (living in a region of $\mathbb{R}^N$) that describe the system's configurations. If the value of each generalized coordinate is fixed the system cannot move. *Essential* means that none can be eliminated: if we fixed $N - 1$ generalized coordinates the system could still move. *Independent* means that knowing $N - 1$ of them leaves the remaining one free, and the latter's value cannot be computed from the others in any way.

- **Configuration space**: the region where the generalized coordinates can vary.
- **Fixed** or **scleronomous constraint**: a constraint whose analytic expression *does not depend* explicitly on time. The set of admissible configurations or velocity distributions *does not* depend on the instant of time.
- **Moving** or **rheonomous constraint**: a constraint whose analytic expression *depends* explicitly on time. The set of admissible configurations or velocity distributions *depends* on the instant of time.
- **Holonomic constraint**: an *a priori* restriction on the set of possible configurations of a system, hence on its displacements.
- **Mobility constraint**: a constraint that restricts the set of possible velocity distributions. It may be due to a holonomic constraint, and if not it is called *pure mobility constraint* or *nonholonomic*.
- **Pure mobility constraint** or **nonholonomic constraint**: a restriction on admissible velocity distributions that is inequivalent to a collection of holonomic constraints. The nonholonomic constraints we consider are those translating into *linear* equations for the $\dot{q}_k$.
- **Virtual displacement**: an infinitesimal displacement compatible with the constraints imagined fixed at the considered instant. In case of *fixed* constraints one special case is the effective elementary displacement; for *moving* constraints this is no longer true.
- **Virtual velocity distribution** and **virtual velocity**: velocity distribution and velocity compatible with the constraints imagined fixed at the considered instant. For systems subject to *fixed* constraints the effective velocity is virtual; this is *not* true in presence of *moving* constraints.
- **Generalized velocities**: independent quantities that allow to express the $\dot{q}_k$ when related by nonholonomic constraints.
- **Reversible/irreversible virtual displacement or velocity distribution**: a virtual displacement or velocity distribution is reversible if its opposite is virtual. If not, we say it is irreversible.
- **Bilateral/unilateral constraints**: a system is subject to bilateral constraints if *every* virtual displacement is *reversible* starting from any possible configuration. If there are configurations admitting *irreversible* virtual displacements we call them *boundary configurations* and the constraints *unilateral*.
- **Degrees of freedom**: the number of degrees of freedom of a non-labile system only subject to holonomic constraints equals the number of generalized coordinates, i.e. the number of essential independent parameters describing the system's configurations. More generally, the degrees of freedom correspond to the number of independent parameters allowing to describe the set of virtual displacements (or velocity distributions). A system with $N$ generalized coordinates and $G$ degrees of freedom admits $\infty^N$ possible configurations and $\infty^G$ virtual displacements.
- **Hyper-constrained system**: a system subject to superfluous constraints: the set of possible configurations or velocity distributions is unaffected if we remove one or more constraints.

- **Isostatic system**: a system not admitting motions nor virtual displacements or velocity distributions. No constraint can be removed or weakened without altering the set of possible configurations or velocity distributions.
- **Hyperstatic system**: a system that cannot move and is hyper-constrained.
- **Labile system**: there are *more* degrees of freedom than generalized coordinates. There exist virtual displacements or virtual velocities not arising from virtual variations of generalized coordinates.

## 4.12 Kinematics: Examples

**Example 4.8** The first picture in Fig. 4.26 shows a disk of radius $2r$ and center $G$ that at the contact point $H$ rolls without slipping along a rod. The rod is constrained to glide horizontally along the $x$-axis with two rollers. An inextensible string, wrapped onto a profile of radius $r$ concentric to the disk and comoving with it, uncoils horizontally until it reaches a point (think of a fixed axle with a pointwise pulley wheel) and drops vertically ending at the point $P$. A second string wraps around the disk and joins it, after a horizontal section $AB$, to the boundary of a second disk of radius $r$ and *fixed* center $Q$, on which it then wraps and continues horizontally to rejoin the center $G$ of the first disk. We assume perfect adherence between disks and strings, and that the strings stay tightly stretched during the motion.

Let us call $\phi$ and $\theta$ the angles of rotation of the two disks, oriented as in the figure, and let $s$ be the $x$-coordinate of one end of the rod.

As we will see below, this system has only one degree of freedom (albeit perhaps not evident at first) and so we can write all velocities of interest using one coordinate among $\phi$, $\theta$, $s$. We set out to examine the kinematics and in particular to determine the relationships between the time derivatives $\dot{s}$, $\dot{\phi}$, $\dot{\theta}$. We also wish to write in terms of $\dot{s}$ the downwards *component* $v$ of the velocity of $P$.

The first thing to observe is that the rod has translational velocity distribution, so the velocity of all of its points, including the contact point with the disk, has component $\dot{s}$ along the $x$-axis, as shown in the middle picture of Fig. 4.26.

The velocities of $G$ and $B$ can be written using the formula characterizing roto-translational velocity distributions:

$$\mathbf{v}(B) = \mathbf{v}(H) + \boldsymbol{\omega} \times HB\,, \qquad \mathbf{v}(G) = \mathbf{v}(H) + \boldsymbol{\omega} \times HG\,. \tag{4.53}$$

Due to the rolling constraint with no slipping the velocity at $H$ is the same both if $H$ is thought of as point of the rod or of the disk's border. We also must not forget that with the choice of $\theta$ indicated in the figure, the disk's angular velocity equals $\boldsymbol{\omega} = -\dot{\theta}\mathbf{k}$ if $\mathbf{k}$ is a unit vector coming out of the plane (hence forming a right-handed basis with $\mathbf{i}$ and $\mathbf{j}$), but it equals $\boldsymbol{\omega} = \dot{\theta}\mathbf{k}$ if $\mathbf{k}$ forms a left-handed basis with $\mathbf{i}$ and $\mathbf{j}$. This is important when computing the cross product in (4.53) using known formulas of linear algebra. In this case however it is much easier to view the velocity of $G$ as

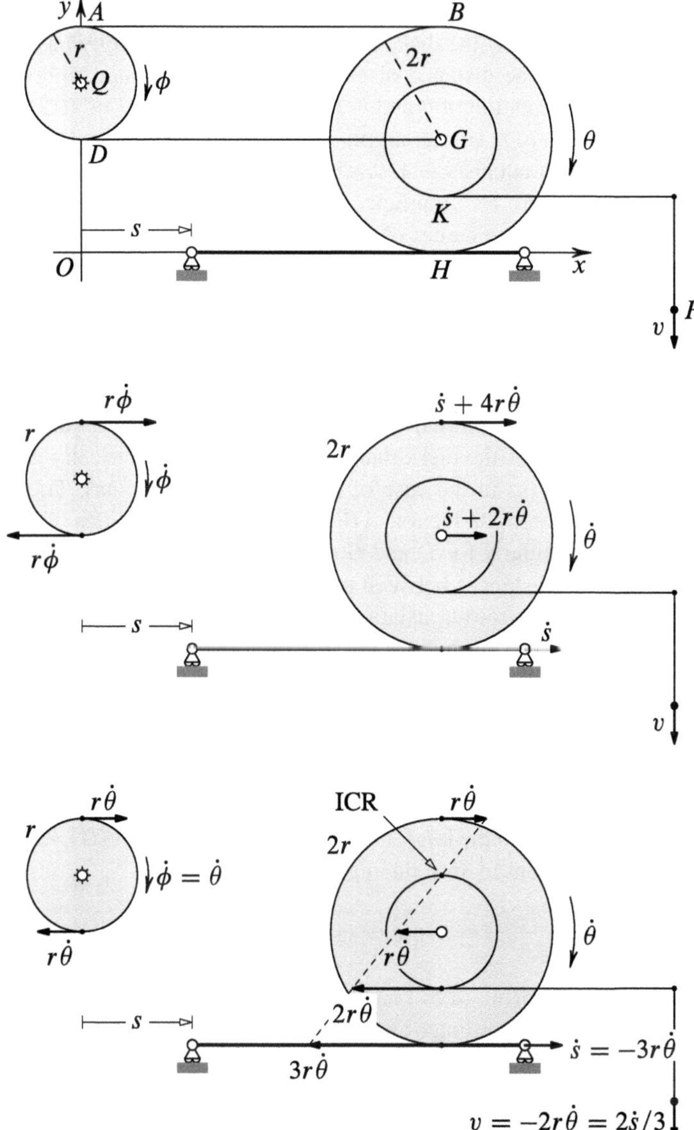


**Fig. 4.26** Two disks and one point connected through inextensible strings

sum of the velocity of $H$ ($\dot{s}$ rightwards) plus the velocity that $G$ would have if $H$ *were* the center of rotation of the disk. The latter part of the velocity has horizontal component $\dot{\theta}$ times $2r$, the distance of $G$ to $H$, and is orientated rightwards, i.e. coherently with the orientation of $\theta$ and $\dot{\theta}$. Hence the component of $\mathbf{v}(G)$ along the positive $x$-axis is $\dot{s} + 2r\dot{\theta}$. A similar argument allows to deduce that the velocity of $B$ in the same direction equals $\dot{s} + 4r\dot{\theta}$. These velocities are shown in the middle picture of Fig. 4.26, where for simplicity we did not plot the axes, the horizontal sections of the strings nor the names of the various points.

The points $A$ and $D$, where the string leaves the border of the disk of radius $r$ and fixed center $Q$, have horizontal velocity expressible via the derivative of $\phi$, which is a priori independent of $\theta$. Here it is obvious that the disk's velocity distribution is rotational about $Q$, so the velocities are horizontal and have component $r\dot{\phi}$, oriented *rightwards* for $A$ and *leftwards* for $D$.

We said the strings wrap around disks with perfect adherence. This means the points of the strings and of the disks that are in contact (geometrically coincident) have the same velocity (as in the case of rolling without slipping). Therefore the points at the ends of the tight section $AB$ have the same velocities as the border points $A$ and $B$, and similarly for $D$ and $G$.

In a tight string the distances between points do not change, so the velocity distribution of this section is roto-translational, i.e. rigid. For $AB$, which clearly has translational velocity distribution, the condition imposes $\mathbf{v}(A) = \mathbf{v}(B)$, and for $DG$ we obtain $\mathbf{v}(D) = \mathbf{v}(G)$. Referring to Fig. 4.26 in the middle, the first condition immediately gives

$$r\dot{\phi} = \dot{s} + 4r\dot{\theta}\,. \tag{4.54}$$

As for $\mathbf{v}(D) = \mathbf{v}(G)$, we need to observe that the components of the two velocities have opposite orientation: to the left for $\mathbf{v}(D)$ and to the right for $\mathbf{v}(G)$. The equality must keep this into account by introducing a minus sign:

$$-r\dot{\phi} = \dot{s} + 2r\dot{\theta}\,. \tag{4.55}$$

The linear system comprising (4.54) and (4.55) gives us the relationships between the first derivatives of the coordinates $s$, $\phi$ and $\theta$ created by the constraints. We solve them for $\dot{s}$ and $\dot{\phi}$ in function of $\dot{\theta}$ to obtain

$$\dot{\phi} = \dot{\theta}\,, \qquad \dot{s} = -3r\dot{\theta}\,. \tag{4.56}$$

The equations are integrable, equivalent to

$$\phi = \theta + \text{cost.} \qquad s = -3r\theta + \text{cost.}$$

where the constants will be determined by the value of the coordinates in a given configuration (here not specified).

The angular velocities of the disks are then equal, while the rod's translational velocity is related to the angular velocity of the disk that rolls on it by $\dot{s} = -3r\dot{\theta}$.

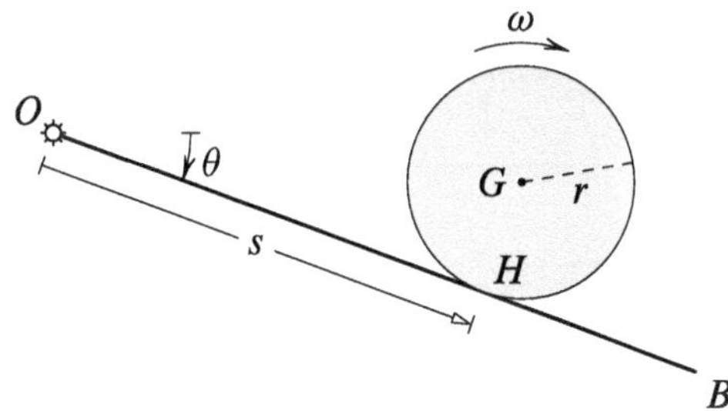


**Fig. 4.27** A disk which is rolling without slipping on a rod pinned at one end and free to rotate around it

It is instructive to note, in the light of what we deduced from (4.56), the velocities of $B$ and $G$ on the larger disk are equal to $r\dot{\theta}$, rightwards, for $B$, and $r\dot{\theta}$, leftwards, for $G$. As a rigid velocity distribution on the plane that is not translational is certainly rotational around an *instantaneous center of rotation*, the previous remark says the center is the midpoint of $B$ and $G$, at distance $r$ from either. In the bottom picture of Fig. 4.26 we see the velocities of $B$, $G$, $K$ and $H$ deduced by viewing the disk's velocity distribution as rotational around its instantaneous center of rotation (i.c.r.). Each velocity is perpendicular to the line joining to this point and is proportional to the distance from it. Looking at what happens at the contact point $H$ we have confirmation that $\dot{s} = -3r\dot{\theta}$ necessarily.

Now we are interested in the velocity at $K$, equal to $2r\dot{\theta}$ to the left. It is identical to that of $P$, except for the different orientation imposed by the fixed point through which the string passes before starting its descent (here too we suppose the string is tight during the motion). Keeping in account the velocities' different orientations, and because of (4.56), we find

$$v = -2r\dot{\theta} \quad \Rightarrow \quad v = 2\dot{s}/3$$

The *descent* velocity of $P$ is therefore two thirds of the velocity of the rod as it shifts to the right. □

**Example 4.9** The system in Fig. 4.27 is made of a rod pinned to the fixed point $O$ along which a disk of radius $r$ rolls without slipping. We want to use generalized coordinates $\theta$ and $s$ and their derivatives to write the velocity of the disk's center $G$ and the component $\omega$ of its angular velocity in the direction shown.

First things first, we use the theorem of composition of angular velocities and decompose the absolute angular velocity $\omega$ of the disk into a sum of two contributions: the first is the angular velocity of the frame rotating with the rod, equal to $\dot{\theta}$ clockwise, while the second summand is $\omega_{\text{rel}}$, the angular velocity *relative* to it. In this frame the rod is fixed and the center of the disk moves with velocity $\dot{s}$ parallel to the rod. Hence the relative angular velocity $\omega_{\text{rel}}$ is $\dot{s}/r$ clockwise. Eventually

$$\omega = \dot{\theta} + \frac{\dot{s}}{r}.$$

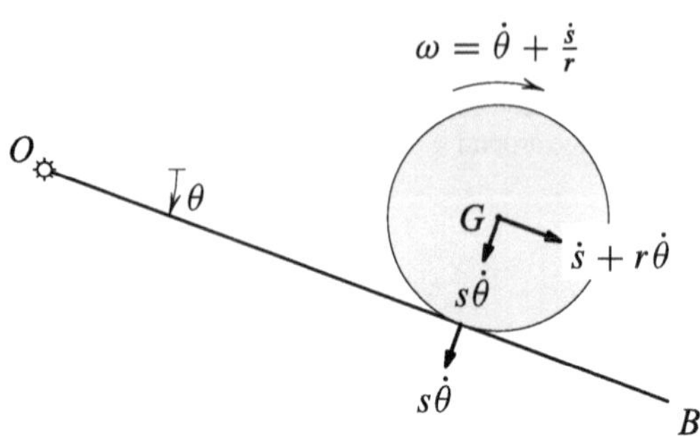


**Fig. 4.28** The velocity of the center $G$ of the disk

Let us write the velocity of $G$ using the formula for the roto-translational velocity distribution of the disk, referred to $H$: $\mathbf{v}(G) = \mathbf{v}(H) + \boldsymbol{\omega} \times HG$. The velocity of the boundary point at $H$ coincides with the velocity of the point on the rod in the same position, because the pure rolling constraint forces the contact points to be the same. But since the rod rotates around $O$ with angular velocity $\dot{\theta}$, the velocity of $H$ is perpendicular to the rod, has modulus $s\dot{\theta}$ and orientation as in Fig. 4.28. To find $\mathbf{v}(G)$ we must add to the previous $\mathbf{v}(H)$ the vector $\boldsymbol{\omega} \times HG$ (which may be interpreted as the velocity $G$ would have if the disk's velocity distribution were purely rotational about $H$). As $HG = r\mathbf{u}$ and $\boldsymbol{\omega} = \omega\mathbf{k} = (\dot{\theta} + \dot{s}/r)\mathbf{k}$, where $\mathbf{u}$ is the unit vector from $H$ to $G$ and $\mathbf{k}$ is the ingoing unit normal to the plane, the cross product $\boldsymbol{\omega} \times HG$ gives, as in Fig. 4.28, a vector of modulus $\dot{s} + r\dot{\theta}$ parallel to the rod.

Naturally, if necessary we could easily project on the fixed Cartesian axes these two components of $\mathbf{v}(G)$. Choosing $x$ oriented left-to-right and $y$ downwards we obtain

$$\dot{x}_G = -s\dot{\theta}\sin\theta + (\dot{s} + r\dot{\theta})\cos\theta\,, \quad \dot{y}_G = s\dot{\theta}\cos\theta + (\dot{s} + r\dot{\theta})\sin\theta\,.$$

We could have reached this conclusion also directly by differentiating the Cartesian coordinates of $G$

$$x_G = s\cos\theta + r\sin\theta\,, \quad y_G = s\sin\theta - r\cos\theta\,,$$

obtained by easy trigonometry. The method used above is anyhow instructive to familiarize ourselves with the formula for the roto-translational velocity distribution. □

# Chapter 5
# Geometry of Masses

The previous chapters were dedicated to Kinematics, which analyzes and describes the motions of systems, both free and constrained, without addressing how they can be achieved and, more importantly without attempting to understand which motion the system will undergo when subject to a certain force.

Experience tells us that to find the dynamical response of a system it is not enough to know its kinematical characteristics (that is, the constraints and the motions allowed by them) and the force applied. It is in fact evident that if we apply the same system of forces to two systems that are geometrically and kinematically identical, the resulting motion might be completely different. The further characteristics that distinguish the behaviour of one system (free or constrained) from another one boil down to the mass and its distribution, and it is this that motivates the title of this chapter.

The ideas we will develop are all associated with masses, whether they are concentrated or uniformly distributed in spatial regions, and they can be divided into two types: *centers of mass* and *moments of inertia*, together with their main properties.

Every particle is characterized by a positive quantity, called *mass*, that determines the dynamical response to the forces we apply to it, as we will see in Chap. 8. For extended systems, the mass is distributed over the entire region occupied by the body. In order to associate a mass with every finite and measurable portion of the system one introduces the notion of *density of mass* $\rho$. This will denote a sufficiently regular function, in general depending on the position. Using $\rho$ we assume to be able to express the mass of any portion $\mathcal{B}$ of the system as

$$m(\mathcal{B}) = \int_{\mathcal{B}} \rho(P)\, d\tau\,. \tag{5.1}$$

The quantity $d\tau$ indicates an infinitesimal *volume* element for three-dimensional bodies, an infinitesimal *area* element for two-dimensional ones, and a *length* element for one-dimensional bodies. In formula (5.1) the function $\rho(P)$ is therefore



P. Biscari et al., *Rational Mechanics*, UNITEXT 177,
https://doi.org/10.1007/978-3-032-07462-1_5

the density of volume, area or length that characterizes the mass distribution within the continuous body. Such density functions have respective physical dimensions $[ML^{-3}]$, $[ML^{-2}]$ and $[ML^{-1}]$. If the density of a body is constant the body is said to be *homogeneous*. For these the uniform value of $\rho$ coincides with the ratio between the mass of any region in the body and the corresponding volume (or area or length, for planar or one-dimensional bodies).

## 5.1 Center of Mass

The *center of mass* $G$ of a discrete system $\mathcal{S}$, made of $n$ particles of masses $m_i$ placed at the points $P_i$, is defined by

$$OG = \frac{\sum_{i=1}^{n} m_i OP_i}{\sum_{i=1}^{n} m_i} = \frac{1}{m} \sum_{i=1}^{n} m_i OP_i \,, \tag{5.2}$$

(where $m = \sum_i m_i$ indicates the total mass and $O$ is an arbitrary origin). This relation can be recast in the useful form

$$m\, OG = \sum_{i=1}^{n} m_i OP_i \,. \tag{5.3}$$

The coordinates of $G$ in a Cartesian frame with origin $O$ are

$$x_G = \frac{1}{m} \sum_{i=1}^{n} m_i x_i \,, \qquad y_G = \frac{1}{m} \sum_{i=1}^{n} m_i y_i \,, \qquad z_G = \frac{1}{m} \sum_{i=1}^{n} m_i z_i \,.$$

For a continuous body $\mathcal{B}$ of density $\rho$, the center of mass is defined to be

$$OG = \frac{\int_{\mathcal{B}} \rho OP d\tau}{\int_{\mathcal{B}} \rho d\tau} = \frac{1}{m} \int_{\mathcal{B}} \rho OP d\tau \,, \tag{5.4}$$

or equivalently

$$mOG = \int_{\mathcal{B}} \rho OP d\tau \tag{5.5}$$

and its Cartesian coordinates are

$$x_G = \frac{1}{m} \int_{\mathcal{B}} \rho x d\tau \,, \quad y_G = \frac{1}{m} \int_{\mathcal{B}} \rho y d\tau \,, \quad z_G = \frac{1}{m} \int_{\mathcal{B}} \rho z d\tau \,. \tag{5.6}$$

For reasons we will explain later (Sect. 7.4.2 in Chap. 7, devoted to forces) the center of mass is also called *center of gravity* or *barycenter*.

For *homogeneous* systems one also uses the word *centroid*, because the position of $G$ is independent of the density: being constant, the density factors through sums and integrals.

In an ensuing Section (Sect. 7.4) we will show the center of mass coincides with the center of a particular system of parallel forces representing the weights $(P_i, m_i\mathbf{g})$, $(i = 1, \dots, n)$, where $\mathbf{g}$ is the acceleration of gravity (see (7.27)) under the assumption it is independent of the point.

Although it is useful to assign the center of mass' position with respect to an origin $O$ in definitions (5.2) and (5.4), the position of $G$ depends exclusively on the positions and masses of the system's points, and not on the choice of origin.

Suppose by contradiction that with a different origin $O'$ we found another center of mass $G'$. In the discrete case (the continuous case is completely similar), from (5.3) we obtain

$$\begin{aligned} mO'G' &= \sum_{i=1}^{n} m_i O'P_i = \sum_{i=1}^{n} m_i (O'O + OP_i) \\ &= mO'O + \sum_{i=1}^{n} m_i OP_i = m(O'O + OG) = mO'G\,, \end{aligned}$$

so $G \equiv G'$. If in particular we pick $O \equiv G$ we find the useful relation

$$\sum_{i=1}^{n} m_i GP_i = m\,GG = \mathbf{0}\,. \tag{5.7}$$

### 5.1.1 Material Symmetries

Material symmetries are an extension of the classical definitions about the notion of geometric symmetry to systems with a mass.

A plane $\pi$ is called *diametral* (or a *diameter*) for system $\mathcal{S}$, and conjugate to the unit vector $\mathbf{u}$, if for any point $P \in \mathcal{S}$ not lying on $\pi$ there is another point $\hat{P} \in \mathcal{S}$ such that (see Fig. 5.1):

(i) $P$ and $\hat{P}$ have the same mass (same density if $\mathcal{S}$ is continuous).
(ii) $P\hat{P} \parallel \mathbf{u}$.
(iii) The midpoint of the segment $P\hat{P}$ belongs to $\pi$.

A plane $\pi$ is a *material symmetry* if it is diametral and $\mathbf{u}$ is orthogonal to $\pi$.

For a system $\mathcal{S}$ contained in a plane $\Pi$, this plane must be a material symmetry, trivially. In this case moreover, as shown in Fig. 5.2, a line $r \subset \Pi$ is called *diametral* (or a *diameter*) and conjugate to the unit vector $\mathbf{u} \in \Pi$, if

(i) $P$ and $\hat{P}$ have equal mass (equal density if $\mathcal{S}$ is continuous).

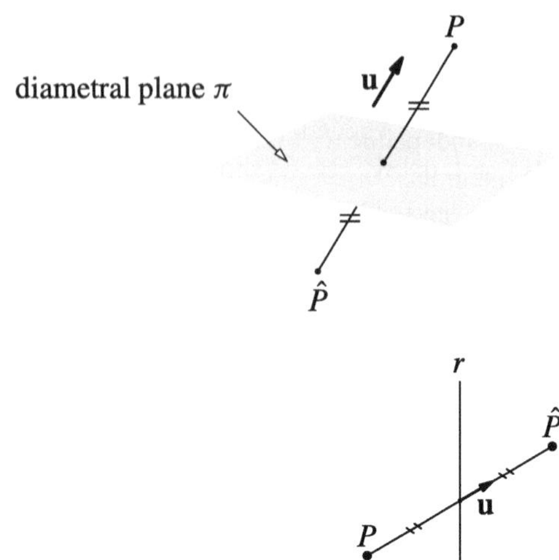


**Fig. 5.1** The diametral plane conjugate to **u**: at points $P$ and $\hat{P}$ the densities (or masses) are equal and the midpoint of the segment $P\hat{P}$ belongs to $\pi$

**Fig. 5.2** The diametral axis $r$ conjugate to **u**: at the points $P$ and $\hat{P}$ the densities (or masses) are equal and the midpoint of the segment $P\hat{P}$ belongs to $r$

(ii) $P\hat{P} \parallel \mathbf{u}$.
(iii) The midpoint of $P$ and $\hat{P}$ belongs to $r$.

We also say $r \subset \Pi$ is a *material symmetry* for the planar system $\mathcal{S}$ if **u** is orthogonal to $r$.

**Example 5.1** (*Material symmetries in a rectangle*) Consider a homogeneous rectangle with edges $a$ and $b$. The lines $r$, $s$ parallel to the edges passing through their midpoints are material symmetries. Moreover, each diagonal is a diameter conjugate to the direction of the other diagonal. Hence the diagonals are material symmetries if and only if $a = b$. □

### 5.1.2 Distributive Property

Consider a material system $\mathcal{S}$, with center of mass $G$ and total mass $m$, made of $k$ subsystems $\{\mathcal{S}_i\,,\ i = 1, \ldots, k\}$ each of which has center of mass $G_i$ and mass $m_i$ (obviously $m_1 + \cdots + m_k = m$). The center of mass of $\mathcal{S}$ coincides with the center of mass of a system of $k$ particles of masses $(m_1, \ldots, m_k)$ placed at $(G_1, \ldots, G_k)$.

This property is an immediate consequence of the distributivity of addition, or of integration in case $\mathcal{S}$ is continuous. In fact, we can write

$$m OG = \int_{\mathcal{S}} \rho OP d\tau = \sum_{i=1}^{k} \int_{\mathcal{S}_i} \rho OP d\tau = \sum_{i=1}^{k} \left( \int_{\mathcal{S}_i} \rho d\tau \right) OG_i = \sum_{i=1}^{k} m_i \, OG_i\,.$$

### 5.1.3 Symmetries

(i) The center of mass $G$ of two particles lies on the segment joining them. In particular, $G$ divides the segment in two parts that are inversely proportional to the masses placed at the endpoints. This is clear if we use a frame system with origin $O \equiv G$ and formula (5.7).

(ii) The center of mass of a planar system is contained in the system's plane. To check this we introduce an axis $z$ orthogonal to the plane. As the coordinates $z_i$ of all points are zero, so will be the coordinate $z_G$ of the center of mass.

(iii) Any diameter (plane or line) contains the center of mass. In fact, by definition of diametral plane and by the additivity seen above, the center of mass can be computed using the centers of gravity of symmetric pairs $\{P, \hat{P}\}$ for the diametral plane. By the (i) above, the center of mass of these pairs lies on the diametral plane, so also the center of mass of the system, by (ii), must belong to this plane.

(iv) If a material system is contained in a region of the plane bounded by a closed convex curve, the center of mass $G$ must belong inside the region. Analogously, if the material system is inside a region bounded by a closed convex surface, its center of mass $G$ belongs to the region.

To prove this property we can consider a point $P_0$ of the closed convex curve (or surface) and the tangent line (tangent plane) at $P_0$. Fix a suitable frame of reference on the plane with origin $P_0$ so that the tangent line is the $y$-axis and the $x$-axis points towards the half-plane containing the system. Immediately, since all points of the system have first coordinate $x_i > 0$, necessarily $x_G > 0$. (In the spatial case, just take a frame in which the $xy$-plane is the tangent plane and the $z$-axis points towards the material system.)

**Example 5.2** (*Center of mass of an annulus*) Let us find the center of mass of a sector of a homogeneous annulus of mass $m$, with angular aperture $2\alpha$ and radii $r_1, r_2$. Consider a frame with origin at the annulus' center so that the system lies on the $xy$-plane, with $x$-axis coinciding with the line bisecting the sector (Fig. 5.3). The $x$-axis is a material symmetry, so the center of mass belongs to the bisectrix: $y_G = z_G = 0$. There remains to find $x_G$, i.e. the distance of the center of mass to the sector's center. Using polar coordinates in $(5.6)_1$ we have

$$x_G = |OG| = \frac{1}{m} \int_{r_1}^{r_2} \int_{-\alpha}^{\alpha} \rho r \cos\theta \, r d\theta \, dr = \frac{2}{3} \frac{\sin\alpha}{\alpha} \frac{r_2^2 + r_2 r_1 + r_1^2}{r_2 + r_1}. \quad (5.8)$$

Formula (5.8) subsumes several interesting particular cases.

- If $2\alpha = 2\pi$, (5.8) proves that the center of mass of an annulus is at its geometric center, as one could prove by noting that the $y$-axis is a material symmetry too.
- Putting $r_1 = 0$ and $r_2 = r$ we find the position of the center of mass of a circular sector of angle $2\alpha$ and radius $r$:

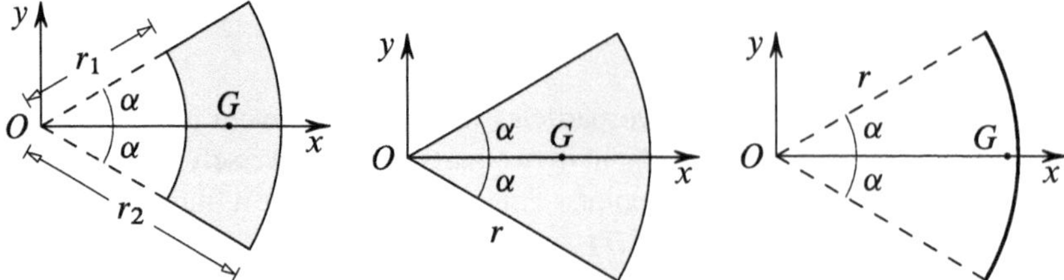


**Fig. 5.3** Centers of mass for circular sectors and a circular arc

$$x_G = |OG| = \frac{2}{3}\frac{\sin\alpha}{\alpha}r\,.$$

In particular, in the limit as $\alpha \to 0$ the center of mass of the sector will have distance $2r/3$ from the origin, and not $r/2$ as we might (wrongly) expect based on the sector's limiting shape being a segment of length $r$. In fact, although in the limit the sector shrinks to a segment, the latter is not homogeneous. One could in fact show that the sector's center of mass as $\alpha \to 0$ coincides with the center of mass of a non-homogeneous segment, whose density grows linearly with the distance to the center of the sector, starting from zero.

Particularly useful are the expressions for $\alpha = \pi/4$ and $\alpha = \pi/2$, respectively representing the center of mass of one quarter of a disk and a semi-disk:

$$\alpha = \frac{\pi}{4} \Rightarrow x_G = \frac{4\sqrt{2}}{3\pi}r\,, \qquad \alpha = \frac{\pi}{2} \Rightarrow x_G = \frac{4r}{3\pi}\,.$$

- Putting $r_1 = r_2 = r$ we find the center of mass of circular arc of angle $2\alpha$ and radius $r$:

$$x_G = |OG| = \frac{\sin\alpha}{\alpha}r\,.$$

Here, as well, we have useful expressions for $\alpha = \pi/4$ and $\alpha = \pi/2$, giving the positions of the center of mass of one quarter of a circle and a semi-circle:

$$\alpha = \frac{\pi}{4} \Rightarrow x_G = \frac{2\sqrt{2}}{\pi}r\,, \qquad \alpha = \frac{\pi}{2} \Rightarrow x_G = \frac{2r}{\pi}\,.$$

If we fix the angle, the center of mass of the arc is clearly farther from the origin than that of the sector. □

**Example 5.3** (*Center of mass of a triangle*) Let us determine the center of mass of a (thin) homogenous plate in the shape of a right triangle $AOB$ with legs of length $a$ and $b$. Consider a frame with origin at $O$, $x$-axis along the leg $OA$ (length $a$) and $y$-axis along $OB$ (length $b$), as in Fig. 5.4. Here the hypotenuse is given by the line

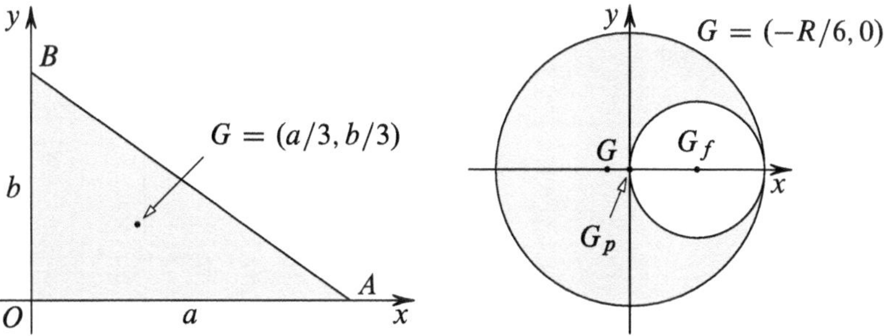


**Fig. 5.4** Center of mass of a triangle and a disk with a circular hole

of equation $y = b(1 - x/a)$. Homogeneity ensures that $m = \rho ab/2$, where $\rho$ is the area density. Applying the definition gives

$$x_G = \frac{1}{m}\int_0^a \int_0^{b(1-x/a)} \rho x \, dy dx = \frac{a}{3}$$

and similarly one shows that $y_G = b/3$. This is true more generally. One can in fact prove that the coordinates of the center of mass of an arbitrary homogeneous triangle (right or not) are given by averaging the vertices' coordinates. □

**Remark 5.4** Rotating the triangle of Fig. 5.4 around the $y$-axis generates a circular cone with base radius $a$ and height $b$. Due to the manifest material symmetry, the cone's center of mass is on the $y$-axis. It is also evident that the centers of gravity of all triangles composing the cone have second coordinate $b/3$. Notwithstanding, and contrary to what we could expect, the reader can compute everything explicitly and show the $y$-coordinate of the cone's center of mass is $y_G^{(\text{cone})} = b/4$. □

**Example 5.5** (*Subtraction property*) Let us determine the position of the center of mass of a homogenous circular plate of radius $R$ with a round hole of radius $R/2$ (Fig. 5.4). This center of mass will lie on the $x$-axis joining the centers as this is a material symmetry. Let $(G_p, m_p)$, $(G_f, m_f)$ denote centers of mass and masses of the plate without the hole and with the hole, respectively. To locate the center of mass we can employ the addition formula by considering the hole as a plane of *negative* density. This gives

$$x_G = \frac{m_p x_{G_p} - m_f x_{G_f}}{m_p - m_f} = -\frac{R}{6}\,.$$

□

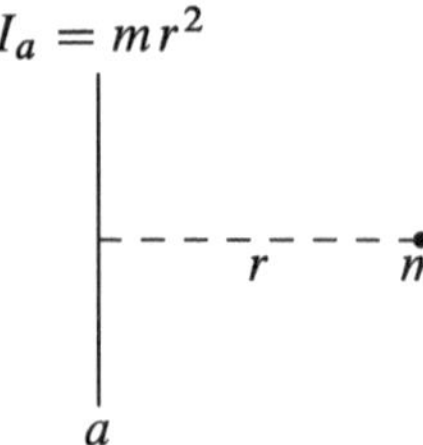


**Fig. 5.5** Moment of inertia of a point with mass $m$ with respect to an axis $a$

## 5.2 Moments of Inertia

The *moment of inertia* of a particle of mass $m$ with respect to an axis $a$ is the number $I_a = mr^2$, where $r$ is the distance of the point to the axis, as in Fig. 5.5. Its physical dimensions are $\left[ML^2\right]$.

For a discrete system of $n$ particles $P_i$ of mass $m_i$, the moment of inertia with respect to an axis $a$ is

$$I_a = \sum_{i=1}^{n} m_i r_i^2 ,$$

where $r_i$ is the distance of $P_i$ to the axis.

For a continuous material system $\mathcal{B}$ we set

$$I_a = \int_{\mathcal{B}} \rho r^2 d\tau \tag{5.9}$$

with the same interpretation for the density $\rho$ and the symbol $d\tau$ used for the center of mass.

For moments of inertia, however, we will provide definitions and main properties only in the case of continuous bodies, because it is to those that they apply more commonly. The concepts are still valid for discrete bodies, though, with the proviso of replacing $\rho d\tau$ with the masses of the single points and replacing integrals with sums over all points forming the discrete system.

Two easy but important properties of moments of inertia are straightforward from the definition, namely:

- the moment of inertia is never negative;
- the moment of inertia vanishes if and only if all points in the body lie on the axis, as shown in Fig. 5.6.

$a$ ——— $I_a = 0$

**Fig. 5.6** Zero moment of inertia: all points of the system belong to the axis

**Example 5.6** Let us compute the moment of inertia of a system with respect to the coordinate axes. At each point set $OP = x\mathbf{i} + y\mathbf{j} + z\mathbf{k}$. It is easy to show that, for instance, the distance of $P$ to the $x$-axis is $r^2 = y^2 + z^2$ (see (A.5) in the Appendix). Similarly,

$$I_x = \int_{\mathcal{B}} \rho(y^2 + z^2)d\tau\,, \quad I_y = \int_{\mathcal{B}} \rho(x^2 + z^2)d\tau\,, \quad I_z = \int_{\mathcal{B}} \rho(x^2 + y^2)d\tau\,.$$

□

The positive quantity

$$\delta_a = \sqrt{\frac{I_a}{m}} \tag{5.10}$$

is called *gyration radius*. Its physical dimensions are those of a length, and clearly $I_a = m\delta_a^2$. The gyration radius is the distance from the axis that one point with mass $m$ equal to the system's total mass should have to produce the same moment of inertia of the system. If a material system is homogeneous, it is useful to define the *geometric moment of inertia*

$$i_a - \frac{I_a}{m} - \frac{1}{V}\int_{\mathcal{B}} r^2 d\tau\,,$$

where $V$ is the body's volume.

The introduction of geometric moments of inertia, whose dimensions are $\left[L^2\right]$, allows to unify the computation of the moments of inertia of all homogeneous bodies with the same shape.

## 5.3 Moments of Inertia with Respect to Parallel Axes

To facilitate the computations it is convenient to establish how the moment of inertia changes when we translate the axis in a parallel way.

**Theorem 5.7** (Huygens-Steiner) *The moment of inertia of a body with respect to an arbitrary axis a equals the sum*

$$I_a = I_{a_G} + md^2$$

*where $I_{a_G}$ is the moment of inertia of the body with respect to the axis $a_G$ parallel to a through the center of mass $G$ of the body, $m$ is the body's total mass and $d$ is the distance of the two axes.*

***Proof*** We use a frame of reference with origin at the center of mass ($x_G = y_G = z_G = 0$) and $z$-axis parallel to the axis $a$. Calling $(x_a, y_a)$ the coordinates of the

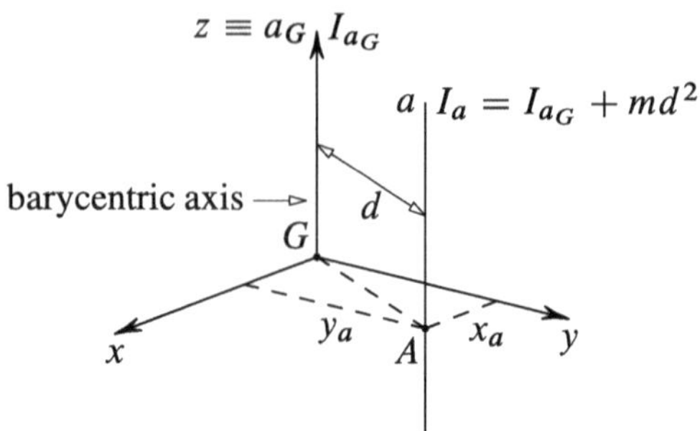


**Fig. 5.7** Moments of inertia for parallel axes

intersection of $a$ with the $xy$-plane, from the definition of moment of inertia we obtain (Fig. 5.7)

$$\begin{aligned} I_a &= \int_{\mathcal{B}} \rho \left( (x - x_a)^2 + (y - y_a)^2 \right) d\tau \\ &= \int_{\mathcal{B}} \rho \left( x^2 + y^2 \right) d\tau - 2x_a \int_{\mathcal{B}} \rho x d\tau - 2y_a \int_{\mathcal{B}} \rho y d\tau + \left( x_a^2 + y_a^2 \right) \int_{\mathcal{B}} \rho d\tau \\ &= I_{a_G} - 2m \left( x_a x_G + y_a y_G \right) + m \left( x_a^2 + y_a^2 \right) = I_{a_G} + md^2 \end{aligned}$$

where we used definitions (5.9) and (5.4), and we took into account the fact that the center of mass is the origin of the frame of reference. □

A consequence of this theorem is that, among all parallel axes, the one passing through the center of mass is the axis for which the body has *smallest* moment of inertia. Moreover, the Huygens-Steiner theorem also says that if $a_1$ and $a_2$ are distinct parallel axes with distances $d_1$ and $d_2$ from the center of mass then

$$I_{a_2} = I_{a_1} + m \left( d_2^2 - d_1^2 \right) .$$

**Example 5.8** (*Rectangle*) Let us determine the moment of inertia of a homogeneous rectangle of edges $a, b$ and mass $m$ with respect to: (1) one of the edges; (2) an axis parallel to the edges through the center of mass. Let $\rho = m/(ab)$ denote the density of mass per unit of area. Consider a frame with origin at one vertex and $(x, y)$-axes respectively parallel to the edges of length $a, b$ (see Fig. 5.8). Then

$$I_x = \rho \int_0^b \int_0^a y^2 dx dy = \rho \frac{ab^3}{3} = \frac{mb^2}{3} . \tag{5.11}$$

Similarly, $I_y = ma^2/3$. Applying now the Huygens-Steiner theorem we find

$$I_{x_G} = I_x - m \left( \frac{b}{2} \right)^2 = \frac{mb^2}{12} \tag{5.12}$$

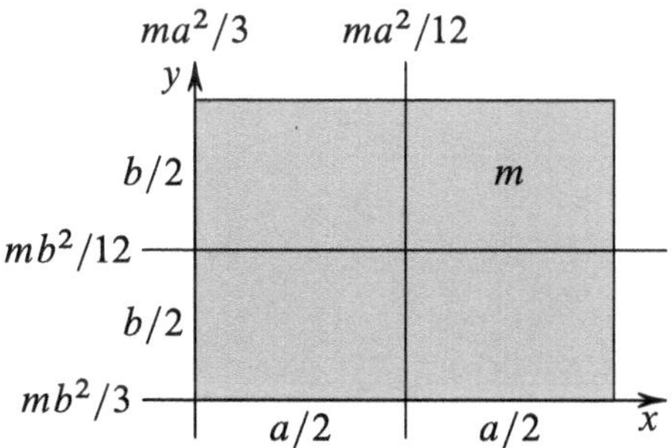


**Fig. 5.8** Moments of inertia for a homogeneous rectangular plate

and also $I_{y_G} = ma^2/12$. Analogous computations allow to show that if $z$, $z_G$ are axes orthogonal to $(x, y)$ and $(x_G, y_G)$, respectively, we have

$$I_z = \frac{m}{3}(a^2 + b^2), \qquad I_{z_G} = \frac{m}{12}(a^2 + b^2).$$

The fact that the moments of inertia with respect to axes orthogonal to the rectangle's axis are equal to the sum of the inertia moments with respect to the plane's axes is actually a property of all planar systems, as we will prove later (see (5.36)).

Let us finally emphasize that for a rectangle with $a = 0$, Eqs. (5.11) and (5.12) provide the moments of inertia of a homogeneous rod, with respect to an axis orthogonal to the rod and an axis passing through one end and the center of mass. Calling $l$ the length we have, in the two cases,

$$I = \frac{ml^2}{3}, \qquad I = \frac{ml^2}{12},$$

as shown in Fig. 5.9. □

**Example 5.9** (*Annular sector*) Let us compute the moment of inertia of a sector of a homogeneous annulus of mass $m$, angle $2\alpha$ and radii $r_1, r_2$ (see Fig. 5.3), with respect to an axis through the center and orthogonal to the plane of the sector. It is convenient to use polar coordinates:

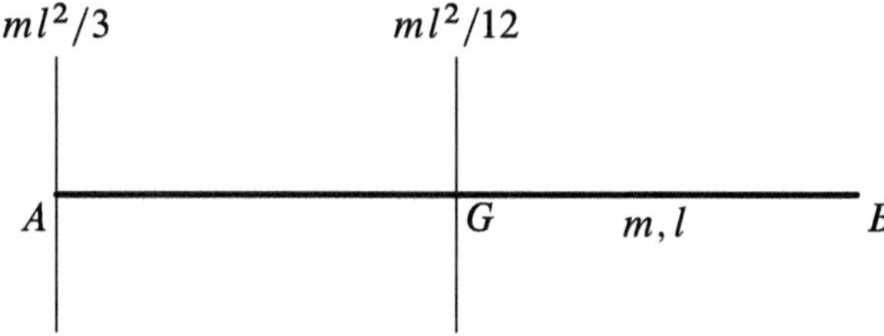


**Fig. 5.9** Two moments of inertia for a homogeneous rod of length $l$ and mass $m$

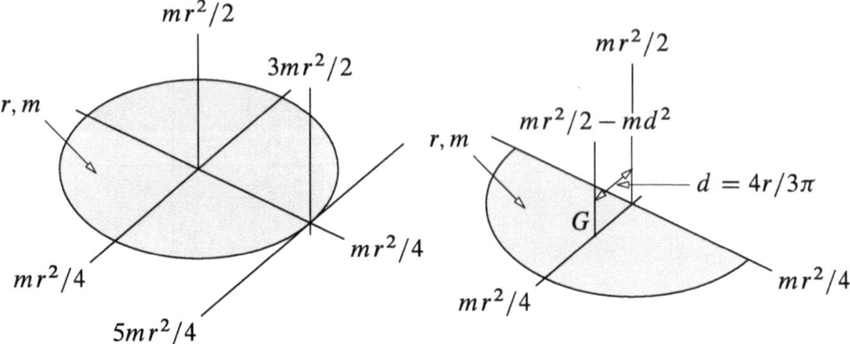


**Fig. 5.10** Moments of inertia for a homogeneous disk and semi-disk

$$I_a = \rho \int_{-\alpha}^{\alpha} \int_{r_1}^{r_2} r^2 \,(r dr d\theta) = \frac{1}{2} m \left(r_2^2 + r_1^2\right), \tag{5.13}$$

where $\rho$ is the area density of the annulus.

The result is independent of the angle $\alpha$. Two useful special cases of (5.13) give us the moments of inertia with respect to the axis through the center of a disk ($r_1 = 0, r_2 = r$) and the center of a circle ($r_1 = r_2 = r$):

$$I = \frac{mr^2}{2}, \qquad I = mr^2 .$$

Figure 5.10 shows the moments of inertia for the disk and semi-disk, both homogeneous, with respect to various axes. These expression follows from calculations similar to (5.13), together with the Huygens-Steiner theorem. □

**Example 5.10** (*Homogeneous cylinder*) Let us determine the moment of inertia of a homogeneous cylinder of height $h$ and radius $R$ with respect to its symmetry axis. In cylindrical coordinates and with $\rho$ denoting the density we have

$$I_a = \rho \int_0^h \int_0^{2\pi} \int_0^R r^2 \,(r dr d\theta dz) = 2\pi h \rho \int_0^R r^3 dr = \frac{\pi h \rho}{2} R^4 = \frac{m R^2}{2} .$$

□

## 5.4 Moments of Inertia with Respect to Concurrent Axes

To understand how the moment of inertia of a body varies as the direction of the axis $a$ varies, we remind that lines through $O$ correspond to unit vectors $\mathbf{u}$ (or the opposite), and conversely. For this reason we will denote by $I_u$ the moment of inertia

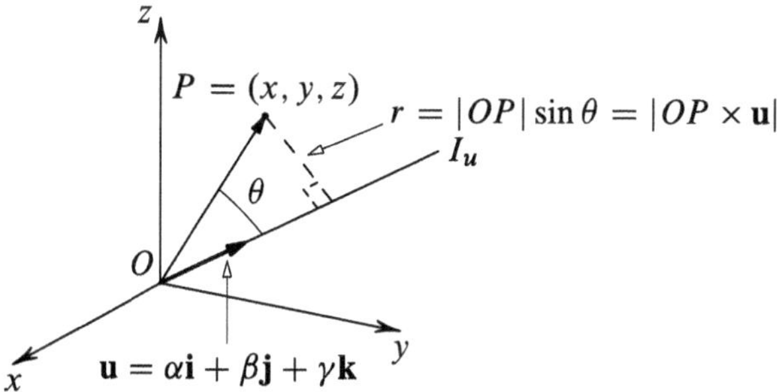


**Fig. 5.11** Moments of inertia for concurrent axes

of a body relative to the line through $O$ and parallel to $\mathbf{u}$. The distance $r$ of the generic element of the body to the axis given by $\mathbf{u}$ is (see (A.7) and Fig. 5.11)

$$r = |OP| \sin\theta = |OP \times \mathbf{u}| \,, \tag{5.14}$$

where $\theta$ is the angle between the radial vector $OP$ and the unit vector $\mathbf{u}$. To compute explicitly the cross product in (5.14) we introduce a frame such that

$$OP = x\mathbf{i} + y\mathbf{j} + z\mathbf{k}\,, \qquad \mathbf{u} = \alpha\mathbf{i} + \beta\mathbf{j} + \gamma\mathbf{k}\,, \tag{5.15}$$

where we recall $(\alpha, \beta, \gamma)$ are the director cosines of $\mathbf{u}$, and $\alpha^2 + \beta^2 + \gamma^2 = 1$. Therefore

$$OP \times \mathbf{u} = (y\gamma - z\beta)\,\mathbf{i} - (x\gamma - z\alpha)\,\mathbf{j} + (x\beta - y\alpha)\,\mathbf{k}$$

and so

$$I_u = \int_{\mathcal{B}} \rho\left[(y\gamma - z\beta)^2 + (x\gamma - z\alpha)^2 + (x\beta - y\alpha)^2\right] d\tau\,. \tag{5.16}$$

Computing the squares in (5.16) and re-arranging terms we obtain

$$\begin{aligned} I_u = {} & \alpha^2 \int_{\mathcal{B}} \rho\left(y^2 + z^2\right) d\tau + \beta^2 \int_{\mathcal{B}} \rho\left(x^2 + z^2\right) d\tau + \gamma^2 \int_{\mathcal{B}} \rho\left(x^2 + y^2\right) d\tau \\ & - 2\beta\gamma \int_{\mathcal{B}} \rho yz d\tau - 2\alpha\gamma \int_{\mathcal{B}} \rho xz d\tau - 2\alpha\beta \int_{\mathcal{B}} \rho xy d\tau\,. \end{aligned} \tag{5.17}$$

Let $I_x$, $I_y$ and $I_z$ be the first three integrals in the formula:

$$I_x = \int_{\mathcal{B}} \rho\left(y^2 + z^2\right) d\tau \quad I_y = \int_{\mathcal{B}} \rho\left(x^2 + z^2\right) d\tau \quad I_z = \int_{\mathcal{B}} \rho\left(x^2 + y^2\right) d\tau \tag{5.18}$$

As $y^2 + z^2$, $x^2 + z^2$, $x^2 + y^2$ are the squared distances of a generic point $P = (x, y, z)$ to the axes $x$, $y$ and $z$, the integrals in (5.18) are nothing but the moments of inertia with respect to the coordinate axes (see Example 5.6).

The remaining three integrals in (5.17) are quantities with the correct physical dimensions (i.e. $[ML^2]$), but they admit no axis for which they are moments of inertia. They are called *inertia products* and denoted

$$I_{xy} = -\int_{\mathcal{B}} \rho xy d\tau\,, \qquad I_{xz} = -\int_{\mathcal{B}} \rho xz d\tau\,, \qquad I_{yz} = -\int_{\mathcal{B}} \rho yz d\tau\,, \tag{5.19}$$

while their opposites $-I_{xy}$, $-I_{xz}$, $-I_{yz}$ are also known as *centrifugal moments*.

These definitions allow to write (5.17) as

$$I_u = I_x\alpha^2 + I_y\beta^2 + I_z\gamma^2 + 2I_{xy}\alpha\beta + 2I_{xz}\alpha\gamma + 2I_{yz}\beta\gamma\,. \tag{5.20}$$

Next we introduce a symmetric matrix, called *inertia matrix*, whose diagonal contains the moments of inertia and the other entries are the inertia products, all computed with respect to the given coordinate axes passing through $O$:

$$\mathbf{I}_O = \begin{bmatrix} I_x & I_{xy} & I_{xz} \\ I_{xy} & I_y & I_{yz} \\ I_{xz} & I_{yz} & I_z \end{bmatrix}. \tag{5.21}$$

In the light of the above, (5.20) reads

$$I_u = \begin{bmatrix} \alpha \\ \beta \\ \gamma \end{bmatrix} \cdot \begin{bmatrix} I_x & I_{xy} & I_{xz} \\ I_{xy} & I_y & I_{yz} \\ I_{xz} & I_{yz} & I_z \end{bmatrix} \begin{bmatrix} \alpha \\ \beta \\ \gamma \end{bmatrix} \tag{5.22}$$

or more compactly,

$$I_u = \mathbf{u} \cdot \mathbf{I}_O \mathbf{u}\,. \tag{5.23}$$

The importance of this fact is the following: once we have computed the inertia matrix via the six integrals (5.18) and (5.19), relations (5.22) and (5.23) allow us to find, via purely algebraic manipulations, the moment of inertia of the body with respect to *any axis* through $O$.

Since the moment of inertia $I_u$ is always positive (except in the special case where the body is one-dimensional and entirely contained in the axis of $\mathbf{u}$), from (5.23) we deduce that the inertia matrix is not only symmetric but also *positive definite* (bar the above exception).

In many treatises on Rational Mechanics or exercise books the moments of inertia $I_x$, $I_y$, $I_z$ with respect to axes are traditionally written $A$, $B$, $C$, and the centrifugal moments $-I_{yz}$, $-I_{xz}$, $-I_{xy}$ are called $A'$, $B'$, $C'$. In this different notation our inertia matrix (5.21) becomes:

$$\mathbf{I}_O \equiv \begin{bmatrix} A & -C' & -B' \\ -C' & B & -A' \\ -B' & -A' & C \end{bmatrix}.$$

Along the lines of the previous proof, one deduces that the inertia product $I_{uv}$ relative to any pair of *orthogonal* axes parallel to unit vectors $\mathbf{u}$ and $\mathbf{v}$ is:

$$I_{uv} = \mathbf{u} \cdot \mathbf{I}_O \mathbf{v}\,. \tag{5.24}$$

### 5.4.1 Inertia Tensor

Now we show that the inertia *matrix* (5.21) is indeed the matrix, for the chosen coordinate axes, of a linear map, called *inertia tensor*. If in fact we *define* the linear mapping (or tensor) $\mathbf{I}_O$ by

$$\mathbf{I}_O \mathbf{v} = \int_{\mathcal{B}} \rho\big[(OP)^2 \mathbf{v} - (OP \cdot \mathbf{v})\, OP\big]\, d\tau \tag{5.25}$$

and then apply (A.23), it is easy to see that the matrix of the inertia tensor $\mathbf{I}_O$ is precisely (5.21). (Here, to simplify notations, we have used the same symbol $\mathbf{I}_O$ for the inertia tensor just defined and for its matrix (5.21), in the hope this does not create misunderstandings).

Recalling that $(OP)^2 = x^2 + y^2 + z^2$ and $OP \cdot \mathbf{i} = x$, $OP \cdot \mathbf{j} = y$, $OP \cdot \mathbf{k} = z$, easy calculations produce

$$\begin{aligned} \mathbf{i} \cdot \mathbf{I}_O \mathbf{i} &= \int_{\mathcal{B}} \rho\big[(OP)^2 - (OP \cdot \mathbf{i})^2\big]\, d\tau \\ &= \int_{\mathcal{B}} \rho\big[(x^2 + y^2 + z^2) - x^2\big]\, d\tau \\ &= \int_{\mathcal{B}} \rho\big[y^2 + z^2\big]\, d\tau \\ &= I_x \end{aligned}$$

and similarly $\mathbf{j} \cdot \mathbf{I}_O \mathbf{j} = I_y$ and $\mathbf{k} \cdot \mathbf{I}_O \mathbf{k} = I_z$.

Moreover,

$$\mathbf{i} \cdot \mathbf{I}_O \mathbf{j} = \mathbf{j} \cdot \mathbf{I}_O \mathbf{i} = -\int_{\mathcal{B}} \rho x y\, d\tau = I_{xy} = I_{yx}$$

and analogously

$$\mathbf{j} \cdot \mathbf{I}_O \mathbf{k} = \mathbf{k} \cdot \mathbf{I}_O \mathbf{j} = -\int_{\mathcal{B}} \rho yz \, d\tau = I_{yz} = I_{zy}$$

$$\mathbf{i} \cdot \mathbf{I}_O \mathbf{k} = \mathbf{k} \cdot \mathbf{I}_O \mathbf{i} = -\int_{\mathcal{B}} \rho xz \, d\tau = I_{xz} = I_{zx}$$

For conciseness let $(\mathbf{e}_1, \mathbf{e}_2, \mathbf{e}_3) = (\mathbf{i}, \mathbf{j}, \mathbf{k})$ denote the coordinate unit vectors, so that

$$I_{ik} = I_{ki} = \mathbf{e}_i \cdot \mathbf{I}_O \mathbf{e}_k$$

where $I_{ik}$ is the matrix of the inertia tensor $\mathbf{I}_O$ in the coordinate system with basis $\mathbf{e}_i$ and origin $O$.

For completeness we also add that, using the tensor product $\mathbf{a} \otimes \mathbf{b}$ of vectors $\mathbf{a}$ and $\mathbf{b}$, one can define the inertia tensor relative to a point $O$ in space to be

$$\mathbf{I}_O = \int_{\mathcal{B}} \rho \big[ (OP)^2 \mathbf{I} - (OP \otimes OP) \big] \, d\tau$$

where the symbol $\mathbf{I}$ in the integral is the *identity tensor* (not the inertia tensor). This definition is clearly equivalent to (5.25).

To sum up, any body $\mathcal{B}$ in a given position and each point in space $O$ define a *symmetric and positive-definite inertia tensor* $\mathbf{I}_O$, such that $\mathbf{u} \cdot \mathbf{I}_O \mathbf{u}$ is the moment of inertia of the body with respect to the axis with unit vector $\mathbf{u}$ through $O$, while $\mathbf{u} \cdot \mathbf{I}_O \mathbf{v}$ is the inertia product with respect to two orthogonal axes with unit vectors $\mathbf{u}$ and $\mathbf{v}$.

### 5.4.2 Discrete Systems of Particles

It is useful to recall the form taken by the previous sections' quantities for the moments and products of inertia, when we deal with discrete distributions of particles. The idea is simply to replace the integrals with sums. For completeness we write the expressions of the components of the inertia tensor, which are deduced from (5.18) and (5.19), in the case of masses $m_i$ $(i = 1, \ldots, n)$ placed at $P_i = (x_i, y_i, z_i)$ in space:

$$I_x = \sum_{i=1}^n m_i (y_i^2 + z_i^2) \quad I_y = \sum_{i=1}^n m_i (x_i^2 + z_i^2) \quad I_z = \sum_{i=1}^n m_i (x_i^2 + y_i^2)$$

$$I_{xy} = -\sum_{i=1}^n m_i x_i y_i \quad I_{xz} = -\sum_{i=1}^n m_i x_i z_i \quad I_{yz} = -\sum_{i=1}^n m_i y_i z_i \tag{5.26}$$

**Remark 5.11** Every definition, property or proof involving moments or product of inertia can be formulated indifferently for a discrete or continuous distribution of mass, with no conceptual difference. The ensuing choice of one approach over the other will depend on the context, and aims at better coherence and expository clarity. □

### 5.4.3 Principal Axes and Principal Moments of Inertia

The matrix of the inertia tensor $\mathbf{I}_O$ is symmetric and hence (see Sect. A.3) diagonalizable. That means there is an orthonormal basis, called *principal axes of inertia*, with respect to which the matrix is diagonal. As proved in (A.31), the elements on the diagonal will be precisely the tensor's eigenvalues while the three *principal axes of inertia* will be given by the corresponding eigenvectors.

If $\hat{\mathbf{i}}, \hat{\mathbf{j}}, \hat{\mathbf{k}}$ are the unit vectors of the principal axes of inertia at $O$, and using a similar notation for the moments and products of inertia relative to them, the inertia matrix in this basis reads

$$\mathbf{I}_O = \begin{bmatrix} \hat{I}_x & 0 & 0 \\ 0 & \hat{I}_y & 0 \\ 0 & 0 & \hat{I}_z \end{bmatrix} . \tag{5.27}$$

Above the inertia products are zero while the quantities $\hat{I}_x$, $\hat{I}_y$, $\hat{I}_z$ are exactly the moments of inertia with respect to the principal axes, that is the *principal moments of inertia*. It is important to observe that these moments are clearly different from the diagonal elements of (5.21), as the latter are referred to the axes of a *generic* basis.

Another simplification in (5.20) occurs when we refer everything to the principal axes of inertia, i.e. when both the components of $\mathbf{u}$ and the inertia matrix are taken with respect to this basis. This produces the simpler expression

$$\mathbf{I}_u = \alpha^2 \hat{I}_x + \beta^2 \hat{I}_y + \gamma^2 \hat{I}_z .$$

We conclude by noting that when the moments and the principal axes refer to axes at the *center of mass*, one speaks of *central* axes and moments.

**Remark 5.12** Henceforth, when no confusion arises, we will not employ specific symbols for principal axes and moments, in contrast to what we did until now. The context should be enough to understand which basis is being used. □

### 5.4.4 Products of Inertia with Respect to Parallel Axes

It is useful to establish a collection of relationships analogous to the ones in the Huygens-Steiner theorem but for the inertia products. Consider barycentric axes $(x, y, z)$ (hence passing through the center of mass $G$), and another parallel triple $(x', y', z')$ through a generic point $Q$. Setting $GQ = \Delta x\mathbf{i} + \Delta y\mathbf{j} + \Delta z\mathbf{k}$, and viewing

$G$ as both origin and center of mass, we compute the inertia products relative to the new axes $x'$, $y'$, $z'$. Starting from $I_{x'y'}$ we have

$$\begin{aligned} I_{x'y'} &= -\int_{\mathcal{B}} \rho x' y' d\tau = -\int_{\mathcal{B}} \rho\,(x - \Delta x)\,(y - \Delta y) d\tau \\ &= -\int_{\mathcal{B}} \rho x y d\tau - \int_{\mathcal{B}} \rho \Delta x \Delta y d\tau + \int_{\mathcal{B}} \rho x \Delta y d\tau + \int_{\mathcal{B}} \rho y \Delta x d\tau \\ &= I_{xy} - m \Delta x \Delta y \end{aligned}$$

where we used

$$\int_{\mathcal{B}} \rho x d\tau = \int_{\mathcal{B}} \rho y d\tau = 0$$

due to the fact that $(x, y, z)$ are barycentric.

All in all, similar calculations for the other inertia products produce

$$\begin{aligned} I_{xy} &= I_{x'y'} + m \Delta x \Delta y \\ I_{xz} &= I_{x'z'} + m \Delta x \Delta z \\ I_{yz} &= I_{y'z'} + m \Delta y \Delta z \end{aligned} \tag{5.28}$$

(recall that here $I_{xy}$, $I_{xz}$, $I_{yz}$ are the inertia products with respect to *barycentrric* axes, not generic ones).

## 5.5 Inertia Ellipsoid

Formula (5.23) can be interpreted in a particularly suggestive manner by defining, as the direction of $\mathbf{u}$ varies in space, the locus of points $P$ satisfying

$$OP = \frac{\mathbf{u}}{\sqrt{I_u}}\,. \tag{5.29}$$

(In the ensuing formulas we must not forget that for dimensional reasons $OP = k\mathbf{u}/\sqrt{I_u}$; $k$ is a constant such that $[k] = L^2 M^{1/2}$ and later we will set it to 1 to simplify the discussion.)

Notice that $I_u > 0$ for any $\mathbf{u}$ (except when the body belongs entirely on an axis through $O$), so (5.29) is not only well defined but it describes a collection of points $P$ at finite distance from $O$, irrespective of the unit vector $\mathbf{u}$.

Essentially, the set (5.29) is built taking, on the half-line through $O$ with direction $\mathbf{u}$, the point at a distance from $O$ equal to $1/\sqrt{I_u}$ (hence we clearly obtain the symmetric point when taking the unit vector $-\mathbf{u}$). The set thus built describes geometrically and somewhat "pictorially" the variation of the body's moment of inertia as the axis through $O$ varies.

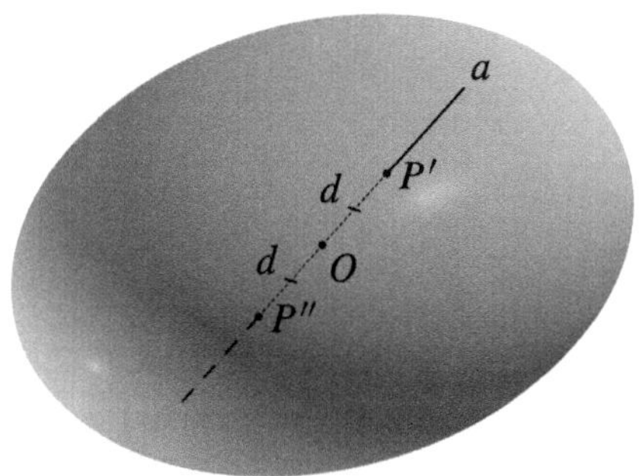


**Fig. 5.12** The inertia ellipsoid of a body with respect to a point $O$ and one axis $a$ passing through $O$

To find the Cartesian equation of this locus it suffices to observe that (5.15) and (5.29) imply

$$x = \frac{\alpha}{\sqrt{I_u}}, \qquad y = \frac{\beta}{\sqrt{I_u}}, \qquad z = \frac{\gamma}{\sqrt{I_u}},$$

and substituting $\alpha = x\sqrt{I_u}$, $\beta = y\sqrt{I_u}$, $\gamma = z\sqrt{I_u}$ in (5.20) yields a quadratic equation

$$I_x\, x^2 + I_y\, y^2 + I_z\, z^2 + 2I_{xy}\, xy + 2I_{xz}\, xz + 2I_{yz}\, yz = 1\,. \tag{5.30}$$

This represents a quadric surface centered at $O$. If we exclude again the case in which the body lies on a line through $O$, from (5.29) we see that the quadric's points are all at a finite distance from the origin, with no points at infinity. We know from geometry that the only quadric surface with such property is the ellipsoid. Hence (5.30) is the equation of an ellipsoid with center $O$, which obviously goes by the name of *inertia ellipsoid* of the body with respect to $O$.

Figure 5.12 shows the inertia ellipsoid of a body relative to a generic point $O$ in space. An axis $a$ intersects the ellipsoid at two points $P'$ and $P''$ at the same distance $d$ to $O$: $d = |OP'| = |OP''|$. The moment of inertia $I_a$ is $1/d^2$ (once units of measure are chosen). As the axis $a$ varies, the moment of inertia $I_a$ becomes bigger in the directions along which the ellipsoid is squashed, and it decreases where the ellipsoid is more elongated.

For a generic body, there is an inertia ellipsoid centered at each point in space, as shown in Fig. 5.13.

If the ellipsoid's center coincides with the center of mass we speak of *central inertia ellipsoid*. Equation (5.30) can be written compactly by putting in (5.23) expression (5.29) for $\mathbf{u}$:

$$OP \cdot \mathbf{I}_O\, OP = 1\,.$$

Finding the central ellipsoid of inertia of a given body allows to compute easily the moment of inertia with respect to any axis in space. Given in fact the inertia ellipsoid at $O$, suppose we wish to find the inertia moment with respect to a line $a_2$ non passing through $O$, as in Fig. 5.14.

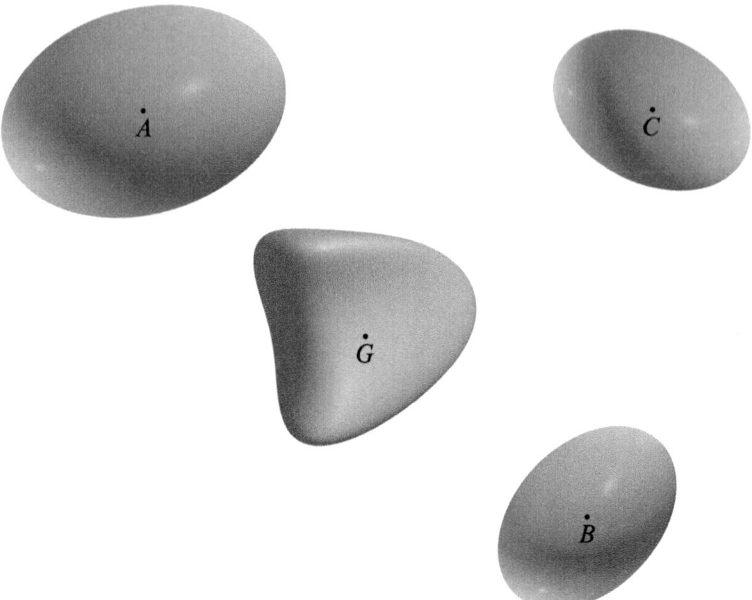


Fig. 5.13 A body with center of mass $G$ and three ellipsoids of inertia centered at points $A$, $B$, $C$ in space

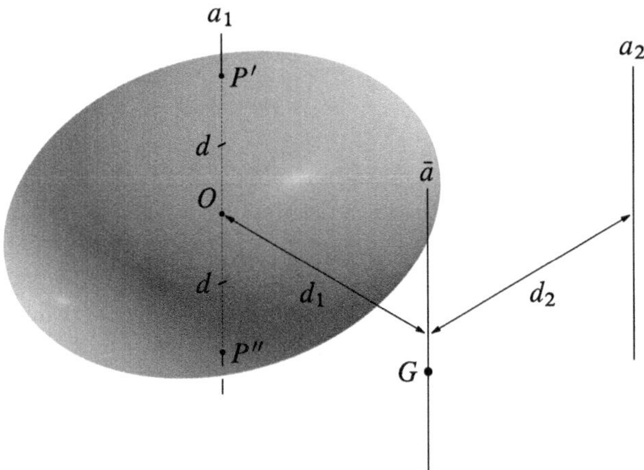


Fig. 5.14 An application of the inertia ellipsoid for computing the moment of inertia with respect to a generic axis

It will be enough to know the position of the center of mass $G$, the mass of the body $m$ and three distances $d, d_1, d_2$ as in the picture, to know the moment

$$I_{a_2} = I_{a_1} + m(d_2^2 - d_1^2) = \frac{1}{d^2} + m(d_2^2 - d_1^2)\,. \tag{5.31}$$

By measuring the distance $d$ we can find the moment of inertia with respect to $a_1$ (parallel to $a_2$ and through $O$). By definition of the ellipsoid we obtain $I_{a_1} = 1/d^2$, and using Huygens-Steiner twice we arrive at formula (5.31).

Equation (5.30) simplifies if we reduce it to canonical form. The theory of quadratic forms guarantees the existence of a triple of axes with respect to which the ellipsoid has canonical equation

$$I_x x^2 + I_y y^2 + I_z z^2 = 1\,.$$

This basis is obviously the same basis that diagonalizes the inertia matrix $\mathbf{I}_O$, its axes are the principal axes of inertia and its moments are the principal moments of inertia with respect to $O$.

**Remark 5.13** The inertia ellipsoid provides a means to understand, even if in a rough way, the shape of the body. If in fact the body is stretched along the principal axis $x$, the relative moment of inertia will be smaller than the other two (as shorter are these distances). Therefore the ellipsoid's semi-axis along $x$ will be longer than the others. Similarly, if the body is squashed along the plane perpendicular to $z$, the moment $I_z$ will be larger and the corresponding semi-axis shorter than the rest.

We further stress that not every ellipsoid is an inertia ellipsoid.

Pick in fact a generic ellipsoid in canonical form

$$I_x x^2 + I_y y^2 + I_z z^2 = 1$$

for which $I_x$, $I_y$, $I_z$ are the principal moments of inertia of a body

$$I_x = \int_{\mathcal{B}} \rho(y^2 + z^2)\, d\tau\,, \quad I_y = \int_{\mathcal{B}} \rho(x^2 + z^2)\, d\tau\,, \quad I_z = \int_{\mathcal{B}} \rho(x^2 + y^2)\, d\tau\,.$$

Immediately,

$$I_y + I_z = \int_{\mathcal{B}} \rho(2x^2 + y^2 + z^2)\, d\tau \geq I_x$$

for instance, so $I_x$, $I_y$, $I_z$ cannot be prescribed randomly. □

**Definition 5.14** (*Gyroscope*) A rigid body is called a *gyroscope* when two central principal moments are equal, say

$$I_x = I_y\,. \tag{5.32}$$

Now the central ellipsoid of inertia is a surface of revolution, i.e. a round ellipsoid. The symmetry axis of the central inertia ellipsoid of a gyroscope, namely the axis $\mathbf{k}$ if we assume (5.32), is called *gyroscopic axis*.

## 5.6 Determining the Principal Axes of Inertia

The principal axes of inertia can be determined analytically by standard diagonalization of real symmetric matrices. However, in many cases, such as barycentric ones, symmetry considerations allow to simplify the task.

(i) Every system possesses at least one triple of orthonormal axes at $O$. There are three situations: (1) if the principal moments of inertia are all distinct, the triple is unique; (2) if two principal moments of inertia coincide and the third is different, all axes in the plane spanned by the first two principal axes are principal; (3) if the principal moments of inertia are all equal, any axis is principal.

These properties are consequences of the multiplicity of the eigenvalues of the inertia matrix $\mathbf{I}_O$. Each time we have two non-orthogonal principal axes, the respective moments coincide, and any axis on the plane they generate is principal as well.

(ii) By examining (5.24) and (5.27) we deduce the following properties. If an axis $a$ is principal, any inertia product involving it is zero, meaning that every inertia product built with $a$ and any other axis orthogonal to $a$ vanishes. Moreover, to prove a given $a$ is principal it suffices to pick any orthogonal basis containing $a$ and check the inertia products involving $a$ and one of the another axes is zero.

(iii) If the system admits a material symmetry plane $\pi$, the axis orthogonal to $\pi$ is principal of inertia with respect to any point $O \in \pi$.

To check this, fix $O \in \pi$ and choose a frame $(x, y, z)$ with the first two axes on the plane of symmetry, and with $z$-axis orthogonal to $\pi$. The points that do not belong on $z = 0$ can be grouped in pairs of equal mass, same $x$- and $y$-coordinates, and opposite $z$-coordinate. Consequently the inertia products $I_{xz}, I_{yz}$ vanish. To guarantee that the $z$-axis is principal we must suppose the plane $\pi$ is a material symmetry: it is not enough to assume it is a diameter.

(iv) Suppose the system has two material symmetry planes $\pi_1, \pi_2$, and they intersect in the line $a$. Choosing any point $O \in a$, and calling $a_1, a_2$ the axes orthogonal to $\pi_1, \pi_2$ at $O$, we have two possibilities. If $\pi_1$ and $\pi_2$ are orthogonal then $\{a_1, a_2, a\}$ is principal at $O$. If $\pi_1$ and $\pi_2$ are not orthogonal, their intersection $a$ is a gyroscopic axis for the body.

In fact $a_1, a_2$ are principal by property (iii). If they are orthogonal they form a principal triple of inertia together with $a$, which is orthogonal to both. If not orthogonal, the plane they span contains infinitely many principal axes by the previous property (i). In the latter case, remembering that $a$ is barycentric in view of the center of mass's symmetry (property (iii) in Sect. 5.1.3), it must be gyroscopic.

(v) Let $a$ be a *central principal* axis of inertia. Then it is principal with respect to any other point. Moreover, every axis $a' \parallel a$ is principal with respect to a unique point on it: the closest point to the center of mass.

To prove this we consider central principal axes of inertia $(x, y, z)$ where $a$ is the $x$-axis, and another triple $(x', y', z')$, *parallel* to it, in which $a'$ is the $x'$-axis, both at a point $Q$ and with $GQ = \Delta x\mathbf{i} + \Delta y\mathbf{j} + \Delta z\mathbf{k}$.

As the first triple is *principal and central* we can use the first two in (5.28) and set $I_{xy} = I_{xz} = 0$. This gives

$$I_{x'y'} = -m\Delta x\Delta y \qquad I_{x'z'} = -m\Delta x\Delta z$$

Hence there are only two possibilities for $x'$ to be principal. The first is that $\Delta y = \Delta z = 0$, and then the $x'$-axis coincides with the $x$-axis, $Q$ belongs to either and ultimately we have shown $x$ is principal with respect to any other point. The other possibility is that $\Delta x = 0$, meaning $Q$ is the point of $x'$ closest to $G$.

(vi) In the collection of all lines through a given point, the principal axes are among those with *largest* and *smallest* moments of inertia.

In fact, formula (5.20) for the principal axes becomes

$$I_u = I_x\,\alpha^2 + I_y\,\beta^2 + I_z\,\gamma^2 \tag{5.33}$$

with $\mathbf{u} = \alpha\mathbf{i} + \beta\mathbf{j} + \gamma\mathbf{k}$. Let us examine how (5.33) varies in function of $\mathbf{u}$, i.e. as the coefficients $(\alpha, \beta, \gamma)$ vary subject to $\alpha^2 + \beta^2 + \gamma^2 = 1$ vary. Without loss of generality let us assume $I_x \le I_y \le I_z$. Then immediately $I_u$ is a bounded and continuous map defined on a compact set. Hence it assumes all values between the minimum $I_x$ and the maximum $I_z$ (attained at $\alpha = \pm 1, \beta = \gamma = 0$ and $\alpha = \beta = 0, \gamma = \pm 1$).

**Example 5.15** (*Round cone*) Let us compute the central principal axes of inertia of a homogeneous right circular cone of mass $m$, base radius $R$ and height $h$.

Consider a reference frame with $z$-axis coinciding with the cone's axis. Clearly the aforementioned property (iv) holds. In fact, the $z$-axis is clearly the intersection of two symmetry planes $\pi_1$ and $\pi_2$ (possibly not orthogonal). To complete the basis of central principal axes of inertia it suffices to choose as $x$- and $y$-axis two lines perpendicular to one another and to the cone's axis through the center of mass $G$. On the other hand the principal axes of inertia relative to $G'$, projection of $G$ on the base plane, are the projections $x'$ and $y'$ of the $x$- and $y$-axis on the same plane together with the $z$-axis. In cylindrical coordinatesÂ the moment of inertia with respect to $z$ equals

$$I_z = \rho\int_0^{2\pi} d\theta \int_0^h dz \int_0^{\frac{R}{h}(h-z)} r^3 dr = \frac{\rho\pi R^4 h}{10} = \frac{3}{10} m R^2,$$

since the cone's total mass is $m = \rho\pi R^2 h/3$. At the same time

$$I_{x'} = \rho\int_0^{2\pi} d\theta \int_0^h dz \int_0^{\frac{R}{h}(h-z)} \left(r^2\sin^2\theta + z^2\right) r dr = \frac{m}{10}\left(\frac{3R^2}{2} + h^2\right).$$

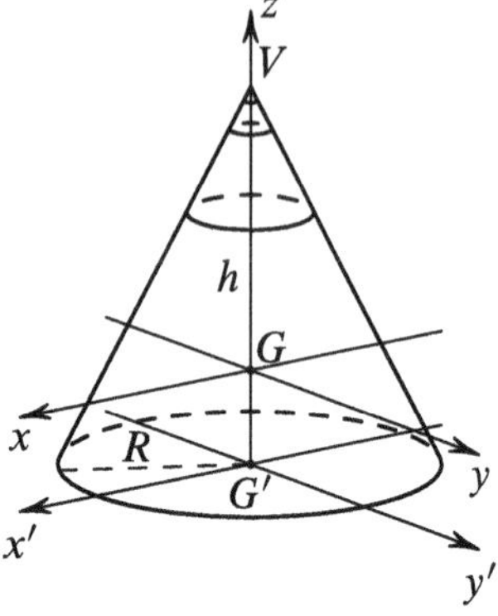


**Fig. 5.15** A homogeneous cone with circular base

Finally, since $|GG'| = h/4$ the Huygens-Steiner theorem gives

$$I_x = I_{x'} - \frac{mh^2}{16} = \frac{3m}{80}\left(4R^2 + h^2\right) ,$$

while clearly $I_y = I_x$ (see Fig. 5.15). Note that for $h = 2R$ we have $I_x = I_y = I_z$.□

## 5.7 Planar Systems

When studying planar bodies it is convenient to choose the figure's plane as coordinate plane on which we place, for instance, the axes $x$, $y$. Let us do just that. As $z = 0$ for all points, we have

$$I_{xz} = I_{yz} = 0 \tag{5.34}$$

and also

$$I_x = \int_{\mathcal{B}} \rho y^2 d\tau , \qquad I_y = \int_{\mathcal{B}} \rho x^2 d\tau . \tag{5.35}$$

Since $I_z = \int_{\mathcal{B}} \rho \left(x^2 + y^2\right) d\tau$, we obtain the important property

$$I_z = I_x + I_y \tag{5.36}$$

valid for all planar bodies.

The $z$-axis is certainly a principal axis of inertia, and finding the other two principal axes, which must lie on the plane of the body, is now simpler.

Consider in fact on the plane $z = 0$ of the system a pair of orthogonal axes $(x, y)$ and a second pair $(\tilde{x}, \tilde{y})$ with the same origin $O$ but obtained rotating by $\theta$ the first pair, as in Fig. 5.16.

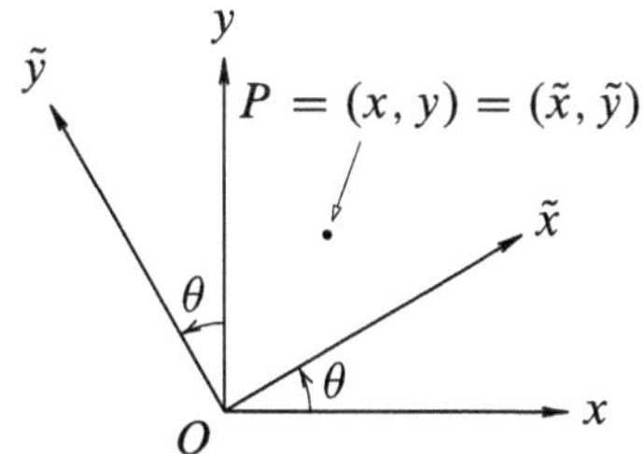


**Fig. 5.16** A rotation of the $(x, y)$ axes by an angle $\theta$

The coordinates of a generic point $P$ of the plane with respect to the two pairs of axes are related by the (inverse) relations

$$\begin{cases} \tilde{x} = x\cos\theta + y\sin\theta \\ \tilde{y} = -x\sin\theta + y\cos\theta \end{cases} \qquad \begin{cases} x = \tilde{x}\cos\theta - \tilde{y}\sin\theta \\ y = \tilde{x}\sin\theta + \tilde{y}\cos\theta \end{cases} \tag{5.37}$$

(while clearly $z = \tilde{z}$).

To determine the angle $\theta$ corresponding to the principal axes of inertia we must express the inertia product $I_{\tilde{x}\tilde{y}}$ in terms of $\theta$, and then set it equal to zero. Using (5.37) together with (5.35) we find

$$\begin{aligned} I_{\tilde{x}\tilde{y}}(\theta) &= -\int_{\mathcal{B}} \rho\tilde{x}\tilde{y}d\tau = -\int_{\mathcal{B}} \rho(x\cos\theta + y\sin\theta)(-x\sin\theta + y\cos\theta)d\tau \\ &= (I_y - I_x)\sin\theta\cos\theta + I_{xy}(\cos^2\theta - \sin^2\theta) \\ &= \frac{1}{2}(I_y - I_x)\sin 2\theta + I_{xy}\cos 2\theta \end{aligned} \tag{5.38}$$

Consider initially the special case where $I_x = I_y$. If $I_{xy} = 0$ as well, the inertia product (5.38) vanishes for any $\theta$, which implies that any two orthogonal axes on the plane are principal. If instead $I_{xy} \neq 0$, for the pair $(\tilde{x}, \tilde{y})$ to be principal it must satisfy $\cos 2\theta = 0$: then the axes form an angle $\pi/4$ with the original axes $(x, y)$ (that is, they are the bisecting lines). At last, in the generic case $I_x \neq I_y$ the inertia product (5.38) vanishes for

$$\tan 2\theta = \frac{2I_{xy}}{I_x - I_y}\,. \tag{5.39}$$

Equation (5.39) gives, for the angle $2\theta$, two solutions that differ by $\pi$, hence two orthogonal solutions for the value $\theta$, which are the directions of the principal axes of inertia.

It is also instructive to compute the inertia moments with respect to the rotated axes $(\tilde{x}, \tilde{y})$. Substituting in (5.37) we obtain

$$
\begin{aligned}
I_{\tilde{x}}(\theta) &= \int_{\mathcal{B}} \rho \tilde{y}^2 d\tau = I_x \cos^2\theta + I_y \sin^2\theta + 2I_{xy} \sin\theta\cos\theta \\
I_{\tilde{y}}(\theta) &= \int_{\mathcal{B}} \rho \tilde{x}^2 d\tau = I_x \sin^2\theta + I_y \cos^2\theta - 2I_{xy} \sin\theta\cos\theta
\end{aligned}
\tag{5.40}
$$

It is easy to check that we would reach the same conclusion using (5.20). In fact, the cosines $\alpha$, $\beta$, $\gamma$ of the angles between this axis and $x$, $y$, $z$ are

$$
\alpha = \cos\theta, \quad \beta = \sin\theta, \quad \gamma = 0\,.
$$

Substituting in (5.20), and recalling (5.34), gives the first relation in (5.40), and a similar argument produces the second one.

From (5.40) and (5.38) we deduce that the derivative in $\theta$ of $I_{\tilde{x}}$ equals $I'_{\tilde{x}}(\theta) = 2I_{\tilde{x}\tilde{y}}(\theta)$, and similarly $I'_{\tilde{y}}(\theta) = -2I_{\tilde{x}\tilde{y}}(\theta)$. Therefore the principal axes, which kill the inertia product, provide at the same time critical values for the moment of inertia, which confirms property (vi) in Sect. 5.6, whereby the principal axes maximize and minimize the moment of inertia.

It is possible and instructive to follows another path. Write the inertia ellipsoid's equation in the axes $(x, y)$:

$$
I_x x^2 + I_y y^2 + I_z z^2 + 2I_{xy} xy = 1 \tag{5.41}
$$

(this is a simplified version (5.30) for the planar case, where some terms are obviously zero). Substitute in (5.41) the second pair of relations in (5.37) so that

$$
\begin{aligned}
&\big(I_x \cos^2\theta + I_y \sin^2\theta + 2I_{xy} \sin\theta\cos\theta\big)\tilde{x}^2 \\
&\quad + \big(I_x \sin^2\theta + I_y \cos^2\theta - 2I_{xy} \sin\theta\cos\theta\big)\tilde{y}^2 + I_z \tilde{z}^2 \\
&\qquad + 2\big(I_{xy} \cos 2\theta + \tfrac{1}{2}(I_y - I_x)\sin 2\theta\big)\tilde{x}\tilde{y} = 1
\end{aligned}
\tag{5.42}
$$

Comparing (5.42) with the ellipsoid's equation in the rotated axes $(\tilde{x}, \tilde{y})$:

$$
I_{\tilde{x}}\tilde{x}^2 + I_{\tilde{y}}\tilde{y}^2 + I_{\tilde{z}}\tilde{z}^2 + 2I_{\tilde{x}\tilde{y}}\tilde{x}\tilde{y} = 1\,,
$$

we deduce that the coefficients of $\tilde{x}^2$, $\tilde{y}^2$, $\tilde{z}^2$ in (5.42) give the moments of inertia relative to the respective axes, while the coefficient of $2\tilde{x}\tilde{y}$ in (5.42) equals the inertia product for the rotated axes, and finally $I_z = I_{\tilde{z}}$. The results found by this procedure clearly confirm what was proved earlier.

Let us further remark that (5.38) can be used to find the inertia product with respect to axes forming an angle $\theta$ with the principal axes. In fact, if we suppose the axes $(x, y)$ are principal then $I_{xy} = 0$, so

$$
I_{\tilde{x}\tilde{y}}(\theta) = (I_y - I_x)\sin\theta\cos\theta\,. \tag{5.43}
$$

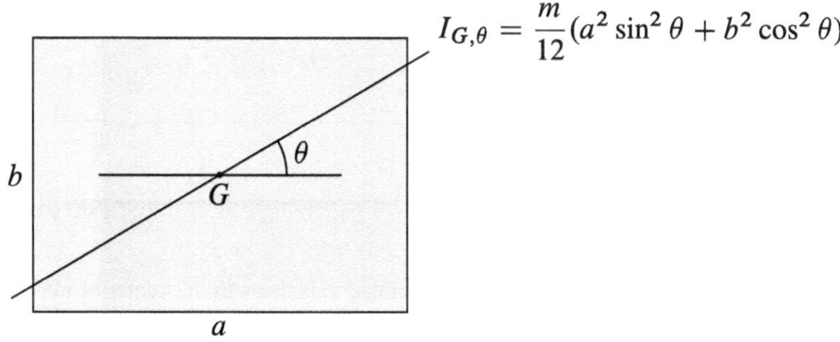


**Fig. 5.17** Moment of inertia of a rectangular homogeneous plate with respect to an oblique axis through the center of mass

Hence we can easily compute the inertia product referred to a generic pair $(\tilde{x}, \tilde{y})$ obtained rotating the principal pair by $\theta$.

Computing the principal moments in the plane, once we know the inertia matrix with respect to a system of axes $(x, y)$, is greatly simplified. In fact the inertia matrix takes the form

$$\mathbf{I}_O = \begin{bmatrix} I_x & I_{xy} & 0 \\ I_{xy} & I_y & 0 \\ 0 & 0 & I_z \end{bmatrix}$$

and to find its eigenvalues, corresponding to the principal moments of inertia, it suffices to solve the characteristic equation, which is easily seen to be

$$(\lambda - I_z)\left(\lambda^2 - \lambda(I_x + I_y) + (I_x I_y - I_{xy}^2)\right) = 0\,.$$

The eigenvalue $\lambda = I_z$ is one of the roots, confirming that $I_z$ is a principal moment, and the remaining solutions

$$\begin{aligned} I_{\tilde{x}} &= \frac{1}{2}\left(I_x + I_y - \sqrt{(I_x - I_y)^2 + 4I_{xy}^2}\right) \\ I_{\tilde{y}} &= \frac{1}{2}\left(I_x + I_y + \sqrt{(I_x - I_y)^2 + 4I_{xy}^2}\right) \end{aligned} \tag{5.44}$$

give us the other two principal moments.

**Example 5.16** Let us determine the moment of inertia of a homogeneous rectangle (with edges $a, b$ and mass $m$) with respect to an axis contained in the plane of the rectangle, through its center of massa and forming an angle $\theta$ with the edge of length $a$ (Fig. 5.17).

Based on the considerations made in Example 5.1 the principal axes at the center of mass are the axes $(x_G, y_G)$, respectively parallel to the edges of lengths $a, b$. If we know the moments of inertia $I_{x_G}, I_{y_G}$ (see Example 5.8), from (5.40) we obtain

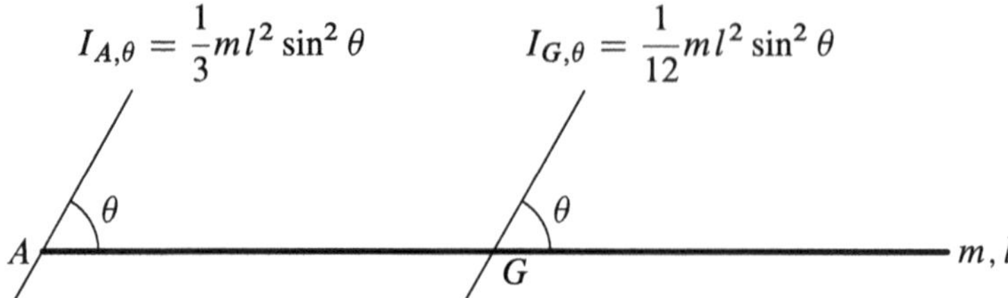


**Fig. 5.18** Moments of inertia with respect to an oblique axis through the center of mass and one through the end of a homogeneous rod

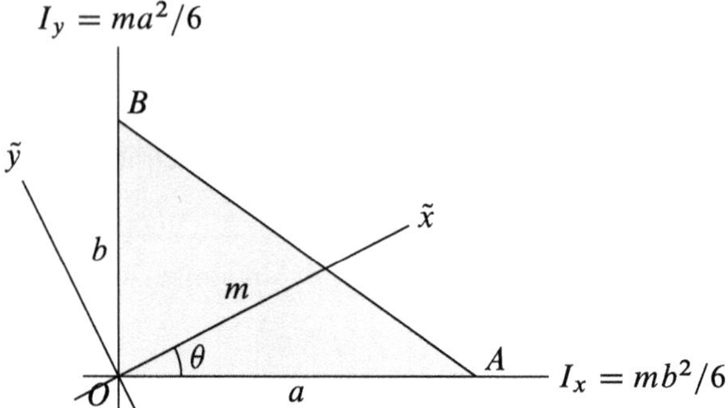


**Fig. 5.19** Moments of inertia with respect to the legs of a homogeneous right-triangular plate and principal axes of inertia with respect to the vertex $O$, whose directions can be computed by (5.46)

$$I_{G,\theta}^{(\text{rect})} = \frac{1}{12}m\left(a^2 \sin^2\theta + b^2 \cos^2\theta\right). \tag{5.45}$$

Note that this moment of inertia does not depend on $\theta$ in the special case of a square plate ($a = b$). At last, if we set $a = l, b = 0$ in (5.45), we obtain the moment of inertia of a homogeneous rod (of length $l$ and mass $m$), with respect to a barycentric axis at an angle $\theta$ with the rod:

$$I_{G,\theta}^{(\text{asta})} = \frac{1}{12}ml^2 \sin^2\theta\,,$$

as shown in Fig. 5.18. The figure shows a parallel axis to the above but passing through an endpoint, and its moment of inertia, obtained using the Huygens-Steiner theorem on two axes at distance $l/2 \sin\theta$. □

**Example 5.17** (*Triangular plate*) Take a plate shaped as a right triangle with right angle at $O$ as in Fig. 5.19. We shall find the principal axes of inertia and the principal moments with respect to $O$.

An integration gives us the moments and products of inertia with respect to the axes $x$ and $y$ through $O$ aligned with the legs:

$$I_x = \frac{1}{6}mb^2, \qquad I_y = \frac{1}{6}ma^2, \qquad I_{xy} = -\frac{1}{12}mab.$$

As the plate is a planar system, we must have

$$I_z = \frac{1}{6}m\left(a^2 + b^2\right), \quad I_{xz} = I_{yz} = 0.$$

Applying formulas (5.44) we obtain the principal moments of inertia

$$I_{\tilde{x}} = \frac{m}{12}\left\{a^2 + b^2 - \sqrt{b^4 - a^2b^2 + a^4}\right\}$$
$$I_{\tilde{y}} = \frac{m}{12}\left\{a^2 + b^2 + \sqrt{b^4 - a^2b^2 + a^4}\right\}$$

while, clearly, $I_{\tilde{z}} = I_z$. Moreover, due to (5.39), solving

$$2\theta = \arctan\left(\frac{ab}{a^2 - b^2}\right) \tag{5.46}$$

for $\theta$ determines the directions of the principal axes $(\tilde{x}, \tilde{y})$. They are represented in Fig. 5.19. □

**Example 5.18** (*Subtraction property*) Let us compute the moment of inertia with respect to the axis $a$ for the plate in Fig. 5.20: this is obtained from a homogeneous square of edge $3l$ and surface density $\rho = m/l^2$ by removing a square corner of edge $l$.

We know that the moment of inertia of a square plate of mass $M$ and edge $L$ with respect to any axis through the center of mass and lying on the plane equals $I_{a_G} = ML^2/12$. We may use this expression for the inertia moment with respect to the axis $a'$ of the whole plate, which would have mass $\rho(3l)^2 = (m/l^2)9l^2 = 9m$. Then

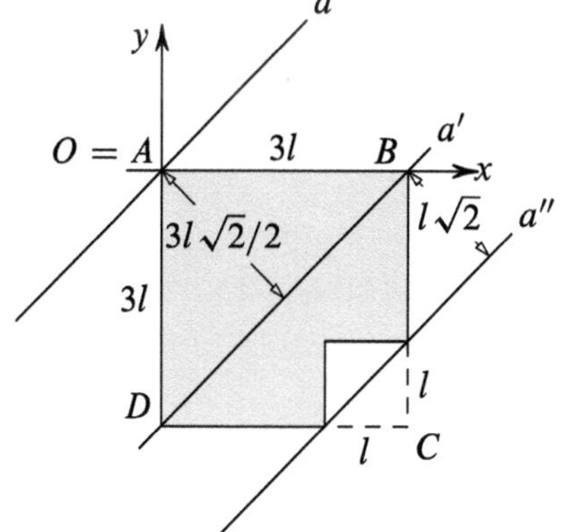


**Fig. 5.20** A chipped square plate

$$I_{a'}^{\text{whole}} = \frac{1}{12} 9m (3l)^2 = \frac{27}{4} ml^2$$

and, using the Huygens-Steiner theorem,

$$I_a^{\text{whole}} = \frac{27}{4} ml^2 + 9m \left( \frac{3\sqrt{2}}{2} l \right)^2 = \frac{189}{4} ml^2 .$$

Let us now compute the moment of inertia of the corner with respect to its diagonal $a''$, as if this square were composed of the same material as the plate, hence with mass $\rho l^2 = m$. We have $I_{a''}^{\text{corner}} = ml^2/12$. By the Huygens-Steiner theorem, and remembering that the distance between the axes $a$ and $a''$ is $5l\sqrt{2}/2$, we find

$$I_a^{\text{corner}} = \frac{1}{12} ml^2 + m \left( \frac{5}{2} \sqrt{2} l \right)^2 = \frac{151}{12} ml^2 .$$

Subtracting from the plate's moment of inertia that of the corner we obtain

$$I_a = I_a^{\text{whole}} - I_a^{\text{corner}} = \frac{189}{4} ml^2 - \frac{151}{12} ml^2 = \frac{104}{3} ml^2 .$$

□

# Chapter 6
# Kinematics of Masses

With a system of masses in motion one associates scalar and vector quantities used to formulate the equations of motion starting from the Cardinal Principles of Mechanics. The topic of this chapter is also known as *kinematics of masses* precisely because it is here where the fundamental ideas of Kinematics converge, in particular those of rigid-body kinematics, and those related to masses, centers of gravity and inertia moments.

The chapter is essentially divided into three parts, corresponding to:

- linear momentum;
- angular momentum;
- kinetic energy.

The importance of these three mechanical quantities will become clear only later, when they will be put to use to formulate the equations of motion of points and systems.

## 6.1 Linear Momentum

**Definition 6.1** The *linear momentum* of a system is defined as

$$\mathbf{Q} = \sum_{i=1}^{n} m_i \mathbf{v}_i \qquad \text{or} \qquad \mathbf{Q} = \int_{\mathcal{B}} \varrho \mathbf{v}\, d\tau\,,$$

depending on whether the system is discrete or continuous.

The following identity relates the linear momentum of a system to the motion of its center of mass.

 
P. Biscari et al., *Rational Mechanics*, UNITEXT 177,
https://doi.org/10.1007/978-3-032-07462-1_6

**Proposition 6.2** (Linear momentum theorem) *The linear momentum of any system satisfies the condition*

$$\mathbf{Q} = m\mathbf{v}_G\,, \tag{6.1}$$

*where m is the system's mass and G its center of mass.*

***Proof*** Equation (6.1) follows from the definition of center of mass. In fact, as (see (5.2))

$$m\,OG = \sum_{i=1}^{n} m_i\,OP_i\,,$$

if we differentiate with respect to time we find

$$m\mathbf{v}_G = \sum_{i=1}^{n} m_i\mathbf{v}_i = \mathbf{Q}\,.$$

□

The linear momentum of a system thus coincides with the linear momentum of its center of mass thought of as a particle where the system's entire mass is concentrated.

## 6.2 Angular Momentum

**Definition 6.3** We call *angular momentum* of a system with respect to the pole $A$ the vector

$$\mathbf{K}_A = \sum_{i=1}^{n} AP_i \times m_i\mathbf{v}_i \qquad \text{or} \qquad \mathbf{K}_A = \int_{\mathcal{B}} AP \times \varrho\mathbf{v}\,d\tau, \tag{6.2}$$

for a discrete system or a continuous body respectively.

This quantity depends in an evident way on the pole $A$ with respect to which we compute it. Because, as we will see later, computing $\mathbf{K}_A$ may simplify by an appropriate choice of pole, it is convenient to know the relationship between angular momenta for different poles $A$ and $B$:

$$\mathbf{K}_B = \mathbf{K}_A + BA \times \mathbf{Q}\,.$$

The proof of this identity, expressing the *angular momentum's dependence on the pole*, follows from the definition and decomposition $BP_i = BA + AP_i$. In fact

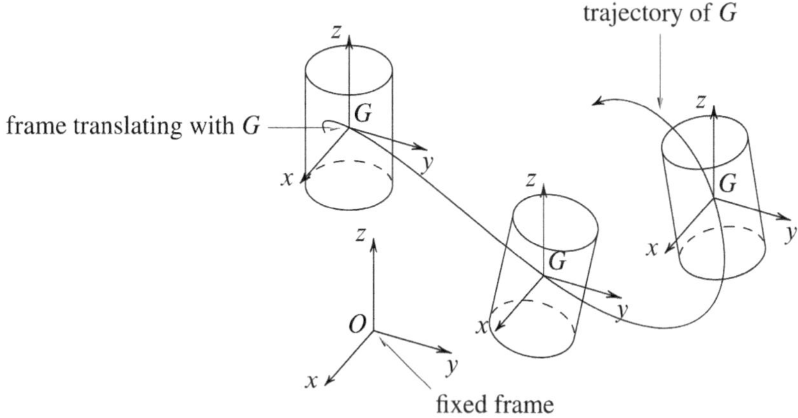


**Fig. 6.1** A fixed frame and a frame translating with the center of mass

$$\mathbf{K}_B = \sum_{i=1}^{n} BP_i \times m_i \mathbf{v}_i = \sum_{i=1}^{n} BA \times m_i \mathbf{v}_i + \sum_{i=1}^{n} AP_i \times m_i \mathbf{v}_i = BA \times \mathbf{Q} + \mathbf{K}_A \tag{6.3}$$

(and similarly for a continuous distribution of masses). An immediate consequence of (6.2) is that for $\mathbf{Q} = \mathbf{0}$ the angular momentum of a system does not depend on the pole: $\mathbf{K}_A = \mathbf{K}_B$.

Computing the angular momentum can be simpler if we use *motions relative to the center of mass*. This refers to the motion of a system's points with respect to an observer that translates with the center of mass, i.e. an observer described by a frame with origin in $G$ and axes with constant direction. In particular, it is convenient to think of a moving frame with axes constantly parallel to the fixed axes, as in Fig. 6.1.

We shall use the tag "r" or "rel" to label the mechanical quantities measured in this frame moving with $G$.

Galilei's theorem (see Sect. 3.2) says the velocity $\mathbf{v}$ of a point seen by the fixed frame equals the sum of $\mathbf{v}_\mathrm{r}$, relative to the frame translating with $G$, and the drag velocity. In this case however the angular velocity of the comoving frame is zero, so the drag velocity simply equals the velocity of $G$ itself. Hence, for the $i$-th point in the system,

$$\mathbf{v}_i = \mathbf{v}_{i,\mathrm{r}} + \mathbf{v}_G \,. \tag{6.4}$$

As first consequence we deduce that in the translating frame the system's linear momentum is always zero. In fact, applying (6.1) to the moving frame,

$$\mathbf{Q}^{\mathrm{rel}} = m\mathbf{v}_G^{\mathrm{rel}} = \mathbf{0} \,, \tag{6.5}$$

since clearly the velocity of $G$ relative to the frame moving with it is zero.

Then (6.2), written in the moving frame, in the light of (6.5) becomes

$$\mathbf{K}_B^{\text{rel}} = BA \times \underbrace{\mathbf{Q}^{\text{rel}}}_{\mathbf{0}} + \mathbf{K}_A^{\text{rel}} = \mathbf{K}_A^{\text{rel}}$$

and it allows to conclude that the angular momentum for the frame moving with $G$ *does not* depend on the pole. For this reason we can just call it $\mathbf{K}^{\text{rel}}$.

**Theorem 6.4** (Decomposition of angular momentum) *The angular momentum of a system can be written as*

$$\mathbf{K}_A = AG \times \mathbf{Q} + \mathbf{K}^{\text{rel}} . \tag{6.6}$$

*In particular,* $\mathbf{K}_G = \mathbf{K}^{\text{rel}}$, *that is: the angular momentum with respect to G computed in the fixed frame coincides with the angular momentum with respect to any pole computed in the frame* that translates with the center of mass.

***Proof*** Equations (6.4) and $\sum m_i AP_i = mAG$ allow to deduce that

$$\begin{aligned}\mathbf{K}_A &= \sum_{i=1}^{n} AP_i \times m_i \mathbf{v}_i = \sum_{i=1}^{n} AP_i \times m_i(\mathbf{v}_G + \mathbf{v}_{i,\text{r}}) \\ &= AG \times m\mathbf{v}_G + \sum_{i=1}^{n} AP_i \times m_i \mathbf{v}_{i,\text{r}} \\ &= AG \times \mathbf{Q} + \mathbf{K}^{\text{rel}} ,\end{aligned}$$

which is (6.6). Putting in it $G$ in place of the generic pole $A$ immediately gives

$$\mathbf{K}_G = \mathbf{K}^{\text{rel}} .$$

Note that the vectors $\mathbf{K}_A$ and $\mathbf{K}^{\text{rel}}$ are equal if and only if $AG \times \mathbf{Q} = \mathbf{0}$, i.e. whenever either $A \equiv G$ or $AG$ is parallel to $\mathbf{v}_G$. □

It is important to comprehend the reason why (6.6) is often useful to compute angular momenta. We interpret the condition as follows: to find $\mathbf{K}_A$ we can add the angular momentum of the center of mass, at which we imagine the total mass of the system is concentrated, to the angular momentum with respect to the pole $G$, *computed in a frame moving with G*.

It is easy to grasp, and we shall see this in the examples, that the frame *translating* with $G$ sees a motion in a simpler way than what the fixed frame sees.

**Remark 6.5** In the special but important case of rigid bodies (cf. Fig. 6.1), for the frame moving with the center of mass the body's motion is always polar around $G$, and $G$ itself is *fixed* in this frame. □

### 6.2.1 *Angular Momentum for a Rotational Velocity Distribution*

Determining the angular momentum of a system subject to a rotational velocity distribution is of great importance due to how commonly it is found in the applications. In the light of Remark 6.5, moreover, it can be useful in more general contexts.

As we will see, we will end up finding a relation that explains the role of the inertia tensor and the inertia matrix.

Consider a body (without loss of generality, continuous) that has a *rotational velocity distribution* with angular velocity $\boldsymbol{\omega}$ around an *instantaneous axis of rotation* on which the pole $H$ lies.

**Proposition 6.6** *The angular momentum with respect to a pole $C$ lying on the instantaneous axis of rotation of a body with rotational velocity distribution is*

$$\mathbf{K}_C = \mathbf{I}_C\boldsymbol{\omega} \tag{6.7}$$

*where $\mathbf{I}_C$ is the inertia tensor relative to $C$.*

***Proof*** The velocity of a generic point $P$ can be expressed as

$$\mathbf{v}_P = \boldsymbol{\omega} \times CP$$

and should be substituted in

$$\mathbf{K}_C = \int_{\mathcal{B}} CP \times \rho\mathbf{v}d\tau\,.$$

This gives

$$\mathbf{K}_C = \int_{\mathcal{B}} CP \times \rho\,(\boldsymbol{\omega} \times CP)\,d\tau = \int_{\mathcal{B}} \rho[(CP)^2\boldsymbol{\omega} - (CP \cdot \boldsymbol{\omega})CP]d\tau$$

where we used (A.8) to compute the triple product. Comparing the latter with the definition of inertia tensor in (5.25) (with $C$ instead of $O$) we see that

$$\mathbf{K}_C = \int_{\mathcal{B}} \rho[(CP)^2\boldsymbol{\omega} - (CP \cdot \boldsymbol{\omega})CP]d\tau = \mathbf{I}_C\boldsymbol{\omega}$$

where $\mathbf{I}_C$ is the inertia tensor relative to point $C$. □

Equation (6.7) subsumes one essential aspect for computing angular momenta of rigid bodies: the linear relations between $\boldsymbol{\omega}$ and $\mathbf{K}$, given by the inertia tensor, when the velocity distribution is rotational. We might even say that here the inertia tensor finds its most important reason to exist and the most relevant application.

This equality can be expressed more directly using matrices and their product:

$$\begin{bmatrix} K_1 \\ K_2 \\ K_3 \end{bmatrix} = \begin{bmatrix} I_1 & I_{12} & I_{13} \\ I_{12} & I_2 & I_{23} \\ I_{13} & I_{23} & I_3 \end{bmatrix} \begin{bmatrix} \omega_1 \\ \omega_2 \\ \omega_3 \end{bmatrix} \tag{6.8}$$

(where for simplicity we dropped the pole to which the angular momentum refers). Therefore, using a generic triple $\mathbf{e}_i$,

$$\begin{aligned} \mathbf{K}_H =& (I_1\omega_1 + I_{12}\omega_2 + I_{13}\omega_3)\mathbf{e}_1 \\ &+ (I_{12}\omega_1 + I_2\omega_2 + I_{23}\omega_3)\mathbf{e}_2 \\ &+ (I_{13}\omega_1 + I_{23}\omega_2 + I_3\omega_3)\mathbf{e}_3 \end{aligned} . \tag{6.9}$$

But if we choose *principal inertia axes* $\hat{\mathbf{e}}_i$ at $C$, formula (6.8) simplifies to

$$\begin{bmatrix} \hat{K}_1 \\ \hat{K}_2 \\ \hat{K}_3 \end{bmatrix} = \begin{bmatrix} \hat{I}_1 & 0 & 0 \\ 0 & \hat{I}_2 & 0 \\ 0 & 0 & \hat{I}_3 \end{bmatrix} \begin{bmatrix} \hat{\omega}_1 \\ \hat{\omega}_2 \\ \hat{\omega}_3 \end{bmatrix} ,$$

obviously equivalent to

$$\hat{K}_1 = \hat{I}_1\hat{\omega}_1 , \qquad \hat{K}_2 = \hat{I}_2\hat{\omega}_2 , \qquad \hat{K}_3 = \hat{I}_3\hat{\omega}_3 ,$$

or even better

$$\mathbf{K}_H = \hat{I}_1\hat{\omega}_1\hat{\mathbf{e}}_1 + \hat{I}_2\hat{\omega}_2\hat{\mathbf{e}}_2 + \hat{I}_3\hat{\omega}_3\hat{\mathbf{e}}_3 , \tag{6.10}$$

where the momenta $\hat{I}_i$ and all components refer to the principal axes of inertia (obviously there is no need to use "hats" when the triple is clear from the context).

Remark 6.5 allows to deduce an important consequence: when we want to compute the angular momentum of a rigid body with respect to its center of mass we can use (6.7), replacing $C$ with $G$. This fact is extremely useful and deserves to be mentioned explicitly.

**Proposition 6.7** *The angular momentum of a rigid body with respect to its center of mass $G$ can be written as*

$$\mathbf{K}_G = \mathbf{I}_G\boldsymbol{\omega} . \tag{6.11}$$

In the sequel we shall therefore call $H$ a pole that is

- either lying on the instantaneous axis of rotation of a body with rotational velocity distribution, or
- coincides with the center of mass of a rigid body,

*indifferently*. In either case

$$\mathbf{K}_H = \mathbf{I}_H\boldsymbol{\omega}$$

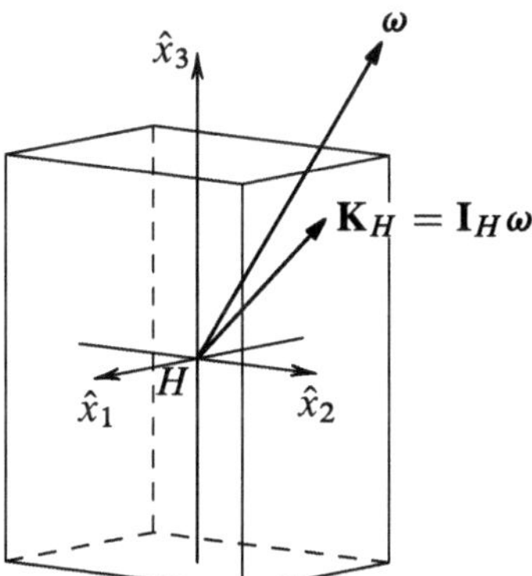


**Fig. 6.2** The angular momentum of a rigid body with respect to a point $H$ which is either the center of mass $G$ or, when the velocity distribution is rotational, a point $C$ on the instantaneous axis of rotation, can be computed by applying the tensor of inertia $\mathbf{I}_H$ to the angular velocity $\boldsymbol{\omega}$. The vectors $\mathbf{K}_H$ and $\boldsymbol{\omega}$ are in general *not* parallel, unless $\boldsymbol{\omega}$ is parallel to a principal axis of inertia, which means it is an eigenvalue of $\mathbf{I}_H$

and Fig. 6.2 shows and explains the relationship of $\boldsymbol{\omega}$ and $\mathbf{K}_H$.

Here is one thing to remember as regards the parallelism between $\mathbf{K}_H$ and $\boldsymbol{\omega}$. From Linear Algebra, in particular the theory of matrix diagonalization or more directly from the analysis of (6.10), we know that $\boldsymbol{\omega}$ and $\mathbf{K}_H$ *are parallel if and only if* $\boldsymbol{\omega}$ *(and so* $\mathbf{K}_H$*) has the same direction as a principal inertia axis.*

Choosing the unit vector $\mathbf{e}_1$ (principal or not) parallel to the angular velocity $\boldsymbol{\omega}$ gives $\boldsymbol{\omega} = \omega_1 \mathbf{e}_1$ and by (6.9) we can prove

$$\mathbf{K}_H \cdot \mathbf{e}_1 = \mathbf{I}_H \boldsymbol{\omega} \cdot \mathbf{e}_1 = I_1 \omega_1 . \tag{6.12}$$

In other words, the component of the angular momentum $\mathbf{K}_H = \mathbf{I}_H\boldsymbol{\omega}$ parallel to $\boldsymbol{\omega}$ coincides with $I_1\omega_1$, namely the product of the moment of inertia for the axis parallel to $\boldsymbol{\omega}$ times the angular velocity itself. (Clearly this does not imply $\mathbf{I}_H\boldsymbol{\omega} = I_1\boldsymbol{\omega}$, which holds if and only if $\boldsymbol{\omega}$ is parallel to a principal axis, as we saw above.)

It is interesting to spell out the role played by the inertia ellipsoid with respect to the pole $H$, by which we can find the direction of $\mathbf{K}_H$ starting from that of $\boldsymbol{\omega}$. Consider the inertia ellipsoid referred to principal axes. The normal vector to the surface, defined by

$$f(x_1, x_2, x_3) = \hat{I}_1 x_1^2 + \hat{I}_2 x_2^2 + \hat{I}_3 x_3^2 - 1 = 0 , \tag{6.13}$$

is parallel to the gradient of $f$

$$\nabla f = 2\hat{I}_1 x_1\, \hat{\mathbf{e}}_1 + 2\hat{I}_2 x_2\, \hat{\mathbf{e}}_2 + 2\hat{I}_3 x_3\, \hat{\mathbf{e}}_3 . \tag{6.14}$$

Let $P_\omega$ be the intersection point of the half-line emanating from $H$ with direction $\boldsymbol{\omega}$ and ellipsoid (6.13). Then $HP_\omega = \lambda\boldsymbol{\omega}$. Using (6.10) and (6.14) we then discover that the *normal vector to the ellipsoid* at $P_\omega$ is *parallel* to the momentum $\mathbf{K}_H$.

In space, $\boldsymbol{\omega}$ and $\mathbf{K}_H$ are in general not parallel, but they become so if the rigid body is planar. We know in fact that in that case the angular velocity $\boldsymbol{\omega}$ is orthogonal to the plane, hence parallel to the principal inertia axis passing though $H$.

This property simplifies even further (6.7). Call $K_H$ and $\omega$ the *components* of $\mathbf{K}_H$ and $\boldsymbol{\omega}$ along a common unit vector orthogonal to the plane where the motion occurs and that contains the rigid body. Then

$$K_H = I_H\omega$$

where $I_H$ denotes the (principal) moment of inertia with respect to the axis at $H$ *perpendicular* to the plane of motion.

We can anticipate that $\mathbf{K}_H$ and $\boldsymbol{\omega}$ being parallel, which is always the case in the planar case, is one of the reasons that make the study of the dynamics of planar systems substantially easier that that of spatial systems.

### 6.2.2 *Angular Momentum of a Rigid Body*

In this section we establish in more generality the relationships that allow to express the angular momentum of a body with rigid velocity distribution, and hence a rigid body, with respect to a generic pole $A$.

In the light of Theorem 6.4 and remembering that in the frame translating with the center of mass the body's velocity distribution is certainly rotational, with instantaneous axis of rotation passing through $G$, we immediately deduce

$$\mathbf{K}_A = AG \times \mathbf{Q} + \mathbf{I}_G\boldsymbol{\omega}\,. \tag{6.15}$$

In other words, to $\mathbf{K}_G = \mathbf{I}_G\boldsymbol{\omega}$ we must add the moment of the linear momentum $\mathbf{Q} = m\mathbf{v}(G)$, applied at $G$, with respect to the pole $A$.

The overall angular momentum coincides therefore with the moment of the linear momentum vector applied at $G$ only if the velocity distribution is translational, as per (6.15):

$$\mathbf{K}_A = AG \times \mathbf{Q} \iff \omega = \mathbf{0}.$$

Here is a useful companion to the above result.

**Proposition 6.8** *The angular momentum of a rigid body with respect to a pole A can be written*

$$\mathbf{K}_A = m\,AG \times \mathbf{v}_A + \mathbf{I}_A\boldsymbol{\omega}\,, \tag{6.16}$$

*where* $\mathbf{v}_A$ *is the velocity of the point transiting through the pole A,* $\boldsymbol{\omega}$ *is the angular velocity of the rigid body and* $\mathbf{I}_A$ *the inertia tensor with respect to A (see (5.21)).*

***Proof*** Law (2.22) for rigid velocity distributions implies $\mathbf{v}_P = \mathbf{v}_A + \boldsymbol{\omega} \times AP$ for all $P$. Substituting in definition $(6.2)_2$ we have

$$\begin{aligned} \mathbf{K}_A &= \int_{\mathcal{B}} AP \times \rho[\mathbf{v}_A + \boldsymbol{\omega} \times AP]d\tau \\ &= \int_{\mathcal{B}} \rho AP \times \mathbf{v}_A d\tau + \int_{\mathcal{B}} \rho AP \times (\boldsymbol{\omega} \times AP)d\tau \,. \end{aligned} \tag{6.17}$$

In the first integral on the right $\mathbf{v}_A$ moves outside the integral by definition (5.4), and in particular (5.5) with $A$ instead of $O$. Hence

$$\int_{\mathcal{B}} \rho AP \times \mathbf{v}_A d\tau = \left( \int_{\mathcal{B}} \rho AP d\tau \right) \times \mathbf{v}_A = mAG \times \mathbf{v}_A \,. \tag{6.18}$$

In analogy to the proof of Proposition 6.6 we use (A.8) to expand the triple product in the second integral of (6.17), so that

$$\int_{\mathcal{B}} \rho AP \times (\boldsymbol{\omega} \times AP)d\tau = \int_{\mathcal{B}} \rho[(AP)^2\boldsymbol{\omega} - (AP \cdot \boldsymbol{\omega})AP]d\tau \,.$$

Comparing the latter with the definition of inertia tensor in (5.25) (here too $A$ replaces $O$) we find

$$\int_{\mathcal{B}} \rho[(AP)^2\boldsymbol{\omega} - (AP \cdot \boldsymbol{\omega})AP]d\tau = \mathbf{I}_A\boldsymbol{\omega}$$

where $\mathbf{I}_A$ is the inertia tensor relative to $A$. Adding up this term to (6.18) we conclude

$$\mathbf{K}_A = m\, AG \times \mathbf{v}_A + \mathbf{I}_A\boldsymbol{\omega} \,.$$

□

When we choose $A = G$ in (6.16) we recover property (6.11) already seen. Choosing instead a pole $A$ on the instantaneous axis of rotation (if the velocity distribution is rotational) we obtain $\mathbf{v}(A) = \mathbf{0}$ and hence (6.7) once again.

### 6.2.3 Time Derivative of the Angular Momentum

Whenever we want to find the rate of variation in time of the angular momentum we should pay attention, because we must account for the motion not only of the system's points, but also of the possible motion of the momenta's pole $A$. In fact, in many situations, and for a number of reasons, it might be convenient to pick a pole $A$ that is *not fixed*, or even one that does not follow one of the system's particles.

Consider a disk that rolls without slipping on a straight guide, and call $C$ the contact point. It can prove useful to find the angular momentum $\mathbf{K}_C$ with respect to this point. Here the pole is not fixed, nor does it follow the motion of any material element, since the point that *passes through* $C$ changes instant by instant. Besides, the pure rolling constraint forces $\mathbf{v}_C = \mathbf{0}$. This does not imply the *geometric* point $C$ is fixed. It merely says the *particle* passing through $C$ is instantaneously at rest (whereas it was moving before reaching the guide, and starts moving again as soon as it leaves the guide).

In the light of the previous example it becomes necessary to introduce a notation able to distinguish the velocity of a geometric point $A$ (such as the pole used in momenta) from the velocity of the particle that at that moment passes through the geometric point. We will denote as follows:

$$\begin{aligned} \dot{A} &: \quad \text{velocity of the geometric point } A \\ \mathbf{v}_A &: \quad \text{velocity of the particle passing through } A. \end{aligned}$$

The above quantities will coincide whenever the particle at $A$ is always the same, as in definition (1.1). This happens for example when we use a letter to indicate the end of a rod, the center of a disk or the vertex of a square. On the other hand, $\dot{A}$ might differ from $\mathbf{v}_A$ when the particle at $A$ varies in time.

**Example 6.9** For a pure rolling constraint (see Sect. 4.6.1, and in particular Fig. 4.16) the velocity of the particle passing through the guide is zero: $\mathbf{v}_C = \mathbf{0}$. Despite that, the contact point changes geometric position: $\dot{C} \neq \mathbf{0}$. Actually it is easy to convince ourselves that since $C$ is the projection of the disk' center $G$ on the guide, its velocity coincides with the center's velocity: $\dot{C} = \mathbf{v}_G$. □

**Proposition 6.10** *The time derivative of the angular momentum (for discrete systems) equals*

$$\dot{\mathbf{K}}_A = -\dot{A} \times m\mathbf{v}_G + \sum_{i=1}^{n} AP_i \times m_i \mathbf{a}_i \,. \tag{6.19}$$

***Proof*** The claim is immediate once we observe

$$\frac{dAP_i}{dt} = \mathbf{v}_i - \dot{A} \,,$$

since $A$ follows the motion of the geometric pole. So we have

$$\dot{\mathbf{K}}_A = \frac{d}{dt}\sum_{i=1}^{n} AP_i \times m_i\mathbf{v}_i = \sum_{i=1}^{n} (\mathbf{v}_i - \dot{A}) \times m_i\mathbf{v}_i + \sum_{i=1}^{n} AP_i \times m_i\mathbf{a}_i \,.$$

Furthermore, as $\mathbf{v}_i \times m_i\mathbf{v}_i = 0$ we obtain (6.19). □

Next we prove an alternative for (6.19), which will be useful when studying rigid body dynamics.

**Proposition 6.11** *The time derivative of the* barycentric *angular momentum* $\mathbf{K}_G$ *in a rigid body is*

$$\begin{aligned}\dot{\mathbf{K}}_G = \mathbf{I}_G\boldsymbol{\omega} + \boldsymbol{\omega} \times \mathbf{I}_G\boldsymbol{\omega} = & \left(\bar{I}_1\dot{\omega}_1 - (\bar{I}_2 - \bar{I}_3)\omega_2\omega_3\right)\bar{\mathbf{e}}_1 \\ & + \left(\bar{I}_2\dot{\omega}_2 - (\bar{I}_3 - \bar{I}_1)\omega_1\omega_3\right)\bar{\mathbf{e}}_2 \\ & + \left(\bar{I}_3\dot{\omega}_3 - (\bar{I}_1 - \bar{I}_2)\omega_1\omega_2\right)\bar{\mathbf{e}}_3\end{aligned} \tag{6.20}$$

*where* $\{\bar{\mathbf{e}}_1, \bar{\mathbf{e}}_2, \bar{\mathbf{e}}_3\}$ *is a basis of central principal axes of inertia (principal axes at the center of mass),* $\{\bar{I}_1, \bar{I}_2, \bar{I}_3\}$ *are the corresponding central principal moments of inertia, and* $\boldsymbol{\omega} = \omega_1\bar{\mathbf{e}}_1 + \omega_2\bar{\mathbf{e}}_2 + \omega_3\bar{\mathbf{e}}_3$.

***Proof*** For a rigid body, once we choose the center of mass as pole we have $\mathbf{K}_G = \mathbf{I}_G\boldsymbol{\omega}$. When differentiating, it is convenient to refer to the comoving triad, because the comoving frame sees the body at rest and so in this frame the time derivative of the inertia tensor vanishes. Using (3.1) to relate the absolute time derivative and the derivative in the comoving frame,

$$\dot{\mathbf{K}}_G = (\mathbf{K}_G)' + \boldsymbol{\omega} \times \mathbf{K}_G\,, \tag{6.21}$$

where the dash indicates the time derivative in the comoving frame.

In (6.21) we have$(\mathbf{K}_G)' = (\mathbf{I}_G\boldsymbol{\omega})' = \mathbf{I}_G\boldsymbol{\omega}'$. As $\boldsymbol{\omega}' = \boldsymbol{\omega}$ (see (3.3)), we obtain $\dot{\mathbf{K}}_G = \mathbf{I}_G\boldsymbol{\omega} + \boldsymbol{\omega} \times \mathbf{I}_G\boldsymbol{\omega}$, or (6.20). To finish the proof it suffices to write the above relation with respect to central principal axes of inertia. □

**Remark 6.12** A completely similar expression to (6.20) holds when all quantities refer to a generic point $A$, rather than the center of mass $G$ as here, as long as the point is *fixed* (in particular, it must belong at each instant to the instantaneous axis of rotation). That is because, by (6.16), the angular momentum with respect to such a point has an expression similar to the barycentric one: $\mathbf{K}_A = \mathbf{I}_A\boldsymbol{\omega}$. □

## 6.3 Kinetic Energy

The final mechanical quantity we present associates with a point, or system of points, in motion a scalar, denoted $T$, called kinetic energy.

**Definition 6.13** The *kinetic energy* is by definition

$$T = \frac{1}{2}mv^2 \qquad T = \frac{1}{2}\sum_{i=1}^{n} m_i v_i^2 \qquad T = \frac{1}{2}\int_{\mathcal{B}} \varrho v^2 d\tau\,, \tag{6.22}$$

for a point, a discrete system or a continuous body respectively.

Note the kinetic energy is always positive, and vanishes when all points have zero velocity. Computing it is made easier by some useful properties, the first of which allows to decompose it into the sum of the kinetic energy of the mass plus the kinetic energy measured in a frame translating with it.

**Theorem 6.14** (König: decomposition of kinetic energy) *The kinetic energy of any material system can be written as*

$$T = \frac{1}{2} m v_G^2 + T_{\text{rel}}^{(G)}, \tag{6.23}$$

*sum of the kinetic energy of the center of mass with the entire system's mass concentrated in it, plus the kinetic energy of the material system in the motion* relative *to the center of mass.*

***Proof*** For brevity we will only discuss a discrete system, though the proof can be easily adapted for extended rigid bodies. We saw in (6.4) that Galilei's theorem implies

$$\mathbf{v}_{i,\text{a}} = \mathbf{v}_{i,\tau} + \mathbf{v}_{i,\text{r}} = \mathbf{v}_G + \mathbf{v}_{i,\text{r}}, \tag{6.24}$$

since the drag velocity distribution in a barycentric frame is translational and in particular the drag velocity equals the velocity of the center of mass. Substituting (6.24) in (6.22) and using (6.5) gives

$$\begin{aligned} T &= \frac{1}{2} \sum_{i=1}^{n} m_i \, (\mathbf{v}_G + \mathbf{v}_{i,\text{r}}) \cdot (\mathbf{v}_G + \mathbf{v}_{i,\text{r}}) \\ &= \frac{1}{2} \sum_{i=1}^{n} m_i (\mathbf{v}_G \cdot \mathbf{v}_G) + \sum_{i=1}^{n} m_i (\mathbf{v}_G \cdot \mathbf{v}_{i,\text{r}}) + \frac{1}{2} \sum_{i=1}^{n} m_i (\mathbf{v}_{i,\text{r}} \cdot \mathbf{v}_{i,\text{r}}) \\ &= \frac{1}{2} \Big( \sum_{i=1}^{n} m_i \Big) v_G^2 + \mathbf{v}_G \cdot \underbrace{\sum_{i=1}^{n} m_i \mathbf{v}_{i,\text{r}}}_{\mathbf{Q}^{(G)} = \mathbf{0}} + \frac{1}{2} \sum_{i=1}^{n} m_i v_{i,\text{r}}^2 \\ &= \frac{1}{2} m v_G^2 + T_{\text{rel}}^{(G)}. \end{aligned}$$

□

### 6.3.1 Kinetic Energy for a Rigid Body

Let us examine in detail, once more, how computing the kinetic energy becomes simpler for rigid systems.

**Proposition 6.15** *The kinetic energy of a rigid body can be written as*

$$T = \frac{1}{2} m v_A^2 + m \mathbf{v}_A \cdot \boldsymbol{\omega} \times AG + \frac{1}{2} \mathbf{I}_A \boldsymbol{\omega} \cdot \boldsymbol{\omega}, \tag{6.25}$$

*where A is any point comoving with the rigid system, $\boldsymbol{\omega}$ and G are the latter's angular velocity and center of mass, and $\mathbf{I}_A$ is the inertia matrix of the system with respect to A (see (5.21)).*

***Proof*** To find the kinetic energy of a rigid system we must insert in definition (6.22) the velocity distribution law (2.22), whereby $\mathbf{v}_P = \mathbf{v}_A + \boldsymbol{\omega} \times AP$ for all $P$ comoving with the system. Then

$$\begin{aligned}
T &= \frac{1}{2} \int_{\mathcal{B}} (\mathbf{v}_A + \boldsymbol{\omega} \times AP)^2 \rho d\tau \\
&= \frac{1}{2} \int_{\mathcal{B}} v_A^2 \rho d\tau + \int_{\mathcal{B}} \mathbf{v}_A \cdot \boldsymbol{\omega} \times AP \rho d\tau + \frac{1}{2} \int_{\mathcal{B}} (\boldsymbol{\omega} \times AP)^2 \rho d\tau \\
&= \frac{1}{2} m v_A^2 + \mathbf{v}_A \cdot \boldsymbol{\omega} \times \int_{\mathcal{B}} AP \rho d\tau + \frac{1}{2} \int_{\mathcal{B}} (\boldsymbol{\omega} \times AP) \cdot (\boldsymbol{\omega} \times AP) \rho d\tau \\
&= \frac{1}{2} m v_A^2 + \mathbf{v}_A \cdot \boldsymbol{\omega} \times mAG + \frac{1}{2} \int_{\mathcal{B}} \underbrace{\boldsymbol{\omega}}_{\mathbf{a}} \times \underbrace{AP}_{\mathbf{b}} \cdot \underbrace{(\boldsymbol{\omega} \times AP)}_{\mathbf{c}} \rho d\tau .
\end{aligned}$$

In the last term with use the mixed product's property $\mathbf{a} \times \mathbf{b} \cdot \mathbf{c} = \mathbf{b} \times \mathbf{c} \cdot \mathbf{a}$ (see (A.10)), so that

$$\begin{aligned}
T = \frac{1}{2} m v_A^2 + \mathbf{v}_A \cdot \boldsymbol{\omega} \times mAG + \frac{1}{2} \left[ \int_{\mathcal{B}} AP \times (\boldsymbol{\omega} \times AP) \rho d\tau \right] \cdot \boldsymbol{\omega} \\
= \frac{1}{2} m v_A^2 + m \mathbf{v}_A \cdot \boldsymbol{\omega} \times AG + \frac{1}{2} \mathbf{I}_A \boldsymbol{\omega} \cdot \boldsymbol{\omega} .
\end{aligned}$$

In the final passage the term in brackets coincides with the second term in (6.17), so it does equal $\mathbf{I}_A \boldsymbol{\omega}$. □

**Remarks**

- Formula (6.25) simplifies if $A$ is chosen so to kill the second summand. In particular, with $A \equiv G$ we have $AG = \mathbf{0}$ and then

$$T = \frac{1}{2} m v_G^2 + \frac{1}{2} \mathbf{I}_G \boldsymbol{\omega} \cdot \boldsymbol{\omega} . \tag{6.26}$$

  If there exists, and we can find, a point $C$ on the instantaneous axis of rotation (i.e. such that $\mathbf{v}_C = \mathbf{0}$, see Sect. 2.8) formula (6.25) simplifies further:

$$T = \frac{1}{2}\mathbf{I}_C\boldsymbol{\omega}\cdot\boldsymbol{\omega} = \frac{1}{2}\mathbf{K}_C\cdot\boldsymbol{\omega} \qquad \left(\mathbf{v}_C = \mathbf{0}\right). \tag{6.27}$$

- Comparing (6.26) with König's theorem (6.23) we discover that for a rigid body

$$T_{\text{rel}}^{(G)} = \frac{1}{2}\mathbf{I}_G\boldsymbol{\omega}\cdot\boldsymbol{\omega} = \frac{1}{2}\mathbf{K}_G\cdot\boldsymbol{\omega}. \tag{6.28}$$

  This should not be a surprise, since the motion of the rigid body relative to its center of mass is polar (see Sect. 2.5.2), as $G$ is fixed by construction in the relative system. In particular, $G$ always belongs on the body's axis of rotation in this relative system, and so (6.28) is nothing but a special case of (6.27).
- Expression (5.23), obtained when studying the moments of inertia for concurrent axes (see Sect. 5.4), allows to simplify the quadratic terms in $\boldsymbol{\omega}$ in the quadratic energy. Define in fact a unit vector $\mathbf{u}$ such that $\boldsymbol{\omega} = \omega\,\mathbf{u}$. Then (5.23) implies

$$\frac{1}{2}\mathbf{I}_A\boldsymbol{\omega}\cdot\boldsymbol{\omega} = \frac{1}{2}\Big(\mathbf{I}_A\mathbf{u}\cdot\mathbf{u}\Big)\omega^2 = \frac{1}{2}I_{A\omega}\omega^2, \tag{6.29}$$

  where $I_{A\omega}$ is the moment of inertia with respect to the axis *parallel to $\boldsymbol{\omega}$ and passing through $A$*. For example, let us reintroduce the *barycentric principal* axes $\{\bar{\mathbf{e}}_1, \bar{\mathbf{e}}_2, \bar{\mathbf{e}}_3\}$ and the corresponding *central principal moments* $\{\bar{I}_1, \bar{I}_2, \bar{I}_3\}$, and also set $\boldsymbol{\omega} = \omega_1\bar{\mathbf{e}}_1 + \omega_2\bar{\mathbf{e}}_2 + \omega_3\bar{\mathbf{e}}_3$. Then (6.26) and (6.29) imply

$$T = \frac{1}{2}mv_G^2 + \frac{1}{2}I_{G\omega}\omega^2 = \frac{1}{2}mv_G^2 + \frac{1}{2}\left(\bar{I}_1\omega_1^2 + \bar{I}_2\omega_2^2 + \bar{I}_3\omega_3^2\right).$$

- Since the kinetic energy is always positive (and zero only when all particles are at rest), from (6.27) we deduce that in a rotational velocity distribution the angle between $\mathbf{K}_C$ and $\boldsymbol{\omega}$ is always less than $\pi/2$, by known properties of the dot product.

### *6.3.2 Kinetic Energy for a Holonomic System*

We have shown that in a holonomic system the position $P_i$ of any point can be expressed in terms of a suitable number of generalized coordinates and possibly time, $P_i = P_i(\mathsf{q}; t)$, where $\mathsf{q} = \{q_1, \ldots, q_N\}$ is the collection of Lagrangian coordinates. As already shown (see (4.21)),

$$\mathbf{v}_i = \sum_{k=1}^{N}\frac{\partial P_i}{\partial q_k}\dot{q}_k + \frac{\partial P_i}{\partial t}$$

and the kinetic energy reads

$$T = \frac{1}{2} \sum_{h,k=1}^{N} a_{hk} \dot{q}_k \dot{q}_h + \sum_{k=1}^{N} b_k \dot{q}_k + c \,, \tag{6.30}$$

where we have introduced quantities

$$\begin{aligned} a_{kh}(\mathsf{q}; t) &= \sum_{i=1}^{n} m_i \frac{\partial P_i}{\partial q_k} \cdot \frac{\partial P_i}{\partial q_h} \,, \\ b_k(\mathsf{q}; t) &= \sum_{i=1}^{n} m_i \frac{\partial P_i}{\partial q_k} \cdot \frac{\partial P_i}{\partial t} \,, \\ c(\mathsf{q}; t) &= \frac{1}{2} \sum_{i=1}^{n} m_i \frac{\partial P_i}{\partial t} \cdot \frac{\partial P_i}{\partial t} \,. \end{aligned} \tag{6.31}$$

The $N \times N$ symmetric matrix

$$\mathsf{A} = [a_{kh}] \quad (k, h = 1, \ldots, N) \tag{6.32}$$

composed by the coefficients in $(6.31)_1$ is called *mass matrix*.

Together with the vectors $\mathsf{q} = \{q_1, \ldots, q_N\}$ and $\dot{\mathsf{q}} = \{\dot{q}_1, \ldots, \dot{q}_N\}$ we further define $\mathsf{b} = \{b_1, \ldots, b_N\}$ and the scalar $c$ defined by $(6.31)_{2,3}$.

In this way the kinetic energy (6.30) can be written succinctly as

$$T(\mathsf{q}, \dot{\mathsf{q}}, t) = \frac{1}{2} \mathsf{A}(\mathsf{q}, t)\dot{\mathsf{q}} \cdot \dot{\mathsf{q}} + \mathsf{b}(\mathsf{q}, t) \cdot \dot{\mathsf{q}} + c(\mathsf{q}, t) \,. \tag{6.33}$$

The coefficients $b_k$ and the scalar $c$ depend on the partial derivatives of the positions with respect to time, as shown in (6.31). In presence of *fixed* constraints only, these derivatives vanish so the kinetic energy is a *quadratic* function of the derivatives of the Lagrangian coordinates

$$T(\mathsf{q}, \dot{\mathsf{q}}, t) = \frac{1}{2} \mathsf{A}(\mathsf{q}, t)\dot{\mathsf{q}} \cdot \dot{\mathsf{q}} \qquad \text{(fixed constraints)}.$$

### 6.3.3 Mass Matrix

The *mass matrix* is symmetric, positive definite, diagonalizable and invertible.

- The mass matrix' symmetry, implying its diagonalizability, is a trivial consequence of the symmetry of the dot product.
- The invertibility of $\mathsf{A}$, which will be crucial for Lagrangian determinism (see Sect. 14.4.1), follows instead from the kinetic energy's positivity. By definition, in fact, the kinetic energy of a system cannot be negative, and vanishes precisely

when the velocities of all points are zero.

If $\mathsf{A}$ had a negative eigenvalue, it would suffice to choose the $n$-tuple of Lagrangian velocities $\dot{\mathsf{q}}$ parallel to the corresponding eigenvector and we would have negative kinetic energy (whose sign is determined by the quadratic term for sufficiently large velocities).

Similarly, if $\mathsf{A}$ had a zero eigenvalue we could build a velocity distribution for which the quadratic part of $T$ vanished. But then the sign of $T$ would be given by the linear term in $\dot{\mathsf{q}}$, and would thus change when the velocities changed direction. The mass matrix must therefore have strictly positive eigenvalues, making it invertible and positive definite.

## 6.4 Example: Calculation of Momenta and Kinetic Energy

**Example 6.16** We set out to compute the linear momentum, angular momentum and kinetic energy of the rod $AB$ in Fig. 6.3, which is homogeneous, has length $l$, mass $m$ and is constrained to move on a plane with the end $A$ gliding along a fixed horizontal guide. Call $s$ the abscissa of $A$ with respect to some origin $O$ and $\theta$ the rod's angle of rotation measured clockwise from the guide's direction. The quantities $(s, \theta)$ are generalized coordinates, since they are independent and from them we can describe any configuration.

We use axes $x$ and $y$ as in the figure and a coherently oriented orthogonal $z$-axis, so that the rod's angular velocity is $\boldsymbol{\omega} = \dot{\theta}\mathbf{k}$.

To find the linear momentum we use $\mathbf{Q} = m\mathbf{v}_G$, namely the *Linear momentum theorem*. Multiply by $m$, the mass of the system, the components of the velocity of the center of mass and write

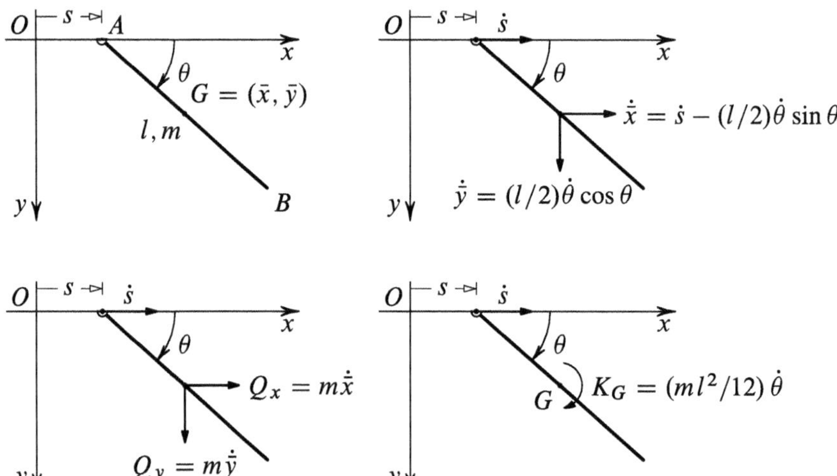


**Fig. 6.3** A rod with one end constrained to move along a horizontal guide

$$Q_x = m\dot{\bar{x}}, \quad Q_y = m\dot{\bar{y}}$$

where $(\bar{x}, \bar{y})$ are the coordinates of the center of mass $G$, placed at the rod's midpoint. Then

$$\bar{x} = s + (l/2)\cos\theta\,, \quad \bar{y} = (l/2)\sin\theta\,.$$

Differentiating in time, and recalling $s$ and $\theta$ are functions of time,

$$Q_x = m[\dot{s} - (l/2)\dot{\theta}\sin\theta]\,, \quad Q_y = m[(l/2)\dot{\theta}\cos\theta]\,.$$

Computing the angular momentum is more delicate and instructive. For the sake of an example we shall use $G$, $A$, $O$ as poles.

Theorem 6.4 allows us to identify $\mathbf{K}_G$ with the angular momentum measured in the relative motion of the center of mass. In other words, to find $\mathbf{K}_G$ it is possible and convenient to refer to the velocity distribution relative to the *frame translating* with the center of mass. In this case this frame sees the rod rotate around $G$, which appears to it fixed. So we can use $\mathbf{K}_G = \mathbf{I}_G\boldsymbol{\omega}$, expressing the angular momentum for a rotational velocity distribution around $G$. The angular velocity of the rod has modulus $\dot{\theta}$, while the inertia moment $I_G$ with respect to the axis of rotation (here a principal inertia axis) is $ml^2/12$. Hence

$$K_G = I_G\dot{\theta} = (ml^2/12)\dot{\theta}$$

where $K_G$ is the orthogonal component to the plane (the only non-zero one).

Now let us find the angular momentum with respect to the end $A$ as pole. It is tempting to use the formula $K_A = I_A\dot{\theta} = (ml^2/3)\dot{\theta}$, but this would be *wrong* because $A$ *is not* the center of mass, *nor* the instantaneous center of rotation. To compute $\mathbf{K}_A$ we must instead use (6.6), which here reads as follows: add to $\mathbf{K}_G$ the momentum with respect to $A$ of the vector $\mathbf{Q}$ applied at $G$.

For $AG \times \mathbf{Q}$ it is useful to think of the components $m\dot{\bar{x}}$ and $m\dot{\bar{y}}$ separately. The moment of the horizontal part, oriented clockwise and perpendicular to the plane, equals $-(l/2)(\sin\theta)m\dot{\bar{x}}$, while the vertical part gives $(l/2)(\cos\theta)m\dot{\bar{y}}$. All in all

$$\begin{aligned}[AG \times \mathbf{Q}]_z &= -(l/2)(\sin\theta)m\dot{\bar{x}} + (l/2)(\cos\theta)m\dot{\bar{y}} \\ &= -(l/2)(\sin\theta)m[\dot{s} - (l/2)\dot{\theta}\sin\theta] + (l/2)(\cos\theta)m[(l/2)\dot{\theta}\cos\theta] \\ &= -m(l/2)\dot{s}\sin\theta + m(l^2/4)\dot{\theta}\end{aligned}$$

To sum up, adding $AG \times \mathbf{Q}$ and $\mathbf{K}_G$ says that the angular momentum with respect to the pole $A$ has modulus

$$K_A = -m(l/2)\dot{s}\sin\theta + m(l^2/3)\dot{\theta}\,.$$

If we now want to find $\mathbf{K}_O$, we can use transformation rule (6.2) for changing pole in two ways. Either from the center of mass, i.e. $\mathbf{K}_O = OG \times \mathbf{Q} + \mathbf{K}_G$, or from $A$: $\mathbf{K}_O = OA \times \mathbf{Q} + \mathbf{K}_A$. In the second case we will first need to know that

$$[OA \times \mathbf{Q}]_z = ms\dot{\bar{y}} = ms(l/2)\dot{\theta}\cos\theta$$

to obtain

$$K_O = -m(l/2)\dot{s}\sin\theta + m(l^2/3)\dot{\theta} + ms(l/2)\dot{\theta}\cos\theta\,.$$

For the kinetic energy König's theorem (6.23) is useful:

$$T = \frac{1}{2}mv_G^2 + T_{\text{rel}}^{(G)}\,,$$

where $T_{\text{rel}}^{(G)}$ is the kinetic energy for the frame translating with the center of mass and center at this point. Here

$$\begin{aligned}\frac{1}{2}mv_G^2 &= \frac{1}{2}m[(\dot{\bar{x}})^2 + (\dot{\bar{y}})^2] = \frac{1}{2}m[(\dot{s} - (l/2)\dot{\theta}\sin\theta)^2 + ((l/2)\dot{\theta}\cos\theta)^2] \\ &= \frac{1}{2}m[\dot{s}^2 - l\dot{s}\dot{\theta}\sin\theta + l^2\dot{\theta}^2/4]\end{aligned}$$

and

$$T_{\text{rel}}^{(G)} = \frac{1}{2}(ml^2/12)\dot{\theta}^2\,,$$

where we could use $T_{\text{rel}}^{(G)} = \frac{1}{2}I_G\dot{\theta}^2$, with $I_G$ being the inertia moment for the $z$-axis through $G$ (instantaneous axis of rotation in the frame translating with the center of mass). It would be *wrong* to write $T = \frac{1}{2}mv_A^2 + \frac{1}{2}I_A\dot{\theta}^2$, because $A$ *is not* the center of mass. We should instead use (6.25).

We exploit this concrete example to make the general expressions of the kinetic energy of a holonomic system, namely (6.30) and (6.33), more concrete. Let us write the overall kinetic energy obtained from König's theorem by arranging terms as follows

$$T = \frac{1}{2}[m\dot{s}^2 + \frac{ml^2}{3}\dot{\theta}^2 - ml\sin\theta\dot{s}\dot{\theta}]\,. \tag{6.34}$$

This can be recast as

$$T = \frac{1}{2}\begin{bmatrix} m & -ml/2\sin\theta \\ -ml/2\sin\theta & ml^2/3 \end{bmatrix}\begin{bmatrix}\dot{s}\\ \dot{\theta}\end{bmatrix}\cdot\begin{bmatrix}\dot{s}\\ \dot{\theta}\end{bmatrix} \tag{6.35}$$

The mass matrix $\mathsf{A}$ in (6.33) is here defined by

$$\mathsf{A} = \begin{bmatrix} m & -ml/2\sin\theta \\ -ml/2\sin\theta & ml^2/3 \end{bmatrix}$$

and the elements $a_{ik}$ in the first sum of (6.30) are

$$a_{11} = m\,, \qquad a_{12} = a_{21} = -m(l/2)\sin\theta\,, \qquad a_{22} = ml^2/3\,.$$

Hence (6.34) and (6.35) are concrete and particular cases of (6.30) and (6.33) respectively, where now $(q_1, q_2) = (s, \theta)$. As the constraints are fixed, the terms $\mathsf{b}(\mathsf{q}, t)$ and $c(\mathsf{q}, t)$ vanish and for the same reason $\mathsf{A}$ here *does not* depend explicitly upon time. □

# Chapter 7
# Forces, Work, Power

In this chapter we will discuss the notion of force in an intuitive manner. We shall think of a force **F**, acting on the point $P$, as the vector that characterizes the interaction of $P$ with other bodies, and implicitly agree that forces have an absolute nature, i.e. independent of a frame. We will return to this in Chap. 8, when we discuss the concept starting from the Laws of Mechanics.

For the time being we will just present several types of vector-valued functions that can represent forces, with the help of concrete examples taken from elementary Physics.

**Constant forces**
There exist forces that are represented by a constant vector $\mathbf{F}_0$, meaning independent not only from the position and the velocity of the point on which it acts, but also on the instant considered. A standard example is the weight, which is a constant force, at least within a suitable approximation that we will explain in Sect. 13.6, Chap. 9.

**Positional forces**
Positional forces are those that depend only on the position of the point on which they act: $\mathbf{F} = \mathbf{F}(P)$. Elastic forces, the gravitational force and Coulomb forces belong in this category.

**Velocity-dependent forces**
Here **F** depends on the velocity distribution of the point of application: $\mathbf{F} = \mathbf{F}(\mathbf{v})$. Resistant forces are of this type. Let us consider what happens if we put our hand outside the window of a car in motion: the pressure (the modulus of the force per unit of area) exerted by the air increases with the car's speed, and the force's direction and orientation is opposite to the velocity. In the light of the absolute character of forces it may seem strange that some forces depend on the velocity (which, as described in Chap. 3, has relative nature). But here by **v** we actually mean the difference between the velocity of the point and that of the air that opposes the motion. The difference between the velocities of two points at the same position is absolute because of

 
P. Biscari et al., *Rational Mechanics*, UNITEXT 177,
https://doi.org/10.1007/978-3-032-07462-1_7

Galilei's theorem (see Sect. 3.2). In our example $\mathbf{v}$ should be thought of as the vehicle's velocity with respect to the frame comoving with the air.

**Time-dependent forces**
These are forces whose value varies in time, even when the points' velocity distribution stays the same: $\mathbf{F} = \mathbf{F}(t)$. A simple example is the water pressure on a point $P$ at the bottom of a full bathtub. If we pull the plug, the pressure decreases as time goes by since in this lapse the water flows out.

**Active forces**
The previous examples are clearly only special cases. In general, there exist forces that depend on all of the aforementioned variables. We will call *active forces* those with known dependency on the velocity distribution (position and velocity) of the application point, and known explicit dependency on time:

$$\mathbf{F} = \mathbf{F}(P, \mathbf{v}, t) .$$

A fundamental feature of any active force is that the dependency on the velocity distribution and time is known *a priori*, namely before we even know if there are other forces acting on the same point, and clearly before we find the motion they will cause.

We will explain in Chap. 8 why we choose to consider active forces that depend on the motion of the application point through current position and velocity at most, and not through time derivatives of order higher than one.

**Elementary work**
Consider a moving point $P$ subject to a force $\mathbf{F}$ in the time interval $[t, t + dt]$, and let $dP$ denote the corresponding elementary displacement. We define the elementary work to be the scalar

$$dL = \mathbf{F} \cdot dP . \tag{7.1}$$

Observe that the elementary work (7.1) is certainly zero when force and displacement are orthogonal, while positive (or negative) when the angle between them is smaller (or greater) than $\pi/2$.

In components, (7.1) reads

$$dL = F_x dx + F_y dy + F_z dz , \tag{7.2}$$

which means the work is a particular *differential form*, for which we have written a small reminder in Appendix A.6.

**Theorem 7.1** *If the elementary work of a force is an exact differential form, the force is positional.*

***Proof*** The proof is immediate because (7.2) is of form (A.44) when we identify:

$$\begin{aligned} &x_1 = x, \quad x_2 = y, \quad x_3 = z; && x_4 = \dot{x}, \quad x_5 = \dot{y}, \quad x_6 = \dot{z}, \quad x_7 = t; \\ &\Psi_1 = F_x, \quad \Psi_2 = F_y, \quad \Psi_3 = F_z; && \Psi_4 = \Psi_5 = \Psi_6 = \Psi_7 = 0. \end{aligned} \tag{7.3}$$

Taking $i = 1$ and $j = 4, 5, 6, 7$, from (A.46) we deduce

$$\frac{\partial F_x}{\partial \dot{x}} = \frac{\partial F_x}{\partial \dot{y}} = \frac{\partial F_x}{\partial \dot{z}} = \frac{\partial F_x}{\partial t} = 0$$

so $F_x$ cannot depend neither on velocity nor on time. Similarly, choosing $i = 2$ or $i = 3$, and $j = 4, 5, 6, 7$, we see that for the compatibility condition to hold the other components of the force **F** must be positional. As soon as **F** depends on the velocity or time, the elementary work cannot be an exact differential. □

**Work along a closed path**

Let us now examine the work of a force in a finite time interval $[t_1, t_2]$ (work along a finite path), i.e. the integral of the elementary work along the curves traced out by the point's orbit.

In case the elementary work is an exact form, property (A.47) guarantees the finite work will only depend on the initial and final points of the trajectory. If on the contrary the work is not an exact differential, only knowing the force and the endpoints will no longer be enough to determine the finite work over the time interval $[t_1, t_2]$. More precisely, we will prove that for positional forces the work will depend also on the particular trajectory $P \equiv P(s)$ joining the initial and end points. In the general case of a force $\mathbf{F}(P, \mathbf{v}, t)$ the work will depend on the arclength time law $s \equiv s(t)$ as well, besides the trajectory $P \equiv P(s)$ and hence the entire motion $P(t)$ (trajectory and time law).

**Work and power**

Suppose we know both the active force $\mathbf{F}(P, \mathbf{v}, t)$ and the corresponding motion $P \equiv P(t)$. The knowledge of the equation of motion allows to reduce the elementary work's differential form, that involves seven independent variables, to a form in one variable only. From (7.1) we have in fact

$$dL = \mathbf{F}\left(P(t), \dot{P}(t), t\right) \cdot \dot{P}(t)\, dt = \Pi(t) dt \tag{7.4}$$

where

$$\Pi = \mathbf{F} \cdot \mathbf{v} \tag{7.5}$$

is called *power*. Integrating (7.4) gives the work along the finite path

$$L = \int_{t_1}^{t_2} \Pi(t) dt\,. \tag{7.6}$$

Hence if between the same endpoints $P_1 = P(t_1)$ and $P_2 = P(t_2)$ the motion changes (either the trajectory, or the time law changes), then the composite map $\Pi(t)$ will change, and therefore the work will be different in general.

It is important to notice that along closed paths (when $P_1 \equiv P_2$) the work is, in general, non-zero. In particular, for resistant-like forces $\mathbf{F} = -\Psi(P, v)\,\mathbf{v}$ with $\Psi(P, v) > 0$, we have $\Pi(t) = -\Psi v^2 \leq 0$. The corresponding work is then strictly negative along any non-trivial closed path (that is, any closed path other than the constant path $P(t) \equiv P_1$).

**Positional forces**
Consider now the case of a positional force $\mathbf{F} = \mathbf{F}(P)$. For the elementary work to be an exact differential, the compatibility conditions (A.46) must be valid. Taking $h, k = 1, 2, 3$ and using (7.3), the following must then hold

$$\frac{\partial F_x}{\partial y} = \frac{\partial F_y}{\partial x}, \qquad \frac{\partial F_x}{\partial z} = \frac{\partial F_z}{\partial x}, \qquad \frac{\partial F_y}{\partial z} = \frac{\partial F_z}{\partial y}, \tag{7.7}$$

or equivalently rot $\mathbf{F} = \mathbf{0}$. When just one in (7.7) is false, the force, even if positional, returns an elementary work that is not exact. Yet it can be shown that to compute the work of a positional force it is enough to have only the trajectory's equation, without knowing the time law. In fact, supposing we are given the trajectory $P = P(s)$, parametrized by arclength $s$, we have

$$dL = \mathbf{F}(P(s)) \cdot \frac{dP(s)}{ds} ds = \mathbf{F}(P(s)) \cdot \mathbf{t}(s) ds = F_t(s) ds\,, \tag{7.8}$$

where we used the fact that $dP/ds$ is the unit tangent vector $\mathbf{t}(s)$ to the trajectory (see Sect. A.2) and we called $F_t = \mathbf{F} \cdot \mathbf{t}$ the tangential component of the force. From (7.8) we immediately obtain

$$L = \int_{s_1}^{s_2} F_t(s) ds\,, \tag{7.9}$$

where $s_1$ and $s_2$ are the arclengths of $P_1$ and $P_2$, i.e. $P_1 = P(s_1)$ and $P_2 = P(s_2)$. From (7.9) we deduce that the work along a finite path depends on the trajectory between $P_1$ and $P_2$, so if we change trajectory the work, in general, will change. On the other hand the work does not change with the time law $s = s(t)$ describing a given trajectory. In this case as well the work along a closed path is, in general, non-zero.

## 7.1 Conservative Forces

Suppose now the elementary work of a positional force is the differential of a function $U(P)$, called *potential*: $dL = dU$, with

$$F_x = \frac{\partial U}{\partial x}, \qquad F_y = \frac{\partial U}{\partial y}, \qquad F_z = \frac{\partial U}{\partial z},$$

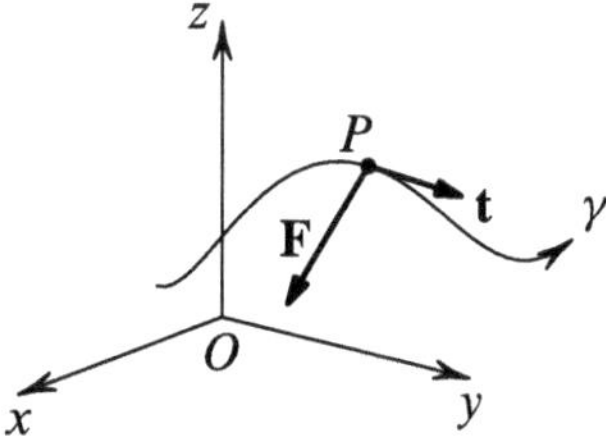


**Fig. 7.1** Directional derivative of the potential

which amounts to saying the force field **F** is the gradient of the potential $U$:

$$\mathbf{F} = \operatorname{grad} U \, . \tag{7.10}$$

Then

$$L = \int_{U_1}^{U_2} dU = U_2 - U_1 \, , \tag{7.11}$$

where $U_1 = U(P_1)$ and $U_2 = U(P_2)$. If so, the force is called *conservative*, and the elementary work is therefore the differential of the potential function $U(P)$.

This is the only case in which the work does not depend on any element of motion but solely on the initial and final positions. In particular, from (7.11) it is straightforward that the work of a conservative force along a closed path vanishes. It is also easy to see the converse: if the work of a positional force is zero along every closed path then the force is conservative.

A sufficient, but not necessary, condition for a positional force to be conservative is that the domain be simply connected, and that (7.7) hold.

**Directional derivative of the potential**

Let $\mathbf{F}(P)$ be a conservative positional force with potential $U(P)$. We want to compute the directional derivative of $U$ along a curve $\gamma$ parametrized by arclength: $P = P(s)$ (see Fig. 7.1). We have:

$$\begin{aligned} \frac{dU}{ds} &= \frac{\partial U}{\partial x}\frac{dx}{ds} + \frac{\partial U}{\partial y}\frac{dy}{ds} + \frac{\partial U}{\partial z}\frac{dz}{ds} \\ &= F_x\frac{dx}{ds} + F_y\frac{dy}{ds} + F_z\frac{dz}{ds} = \mathbf{F}\cdot\mathbf{t} = F_t\,(P(s)) \end{aligned}$$

Hence the potential's directional derivative coincides with the tangential component of the force:

$$\frac{dU}{ds} = F_t \, . \tag{7.12}$$

**Potential**

Several of the forces predicted by the models of Mechanics are conservative. In such cases it is important to find the potential function.

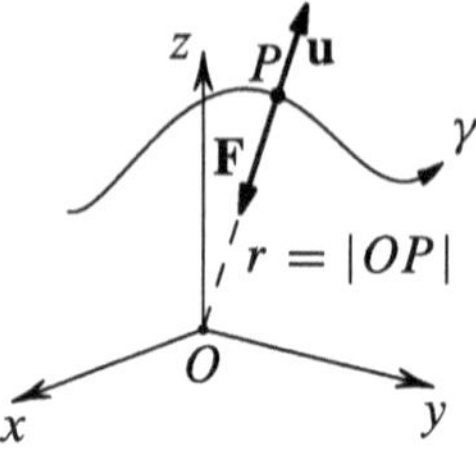


**Fig. 7.2** A central force

**Constant forces**
If $\mathbf{F} \equiv \mathbf{F}_0$ immediately $dL = \mathbf{F}_0 \cdot dP = d\,(\mathbf{F}_0 \cdot OP) = dU$, as long as we introduce the potential

$$U(P) = \mathbf{F}_0 \cdot OP + \text{const.}$$

An example of force that is roughly constant is the weight. Choosing the $y$-axis pointing downward (same orientation as the weight vector) we have

$$U(y) = mgy + \text{const.}$$

Had we chosen $y$ pointing upwards, we would have had $U(y) = -mgy + \text{const.}$ The equipotential surfaces ($U$ = const) are horizontal planes.

**Central forces with modulus depending on the distance**
Consider a central force, namely a force always directed along the line joining $P$ with a fixed point $O$ called *center* (see Fig. 7.2), whose modulus only depends on $r = |OP|$:

$$\mathbf{F} = \Psi(r)\mathbf{u}, \qquad \text{con} \quad \mathbf{u} = \frac{OP}{r} \quad \text{with} \quad r = |OP|\,. \tag{7.13}$$

Central forces are conservative, as we can prove directly by constructing their potential explicitly. Suppose in fact $\mathbf{F}$ is conservative, and call $U$ the potential. Consider a particular curve, a straight line through $O$ and $P$ (oriented as $\mathbf{u}$). Since along such curve $s = r$ and $\mathbf{t} = \mathbf{u}$, from (7.12) we obtain

$$\frac{dU}{dr} = (\Psi(r)\mathbf{u}) \cdot \mathbf{u} = \Psi(r)\,,$$

and therefore the potential $U(r)$ is a primitive of the function $\Psi$:

$$U(r) = \int^r \Psi(\rho)d\rho + \text{constant.} \tag{7.14}$$

Differentiating explicitly confirms that (7.10) holds. In other words the gradient of the potential just found coincides with (7.13).

An important example of central force is given by *springs*, which are particular active forces that yield linear elastic forces

$$\mathbf{F} = -k\,OP\,, \tag{7.15}$$

where $k$ is a positive constant (*elastic constant*). We deduce $\Psi(r) = -kr$ and so one potential function is

$$U(r) = -\frac{1}{2}kr^2\,.$$

A second important example of central force is the *gravitational force*, which we can write as:

$$\mathbf{F} = -\frac{hmM}{r^2}\mathbf{u}\,. \tag{7.16}$$

Here $\Psi(r) = -\dfrac{hmM}{r^2}$ and therefore we can choose

$$U(r) = \frac{hmM}{r}\,.$$

Also the Coulomb force, whose form is similar to (7.16), is central. For these cases the equipotential surfaces are concentric spheres with center $O$.

**Potential energy**
The potential with the opposite sign has the physical meaning of energy and goes by the name of *potential energy*, $V$:

$$V = -U\,.$$

## 7.2 Systems of Forces

We know that a force is described by an *applied vector*, that is, a pair $(P, \mathbf{f})$ made of the point of application $P$, and the vector $\mathbf{f}$ that describes the force by prescribing its direction, modulus and orientation. It is not always necessary to be so formal, though, and in general it is more convenient to speak of "a force $\mathbf{f}$ applied at the point $P$".

**Definition 7.2** (*Line of action*) Given a force $\mathbf{f}$ applied at $P$, we call *line of action* the straight line through $P$ parallel to $\mathbf{f}$.

**Definition 7.3** (*Moment*) We call *moment* or *torque* of a force $\mathbf{f}$ *with respect to the pole O* the vector

$$\mathbf{M}_O = OP \times \mathbf{f}\,.$$

The moment of a force *depends* on its pole. In fact, if we pass from pole $O$ to pole $Q$ the moment changes by the rule

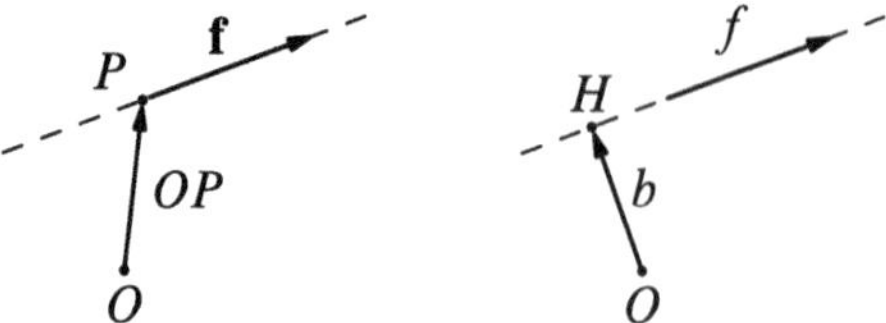


**Fig. 7.3** The moment of a force: $\mathbf{M}_O = OP \times \mathbf{f} = \pm bf\mathbf{k}$, where $\mathbf{k}$ is a unit vector perpendicular to the plane which contains both the point $O$ and the force $\mathbf{f}$

$$\mathbf{M}_Q = QP \times \mathbf{f} = \big(QO + OP\big) \times \mathbf{f} = QO \times \mathbf{f} + OP \times \mathbf{f} = QO \times \mathbf{f} + \mathbf{M}_O\,.$$

Note that the moment *does not* change if we simply translate the force along its line of action, since $\mathbf{M}_Q = \mathbf{M}_O$ if $QO$ is parallel to $\mathbf{f}$. In particular, the moment of a force is zero with respect to the application point, but also with respect to any pole on the line of action.

Let $H$ denote the orthogonal projection of $O$ onto the force's line of action and decompose the vector $OP$ as sum $OH + HP$. Now $HP$ is parallel to $\mathbf{f}$ while $OH$ is the component of $OP$ perpendicular to the force. Thus

$$\mathbf{M}_O = OP \times \mathbf{f} = (OH + HP) \times \mathbf{f} = OH \times \mathbf{f} + \underbrace{HP \times \mathbf{f}}_{\mathbf{0}} = OH \times \mathbf{f}$$

and, for the sole purpose of computing moments we may think the force as applied at $H$, foot of the perpendicular line through $O$ to the line of action.

It is convenient to call *arm* of a force with respect to a pole $O$ the distance $b$ of $O$ to the *line* of action. Indicating by $\mathbf{u}$ the unit vector of $OH$ we may set $OH = b\mathbf{u}$. Now pick a unit vector $\mathbf{k}$ perpendicular to the plane of $\mathbf{f}$ and $O$, let $f$ be the force's modulus and $\mathbf{d}$ its unit vector. Then

$$\mathbf{M}_O = OH \times \mathbf{f} = b\mathbf{u} \times f\mathbf{d} = bf\underbrace{\mathbf{u} \times \mathbf{d}}_{\pm\mathbf{k}} = \pm bf\mathbf{k}\,.$$

In other words, the moment $\mathbf{M}_O$ of a force $\mathbf{f}$ is perpendicular to the plane containing $\mathbf{f}$ and $O$, and has as $\mathbf{k}$-component the product of the arm $b$ times the modulus $f$, with sign determined by whether the cross product $\mathbf{u} \times \mathbf{d}$ agrees or not with $\mathbf{k}$. The construction is shown in Fig. 7.3.

**Definition 7.4** (*Axial moment*) We call *axial moment* or *axial torque* of a force $\mathbf{f}$ *with respect to an axis* $a$, with unit vector $\mathbf{u}$, the quantity

$$M_a = \mathbf{M}_O \cdot \mathbf{u} = OP \times \mathbf{f} \cdot \mathbf{u}$$

where $O$ is *any* point on the axis.

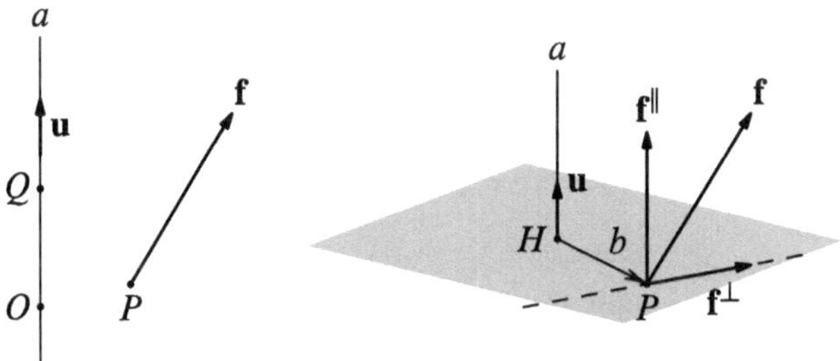


**Fig. 7.4** The axial moment: $M_a = \mathbf{M}_O \cdot \mathbf{u} = \mathbf{M}_Q \cdot \mathbf{u}$. Only the part of the force perpendicular to the axis gives a contribution to $M_a$

The concept is well defined since the axial moment $M_a$ does not depend on the pole $O$ chosen on the axis. In fact if $Q$ is another point on the axis

$$\mathbf{M}_Q \cdot \mathbf{u} = QP \times \mathbf{f} \cdot \mathbf{u} = (QO + OP) \times \mathbf{f} \cdot \mathbf{u} = OP \times \mathbf{f} \cdot \mathbf{u} = M_a$$

by the triple product's properties and the fact $QO$ and $\mathbf{u}$ are parallel.

Let us decompose the force $\mathbf{f}$ as sum of $\mathbf{f}^{\parallel}$, parallel to the axis, and a perpendicular component $\mathbf{f}^{\perp}$, as in Fig. 7.4 right. Immediately, only the part perpendicular to the axis is involved in the axial moment. In fact, let us choose for convenience as pole the point $H$ on the axis coinciding with the orthogonal projection of the force's application point,

$$M_a = HP \times \mathbf{f} \cdot \mathbf{u} = HP \times (\mathbf{f}^{\parallel} + \mathbf{f}^{\perp}) \cdot \mathbf{u} = HP \times \mathbf{f}^{\perp} \cdot \mathbf{u}$$

since $HP \times \mathbf{f}^{\parallel} \cdot \mathbf{u} = 0$, due to the parallelism of $\mathbf{f}^{\parallel}$ and $\mathbf{u}$, unit vector of $a$.

The axial moment in practice corresponds to the $\mathbf{u}$-component of the moment with respect to $H$ of the part $\mathbf{f}^{\perp}$ on the plane through $H$ and perpendicular to $\mathbf{u}$.

More generally, computing the vector $\mathbf{M}_O$ is the same as computing its three components with respect to axes $x$, $y$ and $z$ through $O$. By definition of component

$$M_x = \mathbf{M}_O \cdot \mathbf{i}, \qquad M_y = \mathbf{M}_O \cdot \mathbf{j}, \qquad M_z = \mathbf{M}_O \cdot \mathbf{k}\,,$$

so in the end finding $\mathbf{M}_O$ is equivalent to finding three axial moment.

### *7.2.1 Resultant Force and Resultant Moment*

Let now $\mathcal{S}$ indicate a generic system of forces $\mathbf{f}_i$ $(i = 1, \ldots, n)$ applied at points $P_i$, for which we define the *resultant (force)* and *resultant moment.*

**Definition 7.5** (*Resultant, resultant moment*) Given a system of forces $\mathcal{S}$ we define the *resultant* $\mathbf{R}$ and *resultant moment* $\mathbf{M}_O$ *with respect to to pole* $O$ by:

$$\mathbf{R} = \sum_{i=1}^{n} \mathbf{f}_i \quad \text{and} \quad \mathbf{M}_O = \sum_{i=1}^{n} OP_i \times \mathbf{f}_i \,. \tag{7.17}$$

($\mathbf{R}$ and $\mathbf{M}_O$ will occasionally be referred to as the *characteristic vectors* of the system of forces.)

For brevity one often simply speaks of the "moment" of the system of forces. The moment depends on the pole with respect to which it is computed, and the difference between moments with different poles only depends on the resultant.

**Proposition 7.6** (Change of pole) *If* $\mathbf{M}_O$ *and* $\mathbf{M}_Q$ *are the moments of a system* $\mathcal{S}$ *of forces with respect to the poles* $O$ *and* $Q$*, then*

$$\mathbf{M}_Q = \mathbf{M}_O + QO \times \mathbf{R} \,. \tag{7.18}$$

In fact:

$$\begin{aligned} \mathbf{M}_Q &= \sum_{i=1}^{n} QP_i \times \mathbf{f}_i = \sum_{i=1}^{n} \left(QO + OP_i\right) \times \mathbf{f}_i \\ &= \sum_{i=1}^{n} QO \times \mathbf{f}_i + \mathbf{M}_O = QO \times \mathbf{R} + \mathbf{M}_O \end{aligned} \tag{7.19}$$

A straightforward consequence of the above property is that in a system of forces with zero resultant the moment does not depend on the pole:

$$\mathbf{R} = \mathbf{0} \quad \Rightarrow \quad \mathbf{M}_Q = \mathbf{M}_O \tag{7.20}$$

for any choice of points $O$ and $Q$. Naturally the converse holds too: if $\mathbf{M}_O = \mathbf{M}_Q$ then (7.18) implies $QO \times \mathbf{R} = \mathbf{0}$. For this to hold for any $O$ and $Q$, necessarily $\mathbf{R} = \mathbf{0}$.

From rule (7.18) it further follows that the moment is constant along each line parallel to $\mathbf{R}$, since for any pair of points $A$ and $B$ such that $AB \parallel \mathbf{R}$ we have $\mathbf{M}_A = \mathbf{M}_B$. This property is highlighted in Fig. 7.5 left, where we have lines $r_1, r_2$ parallel to $\mathbf{R}$, moments computed with respect to poles $A, B$ on $r_1$ and poles $C, D$ on $r_2$. The moment changes from one line to the other, but stays constant along each line.

### 7.2.2 *Scalar Invariant, Central Axis and Line of Action of the Resultant*

**Definition 7.7** (*Scalar invariant*) The *scalar invariant* of a system of forces is the dot product of resultant and moment:

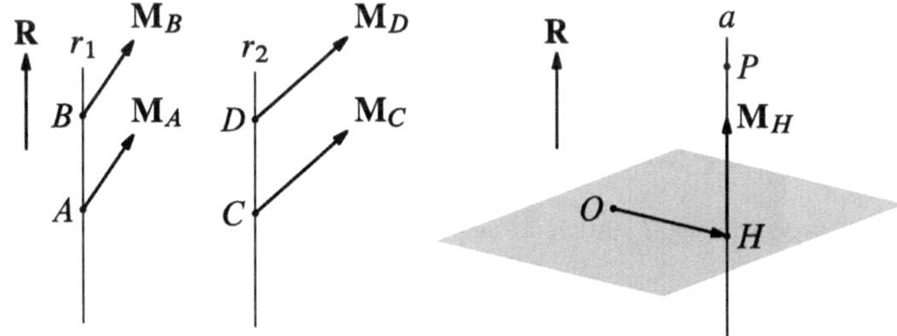


**Fig. 7.5** The moment of a system of forces is constant when computed with respect to points belonging on the same line parallel to **R** (left). When the moment is computed with respect to points on the central axis $a$ it is parallel to **R** (right)

$$I = \mathbf{R} \cdot \mathbf{M}_O. \tag{7.21}$$

It *does not* depend on the moment's pole. In fact, recalling (7.18),

$$\mathbf{R} \cdot \mathbf{M}_Q = \mathbf{R} \cdot (\mathbf{M}_O + QO \times \mathbf{R}) = \mathbf{R} \cdot \mathbf{M}_O + \underbrace{\mathbf{R} \cdot QO \times \mathbf{R}}_{=0} = \mathbf{R} \cdot \mathbf{M}_O\,.$$

This property of the scalar invariant expresses the fact that as the pole varies, *the component of the moment parallel to the resultant does not change, and only the orthogonal part changes.*

When the resultant of the system of forces is zero, as we already know, the moment is independent of the pole. In case $\mathbf{R} \neq \mathbf{0}$ instead, i.e. $\mathbf{M}_O$ depends on $O$, we can prove that there exists a line, called *central axis* of the system of forces, made of the points $P$ for which the moment is *parallel* to **R**.

**Proposition 7.8** (Central axis) *For a system of forces with non-zero resultant* **R** *there exists a line, called* central axis, *with respect to whose points the system's moment is* parallel *to the resultant and has smallest modulus among the moments of any other point in space. In the special case* $I = 0$ *this line consists of the poles whose moment* vanishes, *and is called* line of action of the resultant.

***Proof*** The moment $\mathbf{M}_O$ with respect to an arbitrary point $O$ in space will usually have a component parallel to **R** and a perpendicular one.

As the moment is constant along every line parallel to **R**, as Fig. 7.5, left, clearly shows, to find the central axis it will suffice to find its intersection $H$ with a plane perpendicular to **R** and passing through $O$. The vector $OH$ we seek should then detect a point $H$ such that $OH \cdot \mathbf{R} = 0$ and

$$\mathbf{M}_H \times \mathbf{R} = \mathbf{0}$$

(parallelism of $\mathbf{M}_H$ and **R**). From (7.18) therefore

$$(\mathbf{M}_O + \mathbf{R} \times OH) \times \mathbf{R} = \mathbf{0}$$

and by the vector identity (A.9), the above is equivalent to

$$\mathbf{M}_O \times \mathbf{R} + R^2 OH - \underbrace{(OH \cdot \mathbf{R})}_{=0}\mathbf{R} = \mathbf{0}$$

where we observed that $OH$ and $\mathbf{R}$ are orthogonal. Hence

$$OH = \frac{\mathbf{R} \times \mathbf{M}_O}{R^2} .$$

The vector $OP$ from $O$ to a generic point $P$ on the line through $H$ and parallel to $\mathbf{R}$ (called $a$ in Fig. 7.5 right) decomposes as $OP = OH + HP$, where $HP = \lambda\mathbf{R}$ for some parameter $\lambda$. Therefore

$$OP(\lambda) = \frac{\mathbf{R} \times \mathbf{M}_O}{R^2} + \lambda\mathbf{R} . \tag{7.22}$$

Line (7.22), called *central axis* of the system of forces, is then made of the poles whose moments are constant and is *parallel* to the resultant.

For any other point $Q$ in space

$$\mathbf{M}_Q = \mathbf{M}_P + QP \times \mathbf{R}$$

because $P$ lies on the central axis so $\mathbf{M}_P$ is parallel to $\mathbf{R}$, while the second term is perpendicular to it (by definition of cross product). Pythagoras' theorem applied to orthogonal vectors gives

$$|\mathbf{M}_Q|^2 = |\mathbf{M}_P|^2 + |QP \times \mathbf{R}|^2 \geq |\mathbf{M}_P|^2$$

and the modulus of $\mathbf{M}_Q$ is larger than the modulus of $\mathbf{M}_P$ (with equality only when $Q$ belongs on the central axis as well).

Finally, in the special case $I = 0$, the central axis has a further property. In fact, as $I = \mathbf{R} \cdot \mathbf{M}_O$ is independent of the pole $O$, its vanishing implies the moment is everywhere *orthogonal* to $\mathbf{R}$. But since we know that along the central axis (7.22) the moment is actually *parallel* to $\mathbf{R}$, we are obliged to conclude that for $I = 0$ the central axis consists of points for which *the system's moment is zero*. Line (7.22) is in this case called *line of action of the resultant*. □

## 7.3 Special Systems of Forces

**Definition 7.9** (*Balanced system*) A system of forces is said to be *balanced* if it has zero resultant and moment: $\mathbf{R} = \mathbf{0}, \ \ \mathbf{M}_O = \mathbf{0}$.

The simplest balanced system is clearly the one with no vectors. Moreover, equation (7.19) proves that if a system is balanced when we compute the moment with

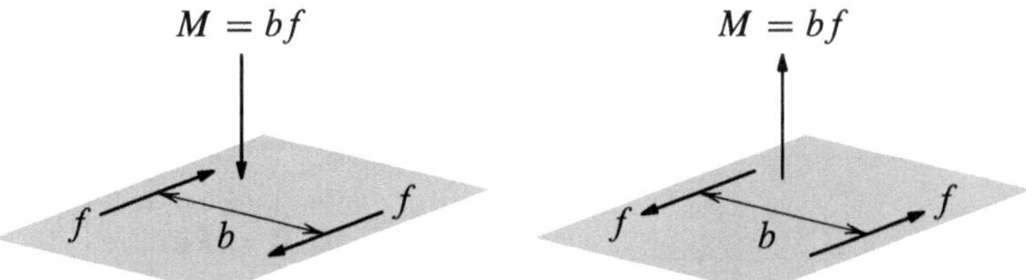


**Fig. 7.6** Two couples with the same arm $b$ and forces of equal intensity $f$. The couple's moment is perpendicular to the plane and its component $M = bf$ with respect to a given unit vector is determined by the orientation of the forces

respect to a pole $O$, it stays balanced if we compute the moment with respect to any point $Q$.

Among systems of forces, those made of two equal and oppositely oriented forces represent the simplest case where the resultant is zero ($\mathbf{R} = \mathbf{0}$) but, in general, the moment is not ($\mathbf{M} \neq 0$).

**Definition 7.10** (*Couple*) A couple is a system of *two* equal and oppositely oriented forces, hence parallel, which do not have the same line of action in general.

By (7.20) the moment of a couple does not depend on the pole. For this reason one speaks of *moment*, or *torque*, of a couple, without specifying or indicating the pole.

The two forces in a couple, if not aligned, still lie on a plane called the *plane of the couple*. Easily, the moment $\mathbf{M}$ is perpendicular to this plane and has as modulus the product of the modulus $f$ of one of the forces times the distance $b$ between the lines of action, called *arm of the couple* (Fig. 7.6).

It is important to notice that for any given value $\mathbf{M}$ of the moment, one can build infinitely many couples $(P_1, \mathbf{f})$ and $(P_2, -\mathbf{f})$ whose moment is $\mathbf{M}$. Let $\mathbf{k}$ and $M$ be the unit vector and modulus of $\mathbf{M}$, so that $\mathbf{M} = M\mathbf{k}$, and choose in any plane orthogonal to $\mathbf{k}$ a point $P_1$ and two orthogonal unit vectors $\mathbf{i}$ and $\mathbf{j}$ such that $\mathbf{k} = \mathbf{i} \times \mathbf{j}$ (this can be done in infinitely many ways). Apply a force $\mathbf{f} = -f\mathbf{j}$ at $P_1$ and a parallel and opposite force $\mathbf{f} = f\mathbf{j}$ at $P_2$ so that $P_1P_2 = b\mathbf{i}$. The moment of the couple, with respect to the pole $P_1$, equals $P_1P_2 \times (\mathbf{f}) = b\mathbf{i} \times (f\mathbf{j}) = bf\mathbf{i} \times \mathbf{j} = bf\mathbf{k}$. This equals $\mathbf{M}$ if $b$ and $f$ are chosen so that $bf = M$ (this, too, is possible in infinitely many ways).

**Definition 7.11** (*Planar system*) A system of forces is *planar* when the forces and the application points belong to a common plane.

Evidently the resultant $\mathbf{R}$ lies on the plane, which we indicate with $\pi$. Note that the moment of *any* force with respect to a pole $O$ on $\pi$ is perpendicular to the plane itself:

$$(OP_i \times \mathbf{f}_i) \perp \pi$$

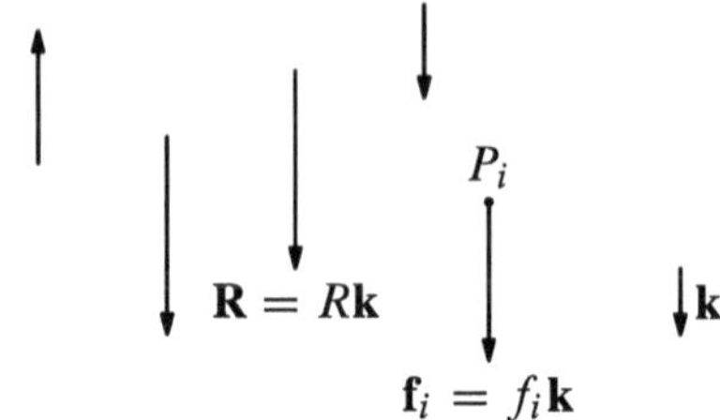


**Fig. 7.7** A system of parallel forces and its resultant $\mathbf{R} = R\mathbf{k}$, where $R = \sum_i f_i$

(the cross product is perpendicular to either factor), and then also the total moment $\mathbf{M}_O$ is perpendicular to $\pi$. All in all, *the scalar invariant* $I = \mathbf{R} \cdot \mathbf{M}_O$ *of a planar system always vanishes.*

**Definition 7.12** (*System of concurrent forces*) The forces of a system are *concurrent* at a point $Q$ if their lines of action all meet at $Q$.

If all forces concur at a given $Q$, every single vector $\mathbf{f}_i$ will have zero momentum with respect to $Q$, so the resultant moment with respect to the point will be zero too. Using that point to compute the scalar invariant,

$$I = \mathbf{R} \cdot \mathbf{M}_Q = 0\,.$$

**Definition 7.13** (*System of parallel forces*) The forces of a system are *parallel* when the vectors $\mathbf{f}_i$ are parallel to one unit vector $\mathbf{k}$, and all $\mathbf{f}_i$ can therefore be written as $\mathbf{f}_i = f_i\mathbf{k}$, where $f_i$ is the component of each along the unit vector $\mathbf{k}$ (the forces need not have the same orientation, as shown in Fig. 7.7).

We show that in this case, too, the system's scalar invariant is zero. The resultant is evidently parallel to $\mathbf{k}$

$$\mathbf{R} = \sum_{i=1}^{n} \mathbf{f}_i = \left( \sum_{i=1}^{n} f_i \right) \mathbf{k} = R\mathbf{k} \qquad (R = \sum_{i=1}^{n} f_i)\,,$$

while for the resultant moment

$$\mathbf{M}_O = \sum_{i=1}^{n} OP_i \times \mathbf{f}_i = \sum_{i=1}^{n} OP_i \times f_i\mathbf{k} = \left( \sum_{i=1}^{n} f_i\, OP_i \right) \times \mathbf{k}\,. \tag{7.23}$$

Hence in this case as well the scalar invariant vanishes

$$I = \mathbf{R} \cdot \mathbf{M}_O = R\,\mathbf{k} \cdot \left( \sum_{i=1}^{n} f_i OP_i \right) \times \mathbf{k} = 0\,.$$

## 7.4 Equivalent Systems and Reduction of a System of Forces

Now we introduce an equivalence relation on systems of forces. Consider two systems $\mathcal{S}, \mathcal{S}'$ described by

$$\mathcal{S} = \{P_i, \mathbf{f}_i\} \quad (i = 1, \dots, n), \qquad \mathcal{S}' = \{P'_j, \mathbf{f}'_j\} \quad (j = 1, \dots, m).$$

In general, $\mathcal{S}$ and $\mathcal{S}'$ consists of a distinct number of forces ($n$ and $m$) $\mathbf{f}_i$ and $\mathbf{f}'_j$ applied at different points in space ($P_i$ and $P'_j$ respectively).

**Definition 7.14** (*Equivalent systems*) Two systems of forces $\mathcal{S}$ and $\mathcal{S}'$ are *equivalent* if they have the same characteristic vectors:

$$\mathbf{R} = \mathbf{R}' \quad \text{and} \quad \mathbf{M}_O = \mathbf{M}'_O. \tag{7.24}$$

Systems that are equivalent with respect to a pole $O$ are equivalent with respect to any other pole $Q$. Suppose in fact that (7.24) hold. Then:

$$\mathbf{M}_Q = \mathbf{M}_O + QO \times \mathbf{R} = \mathbf{M}'_O + QO \times \mathbf{R}' = \mathbf{M}'_Q.$$

Given a system $\mathcal{S}$ the ensuing *elementary operations* always produce an equivalent system.

- Translating a force along its line of action. Replacing a force $\mathbf{f}$ applied at $P$ with the same force applied at $P + \lambda\mathbf{f}$ obviously does not modify the resultant, nor the resultant moment, since the two forces have the same moment with respect to any pole:
$$OP \times \mathbf{f} = \big(OP + \lambda\mathbf{f}\big) \times \mathbf{f}.$$
- Replacing forces applied at one point $P$ with their resultant, applied at the same point. This operation leaves the sum of the forces invariant, but also the resultant moment. In fact, call $\mathbf{f}_i$ the forces at $P$ to be replaced by $\sum_i \mathbf{f}_i$. Then:
$$\sum_{i=1}^{m} OP \times \mathbf{f}_i = OP \times \left(\sum_{i=1}^{m} \mathbf{f}_i\right).$$

Combining the previous elementary operations one can prove that the following do not alter resultant and moment of a system of forces either.

- Including or suppressing a couple with zero moment (two equal and opposite forces with the same line of action).
- Replacing concurrent forces by the resultant, applied at the intersection $P$ of the lines of action.

Now we establish, for any system of forces $\mathcal{S}$, what is the simplest system $\mathcal{S}'$ equivalent to $\mathcal{S}$. Suppose we have a system of forces and we want to construct an equivalent system $\mathcal{S}'$, i.e. one with same resultant and moment, but consisting of the minimum number of forces.

**Theorem 7.15** *Let* $\mathbf{R}$ *and* $\mathbf{M}_O$ *be the resultant and moment of system* $\mathcal{S}$, *and* $I = \mathbf{R} \cdot \mathbf{M}_O$ *the scalar invariant. Then:*

*(i)* $\mathbf{R} = \mathbf{0}$, $\mathbf{M}_O = \mathbf{0}$ *(balanced system)* $\Longrightarrow \mathcal{S}$ *is equivalent to the system with no forces.*
*(ii)* $\mathbf{R} = \mathbf{0}$, $\mathbf{M}_O \neq \mathbf{0} \Longrightarrow \mathcal{S}$ *is equivalent to a couple.*
*(iii)* $\mathbf{R} \neq \mathbf{0}$, $I = 0 \Longrightarrow \mathcal{S}$ *is equivalent to a system* $\mathcal{S}'$ *with only one force, equal to* $\mathbf{R}$, *applied at a point on the* resultant's line of action, *described in* (7.22), *of equation*

$$OP(\lambda) = \frac{\mathbf{R} \times \mathbf{M}_O}{R^2} + \lambda \mathbf{R} . \tag{7.25}$$

*(iv)* $\mathbf{R} \neq \mathbf{0}$, $I \neq 0 \Longrightarrow \mathcal{S}$ *is equivalent to a system* $\mathcal{S}'$ *made of one force, equal to* $\mathbf{R}$ *and applied at any point, plus a suitable couple. Furthermore, if we apply* $\mathbf{R}$ *at a point on line* (7.25), *here called* central axis, *the couple needed to have equivalence will have torque parallel to* $\mathbf{R}$ *and smallest possible modulus.*

***Proof*** • Property *(i)* is evident.
• To prove *(ii)* we must find a couple with given moment $\mathbf{M}_O$. There actually exist infinitely such many couples, as we have seen when we introduced the notion of couple.
• Property *(iii)* characterizes systems of forces equivalent to a system made of one force only. When $\mathbf{R} \neq \mathbf{0}$ and $I = 0$ we know there is a line, called line of action of the resultant, with respect to whose points the system's overall moment is zero. Consider then a system $\mathcal{S}' = \{P, \mathbf{R}\}$ consisting on *only* one force $\mathbf{R}$, equal to the resultant of $\mathcal{S}$, applied at a point $P$ of the resultant's line of action. The equivalence of $\mathcal{S}$ and $\mathcal{S}'$ is straightforward.

In fact the resultants of $\mathcal{S}$ and $\mathcal{S}'$ are clearly equal since $\mathcal{S}'$ only has one force $\mathbf{R}$. Moreover, the moment of $\mathcal{S}$ is null with respect to a pole on the resultant's line of action, and so is the moment of $\mathcal{S}'$ made by $\mathbf{R}$ also applied to the same line. Hence also the moments are equal.
• In the general situation ($\mathbf{R} \neq 0, I \neq 0$), as the scalar invariant is non-zero there is *no* pole for which the forces' moment vanishes, and the system of forces $\mathcal{S}$ is equivalent to a system $\mathcal{S}'$ made of the sole resultant $\mathbf{R}$ applied at any point $Q$, together with a couple $(P_1, \mathbf{f})$, $(P_2, -\mathbf{f})$ with moment $\mathbf{M}_Q$. In fact, $\mathcal{S}'$ has resultant $\mathbf{R}$ (the couple does not contribute to the resultant) and moment with respect to $Q$ equal to $\mathbf{M}_Q$ (the resultant applied exactly at $Q$ does not contribute to the overall moment, which reduces to the couple's torque).

As we can choose to apply one of the couple's vector at $Q$, system $\mathcal{S}'$ can be effectively reduced to two forces: the sum of $\mathbf{R}$ and $\mathbf{f}$ at $Q$, and the second force $-\mathbf{f}$ applied at a suitable point chosen so that the couple's torque becomes $\mathbf{M}_Q$.

Finally, if the equivalent system $S'$ is built applying the resultant at a point $P$ on the central axis (7.22), rather than at a generic $Q$, the couple's moment $\mathbf{M}_Q$ to add will be parallel to $\mathbf{R}$ and of smallest modulus. This property is a natural consequence of what we saw in Sect. 7.2.2. □

**Remark 7.16** In Sect. 2.1 we proved that the following always have *zero scalar invariant*:

- planar systems;
- systems of concurrent forces;
- systems of parallel forces.

All these types, if their resultant is non-zero, admit a resultant line of action and hence fall under case *(iii)* of Theorem 7.15 (and never under *(iv)*).

In particular, when the forces are *concurrent*, it is easy to prove that the line of action of the resultant passes through the common intersection point.

More generally, every system with zero scalar invariant can always be reduced to one force (when $\mathbf{R} \neq 0$) *or* a couple (when $\mathbf{R} = 0$). □

### 7.4.1 Transport Moment

In many situations it is useful to transport a force to another point by keeping it parallel to itself and without altering the overall resultant and moment of the system to which it belongs. Call $A$ the point the force $\mathbf{f}$ is initially applied to and $B$ the new point where we want to apply it. It will suffice to add to the system a couple that generates torque $-AB \times \mathbf{f}$. In this way the resultant does not change (a couple has zero resultant) and the total moment generated by $\mathbf{f}$ and the added couple is zero with respect to the pole $A$, see Fig. 7.8. The moment of the new couple is known for obvious reasons as *transport moment*.

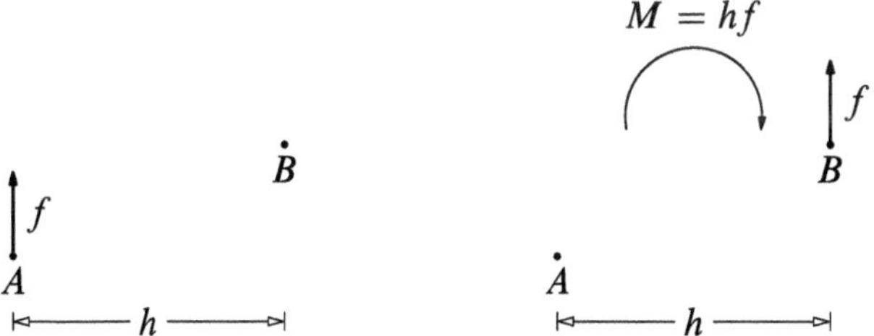


**Fig. 7.8** By adding a couple with an appropriate *torque* it is possible to transport to point $B$ the force which was applied in $A$. Notice that the systems of forces described on the left and on the right have the same resultant and the same total moment, and are thus equivalent. In this picture the couple's torque $M$ is perpendicular to the plane of the drawing, with orientation and intensity as shown

### 7.4.2 Center of Parallel Forces

Let us now examine more carefully a system of parallel forces, for which we assume $\mathbf{R} \neq 0$. Going back to the discussion following Definition 7.13 we recall that in this case $I = 0$ and, as pointed out in Remark 7.16, the system's resultant has a line of action, of which we seek the equation.

Instead of specializing (7.25) to this particular case it is more convenient to note that the line we seek consists of the points $P$ such that the moment with respect to a pole $O$ of the resultant $\mathbf{R} = R\,\mathbf{k}$, as if applied at $P$, equals the total moment $\mathbf{M}_O$, which is given by (7.23). Hence

$$OP \times R\,\mathbf{k} = \sum_{i=1}^{n} f_i OP_i \times \mathbf{k}$$

and so

$$\left( R\,(OP) - \sum_{i=1}^{n} f_i OP_i \right) \times \mathbf{k} = \mathbf{0}\,.$$

The factors of the cross product are therefore parallel and

$$R\,(OP) = \sum_{i=1}^{n} f_i OP_i + \lambda\mathbf{k}$$

for some arbitrary scalar $\lambda$. It is best to divide by $R$, and by the arbitrariness of $\lambda$ write

$$OP(\lambda) = \frac{1}{R}\sum_{i=1}^{n} f_i OP_i + \lambda\mathbf{k} \tag{7.26}$$

where we emphasized the dependence of $P$ upon $\lambda$.

This is the (vectorial) equation of the line of action of the resultant of the system of parallel forces: the locus of points $P(\lambda)$ at which to apply the resultant $\mathbf{R}$ to obtain an equivalent system.

Among all points of this line there is one point that enjoys a particularly important property, for which reason we call it *center of parallel forces*.

Let $\bar{P}$ indicate the point obtained in (7.26) with $\lambda = 0$:

$$O\bar{P} = \frac{1}{R}\left( \sum_{i=1}^{n} f_i\, OP_i \right) \tag{7.27}$$

and observe that it is the unique point on the line that *does not* depend on the forces' direction $\mathbf{k}$.

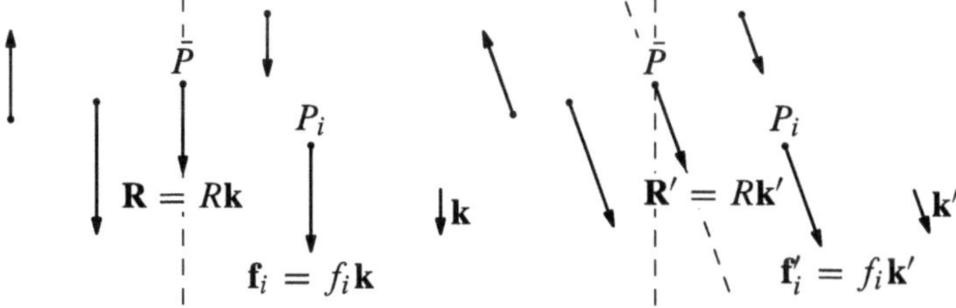


**Fig. 7.9** A system of parallel forces and the line of action of its resultant (left). A system obtained through a rotation of the unit vector **k**, common to all forces, with a new line of action of the resultant is shown on the right. The point $\bar{P}$ is called *center of the system of parallel forces*, and does not depend on the direction of the unit vector **k**

Figure 7.9 left, shows a system of forces parallel to the unit vector **k** and applied at points $P_i$. The line of action of the resultant **R** is dashed and $\bar{P}$ is the center of the parallel forces. On the right we see the system consisting of the same forces, applied at the same point $P_i$ but rotated so to become parallel to some other unit vector $\mathbf{k}'$. The line of action of the resultant of the second system is dashed and now parallel to $\mathbf{k}'$. The center of parallel forces $\bar{P}$ belongs to this line as well, since it is independent of the choice of common unit vector.

## Center of Gravity

The weights of a collection of particles are "reasonably parallel" vectors only when we are in a small region of space because, for example, faraway points—even if in proximity of the earth's surface—are subject to non-parallel weight forces.

Suppose we are in a situation where the weights $\mathbf{p}_i$ of the points $P_i$ can be considered *parallel*, with $\mathbf{p}_i = p_i\mathbf{k}$ ($p_i > 0$ and **k** vertical and pointing downward). Under the additional assumption that the acceleration of gravity $g$ is uniform we will also have $p_i = m_i g$.

The center of parallel forces $\bar{P}$ of the system is now called *center of gravity* (or *barycenter*, standing for "center of the weights").

Denote by $p$ and $m$ the weight and the total mass of the particles, Then (7.27) reads:

$$O\bar{P} = \frac{1}{p}\left(\sum_{i=1}^{n} p_i\, OP_i\right) = \frac{1}{mg}\left(\sum_{i=1}^{n} m_i g\, OP_i\right) = \frac{1}{m}\left(\sum_{i=1}^{n} m_i\, OP_i\right).$$

This mimics precisely the definition of center of mass given in (5.2), for which reason the terms center of gravity and center of mass are used as synonyms.

For theoretical clarity it is best to say that:

- relation (5.2), and its extension (5.4) to the continuous case, define the *center of mass* $G$ of a material system.
- For a (discrete or continuous) system that occupies a small spatial region where it is subject to a uniform gravitational action the weights are parallel, and its center of gravity $\bar{P}$ coincides with the center of mass $G$.
- Based on this it is common to use *center of mass* and *center of gravity* as if they were synonyms, even if the two entities are conceptually distinct.

It follows from these observations that the center of gravity (in the sense of "center of weights") enjoys all of the properties proved in the first part of Chap. 5 regarding the center of mass.

## 7.5 Elementary Work and Power of a System of Forces

The *elementary work* of a force $\mathbf{F}$ along an infinitesimal displacement $dP$ of the application point is $dL = \mathbf{F} \cdot dP$, while the *power* of $\mathbf{F}$ is defined as $\Pi = \mathbf{F} \cdot \mathbf{v}$, where $\mathbf{v}$ is the velocity of the application point $P$, as earlier seen.

The total elementary work and total power of a system of forces $\mathbf{F}_i (i = 1, \ldots, n)$ acting on points $P_i$ are instead defined as sums of the single forces' powers and works:

$$dL = \sum_{i=1}^{n} \mathbf{F}_i \cdot dP_i \qquad \Pi = \sum_{i=1}^{n} \mathbf{F}_i \cdot \mathbf{v}_i \,. \tag{7.28}$$

It is important to establish formulas for the elementary work and the power when the forces are applied to the points of:

- a rigid system;
- a system with $N$ generalized coordinates $\mathsf{q} = (q_1, q_2, \ldots, q_N)$.

### 7.5.1 Systems of Forces Applied to a Rigid Body

Relation (2.26) expresses the infinitesimal displacement of the points of a rigid body as

$$dP_i = dQ + \boldsymbol{\varepsilon} \times QP_i \qquad (\boldsymbol{\varepsilon} = \boldsymbol{\omega} dt)$$

where $\boldsymbol{\varepsilon}$ is the infinitesimal rotation vector of the body with angular velocity $\boldsymbol{\omega}$ at the instant considered.

The elementary work is obtained by substitution in definition (7.28):

$$
\begin{aligned}
dL &= \left(\sum_{i=1}^{n} \mathbf{F}_i\right) \cdot dQ + \sum_{i=1}^{n} \mathbf{F}_i \cdot \varepsilon \times QP_i \\
&= \left(\sum_{i=1}^{n} \mathbf{F}_i\right) \cdot dQ + \left(\sum_{i=1}^{n} QP_i \times \mathbf{F}_i\right) \cdot \varepsilon ,
\end{aligned}
\tag{7.29}
$$

where on the right we permuted cyclically the mixed product. Introducing the resultant vector $\mathbf{R}$ and the resultant moment $\mathbf{M}_Q$ of the system of applied forces $(P_i, \mathbf{F}_i)$ (see (7.17)) we obtain

$$dL = \mathbf{R} \cdot dQ + \mathbf{M}_Q \cdot \varepsilon . \tag{7.30}$$

In a similar way, substituting in $(7.28)_2$ the velocity $\mathbf{v}_i = \mathbf{v}_Q + \omega \times QP_i$ of a generic point $P_i$ in a rigid system, gives

$$
\begin{aligned}
\Pi &= \sum_{i=1}^{n} \mathbf{F}_i \cdot \mathbf{v}_i = \sum_{i=1}^{n} \mathbf{F}_i \cdot (\mathbf{v}_Q + \omega \times QP_i) \\
&= \sum_{i=1}^{n} \mathbf{F}_i \cdot \mathbf{v}_Q + \sum_{i=1}^{n} \mathbf{F}_i \cdot \omega \times QP_i = \mathbf{R} \cdot \mathbf{v}_Q + \sum_{i=1}^{n} \omega \cdot QP_i \times \mathbf{F}_i \\
&= \mathbf{R} \cdot \mathbf{v}_Q + \omega \cdot \mathbf{M}_Q
\end{aligned}
$$

which we shorten to

$$\Pi = \mathbf{R} \cdot \mathbf{v}_Q + \mathbf{M}_Q \cdot \omega . \tag{7.31}$$

From (7.30) and (7.31) we infer that to measure the elementary work or the power of the forces acting on a rigid system, nothing changes if we replace the system of forces $\{(P_i, \mathbf{F}_i)\}$ with an equivalent system in the sense of definition (7.24). In fact, by (7.30) and (7.31) both the elementary work and the power of a system of forces acting on a rigid body only depend on the system's characteristic vectors. So, to compute the work or power of a system of forces acting on a rigid body, we may replace the system by a simpler one, provided that the new system is equivalent. Observe that $Q$ can be chosen randomly; but it is crucial that the point whose infinitesimal displacement appears in (7.30) (or whose velocity appears in (7.31)) coincides with the point with respect to which one takes the moment $\mathbf{M}_Q$.

Two useful and easy consequences of (7.30) and (7.31) are the formulas for the elementary work and power of a couple, or an equivalent system for which $\mathbf{R} = \mathbf{0}$, and hence the moment $\mathbf{M}$ is *independent* of the pole:

$$dL = \mathbf{M} \cdot \varepsilon \qquad \Pi = \mathbf{M} \cdot \omega .$$

**Example 7.17** (*Couples acting on planar rigid bodies*) Consider a rigid body in planar motion, or more generally constrained to move by roto-translations with axis parallel to $\mathbf{u}$ (see Sect. 2.5.2). Calling $\theta$ the angle of rotation, the rigid body's infinitesimal rotations will satisfy $\varepsilon = \omega dt = d\theta\,\mathbf{u}$ (see (2.29)). Consider now a couple with torque $\mathbf{M}$, acting on rigid body. Equation (7.30) shows that the infinitesimal work it does is $dL = \mathbf{M} \cdot \varepsilon = (\mathbf{M} \cdot \mathbf{u})d\theta$. Let $M_u = \mathbf{M} \cdot \mathbf{u}$ be the component of $\mathbf{M}$ parallel to the axis of rotation, and suppose $M_u$ depends only on $\theta$, so that

$$dL = M_u(\theta)\, d\theta\,. \tag{7.32}$$

Then (7.32) shows that the couple's work is an exact differential. It is in fact possible to find a potential $U$, which will be an arbitrary primitive of $M_u(\theta)$. In fact,

$$M_u(\theta) = U'(\theta) \quad \Longrightarrow \quad dL = dU = U'(\theta)\, d\theta\,.$$

In the particular case where the couple is constant ($\mathbf{M}_0 = M_0\mathbf{u}$) the potential is $U_0 = M_0\theta + \text{constant}$. Another interesting example is the *torsion spring*, the rotational analogue of the elastic force (7.15). A torsion spring produces a couple whose torque is proportional to the rotation angle $\theta$ with respect to its rest configuration: $\mathbf{M}_\mathrm{k} = M_\mathrm{k}(\theta)\mathbf{u}$, with $M_\mathrm{k}(\theta) = -\mathrm{k}\theta$ and $\mathrm{k} > 0$. The potential of a torsion spring is

$$U_\mathrm{k}(\theta) = -\frac{1}{2}\mathrm{k}\theta^2\,, \tag{7.33}$$

in substantial similarity to the linear spring's. □

### 7.5.2 Forces Applied to a Holonomic System

Consider a holonomic system whose configurations are described by $N$ generalized coordinates $\mathsf{q} = (q_1, \ldots, q_N)$:

$$P_i = P_i(q_1, \ldots, q_N; t) \tag{7.34}$$

(the functions assigning the positions of the $P_i$ depend explicitly on time in the case of moving constraints).

Let us calculate the elementary work of a system of forces applied to the $P_i$, deducing from formula (7.34) that the elementary displacements are:

$$dP_i = \sum_{k=1}^{N} \frac{\partial P_i}{\partial q_h} dq_h + \frac{\partial P_i}{\partial t} dt\,. \tag{7.35}$$

Substituting (7.35) in the first of (7.28), the elementary work becomes

$$
\begin{aligned}
dL &= \sum_{i=1}^{n} \mathbf{F}_i \cdot dP_i = \sum_{i=1}^{n} \mathbf{F}_i \cdot \Big( \sum_{h=1}^{N} \frac{\partial P_i}{\partial q_h} dq_h + \frac{\partial P_i}{\partial t} dt \Big) \\
&= \sum_{i=1}^{n} \mathbf{F}_i \cdot \left( \sum_{h=1}^{N} \frac{\partial P_i}{\partial q_h} dq_h \right) + \sum_{i=1}^{n} \mathbf{F}_i \cdot \frac{\partial P_i}{\partial t} dt \\
&= \sum_{h=1}^{N} \left( \sum_{i=1}^{n} \mathbf{F}_i \cdot \frac{\partial P_i}{\partial q_h} \right) dq_h + \sum_{i=1}^{n} \mathbf{F}_i \cdot \frac{\partial P_i}{\partial t} dt \\
&= \sum_{h=1}^{N} Q_h \, dq_h + Q_t \, dt
\end{aligned}
\tag{7.36}
$$

where $Q_h$ and $Q_t$ are defined to be

$$
Q_h = \sum_{i=1}^{n} \mathbf{F}_i \cdot \frac{\partial P_i}{\partial q_h}, \qquad Q_t = \sum_{i=1}^{n} \mathbf{F}_i \cdot \frac{\partial P_i}{\partial t}. \tag{7.37}
$$

Note $dL$ is a differential form in the $N + 1$ variables $(q_1, \ldots, q_N; t)$, and in general it will not be exact. In other words there might not exist a function whose differential is $dL$.

As regards the power $\Pi$ of the same forces, it is easy to deduce, from the velocity

$$
\mathbf{v}_i = \frac{dP_i}{dt} = \sum_{h=1}^{N} \frac{\partial P_i}{\partial q_h} \dot{q}_h + \frac{\partial P_i}{\partial t}
$$

of $P_i(q_1, \ldots, q_N; t)$, that

$$
\Pi = \sum_{i=1}^{n} \mathbf{F}_i \cdot \mathbf{v}_i = \sum_{i=1}^{n} \mathbf{F}_i \cdot \left( \sum_{h=1}^{N} \frac{\partial P_i}{\partial q_h} \dot{q}_h + \frac{\partial P_i}{\partial t} \right) = \sum_{h=1}^{N} Q_h \dot{q}_h + Q_t \,.
$$

The passages (here omitted) are completely similar to those in (7.36).

## 7.6 Virtual Work and Virtual Power

Recall that a virtual displacement $\delta P$ is an infinitesimal displacement compatible with the constraints, imagined fixed at the instant of concern (this is essential in presence of moving constraints). A virtual velocity $\mathbf{v}'$ is, similarly, a velocity compatible with the constraints, again thought of as fixed at the instant considered.

The *virtual work* of a force is the dot product of the force itself and a *virtual* displacement of its point of application:

$$\delta L = \mathbf{F} \cdot \delta P\,.$$

Note the use of the letter "$\delta$" (instead of "$d$") to distinguish virtual work from elementary work. The latter is done under an effective elementary displacement.

In total analogy to (7.5) we define the *virtual power* of a force as

$$\Pi' = \mathbf{F} \cdot \mathbf{v}'$$

where the dash on $\Pi$ distinguishes the virtual power from the effective power, which corresponds to the effective velocity $\mathbf{v}$ and not the virtual one $\mathbf{v}'$.

Obviously the virtual work and virtual power of a system of forces $\mathbf{F}_i$ applied to points $P_i$ are

$$\delta L = \sum_{i=1}^{n} \mathbf{F}_i \cdot \delta P_i \qquad \Pi' = \sum_{i=1}^{n} \mathbf{F}_i \cdot \mathbf{v}_i' \tag{7.38}$$

As we saw in Sect. 4.1.4, *for moving constraints* the effective velocity is usually not a virtual velocity, whence the effective elementary work and effective power will not normally be special instances of virtual work and virtual power.

Next we explain how to write the virtual work and power of forces acting on a rigid body and on a system with $N$ generalized coordinates.

### 7.6.1 Rigid Bodies

Take (4.19), giving the virtual displacement of the points of a rigid body

$$\delta P_i = \delta Q + \varepsilon' \times QP_i\,.$$

Substituting in the first of (7.38), and using arguments similar to those of (7.29), we eventually attain

$$\delta L = \mathbf{R} \cdot \delta Q + \mathbf{M}_Q \cdot \varepsilon'\,. \tag{7.39}$$

Analogously, for the virtual power, substituting

$$\mathbf{v}_i' = \mathbf{v}_Q' + \boldsymbol{\omega}' \times QP_i$$

in the second of (7.38) produces

$$\Pi' = \mathbf{R} \cdot \mathbf{v}_Q' + \mathbf{M}_Q \cdot \boldsymbol{\omega}'\,. \tag{7.40}$$

Here, too, it is worth recalling the virtual work and virtual power of a couple or an equivalent system:

$$\delta L = \mathbf{M} \cdot \varepsilon' \qquad \Pi' = \mathbf{M} \cdot \boldsymbol{\omega}'\,, \tag{7.41}$$

due to their usefulness in the applications, Above, $\varepsilon'$ and $\omega'$ are respectively the vector of infinitesimal rotation and the angular velocity, both virtual, of the rigid body on which the couple acts.

Therefore, in analogy to what we observed at the end of Sect. 7.5.1, to compute the virtual work or virtual power of a system of forces acting on a rigid body we can replace the system itself with a simpler system, clearly if equivalent.

### 7.6.2 Holonomic Systems

Equation (4.23) allows to write the virtual displacement of the points of a system with $N$ generalized coordinates as

$$\delta P_i = \sum_{h=1}^{N} \frac{\partial P_i}{\partial q_h} \delta q_h \,,$$

where we note that, in contrast to the effective displacement (7.35), here the last term involving time derivative and infinitesimal variation $dt$ is absent since the constraints and time are imagined fixed. Hence the virtual work is

$$\delta L = \sum_{i=1}^{n} \mathbf{F}_i \cdot \Big( \sum_{h=1}^{N} \frac{\partial P_i}{\partial q_h} \delta q_h \Big) = \sum_{h=1}^{N} \Big( \sum_{i=1}^{n} \mathbf{F}_i \cdot \frac{\partial P_i}{\partial q_h} \Big) \delta q_h = \sum_{h=1}^{N} Q_h \, \delta q_h \,, \tag{7.42}$$

where, as in the first of (7.37), we set

$$Q_h = \sum_{i=1}^{n} \mathbf{F}_i \cdot \frac{\partial P_i}{\partial q_h} \,. \tag{7.43}$$

At this point, as mentioned in Sect. 4.3, beside $\mathsf{q} = (q_1, q_2, \dots, q_n)$ it is useful to introduce the vectors of $\mathbb{R}^N$

$$\begin{aligned} \delta\mathsf{q} &= (\delta q_1, \delta q_2, \dots, \delta q_N) \\ \mathsf{Q} &= (Q_1, Q_2, \dots, Q_N) \end{aligned} \tag{7.44}$$

so that now the virtual work (7.42) reads

$$\delta L = Q_1 \delta q_1 + \cdots Q_N \delta q_N = \mathsf{Q} \cdot \delta \mathsf{q} \,. \tag{7.45}$$

The last dot product should be thought of as taken in the configuration space, whose dimension equals the number of generalized coordinates of the system.

Equation (7.45), which we will see has a major importance, is much similar to (7.1), which in turn describes the elementary work of a force $\mathbf{F}$ applied to a point

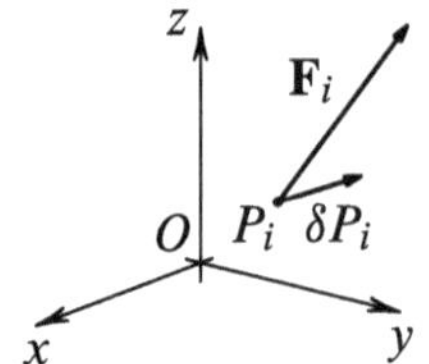

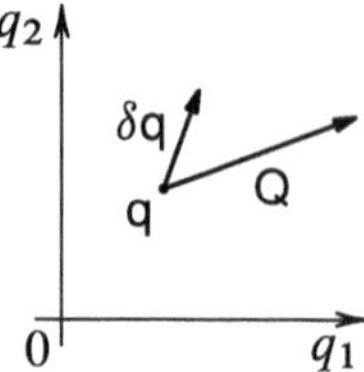


**Fig. 7.10** The total virtual work of the forces applied to the points $P_i$ of a system equals the dot product $\mathsf{Q} \cdot \delta\mathsf{q}$ (the picture corresponds to a system with two generalized coordinates: $\mathsf{q} = (q_1, q_2)$)

$P$: instead of $\mathbf{F}$ and $dP$, in (7.45) we have $\mathsf{Q}$ and $\delta\mathsf{q}$ respectively. Due to this, the vector $\mathsf{Q}$ is called *generalized force* and its components $Q_h (h = 1, 2, \ldots, N)$ are the *Lagrangian* (or *generalized) components of the force*.

The total virtual work of forces $\mathbf{F}_i$, by (7.42) and (7.45), is then given by the dot product of the vector of Lagrangian components $Q_h$ and the vector of the $\delta q_h$, virtual variations of the generalized coordinates.

This analogy between points and holonomic systems will turn out to be extremely important both to Statics and to Dynamics. In particular, the Lagrangian components of the active forces will play a fundamental role both in detecting the equilibrium configurations, and in steering the motion of holonomic systems.

Figure 7.10 indicates a way to visualise these concepts: on the left we see in three-dimensional physical space the force $\mathbf{F}_i$ applied to the generic point $P_i$ of a system here imagined having two degrees of freedom.

## 7.7 Systems of Conservative Forces

Consider a system of forces $\mathbf{F}_i$, applied to points $P_i$ of a system with $N$ generalized coordinates. Suppose each is conservative, hence arising from a potential $U_i(P)$ (perhaps the same for some, or all, forces), so $\mathbf{F}_i = \operatorname{grad} U_i$.

In these hypotheses the formula to find the $Q_h$ gives

$$Q_h = \sum_{i=1}^{n} \mathbf{F}_i \cdot \frac{\partial P_i}{\partial q_h} = \sum_{i=1}^{n} \operatorname{grad} U_i \cdot \frac{\partial P_i}{\partial q_h} = \frac{\partial \tilde{U}}{\partial q_h} \tag{7.46}$$

where we used the chain rule on $\tilde{U}$, the total potential of the system of forces as function of the generalized coordinates

$$\tilde{U}(q_1, \ldots, q_N, t) = \sum_{i=1}^{n} U_i(P_i(q_1, \ldots, q_N, t)) . \tag{7.47}$$

It is important then to note that while the potential $U_i$ of each conservative force does not depend on time, it may happen (essentially because of the possible presence

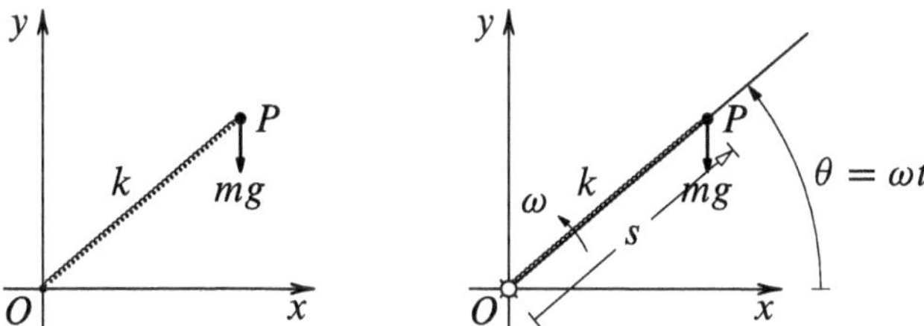


**Fig. 7.11** An elastic force acting on an unconstrained particle $P$ (left) and on a particle constrained to move on a rotating guide (right)

of moving constraints) that $\tilde{U}$ becomes dependent on time. Its derivatives with respect to the $q_h$ are the force's Lagrangian components $Q_h$.

Similarly,

$$Q_t = \sum_{i=1}^{n} \mathbf{F}_i \cdot \frac{\partial P_i}{\partial t} = \sum_{i=1}^{n} \operatorname{grad} U_i \cdot \frac{\partial P_i}{\partial t} = \frac{\partial \tilde{U}}{\partial t}.$$

This last formula agrees with the fact that, for fixed constraints, $Q_t = 0$ and $\tilde{U}$ does not explicitly depend on time.

**Example 7.18** To understand these details better let us look at a point of mass $m$ on a vertical plane, subject to an elastic force given by an ideal spring of constant $k$ connecting it to the origin. Referring to Fig. 7.11, the total potential $U(P)$ of the elastic force and the weight is $U(x, y) = -k(x^2 + y^2)/2 - mgy$. But if the point is constrained to move on a guide that rotates uniformly, as in the right picture, there is only one generalized coordinate, say the distance $s$ to the origin. Hence the potential is

$$\tilde{U}(s, t) = -ks^2/2 - mgs \sin(\omega t).$$

The dependence on time is explicit, although both the spring's force and the weight are conservative. In presence of the spring's force *only*, $\tilde{U}$ would not depend on $t$, meaning that having moving constraints is a necessary but not sufficient condition for the potential $\tilde{U}$ to depend on time explicitly. □

From now on, for the sake of simplicity, we will write $U$ for the potential understood as function of both the points' positions and of the generalized coordinates, and possibly time as well. In other words, we will not formally distinguish between the $U$ and $\tilde{U}$ introduced above and related by (7.47).

With this convention let us then summarize what we have discovered: for a system of conservative forces there is a function $U(q_1, \ldots, q_N, t)$ such that

$$Q_h = \frac{\partial U}{\partial q_h} \qquad Q_t = \frac{\partial U}{\partial t}. \tag{7.48}$$

The effective elementary work now becomes

$$dL = \sum_{h=1}^{N} Q_h dq_h + Q_t dt = \sum_{h=1}^{N} \frac{\partial U}{\partial q_h} dq_h + \frac{\partial U}{\partial t} dt = dU \, ,$$

while the effective power of the system of forces is

$$\Pi = \sum_{h=1}^{N} Q_h \dot{q}_h + Q_t = \sum_{h=1}^{N} \frac{\partial U}{\partial q_h} \dot{q}_h + \frac{\partial U}{\partial t} = \frac{dU}{dt} \, . \tag{7.49}$$

Finally, for the virtual work

$$\delta L = \sum_{h=1}^{N} Q_h \delta q_h = \sum_{h=1}^{N} \frac{\partial U}{\partial q_h} \delta q_h = \delta U \, .$$

Therefore, if a system of conservative forces is applied to a holonomic system (with fixed or moving constraints) the elementary work and the virtual work become the exact differentials of the total potential $U(q_1, q_2, \ldots, q_N; t)$ seen as function of the generalized coordinates and of time (as already noted, the latter dependence appears with moving constraints).

It is worth stressing that the presence of constraints might turn exact the elementary work of systems of forces that would not be otherwise conservative.

**Example 7.19** Consider the planar force field

$$\mathbf{F}(x, y) = k[-xy\mathbf{i} + y\mathbf{j}] \, ,$$

where $k$ is a constant. As

$$\frac{\partial F_x}{\partial y} \neq \frac{\partial F_y}{\partial x}$$

we know the field is not conservative, so there exists *no* map $U(x, y)$ such that

$$\delta U = k[-xy \, \delta x + y \, \delta y] = \mathbf{F} \cdot \delta P = \delta L \, .$$

But suppose the force of this field is applied to a point $P$ *constrained* to the line $y = 1$, so that the only possible displacements are those where $\delta P = (\delta x, 0)$. Then

$$\delta L = -kx \, \delta x = \delta(-k \, x^2/2) \, ;$$

while the point that is free to move on the plane is not subject to a conservative force, the constrained point is. □

The condition $\delta L = \delta U$ requires in fact that there exists a function $U$ such that the virtual work (7.45) coincides with its virtual variation *for any collection of virtual displacements allowed by the constraints*. Evidently, the fewer the virtual displacements allowed (the more constraints are present), the easier it will be to find such a potential.

Another example of what we have explained appeared when we computed the potential of a couple acting on a rigid body constrained to planar motions (or roto-translational ones). If so, we saw that the constraint on the angular velocity's direction implies that it suffices that one of the couple's components is positional to ensure the elementary work is exact. No condition whatsoever is imposed on the remaining components. We will encounter further similar examples later in the book (see Sects. 10.2.1, 10.4 and 12.5).

**Example 7.20** (*Computing Lagrangian components*) Consider a rod $AB$ of length $\ell$, constrained as in Sect. 4.1.3: a roller constrains the end $A$ to glide on the $x$-axis, and so the system has two degrees of freedom. Choose as generalized coordinates the abscissa $s$ of $A$ and the angle $\theta$ (see Fig. 4.6).

- Let us compute the Lagrangian components of a force $\mathbf{F}_B = F_{Bx}\mathbf{i} + F_{By}\mathbf{j}$ applied at the end $B$. In order to use definition (7.43) we should write the position of the application point $B$ in function of the generalized coordinates: $OB = (s + l\cos\theta)\mathbf{i} + l\sin\theta\mathbf{j}$. Then

$$
\begin{aligned}
Q_{Bs} &= \mathbf{F}\cdot\frac{\partial B}{\partial s} = \mathbf{F}\cdot\mathbf{i} = F_{Bx} \\
Q_{B\theta} &= \mathbf{F}\cdot\frac{\partial B}{\partial\theta} = \mathbf{F}\cdot\big(-l\sin\theta\mathbf{i} + l\cos\theta\mathbf{j}\big) = (-F_{Bx}\sin\theta + F_{By}\cos\theta)l
\end{aligned}
$$

  That $s$ is an $x$-coordinate forces the Lagrangian component $Q_s$ to coincide with the force's $x$-component. We point out that also the component $Q_\theta$ is non-zero.
- Consider now a couple with torque $\mathbf{M} = M_z\mathbf{k}$. By (7.39) the virtual work it does is

$$
\delta L = \mathbf{M}\cdot\varepsilon' = M_z\mathbf{k}\cdot\delta\theta\mathbf{k} = M_z\delta\theta\,.
$$

  In view of (7.42) the Lagrangian components are the coefficients of the virtual work as function of the virtual displacements. Hence $Q_s = 0$ and $Q_\theta = M_z$. The Lagrangian component relative to $\theta$ coincides with the component of the moment along the axis that defined the angle. □

# Chapter 8
# The Laws of Mechanics

The objective of Mechanics is to relate the motion of bodies to the forces that can act on them. For this one needs to introduce general principles whose role is to explain the relationships between forces and motions. These principles are the so-called Laws of Mechanics: a collection of postulates whose acceptance leads, under the hypothetical-deductive process typical of mathematical sciences, to the desired *cause-effect* relations.

The Laws of Mechanics are the product of a long and delicate activity of abstraction from experimental observations which historically were mainly astronomical. In order to formulate absolute laws starting from experimental evidence it is necessary to make an *idealization process*. This means that we must accept as verified in an absolute and rigorous way certain properties, labelled *special*, that come from the experience. It is patent that in reality these special properties, which we accept as absolute, only hold in an approximate way. Notwithstanding, the more we eliminate any accidental circumstance interpreted as perturbation, the higher is the degree of approximation of those properties we call special, among all other properties arising in the experience.

Usually, in elementary textbooks, the Laws of Mechanics are presented from a perspective going back to Galileo Galilei (Pisa 1564–Florence 1642), Sir Isaac Newton (Woolsthorpe, Lincolnshire 1643–London 1727) and Leonhard Euler (Basel 1707–St.Petersburg 1783). This approach rests on three cardinal laws: the Principle of inertia, the proportionality of force and acceleration, the action-reaction principle.

Indeed, while motion has a relative nature, the same cannot be said for the forces, to which we must give a character that is independent of the frame of reference (as is easy to understand by thinking about the most common daily examples). Hence any attempt to relate the motion of bodies to the forces acting on them involves laws that, if valid in a certain frame, might be false in a frame that moves with respect to the first one. The first problem to solve is therefore the choice of the frame of reference.

 
P. Biscari et al., *Rational Mechanics*, UNITEXT 177,
https://doi.org/10.1007/978-3-032-07462-1_8

## 8.1 The Principles of Mechanics

### 8.1.1 Inertial Frames of Reference

***Postulate*** (*Principle of inertia*) There exist reference frames, called *inertial frames*, for which isolated points are either at rest, or they move with zero acceleration.

The need to refer to a specific class of reference frames (inertial ones) is best understood by noting that the Principles of Mechanics aim to relate forces, which cause a motion, with the motion they generate. This goal clashes with the observation that forces have absolute nature (all frames measure the same forces) whereas motion, specified by velocity and acceleration, has a relative character, as we studied in Chap. 3. The laws we will propose in the sequel therefore will not be valid in all reference frames, but only in the inertial frames defined above. In non-inertial reference frames it will be possible to observe phenomena that only supposedly contradict the Principles introduced here. We will see in Chap. 13 that one can study a body's motion in any frame, as long as one introduces additional forces, called *apparent forces*.

There exist infinitely many inertial frames. Using the Coriolis theorem (Sect. 3.3) one sees that the inertial class consists of $\infty^6$ frames that differ by a Galilean transformation ($\mathbf{a}_O = \mathbf{0}, \boldsymbol{\omega} = \mathbf{0}$), where $\mathbf{a}_O$ is the origin's acceleration and $\boldsymbol{\omega}$ is the angular velocity of the basis describing one frame with respect to the other. In other words: any frame in linear rigid motion with respect to a given inertial frame is itself inertial.

### 8.1.2 Newton's Second Law

Galilei established the law governing the phenomenon of free falling by observing the motion of massive bodies. This law says that in a given time interval $\Delta t$, the variation of a massive body's velocity along the vertical direction is constant. In this way Galilei managed to deduce that what causes a variation in a body's velocity (and hence produces an acceleration) is the action of the weight, and that this variation is independent of the velocity itself.

Newton eventually made this fundamental experimental observation explicit.

***Postulate*** (*Newton's second law*) In an inertial frame, a body subject to a force $\mathbf{F}$ acquires an acceleration parallel to and with the same orientation as, the force, and of modulus proportional to that of the force.

In formulas, Newton's second law is the well-known fundamental equation of Dynamics (a.k.a. Newton's equation):

$$\mathbf{F} = m\mathbf{a}\,, \tag{8.1}$$

where $m$ denotes the aforementioned positive proportionality constant, called *mass* (although to be precise we should say *inertial* mass). The mass may be understood as a measure of the tendency of a point to stay at rest or in uniform linear motion. In fact, from (8.1) we have $\mathbf{a} = \mathbf{F}/m$, meaning that the larger $m$ is, the smaller the acceleration's modulus. This fact is an (imprecise but suggestive) measure of the fact that the bigger a point's mass is, the harder it will be for its motion to differ from an inertial motion.

### 8.1.3 Action-Reaction Principle

***Postulate*** (*Newton's third law*) For any pair of particles $P_1$ and $P_2$ the force $\mathbf{F}_{12}$ that $P_2$ exerts on $P_1$ is equal and opposite to the force $\mathbf{F}_{21}$ that $P_1$ exerts on $P_2$ and has the same line of action. Put otherwise, the interactions of two particles are representable by the forces of a couple with zero arm:

$$\mathbf{F}_{12} + \mathbf{F}_{21} = \mathbf{0}, \qquad OP_1 \times \mathbf{F}_{12} + OP_2 \times \mathbf{F}_{21} = \mathbf{0}.$$

This postulate, also known as *action-reaction principle*, is often stated by saying that *with each action there corresponds an equal and opposite reaction*.

Under the assumption that two points $P_1$, $P_2$, of respective masses $m_1, m_2$, are isolated, from

$$\mathbf{F}_{12} = m_1\mathbf{a}_1, \qquad \mathbf{F}_{21} = m_2\mathbf{a}_2,$$

we deduce

$$m_1\mathbf{a}_1 + m_2\mathbf{a}_2 = \mathbf{0}$$

so *the accelerations of the isolated particles are parallel to the lines joining them, oppositely oriented, and the ratio of their moduli is a constant, equal to the inverse ratio of the masses of the points.*

### 8.1.4 Principle of Superposition of Forces

Galilei, while observing the non-vertical motion of bodies with mass, also had the intuition that the circumstances that determine the motion of such bodies do not interfere with one another. Once again it was Newton that generalized this intuition and turned it into a proper principle.

***Postulate*** Let $\mathbf{F}_1$ be the force that would produce on a given particle (if applied alone) acceleration $\mathbf{a}_1$, and $\mathbf{F}_2$ the force that on the same particle would produce acceleration $\mathbf{a}_2$. Then the acceleration produced by applying the two forces simultaneously is $\mathbf{a}_1 + \mathbf{a}_2$, and so

$$m\,(\mathbf{a}_1 + \mathbf{a}_2) = \mathbf{F}_1 + \mathbf{F}_2\,.$$

Clearly this principle extends to an arbitrary number of forces acting on the same particle, and they can always be replaced by their resultant.

## 8.2 Principle of Mechanical Determinism

Classical Mechanics is the mathematical theory that deals with the motion of macroscopic bodies with astonishing success and a wide application range, going from Molecular Mechanics to celestial motions to the dynamics of terrestrial and space vehicles. Despite that, we must remember that Classical Mechanics is not suited to represent the motion of sub-atomic particles, and that its applicability is questionable even for macroscopic bodies that move at extremely high speeds, close to the speed of light.

If we take into account the experimental observations of real bodies within the descriptive reach of Classical Mechanics, we notice that the motion of material systems is determined by the initial conditions (the configuration and velocity distribution of the system at the time $t_0$ when the observation of the motion begins). In analytical terms this observation, in the case of one particle, translates into the request that in Newton's law (8.1) the forces depend exclusively on position, velocity and time:

$$\mathbf{F} = \mathbf{F}\,(P, \mathbf{v}, t) \tag{8.2}$$

In fact, if the function in (8.2) is known explicitly and the particle's motion is unknown, then (8.1), projected along the axes of frame's basis, furnishes three scalar differential equations of order two in the three unknowns $\{x(t), y(t), z(t)\}$. Moreover, (8.2) warrants that this system is already in *normal form*, i.e. solved for the highest derivatives, which by (8.2) only appear in the acceleration.

But we must emphasize that assumption (8.2) is not enough to guarantee, given initial conditions

$$\begin{array}{l} x(t_0) = x_0\,,\ y(t_0) = y_0\,,\ z(t_0) = z_0\,, \\ \dot{x}(t_0) = \dot{x}_0\,,\ \dot{y}(t_0) = \dot{y}_0\,,\ \ \dot{z}(t_0) = \dot{z}_0 \end{array} \tag{8.3}$$

and the map $\mathbf{F}(P, \mathbf{v}, t)$, that Eq. (8.1) has exactly one solution over some interval $[t_0, t_1]$, thus warranting the motions' determinism. Indeed, let us recall that to ensure the existence and uniqueness of solutions to the initial-value problem (8.1)–(8.3), expression (8.2) must be compatible with the hypotheses in Cauchy's theorem. We state the result without proof in the general case of $n$ differential equations in $n$ unknowns.

**Theorem 8.1** (Cauchy) *The initial-value problem*

$$\ddot{x}_i = \varphi_i\,(x_1, \ldots, x_n; \dot{x}_1, \ldots, \dot{x}_n; t)\,, \qquad x_i(t_0) = x_{i0}\,, \\ \dot{x}_i(t_0) = \dot{x}_{i0}\,, \quad i = 1, \ldots, n \tag{8.4}$$

*admits at least one solution in a time interval $t_0 \le t < t_1$ if the maps $\varphi_i$ in the right-had side of (8.4) are* continuous. *This solution is unique if the $\varphi_i$ are* Lipschitz *on a set of values $(x_i, \dot{x}_i, t)$ containing the initial datum.*

To this end let us recall the definition of Lipschitz map.

**Definition 8.2** A function of several variables $f : D \subset \mathbb{R}^n \to \mathbb{R}$ is Lipschitz in one variable (say, $x_1 \in [a, b]$) if there exists a constant $K$ such that

$$\left| f(x_1'', x_2, \ldots, x_n) - f(x_1', x_2, \ldots, x_n) \right| \le K \left| x_1'' - x_1' \right| \qquad \forall\, x_1', x_1'' \in [a, b] \\ \forall\, x_2, \ldots, x_n\,.$$

To guarantee that $f$ is Lipschitz when the variable of concern ranges in a closed interval it suffices that the partial derivative of $f$ in that variable is defined and continuous on the interval.

Cauchy's theorem does not ensure (and obviously does not exclude) that the solution to a Cauchy problem is defined on an unbounded interval ($t_1 = +\infty$).

**Remark 8.3** A simple example that does not satisfy Cauchy's theorem is the one-dimensional force $F(x) = a\sqrt[3]{x}$, with $a > 0$. If we examine the problem

$$m\ddot{x} = a\sqrt[3]{x}\,; \qquad x(0) = 0\,, \quad \dot{x}(0) = 0\,, \tag{8.5}$$

it is trivial to check that it admits the rest solution $x(t) \equiv 0$. But this is not the only solution, since also

$$x(t) = \left(\frac{a}{6\,m}\right)^{3/2} t^3$$

solves (8.5).

This remark shows how imposing mechanical determinism renders *non-physical* certain specific constitutive laws that would define forces disobeying the hypotheses of Cauchy's theorem.

**Remark 8.4** Newton's second law can be interpreted in two ways. In problems where the total force is a datum of the problem, the law is a vector-valued differential equation of order two in the unknown function $OP(t)$. We could write this dynamical problem symbolically as $m\mathbf{a} = \mathbf{F}(OP, \mathbf{v}, t)$, called a *direct* problem. When it is the motion that is known, and we seek the force or forces that allow to reach the desired objective, one speaks of an *inverse* problem, which would symbolically read $\mathbf{F} = m\mathbf{a}$. Such problems are not differential equation of order two, since now the motion (and hence the acceleration) are known.

A classical inverse problem is Kepler's problem (see Sect. 10.6), in which observations on the orbits of the planets led to the law of universal gravitation that generates those orbits.

## 8.3 Internal and External Forces

When partitioning, ideally, an isolated material system $\mathcal{S}$ into subsets it is often useful to find a subdivision dictated by the property that certain particles can affect the motion of other points, which in turn have a negligible effect on the former, and all of this despite the action-reaction principle.

Think for instance of the effect of Earth on an apple in free fall, and of the far less consequential effect the apple has on Earth. More generally, we can imagine situations where $\mathcal{S} = \mathcal{S}_1 \cup \mathcal{S}_2$ and that the mass of each point of $\mathcal{S}_1$ is far smaller than the mass of each point of $\mathcal{S}_2$.

In such a case the motion of the particles of $\mathcal{S}_2$ will, in a first rough approximation, be influenced by the motion of the particles of $\mathcal{S}_1$. So it may well happen that, although all points interact among themselves, the motion of the points of $\mathcal{S}_2$ can be considered known even if that of the points of $\mathcal{S}_1$ is not. In studying the dynamics of the points of $\mathcal{S}_1$ it is then possible to introduce the following approximation. Take the law that defined the interaction force of a point of $\mathcal{S}_1$ and a point of $\mathcal{S}_2$. It will in general depend on the position and velocity of both. But we can replace the quantities relative to the point of $\mathcal{S}_2$ with those obtained by neglecting the effect of the force on its motion.

For example, if the isolated system consists only of two particles $P$, $Q$, with very different masses $m_P \ll m_Q$, a similar subdivision allows to assume, as a rough approximation, that $Q$ is at rest or in uniform linear motion.

With this approximation, for any $P_i \in \mathcal{S}_1$ we can write

$$m_i \mathbf{a}_i = \sum_{j \neq i \in \mathcal{S}_1} \mathbf{F}_{ij}^{(\mathrm{i})} \left(P_i, P_j, \mathbf{v}_i, \mathbf{v}_j\right) + \sum_{h \in \mathcal{S}_2} \mathbf{F}_{ih}^{(\mathrm{e})} \left(P_i, \mathbf{v}_i, t\right) , \tag{8.6}$$

where the first sum gives the *internal forces*' resultant, while the second sum is the resultant of the *external forces* acting on point $P_i$. In the latter the dependence of $\mathbf{F}_{ih}^{(\mathrm{e})}$ on the position and velocity of the points $Q_h \in \mathcal{S}_2$ has been replaced by their time law, which then generates the explicit dependence of these forces on time. The set of equations (8.6) is called here *fundamental system of dynamics* and is written compactly as

$$m_i \mathbf{a}_i = \mathbf{F}_i^{(\mathrm{i})} + \mathbf{F}_i^{(\mathrm{e})} , \tag{8.7}$$

where $\mathbf{F}_i^{(\mathrm{i})}$ and $\mathbf{F}_i^{(\mathrm{e})}$ are the resultants of the *internal* and *external* forces, respectively, acting on point $P_i$.

In case system $\mathcal{S}_1$ is made of a unique particle, so there are no internal forces, (8.6) reduces to the vectorial equation

$$m\mathbf{a} = \mathbf{F}(P, \mathbf{v}, t) \ ,$$

giving back *Newton's second law for one particle.*

## 8.4 Internal Forces

Newton's third law, formulated as in Sect. 8.1.3, allows us to deduce immediately a few simple but very remarkable consequences:

- the resultant and moment of the internal forces are always zero;
- the work and power of the internal forces are zero in any *rigid* system.

### 8.4.1 Resultant and Moment

Internal forces are pairwise equal and opposite, with the same line of action, so we expect that their sum and the sum of their moments will vanish. The formal proof makes this idea rigorous.

**Theorem 8.5** *The resultant and moment of the internal forces always vanish:*

$$\mathbf{R}^{\text{int}} = \mathbf{0} \ , \qquad \mathbf{M}_O^{\text{int}} = \mathbf{0} \ . \tag{8.8}$$

***Proof*** Let $\mathbf{f}_{ij}$ $(i \neq j)$ be the force that point $P_j$ exerts on point $P_i$ in a system $\{P_h\}$ $(h = 1, \ldots, N)$. From the action-reaction principle we know that

$$\mathbf{f}_{ij} + \mathbf{f}_{ji} = \mathbf{0} \ , \qquad OP_i \times \mathbf{f}_{ij} + OP_j \times \mathbf{f}_{ji} = \mathbf{0} \ .$$

The resultant of the internal forces is

$$\mathbf{R}^{\text{int}} = \sum_{\substack{i,j=1 \\ (i \neq j)}}^{N} \mathbf{f}_{ij} = \sum_{\substack{i,j=1 \\ (i \neq j)}}^{N} \mathbf{f}_{ji} = -\sum_{\substack{i,j=1 \\ (i \neq j)}}^{N} \mathbf{f}_{ij} = -\mathbf{R}^{\text{int}}$$

and therefore $\mathbf{R}^{\text{int}} = \mathbf{0}$.

For the total moment of the internal forces we similarly find:

$$\mathbf{M}_O^{\text{int}} = \sum_{\substack{i,j=1 \\ (i \neq j)}}^{N} OP_i \times \mathbf{f}_{ij} = \sum_{\substack{i,j=1 \\ (i \neq j)}}^{N} OP_j \times \mathbf{f}_{ji} = -\sum_{\substack{i,j=1 \\ (i \neq j)}}^{N} OP_i \times \mathbf{f}_{ij} = -\mathbf{M}_O^{\text{int}}$$

so as before $\mathbf{M}_O^{\text{int}} = \mathbf{0}$. □

**Remark 8.6** Recalling Definition 7.9, we say the system of internal forces is always *balanced.* □

## 8.4.2 Work and Power of Internal Forces in a Rigid Body

A further important consequence of (8.8) concerns the vanishing of the infinitesimal work and of the power of the internal forces of a rigid body, or more generally a system in rigid motion.

**Theorem 8.7** *The infinitesimal work and the power of the internal forces of a* rigid *body are always zero:*

$$dL^{\text{int}} = 0\,, \qquad \Pi^{\text{int}} = 0\,. \tag{8.9}$$

***Proof*** Recalling (7.30), we can write the infinitesimal work of a generic system of forces applied to a rigid body as

$$dL = \mathbf{R} \cdot dO + \mathbf{M}_O \cdot \boldsymbol{\varepsilon}\,.$$

When the forces are internal, from (8.8) we will have

$$dL^{\text{int}} = \mathbf{R}^{\text{int}} \cdot dO + \mathbf{M}_O^{\text{int}} \cdot \boldsymbol{\varepsilon} = 0\,.$$

Equation (7.31) also allows us to express the power of the forces applied to a rigid system as

$$\Pi = \mathbf{R} \cdot \mathbf{v}_O + \mathbf{M}_O \cdot \boldsymbol{\omega}\,,$$

so if the forces are internal, again by (8.8)

$$\Pi^{\text{int}} = \mathbf{R}^{\text{int}} \cdot \mathbf{v}_O + \mathbf{M}_O^{\text{int}} \cdot \boldsymbol{\omega} = 0\,,$$

ending the proof. □

**Remark 8.8** It is important to recall that while the internal forces' resultant and moment are zero for *every* system, their work and power are in general zero *only* for *rigid systems*. The proof is based on (7.30) and (7.31), which can be deduced using rigid-body kinematics.

For a simple counterexample think of a system of two points, joined by a spring, that move towards one another. The powers of the (internal) forces that the spring generates on the points are both positive, making the total power non-zero. □

**Remark 8.9** The statement of Theorem 8.7 still holds when we refer to the *virtual work* and the *virtual power* of the *internal forces* of a rigid body. □

It can be shown that in a rigid body $\Pi^{\text{int}} = 0$ by using an alternative approach to that of Theorem 8.7. The idea is to show the internal forces' power is related to the rate of variation of the distances among the system's points. As in a rigid body the distances are constant in time, as immediate corollary we recover that $\Pi^{\text{int}} = 0$ in a rigid body.

**Proposition 8.10** (Power of internal forces) *Let* $\mathbf{f}_{ij}^{(\mathrm{i})} = \varphi_{ij}^{(\mathrm{i})}\mathbf{e}_{ji}\big( = -\mathbf{f}_{ji}^{(\mathrm{i})}\big)$ *be the internal force that point* $P_j$ *generates on* $P_i$*, where* $\mathbf{e}_{ji}$ *is the unit vector of* $P_jP_i$*. The power of the system of internal forces reads*

$$\Pi^{\text{int}} = \sum_{i<j} \varphi_{ij}^{(\mathrm{i})} \frac{d|P_jP_i|}{dt}$$

*In particular, the power of the internal forces is zero in any rigid velocity distribution.*

***Proof*** Let us initially split the power of the internal forces according to interacting pairs of points

$$\begin{aligned}
\Pi^{\text{int}} &= \sum_{i<j} \left( \mathbf{f}_{ij}^{(\mathrm{i})} \cdot \mathbf{v}_i + \mathbf{f}_{ji}^{(\mathrm{i})} \cdot \mathbf{v}_j \right) = \sum_{i<j} \varphi_{ij}^{(\mathrm{i})} \mathbf{e}_{ji} \cdot (\mathbf{v}_i - \mathbf{v}_j) \\
&= \sum_{i<j} \varphi_{ij}^{(\mathrm{i})} \frac{P_jP_i}{|P_jP_i|} \cdot (\mathbf{v}_i - \mathbf{v}_j) = \sum_{i<j} \frac{\varphi_{ij}^{(\mathrm{i})}}{|P_jP_i|} P_jP_i \cdot \frac{dP_jP_i}{dt} \\
&= \frac{1}{2} \sum_{i<j} \frac{\varphi_{ij}^{(\mathrm{i})}}{|P_jP_i|} \frac{d}{dt} \left( P_jP_i \cdot P_jP_i \right) = \frac{1}{2} \sum_{i<j} \frac{\varphi_{ij}^{(\mathrm{i})}}{|P_jP_i|} \frac{d(P_jP_i)^2}{dt} \\
&= \sum_{i<j} \varphi_{ij}^{(\mathrm{i})} \frac{d|P_jP_i|}{dt} .
\end{aligned}$$

In particular, if the distances between points are constant (as for a rigid velocity distribution) the derivatives of the distances $|P_jP_i|$ vanish, and with them also the power of the internal forces. □

## 8.5 Non-inertial Reference Frames

Whenever the laws of dynamics are not formulated and employed in an inertial frame, (8.1) must be adapted on the basis of the theorems of relative motion, obtained in Chap. 3.

Newton's equation (8.1) contains on the right the acceleration $\mathbf{a}$ of the point with respect to an inertial frame. Introducing a second frame in arbitrary motion with respect to the previous one (hence in general no longer inertial) we know that

$$\mathbf{a} = \mathbf{a}_{\mathrm{r}} + \mathbf{a}_{\tau} + \mathbf{a}_{\mathrm{c}} \tag{8.10}$$

holds, where $\mathbf{a}_r$ is the point's acceleration *relative* to the new frame, $\mathbf{a}_\tau$ is the drag acceleration and $\mathbf{a}_c$ is a complementary acceleration, as proved in the Coriolis Theorem 3.4, Sect. 3.3.

Substituting (8.10) in (8.1) produces

$$\mathbf{F} = m(\mathbf{a}_r + \mathbf{a}_\tau + \mathbf{a}_c)$$

and moving to the left two terms that are on the right (of the dimensions of forces),

$$\mathbf{F} - m\mathbf{a}_\tau - m\mathbf{a}_c = m\mathbf{a}_r \,. \tag{8.11}$$

Let us now define the *drag force* and the *Coriolis force*

$$\begin{cases} \mathbf{F}_\tau = -m\mathbf{a}_\tau & (\textit{drag force}) \\ \mathbf{F}_c = -m\mathbf{a}_c & (\textit{Coriolis force}) \end{cases} \tag{8.12}$$

In this way (8.11) reads

$$\mathbf{F} + \mathbf{F}_\tau + \mathbf{F}_c = m\mathbf{a}_r \,.$$

This relation allows to write Newton's law (8.1) in a *non-inertial* frame, as long as we add to the *effective* force $\mathbf{F}$ the two forces $\mathbf{F}_\tau$, $\mathbf{F}_c$.

The drag and Coriolis forces are called *apparent* because they do not arise from a physical interaction (so they do not obey the action-reaction principle) but rather from the attempt of a *non-inertial* frame to associate every type of acceleration measured with a force.

Apparent forces, defined in (8.12), depend on the position and velocity of the particle they refer to, but also on the kinematic quantities that characterize the frame as non-inertial (acceleration of the origin, angular velocity and acceleration of frames).

As we will see in Chap. 13, which is entirely devoted to the subject, the great utility of apparent forces is to grant us to tackle the problems of Mechanics by putting ourselves in a non-inertial frame, simply by adding to the effective forces the apparent forces that act on the single points in the system.

## 8.6 The Constraints Postulate

Consider a particle constrained to a surface. The motion of this element is profoundly altered by the presence of the constraint. Not only the point's admissible positions are not all possible points in the reference Euclidean space, but as we saw in kinematics also the velocity and acceleration vectors of the element cannot be completely arbitrary. More precisely, in our example the set of configurations compatible with the constraint at a given instant is a regular surface $\mathcal{S}_t \subset \mathcal{E}^3$. The set of possible velocities $\mathcal{V}(P, t)$ at each point $P \in \mathcal{S}_t$ is described by

$$\mathbf{v} = \mathbf{v}_t + \mathbf{v}_n(P, t) \tag{8.13}$$

where $\mathbf{v}_t$ is an arbitrary vector in the tangent plane to $\mathcal{S}_t$ at $P$ (said better, any virtual velocity), while $\mathbf{v}_n(P, t)$ belongs to the orthogonal complement of the tangent plane, and is completely determined when we assign $P$ and $t$. In particular, when the constraint is fixed we have $\mathbf{v}_n(P, t) \equiv \mathbf{0}$.

Similarly, having fixed $P \in \mathcal{S}_t$ and $\mathbf{v} \in \mathcal{V}(P, t)$, the set $\mathcal{A}(P, \mathbf{v}, t)$ of possible accelerations is a collection of vectors of the form

$$\mathbf{a} = \mathbf{a}_t + \mathbf{a}_n(P, \mathbf{v}, t)$$

where $\mathbf{a}_t$ is an arbitrary vector of the tangent plane to $\mathcal{S}_t$ at $P$, and $\mathbf{a}_n(P, \mathbf{v}, t)$ belongs to the orthogonal complement to the tangent plane and is determined by assigning $P$, $\mathbf{v}$ and $t$.

To understand these considerations let us go back to the example in Sect. 4.1.1, where we studied a point constrained to a circle centered at the origin on the plane $z = 0$. Suppose now the radius varies in time, with time law $R = R(t)$. Introduce the generalized coordinate $\theta$ such that

$$x(t) = R(t)\cos\theta(t)\,, \qquad y(t) = R(t)\sin\theta(t)\,, \qquad z(t) = 0\,.$$

Differentiating in time gives $\dot{x} = \dot{R}\cos\theta - R\dot{\theta}\sin\theta$, $\dot{y} = \dot{R}\sin\theta + R\dot{\theta}\cos\theta$, $\dot{z} = 0$. If we introduce the unit radial vector $\mathbf{e}_r = (\cos\theta, \sin\theta, 0)$ and the unit tangent vector $\mathbf{e}_\theta = (-\sin\theta, \cos\theta, 0)$ we can write the velocity of $P$ as $\mathbf{v} = R\dot{\theta}\,\mathbf{e}_\theta + \dot{R}\,\mathbf{e}_r$. This decomposition coincides with (8.13), and indeed

$$\mathbf{v}_t = R\dot{\theta}\,\mathbf{e}_\theta\,, \qquad \mathbf{v}_n = \dot{R}\,\mathbf{e}_r\,.$$

If we fix the constraint's equation (the function $R = R(t)$) for a given value of the angle $\theta$, the particle finds itself at a precise geometric point that is compatible with the constraint. The vector $\mathbf{v}_t$ is arbitrary (as we can choose $\dot{\theta}$ arbitrarily) while $\mathbf{v}_n$ is completely determined. (In particular $\mathbf{v}_n$ is zero if the radius $R$ is constant.)

As regards the accelerations,

$$\mathbf{a}_t = \left(R\ddot{\theta} + 2\dot{R}\dot{\theta}\right)\mathbf{e}_\theta, \qquad \mathbf{a}_n = \left(\ddot{R} - R\dot{\theta}^2\right)\mathbf{e}_r\,.$$

In particular, while $\mathbf{a}_t$ is arbitrary ($\ddot{\theta}$ can be chosen at will), $\mathbf{a}_n$ is determined. (Note $\mathbf{a}_n$ is not zero even when the radius $R$ is constant.)

The discussion clarifies the issue we face when we write Newton's law for a constrained point. The (experimental) law of the active forces acting on the element is a priori totally arbitrary (except for the fact it must have some regularity, as explained). The acceleration vector on the other hand must obey the restrictions introduced by the constraint, namely $\mathbf{a} \in \mathcal{A}(P, \mathbf{v}, t)$.

As, in general, $\mathbf{F}(P, \mathbf{v}, t) \notin \mathcal{A}(P, \mathbf{v}, t)$, the algebra forces Newton's law to change form. It is necessary to introduce a vector that, irrespective of the chosen active force, allows the acceleration to satisfy the restrictions imposed by the constraint:

$$m\mathbf{a} = \mathbf{F}(P, \mathbf{v}, t) + \mathbf{\Phi} . \tag{8.14}$$

From a dynamical viewpoint the vectors $\mathbf{\Phi}$ are naturally interpreted as reactions. Hence it is natural to *postulate* that the action a constraint exerts on a particle is representable by a force.

**Remark 8.11** In the real world the constraints are idealized by mechanical devices that force the particle to stay on a given surface or curve by contact forces associated with the devices' deformations, and finding the reactions' intensity is crucial to ensure the devices are really capable of constraining the motion without failures.

## 8.7 Ideal Constraints

Law (8.2) should be extrapolated from experimental observations of the motion of particles. These laws do not therefore have a general validity, but refer rather to a particular category of phenomena that correspond to a specific physical state in which the points of the system find themselves. In fact it is not always possible to interpret the mutual actions between the material systems of interest in technological applications on the basis of what today's Physics deems to be the fundamental interactions (strong, electromagnetic, weak and gravitational). It is therefore imperative to resort to phenomenological descriptions for the forces. In particular, the values these forces take when the velocity distribution is zero can be measured by static processes. The latter consist in considering devices built to maintain the body at rest and measure the action necessary to achieve that. These measuring processes enable us to connect the concept of force, here introduced in a purely abstract manner, to the intuitive notion of strength of a muscle. One such device is the mechanical dynamometer, an instrument made of a steel coil spring fixed attached to a ring above it. The force to be measured is applied at the bottom end. The device is calibrated using known forces so to operate within, and compare with, a range of measures of the intensity of the force.

The principle of mechanical determinism cannot hold for equation (8.14), for the latter clearly has an overdetermined number of unknowns. To re-establish mechanical determinism also for a constrained element it is necessary to introduce a dynamical characterization of the constrain, namely a phenomenological law that furnishes a priori information on the reaction produced by the physical constraint.

In such a situation the simplest assumption is that the constraint be *smooth*.

Consider a point $P$ constrained to a regular surface $\Sigma$ with implicit Cartesian equation $f(x, y, z) = 0$. The contraption that realizes such a constraint is smooth if and only if it is able to generate a reaction $\mathbf{\Phi}$ that is at all points normal to the surface.

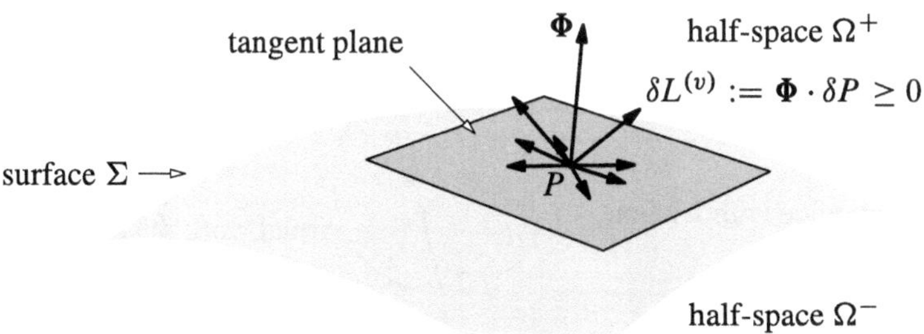


**Fig. 8.1** A point subject to a bilateral constraint with respect to a smooth surface

As $\nabla f$ is parallel to the normal of $\Sigma$, necessarily

$$\mathbf{\Phi} = \lambda \nabla f . \tag{8.15}$$

where $\lambda$ is an unknown factor.

In this case (8.14) becomes

$$m\mathbf{a} = \mathbf{F}(P, \mathbf{v}, t) + \lambda \nabla f . \tag{8.16}$$

Projecting the equation onto the tangent plane gives two scalar equations in the kinematic unknowns. Due to $f(x, y, z) = 0$ there exists two kinematic unknowns, locally. After long but straightforward linear-algebraic considerations, the equation can be locally put in normal form, and determinism is guaranteed once again. Once the kinematic unknowns are found, in order to compute the dynamical unknowns it suffices to dot multiply (8.16) by $\nabla f/|\nabla f|^2$ to obtain

$$\lambda = (m\mathbf{a} - \mathbf{F}) \cdot \frac{\nabla f}{|\nabla f|^2} .$$

It is easy to extend the present discussion to deal with a point constrained to a curve rather than a surface, since a curve can be seen as the intersection of two surfaces.

We point out that both for a particle $P$ constrained to a surface $\Sigma$ or to the half-space $\Omega^+$ bounded by $\Sigma$ (see Fig. 8.1), the reaction $\mathbf{\Phi}$ is normal to $\Sigma$ (see (8.15)), and pointing towards $\Omega^+$. For any virtual displacement $\delta P$ initiating at $P$, therefore,

$$\delta L^{(v)} = \mathbf{\Phi} \cdot \delta P \geq 0 \qquad \forall\, \delta P . \tag{8.17}$$

In fact all reversible virtual displacements belong on the tangent plane $\Pi$ to $\Sigma$, while irreversible ones form an acute angle with $\mathbf{\Phi}$ Fig. 8.1.

This remark enables us to characterize dynamically a family of more general constraints than smooth ones, called *ideal*.

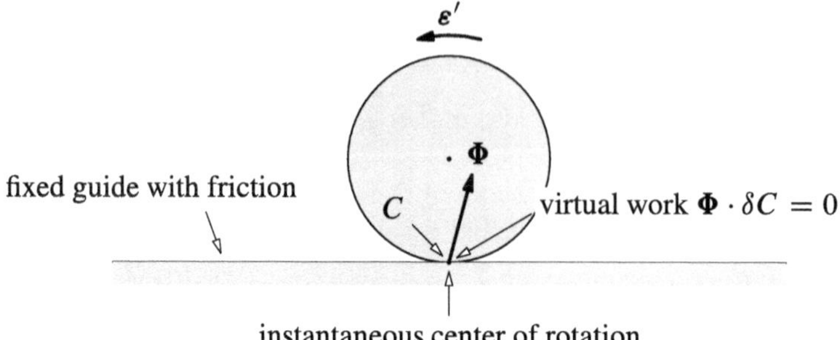


**Fig. 8.2** Pure rolling constraint of a disk on a straight line

**Definition 8.12** (*Ideal constraints*) A constraint is *ideal* if it is capable of generating *all* reactions with non-negative virtual work, or equivalently non-negative virtual power

$$\delta L^{(\mathrm{v})} \geq 0, \qquad \Pi^{(\mathrm{v})} \geq 0 \tag{8.18}$$

(and only these) for *any* virtual displacement or velocity distribution, starting from a generic configuration.

If the ideal constraints are also *bilateral* (the opposites of all virtual displacements or velocity distributions are virtual), then (8.18) become

$$\delta L^{(\mathrm{v})} = 0\,, \qquad \Pi^{(\mathrm{v})} = 0\,. \tag{8.19}$$

When the set of reactive forces associated with a constraint *does not* satisfy (8.18) we say we are in presence of *dissipation* or *dissipative constraints*. For this reason ideal constraints are also known as *non-dissipative*.

Let us immediately remark that the dynamical characterization in (8.18) holds for all smooth constraints, but in certain cases also for *non-smooth* ones.

Consider for instance a disk constrained to roll without slipping on a fixed straight guide, and call $C$ the contact point, as in Fig. 8.2. It is easy to imagine, in agreement with the experience, that to have rolling without slipping it is not enough to have a reaction orthogonal to the guide, because it would be impossible to guarantee $C$ does not slip. So we need a tangential component, naturally caused by some friction (rolling without slipping on a perfectly smooth surface is known to be impossible, as one experiences when driving on an icy road).

Yet in these hypotheses we still have a reaction that, albeit *not* coming from a smooth constraint, produces *zero* virtual work:

$$\delta L^{(\mathrm{v})} = \mathbf{\Phi} \cdot \delta C = 0\,,$$

since the pure rolling constraint forces $\delta C = \mathbf{0}$ (see Fig. 8.2).

## 8.8 Examples of Ideal Constraints

The panorama of possible constraints is very wide and their proper understanding is fundamental for all the applications of Mechanics.

Definition 8.12 makes the concept of ideal constraint precise. Informally, we may think of an ideal constraint as a constraint that produces the collection of forces strictly necessary to restrict the possible motions (or velocity distributions) to those imposed by the constraint itself, and nothing more, without generating any virtual work or power.

Conditions (8.18) and (8.19) allow, as we will see, to deduce the type of reaction to associate with a generic constraint, if thought of as ideal, starting from the kinematic restriction it imposes.

Dissipative constraints, on the other hand, exert an additional reaction component: although this is not necessary to produce the corresponding kinematic restriction, it formulates mathematically the presence of some sort of friction and models its action. The laws of dissipative constraints are therefore inferred based on empirical observations.

We shall discuss one by one the following constraints:

- unilateral and bilateral smooth contacts;
- rolling without slipping;
- screw constraints;
- constraints for planar systems: pins, rollers, sliders, fixed joints.

### 8.8.1 *Smooth Contact Constraint*

The bilateral contact constraint, presented and discussed from a mere kinematic viewpoint in Sect. 4.6, Chap. 4, prevents the detachment of a point, or surface, from another surface. It translates into the restriction that the *normal* components of the velocities (or infinitesimal displacements) of points instantaneously in contact are equal, as shown in Sect. 4.6, in particular in Fig. 4.14, right.

The simplest example, partly already discussed, is a particle $P$ constrained to a surface $\mathcal{S}$. Consider for the moment the constraint as bilateral, so that the point cannot move away from the surface.

Evidently the admissible velocities or virtual displacements generate the tangent space to the surface, both if the surface is fixed or moving, because by definition, when speaking of "virtual" quantities, we must imagine the constraints as fixed or frozen at the time considered. Since the definition of ideal constraint forces the reaction's virtual work or power to vanish, the reaction $\mathbf{\Phi}$ is necessarily perpendicular to the surface, with no restriction on its orientation.

If, instead, with reference to Fig. 8.1, we suppose the constraint only imposes that the point *does not penetrate* the surface, i.e. the constraint should be seen as

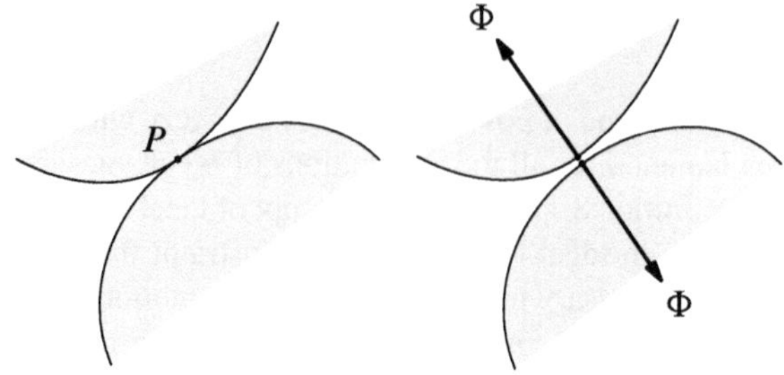


**Fig. 8.3** The reactive force associated with a *smooth* contact constraint between two surfaces is perpendicular to them. For a *unilateral* constraint the reactive force on each surface is oriented towards the body on which it is acting ($\Phi > 0$, where $\Phi$ is the component of each force as shown on the right)

unilateral, then (8.19) only holds when the reaction is orthogonal to the surface and points towards the spatial region allowed for the motion.

The above considerations extend, beyond isolated points, to points in rigid bodies and describe a *smooth* contact constraint. This is in contrast to the next section, when we will allow for friction.

In a *smooth contact* therefore, the reaction consists of a force orthogonal to the surface on which a point moves, or to the bodies' mutually tangent surfaces. For a bilateral constraint there are no restrictions on the orientation, while in the unilateral case one asks that the reaction $\mathbf{\Phi}$ is "repulsive", as depicted in Fig. 8.3.

In either case, in the light of (4.25) and (4.26) it is possible to prove that the system of reactions just described and associated with a *smooth contact constraint*, be it fixed or moving, external to the system or internal, satisfies inequality (8.18), or equality (8.19) for bilateral constraints. This guarantees the constraint is ideal.

### *8.8.2 Pure Rolling*

In Chap. 4, especially Sect. 4.6, we introduced the important constraint called *rolling without slipping*, a.k.a. *pure rolling*, for which the points in contact must have *same* velocity, and which can be, as seen, holonomic or nonholonomic, in different contexts.

Let us find which reaction is created when treating this constraint as ideal. For simplicity we first look at a body rolling without slipping on a fixed surface. As the constraint kills the virtual velocity of the contact point, any reaction $\mathbf{\Phi}$ applied to it has zero virtual power (and similarly zero virtual work), as shown in Fig. 8.4.

The remarkable consequence is that with this constraint we can, and we must, associate a constraint reaction that is *not completely orthogonal* to the contact surface; there must be a tangential part as well, which we must interpret as caused by some static friction. This is totally understandable, since physically it is evident that perfectly smooth surfaces cannot roll without slipping on one another, in general.

The law the reaction must obey is the law of a *static* friction, rather than a dynamic friction: this is due to the fact the velocities of the contact points are equal, so the points are at instantaneous relative rest (on this, see the discussion of the previous

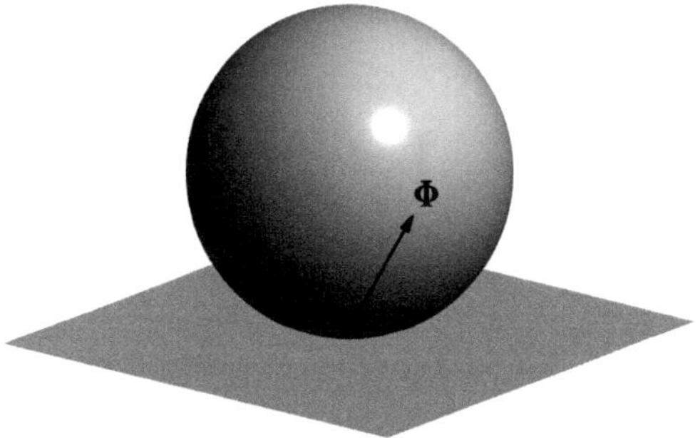


**Fig. 8.4** The reactive force $\mathbf{\Phi}$ exerted on the contact point of a sphere which is rolling without slipping on a plane

sections). Note though that this holds even when the constraint is applied to a moving system, not only when it is in equilibrium.

For a more exhaustive discussion we shall take two moving surfaces. In this case it is useful to separate the case of rolling rigid bodies that are both part of the system (an internal constraint, that is) from the case of a body rolling on a surface in given motion that does not belong to the system (an external constraint).

In the first situation each one of the reactions $\mathbf{\Phi}$ and $-\mathbf{\Phi}$ the surfaces exchange produce virtual power different from zero. But as the (effective) virtual velocities of the contact points $P_1$ and $P_2$ on which the forces act are equal, their *total* virtual power is still zero

$$\mathbf{v}_1' = \mathbf{v}_2' \quad \Rightarrow \quad \Pi' = \mathbf{\Phi} \cdot \mathbf{v}_1' + (-\mathbf{\Phi}) \cdot \mathbf{v}_2' = 0\,. \tag{8.20}$$

For a body in pure rolling on an external surface in given motion, which acts as moving constraint, the reaction's effective power is in general non-zero: $\mathbf{\Phi} \cdot \mathbf{v} \neq 0$, where $\mathbf{v}$ is the effective velocity of the point in contact with the moving constraint. But we must notice that the definition of ideal constraint is based of the demand that the *virtual* power (or work) of the reaction is zero, and the latter is computed imagining the constraint as *fixed*. Hence, for pure rolling, it is found by imposing the contact point's *virtual* velocity vanishes: $\mathbf{\Phi} \cdot \mathbf{v}' = 0$.

To sum up:

- Pure rolling is an ideal constraint, for which the reaction is a force with arbitrary direction applied at the contact point.
- This constraint admits, and supposes the presence of, a tangent part of the reaction, due to some static friction.
- In case of pure rolling on a moving surface external to the system (moving constraint) the reaction's virtual power is zero, while the effective power is not.

Some of these remarks are shown in Fig. 8.5. The system on the left has two generalized coordinates $\theta$ and $\phi$ and the *total* virtual power of the reactions due to the constraint at $P$ is zero, as per (8.20) ($P_1$ and $P_2$ are the physically distinct points that coincide geometrically with $P$). On the right we suppose the rod moves in a given way, thus representing a moving constraint for the disk. The system has one

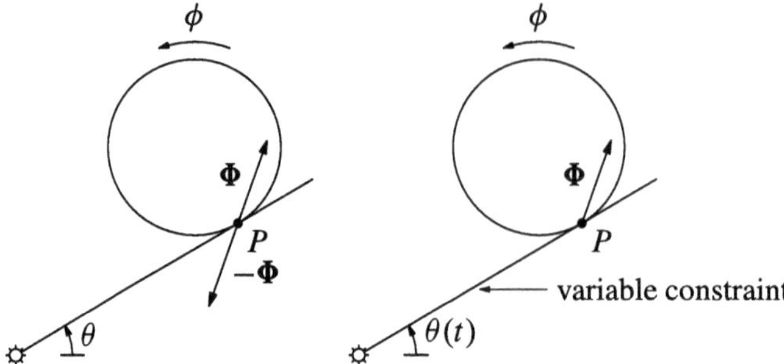


**Fig. 8.5** The pure rolling constraint for a disk. Left: the system is formed by a disk and a rod, with two generalized coordinates and the reactive force is thus internal to the system. Right: the rod is rotating with a prescribed law $\theta(t)$; it plays the role of variable constraint for the disk and this system has only one generalized coordinate

generalized coordinate $\phi$ and the *virtual* power of the rod' reaction $\mathbf{\Phi}$ is equal to zero, since the virtual velocity of the contact point is here zero. Note however that the *effective* power of this reaction does not vanish, because the effective velocity of $P$ is non-zero.

### 8.8.3 Screw Constraint

A *screw* forces the infinitesimal movement $\delta z$ of a rigid body along a comoving axis to be proportional to the infinitesimal variation $\delta\theta$ of the angle of rotation around the axis: $\delta z = h\delta\theta$, where the constant $h$ has the dimension of a length and is related to the *pitch* $p = 2\pi h$, the linear distance the rigid body advances by along the axis after a complete turn Fig. 8.6.

The constraint should be thought of as caused by suitable threads on the body or a part of it, and inside a fixed cavity. The classical example is naturally a bolt that penetrates a threaded hole as we fasten it.

The reaction corresponding to such a constraint is described by a collection of forces distributed along the threads, the resultant of which we call $\mathbf{\Phi}^{\mathrm{v}}$. Let $\mathbf{M}^{\mathrm{v}}_O$ be the moment with respect to a pole on the axis.

If the constraint is to be thought of as ideal we must impose that the reaction's virtual work is zero. Recall that the virtual work and the virtual power of a system of forces applied to a rigid body are given by (7.39) and (7.40). We further denote with $\mathbf{e}$ the axis' unit vector, with $\delta O = \delta z\mathbf{e} = h\delta\theta\mathbf{e}$ the infinitesimal displacement of $O$ along the axis, and finally with $\boldsymbol{\varepsilon}' = \delta\theta\mathbf{e}$ the body's infinitesimal rotation vector about the axis.

Let us write the condition whereby the reactions' virtual work vanishes as

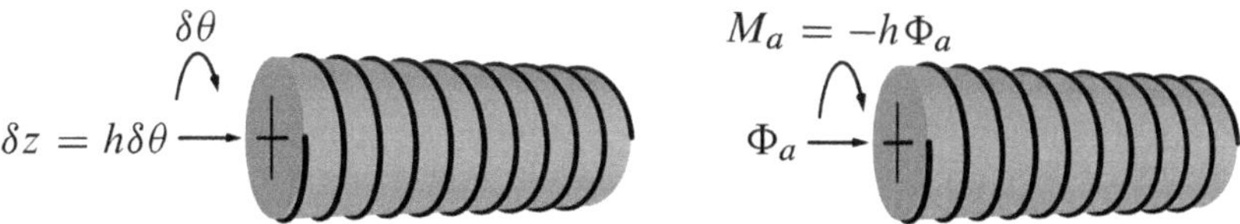


**Fig. 8.6** The screw constraint: the constant $h$ is related to the pitch $p = 2h\pi$ corresponding to the displacement of the screw axis when the body undergoes a full turn

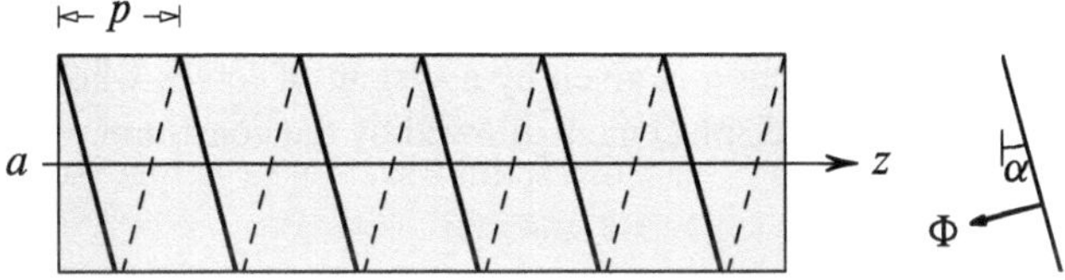


**Fig. 8.7** Lateral view of a screw with circular section when the thread has pitch $p$ and makes an angle $\alpha$ with the screw axis

$$\begin{aligned} \delta L^{\text{v}} &= \boldsymbol{\Phi}^{\text{v}} \cdot \delta O + \mathbf{M}_O^{\text{v}} \cdot \boldsymbol{\varepsilon}' = \left(h\boldsymbol{\Phi}^{\text{v}} + \mathbf{M}_O^{\text{v}}\right) \cdot \mathbf{e}\,\delta\theta \\ &= (h\Phi_a + M_a)\,\delta\theta = 0 \quad \text{for any } \delta\theta \end{aligned} \tag{8.21}$$

where $\Phi_a = \boldsymbol{\Phi}^{\text{v}} \cdot \mathbf{e}$ and $M_a = \mathbf{M}_O^{\text{v}} \cdot \mathbf{e}$ are the axial components of the resultant of reactions and their moment.

By (8.21), in an ideal screw (a "perfectly lubricated" screw) the components $\Phi_a$ and $M_a$ obey

$$M_a = -h\Phi_a \tag{8.22}$$

since this is necessary and sufficient for the reactions' virtual work (and power) to vanish, as indicated in Fig. 8.6 right.

Relation (8.22) between $\Phi_a$ and $M_a$ can be found directly, with the help of Fig. 8.7, for a screw with circular section of radius $R$ with perfectly smooth threading. The reaction $\Phi$ is at each point orthogonal, as in the right figure, so its axial component is $-\Phi\cos\alpha$, whereas the axial moment equals $R\Phi\sin\alpha$. The component $\Phi_a$ of the resultant along the fastening axis is therefore related to the axial moment $M_a$ by $M_a = -R\tan\alpha\,\Phi_a$, which coincides with (8.22) for $h = R\tan\alpha$.

### 8.8.4 Constraints for Planar Systems

We dedicate a specific section to a number of ideal constraints for planar systems, which are the most common in several applications and bear a particular importance form a didactical point of view. We now refer to Fig. 8.8, showing from top to bottom examples of constraints applied to a rigid body that *a priori* can move on the plane.

In the figure the rigid body is represented by a rod, but obviously we should think of a more general situation where the rod is replaced by any other element.

When we talk about planar systems we mean that all applied forces, both active and reactive, belong to the system's plane, while their moments and other possible couples, both active and reactive, are orthogonal to it. This is not a simple geometric feature of the system (to consist of points that belong to a plane, a priori) but also a mechanical property, because it involves all applied forces, external and internal, active and reactive.

This short list also aims to check, via significant and useful examples, how for any *ideal* constraint the reaction is given by a system of forces whose *total virtual work* is *zero* for any virtual displacement allowed by the constraint.

### Pin, or Hinge, Constraint

A planar hinge, depicted in the first figure on the left in Fig. 8.8, constrains a point of the rigid body to a fixed point on the plane (*fixed hinge*) or to a point comoving with a second rigid body that may itself move (*moving hinge*). In either case the hinge allows the constrained body to rotate freely, as shown in the figure.

The reaction corresponding to such a hinge is simply equivalent to a force on the plane, of which the right figure shows the horizontal and vertical components. Note the reaction's virtual work is clearly zero for a fixed hinge, as the constrained point cannot move. But the total virtual work is also zero for moving hinges, because we will need to add up the works of the two reactions acting on the bodies that constrain each other.

**Remark 8.13** Figure 8.9 shows two hinged bodies and the reactions $\boldsymbol{\Phi}_1, \boldsymbol{\Phi}_2$ they exchange due to the pin, applied respectively at $Q_1, Q_2$ (geometrically coinciding but physically distinct), for which the constraint clearly imposes that virtual velocities and displacements are equal: $\delta Q_1 = \delta Q_2$.

The action-reaction principle (Sect. 8.1.3) implies $\boldsymbol{\Phi}_1 + \boldsymbol{\Phi}_2 = \mathbf{0}$ so the total virtual work of these forces vanishes

$$\delta L^{(\mathrm{v})} = \boldsymbol{\Phi}_1 \cdot \delta Q_1 + \boldsymbol{\Phi}_2 \cdot \delta Q_2 = \delta Q_1 \cdot (\boldsymbol{\Phi}_1 + \boldsymbol{\Phi}_2) = 0\,.$$

Notice that it is the *total* virtual work that vanishes, while here the work of each reaction alone is non-zero. □

### Roller

One constraint we have already met (Sect. 4.1.3) is a *roller*, shown in the second picture of Fig. 8.8 (for its representation several drawing conventions are in use).

The planar roller constrains a point of the system to a guide, fixed or moving, or to a profile comoving with a second rigid body, without impeding its rotation as

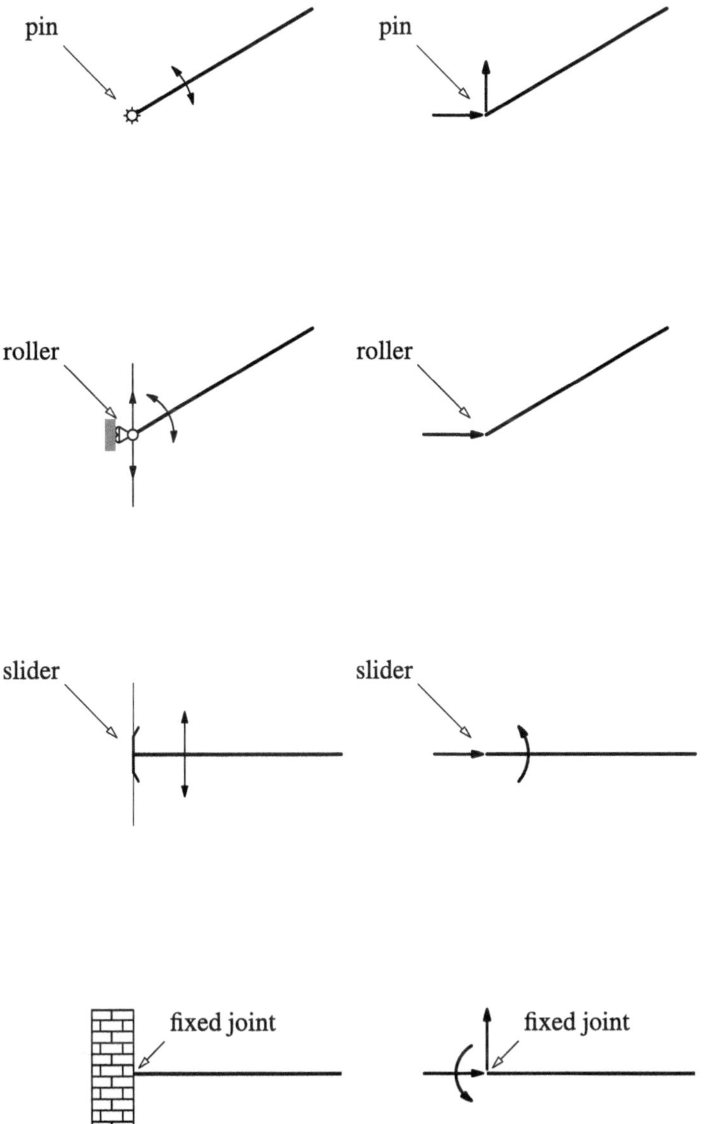


**Fig. 8.8** A collection of the most typical ideal constraints for planar systems: (left) the set of displacements which are possible; (right) the reactive forces and moments, which do zero virtual work. The fixed-joint constraint welds two rigid bodies into one and corresponds to both a reactive force and a reactive moment

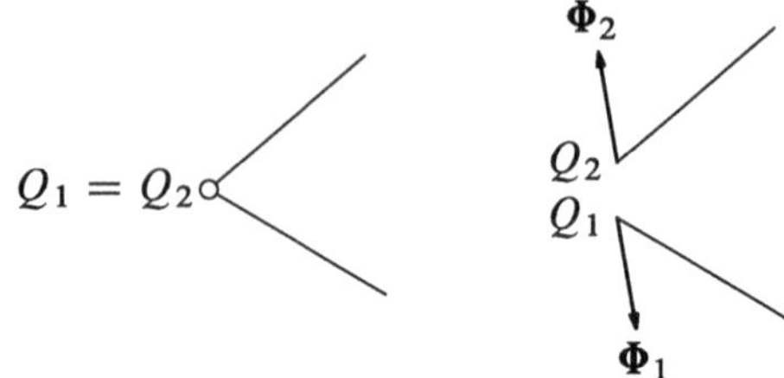


**Fig. 8.9** Two rods linked by a pin and the forces exchanged: $\mathbf{\Phi}_1 = -\mathbf{\Phi}_2$

the figure suggests. The reaction associated with a roller is just given by a force pointwise orthogonal to the guide on which the roller is constrained to move, as in Fig. 8.8 right. Here, as well, we see that the reaction's virtual work is zero, being orthogonal to every virtual displacement of the constrained point.

By definition the roller is a bilateral constraint, hence different from the so-called *smooth support*. In the latter the body could detach itself from the guide and the corresponding reaction therefore has a preferred orientation, the one forcing the point to stay on the guide.

### Slider

The *slider*, as in the third drawing in Fig. 8.8, constrains a point of the rigid body to move on a guide just as a roller would do; in addition, it *does not* allow any relative rotation (the picture is only meant to suggest one way to construct the constraint, and can be replaced by other conventions). This constraint as well is by definition *bilateral*, and the corresponding reaction consists of two elements that subsume the resultant and moment of a system of distributed forces. There is a component along the orthogonal direction to the guide, and an orthogonal moment to the plane. The latter quantity can be interpreted as the moment, with respect to the point where the slider finds itself, of the reactions' planar system. Here, too, the virtual work of the reactions is null.

### Fixed Joint

The last picture in Fig. 8.8 represents a planar joint, which in practice fastens a body to another, thus impeding all relative motions. The figure shows the very common situation where the joint is employed to make one rigid element of the system a comoving part of the external wall.

The reaction associated with a joint produces a force in the system's plane and an orthogonal moment. Referring for concreteness to Fig. 8.8, we may view the reaction as due to the actions exerted by the wall on the invisible portion of rod embedded into it. These actions form a system equivalent to a force and a couple with suitable torque. The force, whose horizontal and vertical components are shown on the right, must

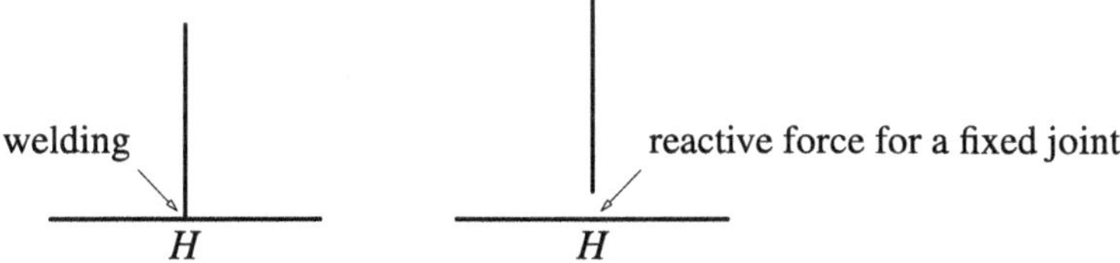


**Fig. 8.10** A T-rod is a rigid body which can be seen as a set of two rods "welded" at $H$. This welding is described quite naturally through the fixed-joint constraint. We can ideally separate the two rods at $H$ (right) introducing an internal action, which is not represented in this picture but can be seen in Fig. 8.11)

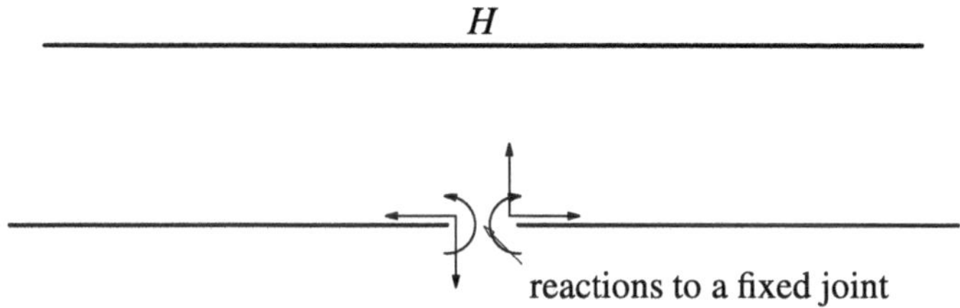


**Fig. 8.11** A rod formed by two pieces welded together through a fixed-joint constraint at $H$. It is natural to think that at $H$ there is an internal action, given by a force and a couple, as in the bottom picture

then be thought of as the resultant of these actions, while the moment, orthogonal to the plane, corresponds to the total moment of those distributed forces.

Interestingly, joints can be used to model "welded" rigid bodies (see Fig. 8.10 for a T-shaped rod).

Even more significantly, we can view a rigid rod as two parts joined together. In this case, thinking of the two parts separately, we can see the equal and opposite reaction forces each part exerts on the other at the joint, as shown in Fig. 8.11.

These considerations will lead us to discuss the concept of "stress" in Chap. 15.

## 8.9 Dissipative Constraints

Real constraints do not simply prescribe the components imposed by the limited motions of the points on which they act. They also present components that affect, and typically hamper, the motion that would otherwise be allowed by the constraint. In this section we will analyze the three most employed models to describe dissipative constraints. They are

- the static rough-contact constraint;
- the dynamic rough-contact constraint;
- the law of rolling friction.

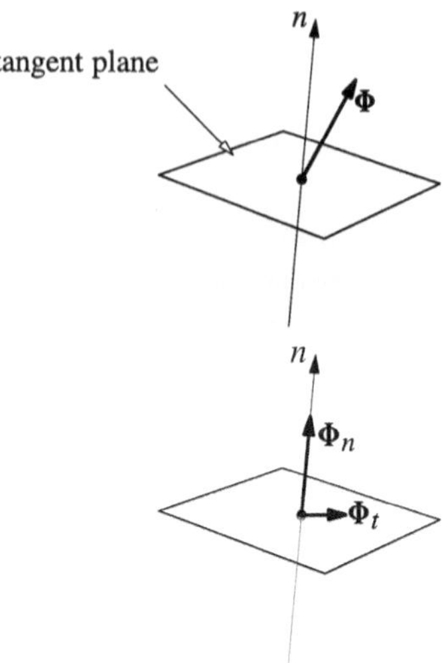


**Fig. 8.12** A point constrained to a rough surface. The constraint reaction $\mathbf{\Phi}$ is decomposed into a normal part $\mathbf{\Phi}_n$ and a tangential part $\mathbf{\Phi}_t$

### 8.9.1 Rough Contact: Law of Static Friction

The existence of some friction associated to a contact constraint translates into the presence of a grazing component of the reaction $\mathbf{\Phi}$, meaning a tangential component to the surface that can (in principle) do virtual work, thus making the constraint non longer ideal. In this case one speaks of a *rough-contact constraint.*

It is important to distinguish the case where the points in contact are at rest, or at rest with respect to one another, to that in which they have different velocities. Here we discuss the first situation, which is always present in static problems although it is not confined to such cases, as we will see later.

Empirical observations have led to deduce that, in case the relative velocity of the contact points is zero, the tangential part $\mathbf{\Phi}_t$ and the normal part $\mathbf{\Phi}_n$ of the reaction $\mathbf{\Phi}$ (represented in Fig. 8.12) must always satisfy the *Coulomb-Morin inequality*

$$|\mathbf{\Phi}_t| \leq \mu_s |\mathbf{\Phi}_n| \tag{8.23}$$

where $\mu_s$ is a dimensionless quantity, called *coefficient of static friction*, sometimes denoted simply by $\mu$ or $f_s$.

Note that (8.23) holds when the points in contact have the same velocity, and this does not only happen in static conditions, as we will see in the sequel regarding rolling without slipping.

The value of the coefficient of static friction $\mu_s$ depends on the physical nature of the touching points or surfaces, and must therefore be deduced from suitable experiments. Although inequality (8.23) is empirical, it does hold in the most common situations.

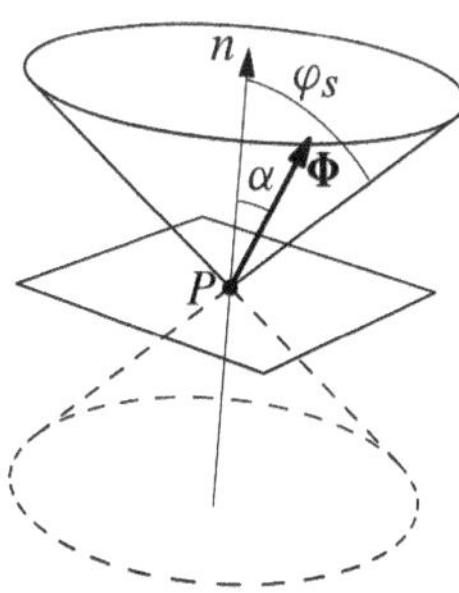


**Fig. 8.13** The friction cone: $\tan\alpha \le \mu_s = \tan\varphi_s$

Putting $\mu_s = 0$ in (8.23) gives $\mathbf{\Phi}_t = \mathbf{0}$, taking us back to a smooth contact. The vanishing of the coefficient of static friction is therefore equivalent to smoothing out the roughness.

There is a useful geometric interpretation of the coefficient of static friction based on the friction cone. This is the cone with axis perpendicular to the surface and angle $\varphi_s$ such that $\tan\varphi_s = \mu_s$. Elementary geometric considerations tell us that in order to satisfy the Coulomb-Morin inequality (8.23) the reaction $\mathbf{\Phi}$ must stay inside the cone, since the angle $\alpha$ formed with the normal must remain less than or equal to $\varphi_s$, as in Fig. 8.13.

Finally, as already mentioned regarding Fig. 8.2 in Sect. 8.7, and as we will see better later on, the presence of static friction by itself does not imply the constraint is necessarily dissipative.

### 8.9.2 Rough Contact: Law of Dynamic Friction

When the contact points are not at relative rest, in view of the kinematical discussion carried out in Sect. 4.6 one speaks of *slipping* contact, where by *slip velocity* we mean the difference between the two points' velocities.

We can have slipping contact in a variety of situations:

(i) a point that slips on a surface, fixed or in motion;
(ii) two touching surfaces that slip on one another;
(iii) a point that slips along a fixed or moving guide.

To fix ideas, in the sequel one should have in mind case (i), remembering though that the discussion naturally extends without any problem to case (ii). For case (iii) instead we will need to make additional remarks.

The components of the velocities or the elementary displacements along the surface normal must be the same, to prevent detachment, as shown in Fig. 4.14. Hence the slip velocity is necessarily tangential.

The presence of friction manifests itself through the tangential component $\mathbf{\Phi}_t$ of the constraint reaction. In presence of non-zero slip velocity, this tangential part is

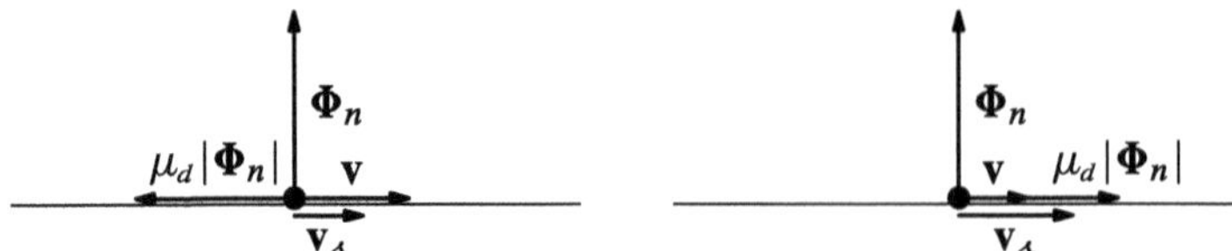


**Fig. 8.14** The dynamic friction force acting on a point moving with velocity $\mathbf{v}$ on a rough surface which has velocity $\mathbf{v}_s$ has a tangential component *opposite* to the slip velocity $\mathbf{u}_s = \mathbf{v} - \mathbf{v}_s$ of the point on the surface and proportional to the normal part of the constraint reaction. Notice that if $\mathbf{v}$ is oriented as $\mathbf{v}_s$ but has *smaller* magnitude, as shown on the right, the dynamic friction force is oriented as $\mathbf{v}$ itself

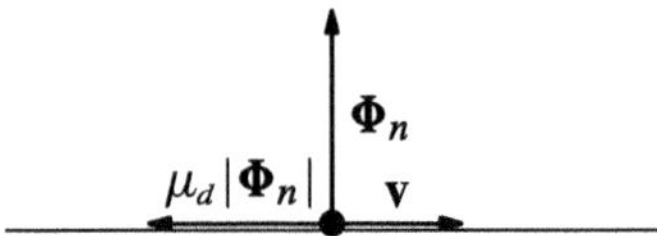


**Fig. 8.15** The dynamic friction force exerted by a *fixed* surface has a tangential component opposite to the velocity of the point and proportional to the normal component of the constraint reaction

related to the normal component $\boldsymbol{\Phi}_n$ by the law of *dynamic friction*, whose nature is empirical exactly as (8.23).

Let $P$ denote a particle and $P_s$ the point (geometrically coinciding with $P$) on the surface, fixed or moving, on which the slip contact occurs.

The law of dynamic friction establishes that the modulus $|\boldsymbol{\Phi}_t|$ of the tangential reaction acting on point $P$ is proportional, with a certain constant $\mu_d$, to the modulus $|\boldsymbol{\Phi}_n|$ of the normal component

$$|\boldsymbol{\Phi}_t| = \mu_d |\boldsymbol{\Phi}_n| \,. \tag{8.24}$$

The *coefficient of dynamic friction* $\mu_d$ is sometimes written $f_d$ and depends on the material properties of the surfaces or contact points, but not, to a good degree, on the slip velocity.

The coefficient of dynamic friction $\mu_d$ has been measured experimentally and is always less than the coefficient of static friction $\mu_s$, so that if a force manages to set a body in motion on a rough guide, the same force keeps it moving.

In contrast to the static Coulomb-Morin law (8.23), the constitutive relation (8.26) is an *equation*, not an inequality. In other words, the moduli of the reactions' normal components determine the friction modulus, and not just an upper bound as we would have at equilibrium.

The tangential part $\boldsymbol{\Phi}_t$ is oppositely oriented to the relative velocity of $P$ with respect to the point $P_s$, i.e. $\mathbf{v} - \mathbf{v}_s$ (as the points are the same geometrically, the difference in velocity does not depend on the frame). Let us define more precisely the *slip velocity* $\mathbf{u}_s$ of the point on the surface to be

$$\mathbf{u}_s = \mathbf{v} - \mathbf{v}_s \tag{8.25}$$

and then write the law of dynamic friction as

$$\boldsymbol{\Phi}_t = -\mu_d |\boldsymbol{\Phi}_n| \frac{\mathbf{u}_s}{|\mathbf{u}_s|} \tag{8.26}$$

where $\boldsymbol{\Phi}_t$ is the friction produced *by* $P_s$ *on* $P$, and the ratio on the right is the unit vector of the slip velocity $\mathbf{u}_s = \mathbf{v} - \mathbf{v}_s$ (see Fig. 8.14). The relation in (8.26), also known as *dynamic Coulomb-Morin law*, becomes meaningless when the contact points have the same velocity, i.e. when the slip velocity is zero.

The minus sign on the right in (8.26) expresses the demand that the dynamic friction *is opposite* to the *relative velocity* of the point with respect to the rough surface on which it moves (see again Fig. 8.14 and the considerations in the caption).

The meaning of (8.26) becomes clearer when looking at a point $P$ moving on a fixed surface, for which then $\mathbf{v}_s = \mathbf{0}$. In this case (8.26) reduces to

$$\boldsymbol{\Phi}_t = -\mu_d |\boldsymbol{\Phi}_n| \frac{\mathbf{v}}{|\mathbf{v}|} \tag{8.27}$$

where $\mathbf{v}$ is the velocity of the point $P$ that the friction opposes, see Fig. 8.15.

In the light of the above considerations the dynamic friction can produce power $\Pi = \boldsymbol{\Phi} \cdot \mathbf{v}$ that can be both negative (always true when the surface is fixed) or positive. The latter can happen when the surface is moving.

**Remark 8.14** Case 8.9.2, that of a point slipping on a fixed or moving *guide*, requires some comments. The reaction $\boldsymbol{\Phi}$ decomposes into a *tangential* part $\boldsymbol{\Phi}_t$ and a part $\boldsymbol{\Phi}_\perp$ on the orthogonal plane to the guide. From the theory of curves we know (Sect. A.2) that the orthogonal plane to a curve contains the unit normal $\mathbf{n}$ and the binormal $\mathbf{b}$ (which together with the unit tangent $\mathbf{t}$ form a Frenet basis). Hence we can write the modulus of $\boldsymbol{\Phi}_\perp$ as

$$|\boldsymbol{\Phi}_\perp| = \sqrt{\Phi_n^2 + \Phi_b^2}\,.$$

Then (8.24) becomes

$$|\boldsymbol{\Phi}_t| = \mu_d |\boldsymbol{\Phi}_\perp|$$

i.e.

$$\left|\boldsymbol{\Phi}_t\right| = \mu_d \sqrt{\Phi_n^2 + \Phi_b^2}\,. \tag{8.28}$$

(Note that here $\Phi_n$ has a different meaning from that in (8.24) or (8.27), where the subscript $n$ denoted the *normal to the surface* of contact, while here it indicates the *unit normal* to the guide.) □

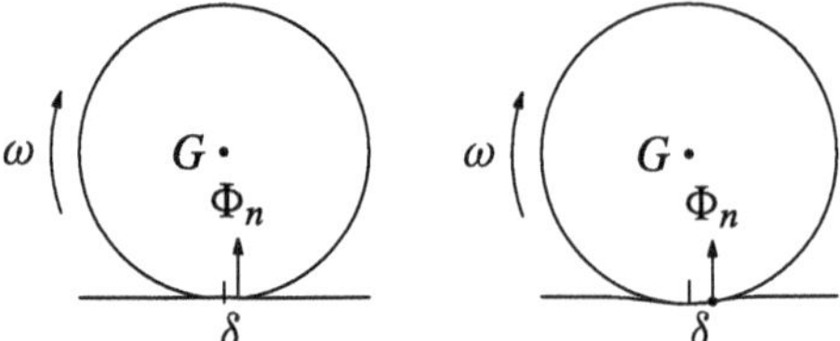


**Fig. 8.16** An intuitive explanation of rolling friction: the normal part of the constraint reaction is applied at distance $\delta$ after the contact point, to take into account an *infinitesimal* deformation of the disk or guide on which it is rolling

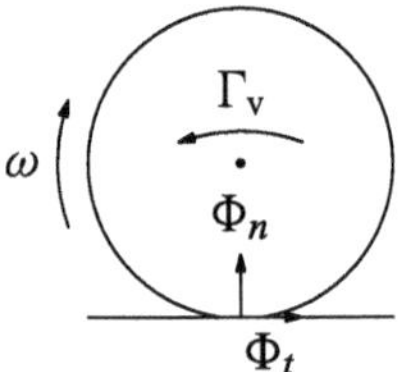


**Fig. 8.17** The torque of rolling friction $\Gamma_v$ accounts for the small distance between the line of action of $\Phi_n$ and the ideal contact point (see Fig. 8.16)

### *8.9.3 Law of Rolling Friction*

The rolling-without-slipping constraint is therefore ideal, even in presence of static friction. In reality the empirical observations detect an additional form of friction, called *rolling friction*, that describes in a more realistic fashion the motion of a body (say, a disk) that rolls without slipping on a fixed or moving surface.

This kind of friction translates into a couple that produces a torque $\mathbf{\Gamma}_v$, proportional to the normal component of the reaction and opposite to the angular velocity of the body upon which it acts (or opposite to the angular velocity relative to the body on which it rolls, more generally).

One motivation for the presence of rolling friction is the concrete possibility that one of the surfaces involved in the rolling is not 100% rigid and undergoes a small deformation, similar to the one represented in Fig. 8.16: on the left, a tire touches the ground not at one point but on an entire patch, technically called *footprint*. On the right we see a guide, along which the rolling occurs, that is not perfectly rigid: the figure highlights the small deformation of the guide due to the load exerted by the disk.

Because of the (infinitesimal) elasticity of the disk or guide on which the motion happens, the normal part of the reaction is applied at a small distance $\delta$ *after* the ideal contact point. This is caused by the resistance to the deformation of the part of disk or guide that gets compressed, which is *not* counterbalanced by the force the disk receives from the successive section of guide. The distance $\delta$ therefore represents a non-zero arm for $\Phi_n$ with respect to the ideal contact point that is right below the center $G$. This arm produces a rolling torque $\mathbf{\Gamma}_v$ to be accounted for.

So, we can describe the reaction as if it were applied at the point of ideal contact Fig. 8.17 as long as we add a torque $\mathbf{\Gamma}_\mathrm{v}$ corresponding to the couple generated by the effective displacement of the line of action of $\Phi_n$ (this boils down to introducing a *transport moment*, as thoroughly explained in Sect. 7.4.1, Chap. 7).

The laws describing rolling friction are codified by the relations

$$\begin{aligned} \mathbf{u}_s \neq \mathbf{0} \quad &\Rightarrow \quad \mathbf{\Gamma}_\mathrm{v} = -\delta_d |\Phi_n| \frac{\boldsymbol{\omega}}{|\boldsymbol{\omega}|} \\ \mathbf{u}_s = \mathbf{0} \quad &\Rightarrow \quad |\mathbf{\Gamma}_\mathrm{v}| \leq \delta_s |\Phi_n| \end{aligned} \tag{8.29}$$

where $\mathbf{u}_s$ is the slip velocity defined in (8.25), and $\boldsymbol{\omega}$ is the angular velocity of the body on which the rolling friction torque $\mathbf{\Gamma}_\mathrm{v}$ acts (relative to the rigid surface on which it rolls, if the surface is moving). The similarity between (8.29) and the laws of static and dynamic friction should not go amiss.

The quantities $\delta_s$ and $\delta_d$ appearing in (8.29) are the *coefficients of rolling friction*. Dimension-wise they are both lengths, and in the most common situations they may be taken to be equal, and written $\delta$. We also known that they are approximately independent of the modulus of $\Phi_n$. Moreover, in the case of wheels, there are measurements showing how $\delta_s$ and $\delta_d$ decrease with the radius.

Clearly, in the light of the above, a meticulous analysis of the rolling friction phenomenon requires we frame it within the theory of elasticity, which is beyond the reach of this treatise.

Finally, note that the presence of dynamic friction turns *dissipative*, hence no longer ideal, the rolling-without-slipping constraint, as one deduces from the virtual work (7.41) of a couple applied to a rigid body.

## 8.10 Classification of Forces

The forces applied to a *system* can be classified using two separate schemes:

- active forces and reactions;
- external forces and internal forces

(reactions are also called “reactive forces”).

Another way to describe this classification is to divide forces in four mutually exclusive categories:

- internal active forces;
- external active forces;
- internal reactions;
- external reactions.

*Active forces*, as introduced and discussed in Chap. 7, are given in terms of the configuration and velocity distribution of the system, plus possibly time.

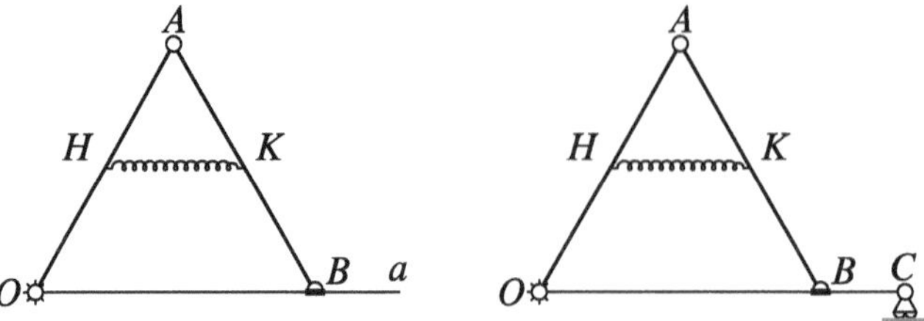


**Fig. 8.18** Two mechanical systems

*Reactions* are instead unknown *a priori*, yet not completely arbitrary, since as we have seen, for ideal constraints their virtual work must obey certain conditions. For rough or dissipative constraints there exist further relations, such as the laws of static, dynamic and rolling friction; these typically function as *constitutive equations* and characterize the reactive forces associated with each constraint.

*External forces* are caused by the action *on* points of the system of external portions of bodies or particles, while *internal forces* are due to the action of other points or bodies included in the system.

It is important to observe that the internal vs. external distinction is subordinated to the choice of mechanical system under exam.

To distinguish the quantities (resultants, moments, work, power) referring to the various types of forces we will deploy superscripts, like: "ext", "e", "int", "i", "a", "r", whose meaning is patent.

These will be omitted whenever it is clear from the context which forces we are referring to, without ambiguity.

**Example 8.15** The system in Fig. 8.18, left, is placed on a vertical plane and consists of two rods joined by a hinge at $A$. Rod $OA$ is further pinned to a point fixed at $O$, while the end $B$ moves along a horizontal guide called $a$. The midpoints of the rods are connected by a spring.

The *external active* forces are the weights of the rods, while the forces the spring exerts on $H$ and $K$ are *active but internal*. The *external reactions* act on $O$ and $B$. There also is an exchange of forces of *internal reactive* type happening at $A$.

If we focus on rod $OA$ and restrict the "system" to it, the spring's force at $H$ is external, and so is the force that the rod (now, the system) receives from $AB$ at the point $A$.

Look at Fig. 8.18, right, where the horizontal guide has been replaced by a rod $OC$ constrained by a roller at $C$.

If now by "system" we mean the three rods $OA$, $AB$, $OC$, we must declare the forces that $AB$ and $OC$ exchange at $B$ as reactive and internal. On the other hand in this new "system" there is a further external reaction at $C$. □

## 8.11 Mach's Point of View on the Foundations of Mechanics

To close the chapter we return to the foundations of Classical Mechanics, for cultural completeness.

We owe Ernst Mach (Chirlitz-Turas, Moravia 1838–Haar 1916) an alternative path to the construction of the axioms and fundamental laws (equivalent to what we presented at the start of the chapter). This road is more gratifying, logically speaking, that the one adopted before, even if it is more complex and artificial.

Once we have defined inertial systems with the scheme of Sect. 8.1.1, we may examine the following axiom, originally due to Mach.

***Postulate*** (*Mach's postulate for isolated pairs*) When two isolated points $P_1$, $P_2$ move in an inertial frame, their accelerations have the same direction as the line $P_1 P_2$, opposite orientations and intensities in constant ratio.

The concept of isolated pair arises with the purpose of idealizing two real bodies that are close to each other and far from all other bodies. By Newton's law we can associate with each pair of points a positive real number, the ratio of their accelerations:

$$\frac{a_1}{a_2} = \sigma\left(P_1, P_2\right) . \tag{8.30}$$

We posit that $\sigma\left(P_1, P_2\right)$ is constant, i.e. independent of the position and velocity of the points, and of time. This makes the ratio the characteristic element of the isolated pair.

Now imagine each particle $P_i$ enters into an isolated pair with every other particle $P_j$. We suppose that for any triple of particles $\left\{P_i, P_j, P_h\right\}$

$$\sigma\left(P_i, P_j\right)\sigma\left(P_j, P_h\right) = \sigma\left(P_i, P_h\right) . \tag{8.31}$$

This allows to use the characteristic ratio $\sigma\left(P_i, P_j\right)$ to define an equivalence class: particles $P_i$ and $P_j$ are equivalent if

$$\sigma\left(P_i, P_j\right) = 1 . \tag{8.32}$$

Relation (8.32) is reflexive, symmetric and transitive, as is easily proved with (8.30) and (8.31).

Call $m_i$ the equivalence class of particle $P_i$. Then Newton's law can be formulated as

$$\begin{aligned} &m_1 \mathbf{a}_1 + m_2 \mathbf{a}_2 = \mathbf{0} , \\ &O P_1 \times m_1 \mathbf{a}_1 + O P_2 \times m_2 \mathbf{a}_2 = \mathbf{0} , \end{aligned} \tag{8.33}$$

provided we *define*

$$\frac{m_2}{m_1} = \sigma\left(P_1, P_2\right) .$$

The quantity $m_i$ is *by definition* the *mass* of particle $P_i$.

Each point of the isolated pair can have a non-zero acceleration, so it is only natural to look at the accelerations $\mathbf{a}_1$ and $\mathbf{a}_2$ as if they were due to the interactions occurring between the particles. These can be expressed by vectorial quantities

$$\mathbf{F}_1 = m_1 \mathbf{a}_1 , \quad \mathbf{F}_2 = m_2 \mathbf{a}_2 , \tag{8.34}$$

which *by definition* are named *forces*. In terms of forces then, (8.33) implies

$$\mathbf{F}_1 = \varphi \mathbf{e}_{21} , \quad \mathbf{F}_2 = \varphi \mathbf{e}_{12} , \tag{8.35}$$

where $\varphi$ is a scalar quantity denoting the forces' intensity. Equations (8.35) are nothing but the action-reaction principle.

The next step is to pass to systems of particles.

***Postulate*** (*Mach's postulate for systems of isolated points*) In the motion of an isolated system of arbitrarily many particles, the forces $\mathbf{F}_{ij}$ and $\mathbf{F}_{ji}$ that any two points $P_i$ and $P_j$ exerts on one another in an inertial frame do not depend on the other points, and agree with (8.35):

$$\mathbf{F}_{ij} = \varphi_{ij} \mathbf{e}_{ji}, \quad \mathbf{F}_{ji} = \varphi_{ij} \mathbf{e}_{ij} .$$

That is, for any point $P_i$ in the system

$$\sum_{j \neq i} \mathbf{F}_{ij} = m_i \mathbf{a}_i . \tag{8.36}$$

The notion of isolated system is the natural extension of isolated pairs to sets of arbitrary cardinality, and (8.36) generalize equations (8.34).

# Chapter 9
# Statics

The Laws of Mechanics enable us to study any kind of motion once the forces and the initial conditions are known. There is however a special case that deserves to be examined first, also in view of its numerous applications: the static solutions, or equilibrium solutions, to the equations of motion. We shall then devote the present chapter to the conditions that allow, and guarantee, a system to be at rest. After that, we will pass to examine the full equations of motion.

## 9.1 Equilibrium and Rest

Consider a system of active forces that do not depend on time: $\mathbf{F} = \mathbf{F}(P, \mathbf{v})$. Among the solutions to Newton's equation

$$m\,\mathbf{a} = \mathbf{F}(P, \mathbf{v}) \tag{9.1}$$

there may be the rest solution, which is tightly related to the concept of equilibrium position.

**Definition 9.1** (*Rest*) A constant solution to (9.1), $P(t) \equiv P^*$ for all $t \geq t_0$, corresponding to initial data $P(t_0) = P^*$, $\mathbf{v}(t_0) = \mathbf{0}$, is called a rest solution.

**Definition 9.2** (*Equilibrium*) A configuration $P^*$ is called an equilibrium if it solves

$$\mathbf{F}(P^*, \mathbf{0}) = \mathbf{0}\,,$$

i.e. if it is a position at which the force vanishes when evaluated at all non-zero values of the velocity.

It is immediate to check that if a position $P^*$ is a rest position, because of (9.1) it must be an equilibrium. Vice versa, suppose we take an equilibrium initial position



P. Biscari et al., *Rational Mechanics*, UNITEXT 177,
https://doi.org/10.1007/978-3-032-07462-1_9

$P(t_0) = P^*$ with $\mathbf{v}(t_0) = \mathbf{0}$: it is obvious that there is the rest solution, but it might not be the only one. More precisely, it will be the only solution if the Cauchy problem associated with (9.1) admits a unique solution (an example when this is not the case is given by (8.5) in Remark 8.3).

At this juncture we must recall that the analytical form of the forces comes from mathematical models. By accepting the Principle of causality (determinism), therefore, we will assume the active forces acting on the system obey the hypotheses of Cauchy's theorem. In this way *rest* and *equilibrium* become synonyms, meaning that each rest solution is an equilibrium and if at time $t = t_0$ the system is in an equilibrium position with zero velocity, it will stay at rest.

## 9.2 Equilibrium of a Particle

The problem of the equilibrium of a free point is conceptually simple. It boils down to finding the zeroes of a vector-valued function:

$$\mathbf{F}(P) = \mathbf{0} \qquad \Longleftrightarrow \qquad \begin{cases} F_x(x, y, z) = 0 \\ F_y(x, y, z) = 0 \\ F_z(x, y, z) = 0 \end{cases}$$

Note that without loss of generality we can just consider positional forces. In fact if there were velocity-dependent active forces acting on the system, in Statics they would have to be evaluated at the value zero of the velocity.

Consider now a point constrained on a curve or surface, subject to an active force $\mathbf{F}$ (see Fig. 9.1). By the Constraint postulate (Sect. 8.6) and the Principle of superposition of forces (Sect. 8.1.4), the equilibria will be those for which the total force acting on $P$ is zero:

$$\mathbf{F}(P) + \mathbf{\Phi} = \mathbf{0}\,. \tag{9.2}$$

As is to be expected, problem (9.2) is underdetermined, since the constraint reaction $\mathbf{\Phi}$ is an unknown of the problem. In fact, up to this moment we have not provided any information regarding the physical nature of the constraint's material.

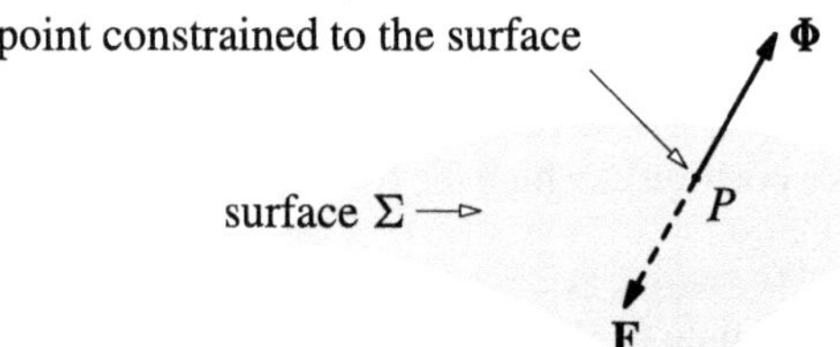


**Fig. 9.1** Equilibrium of a point on a surface

As we saw in Sect. 8.8, Chap. 8 and the subsequent sections, the constraints and the constraint reactions can be smooth or rough, and the corresponding constraint reactions have different descriptions.

Now let us show which are the equilibrium conditions for a point subject to a resultant of the active forces $\mathbf{F}$ and constrained to a surface, which will be first smooth and later rough.

For a smooth surface, as discussed in Sect. 8.8.1, we know the constraint reaction is orthogonal to the surface: $\boldsymbol{\Phi} = \Phi \mathbf{n}$, where $\mathbf{n}$ is the unit normal.

Equation (9.2) can be decomposed along the surface's normal and tangent directions

$$\mathbf{F}_t + \boldsymbol{\Phi}_t = \mathbf{0} \quad \mathbf{F}_n + \boldsymbol{\Phi}_n = \mathbf{0}\,. \tag{9.3}$$

Since for a smooth constraint $\boldsymbol{\Phi}_t = \mathbf{0}$, we deduce that a point $P$ of the surface is an equilibrium if and only if

$$\mathbf{F}_t = \mathbf{0}\,. \tag{9.4}$$

with normal component of the constraint reaction $\boldsymbol{\Phi}$ given by $\boldsymbol{\Phi}_n = -\mathbf{F}_n$.

One says that condition (9.4) is the *free equilibrium equation* of a point constrained to a smooth surface (this easily extends to smooth guides). The word "free" refers to the fact the equation does not involve the constraint reaction but only restricts the active force.

But we known that the constraint can be *bilateral* or *unilateral*, as explained in Sect. 8.8.1: in the former case the constraint forces the point to stay *on the* surface, while in the latter case it prevents the point from *penetrating* the surface, thus making it stay on one side.

In the first case (bilateral constraint) the constraint reaction $\boldsymbol{\Phi}$ has arbitrary orientation, so long as it stays perpendicular to the surface. In this case there is nothing more to add to equation (9.4), which then guarantees the equilibrium.

But if the constraint is unilateral, and $\mathbf{n}$ points toward the part of space containing the point, from the discussion of Sect. 8.8.1 we will have

$$\Phi_n = \boldsymbol{\Phi} \cdot \mathbf{n} \geq 0\,. \tag{9.5}$$

Hence $(9.3)_2$ together with inequality (9.5) forces

$$F_n = \mathbf{F} \cdot \mathbf{n} \leq 0\,.$$

Consequently a point subject to an active force $\mathbf{F}(P)$ and a *unilateral* constraint given by a surface is an equilibrium if and only if at $P$

$$\mathbf{F}_t = 0 \quad \mathbf{F} \cdot \mathbf{n} \leq 0$$

(with $\mathbf{n}$ unit normal and the aforementioned convention).

Next we discuss and deduce the free equilibrium condition for a point constrained to a rough surface, with law of static friction given as in Sect. 8.9.1.

Equation (9.2), projected along the normal and onto the tangent plane to the surface, is equivalent to (9.3). By the Coulomb-Morin inequality (8.23), this implies that to ensure equilibrium we need

$$|\mathbf{F}_t| \leq \mu_s |\mathbf{F}_n| \tag{9.6}$$

since only then the tangent and normal components of the constraint reaction $\boldsymbol{\Phi}$ satisfy (8.23).

We therefore say (9.6) is the *free equilibrium condition* for a point constrained to a rough surface.

Clearly, if the constraint is unilateral, we will accompany (9.6) with $F_n \leq 0$ (always under the assumption that $\mathbf{n}$ is the unit normal oriented towards the half-space containing the point).

In more mathematical terms, suppose we are dealing with an active force $\mathbf{F}$ depending on point $P(x, y, z)$, and let $f(P) = 0$ be the implicit equation describing the constraint surface. Then we may set

$$\psi(P) = |\mathbf{F}_t| - \mu_s |\mathbf{F}_n|$$

and write the collection of equilibrium conditions for a bilateral constraint as

$$\mathbf{F}(P) + \boldsymbol{\Phi} = \mathbf{0} \qquad \psi(P) \leq 0 \qquad f(P) = 0\,.$$

The equilibrium condition (9.6) can also be interpreted geometrically using the *cone of static friction* introduced in Sect. 8.9.1 (see Fig. 8.13). This is the cone with vertex at $P$ and semi-aperture $\varphi_s$ such that $\tan\varphi_s = \mu_s$. Since by (9.2) the active force is equal and opposite to the constraint reaction, *we have equilibrium if the active force does not lie outside the cone of static friction* (see Fig. 9.2). For a *unilateral* rough constraint $\mathbf{F}$ must further belong to the half-cone in whose interior the opposite of $\mathbf{n}$ belongs.

**Remark 9.3** Observe how the friction tends to attain equilibrium. In fact, the larger the coefficient $\mu_s$, the easier (9.6) will hold. We may even see that in the limit as

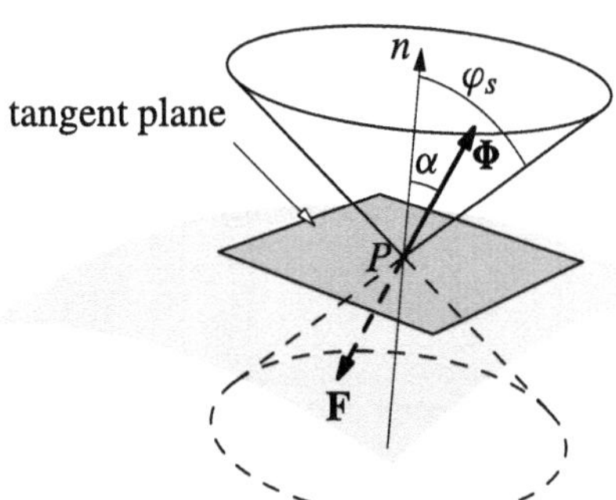


**Fig. 9.2** Static friction cone

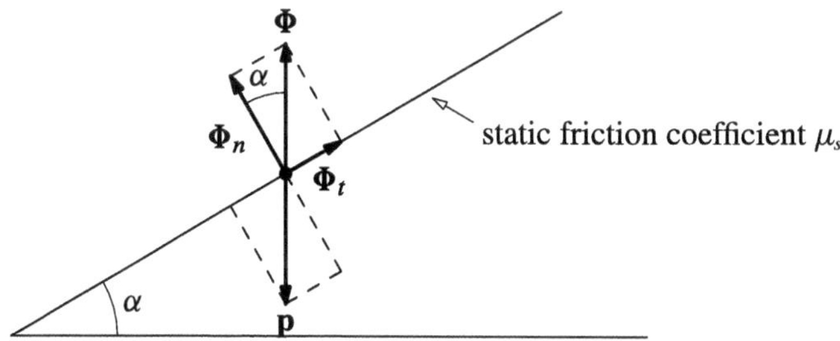


**Fig. 9.3** A point with weight $p$ on a slanted rough guide. Equilibrium is possible only if $\tan\alpha \le \mu_s$

$\mu_s \to \infty$ almost all positions (more precisely, all except those where $F_n = 0$) become equilibria. Once we have found the equilibria we can then use (9.3) to compute the value of the corresponding constraint reaction $\mathbf{\Phi}$. □

**Remark 9.4** Condition (9.6) provides a simple method to estimate the coefficient of static friction $\mu_s$.

Consider a particle $P$ on an inclined plane with varying angle $\alpha$, subject to one vertical active force only, equal to its weight Fig. 9.3. Decompose the weight $p$ in a tangent part $p\sin\alpha$ and a normal part $p\cos\alpha$. Since there is equilibrium until (9.6) holds, this happens if and only if

$$|p\sin\alpha| \le \mu_s |p\cos\alpha|$$

i.e.

$$\tan\alpha \le \mu_s\,.$$

Since the tangent function is monotone, until $\alpha$ is less than a critical value $\varphi_s$ ($0 \le \varphi_s < \pi/2$), called *angle of static friction*, the equilibrium condition holds. The latter is violated when $\alpha$ exceeds that value. So we can deduce an empirical rule to find the coefficient of static friction by setting

$$\mu_s = \tan\varphi_s \ge 0\,.$$ □

**Example 9.5** A point $P$ of mass $m$ is constrained to a smooth circle of radius $R$ and center $O$ lying on a vertical plane. On the point are acting its weight and an elastic force $\mathbf{f}_e = -k(CP)$ with center $C$ at the highest point of the circle, as shown in Fig. 9.4.

In this case the equilibrium criterion says that

$$\mathbf{\Phi} + m\mathbf{g} - kCP = \mathbf{0}\,, \tag{9.7}$$

and the dynamic characterization of the constrain, if we look at Fig. 9.4, is expressed by asking that the reaction $\mathbf{\Phi}$ is parallel to the vector $OP$. Projecting along the Cartesian axes and using well-known geometric and trigonometric properties gives

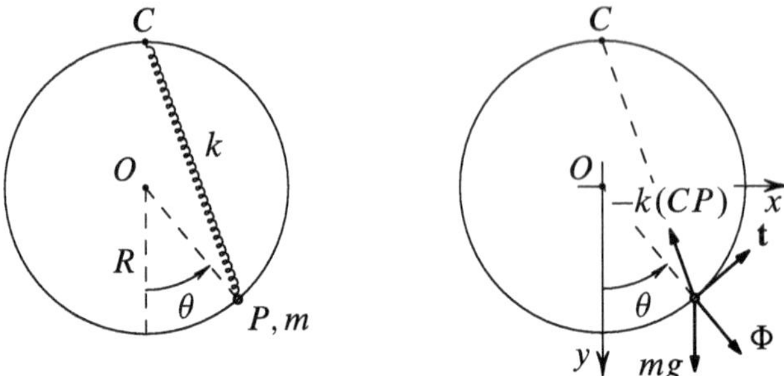


**Fig. 9.4** A particle constrained to a smooth circle and subject to the action of a spring

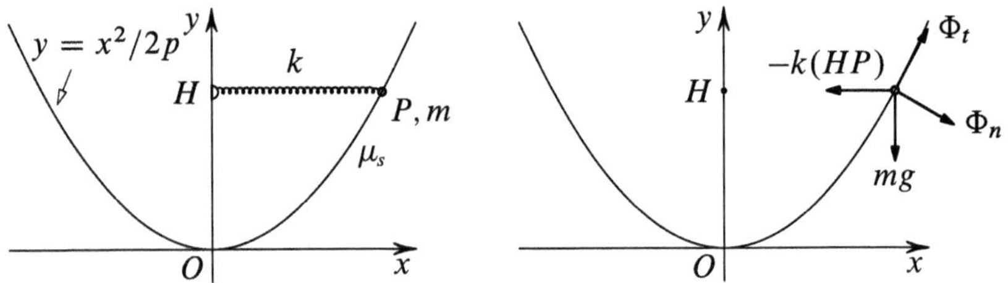


**Fig. 9.5** A particle $P$ constrained to a rough parabolic guide and subject to the action of a spring

$$\Phi \sin\theta - 2kR\sin\tfrac{\theta}{2}\cos\tfrac{\theta}{2} = 0\,, \quad \Phi\cos\theta + mg - 2kR\cos^2\tfrac{\theta}{2} = 0\,. \tag{9.8}$$

Now multiply the first of (9.8) by $\cos\theta$ and the second by $\sin\theta$ then subtract: this gives the pure equation in the kinematic unknown

$$(kR - mg)\sin\theta = 0\,. \tag{9.9}$$

This holds for all $\theta$ if $mg = kR$, or for $\theta = 0, \pi$. On the other hand if we multiply the first equation by $\sin\theta$ and the second by $\cos\theta$ and add, we find the modulus of the constraint reaction (of which we only knew the direction in principle)

$$\Phi = kR + (kR - mg)\cos\theta\,. \tag{9.10}$$

Hence $\Phi = kR$ for $mg = kR$; $\Phi = 2kR - mg$ when $\theta = 0$, and lastly $\Phi = mg$ for $\theta = \pi$.

Note that equations (9.9) and (9.10) could have been found directly by projecting (9.7) along the tangent and normal directions to the circle. □

**Example 9.6** A point $P$ of mass $m$ is constrained to a rough parabola of equation $2py = x^2$ and coefficient of static friction $\mu_s$ lying on a vertical plane. Acting on the point are the weight and an elastic force $\mathbf{f}_e = -k(HP)$ with center $H$ at the projection of the point on the $y$-axis, as in Fig. 9.5.

In this case the equilibrium criterion imposes

$$\mathbf{\Phi} + m\mathbf{g} - k\,HP = \mathbf{0}\,, \tag{9.11}$$

where the constraint reaction is characterized by the Coulomb-Morin law. Set $\mathbf{F} = m\mathbf{g} - k\,HP$, so from (9.11) we have $\mathbf{\Phi} = -\mathbf{F}$ and therefore the Coulomb-Morin law holds whenever

$$|F_t| \le \mu_s\,|F_n|\,,$$

where $F_t = \mathbf{F}\cdot\mathbf{t}$ and $F_n = \mathbf{F}\cdot\mathbf{n}$. Computing the unit tangent and unit normal

$$\mathbf{t} = \frac{1}{\sqrt{p^2+x^2}}(p\,\mathbf{i} + x\,\mathbf{j})\,, \qquad \mathbf{n} = \frac{1}{\sqrt{p^2+x^2}}(-x\,\mathbf{i} + p\,\mathbf{j})\,,$$

we see that we are in equilibrium when the inequality

$$(kp + mg)\,|x| \le \mu_s|kx^2 - mgp| \tag{9.12}$$

holds. Define the dimensionless quantity $\kappa = kp/(mg)$, so (9.12) reads

$$\mu_s^2\kappa^2x^4 - \big(\kappa^2 + 2(1+\mu_s^2)\kappa + 1\big)p^2x^2 + \mu_s^2p^4 \ge 0\,. \tag{9.13}$$

In the smooth case ($\mu_s = 0$) inequality (9.13) admits the only solution $x = 0$. For any positive $\mu_s$, it is possible to find the four real roots (two positive and two negative) of the quartic equation associated with (9.13). Studying the expression's sign allows to prove that all points such that

$$\frac{|x|}{p} \le \frac{\sqrt{(1+\kappa)^2 + 4\mu_s^2\kappa} - (1+\kappa)}{2\mu_s\kappa}$$

or

$$\frac{|x|}{p} \ge \frac{\sqrt{(1+\kappa)^2 + 4\mu_s^2\kappa} + (1+\kappa)}{2\mu_s\kappa}$$

are equilibria. While for a smooth constraint we have just one equilibrium, in presence of friction there exist infinitely many such. This is to be expected as we are solving an inequality. □

## 9.3 Equilibrium Balance Equations

*We will say a system of particles $\mathscr{I} = \{P_i,\ i = 1,\dots,n\}$ is in equilibrium when each point is*. Hence, the total force acting on each point must vanish.

In the general case it will be possible to divide $\mathscr{I}$ into the set $\mathscr{I}_0$ of free points, on which only active forces act, and the set $\mathscr{I}_\varphi$ formed by constrained points. In an

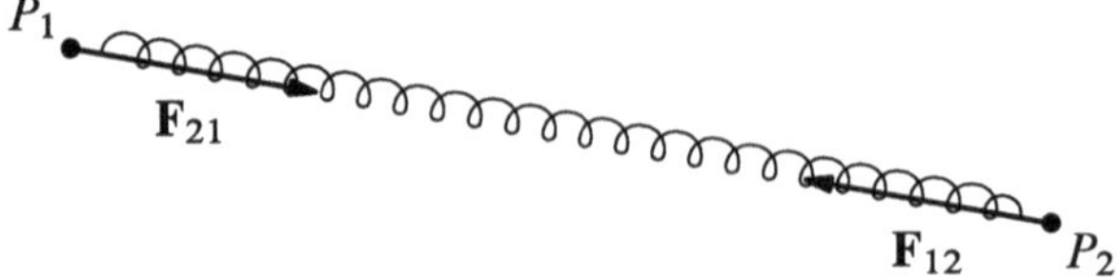


**Fig. 9.6** An example where the balance equations, applied to the external forces acting on a system, are not enough for an equilibrium

equilibrium configuration $\mathscr{C}^*$:

$$\begin{cases} \mathbf{F}_i = \mathbf{0} & \forall\, P_i \in \mathscr{I}_0 \\ \mathbf{F}_i + \mathbf{\Phi}_i = \mathbf{0} & \forall\, P_i \in \mathscr{I}_\varphi \,. \end{cases} \tag{9.14}$$

**Proposition 9.7** (Equilibrium balance equations) *Consider a system of points in $\mathscr{I}_0$ and let us divide the acting forces into* internal forces *and* external forces *(see Sect. 8.3). A* necessary *condition to have equilibrium is that the system of external forces is balanced, i.e. both the resultant and the moment with respect to a generic pole $O$ are zero*

$$\mathbf{R}^{(e)} = \mathbf{0}\,, \qquad \mathbf{M}_O^{(e)} = \mathbf{0}\,. \tag{9.15}$$

*Equations* (9.15) *are called* first *and* second equilibrium balance equation *(the latter with respect to pole $O$).*

***Proof*** In an equilibrium condition the resultant of the forces acting on each point must be zero. Therefore an equilibrium forces the system of all forces acting on the system to be balanced (zero resultant and zero resultant moment, as per Definition 7.9). On the other hand the system of internal forces is always balanced because of the action-reaction principle, as proved by Theorem 8.5. Subtracting those, a necessary condition for the equilibrium of a system is that the system of external forces is balanced. □

It is fundamental to stress that *for a generic system the equilibrium balance equations are typically not sufficient to ensure equilibrium.* To see this consider the counterexample in Fig. 9.6, where a spring joins two isolated points $P_1$, $P_2$. In any configuration of the system the balance equations hold, since the two elastic forces $\mathbf{F}_{21}$, $\mathbf{F}_{12}$ are internal. However the system is evidently not in equilibrium since acting on either point there is a non-zero force. At the same time it is also obvious that, in general, having zero resultant and zero resultant moment, as required by (9.15), cannot imply that on *each* point the forces are zero, which is what (9.14) asks for.

This example confirms that a generic system of points is in equilibrium *if and only if* each point is in equilibrium.

## 9.4 Principle of Virtual Work

For systems subject to ideal constraints it it is possible to prove a fundamental result, called *Principle of virtual work*, that allows to characterize the equilibria of a mechanical system using a simple principle of global nature. In the statement we will for simplicity write $\delta P$ (without subscripts) for a generic virtual displacement of the points of a system, i.e. an infinitesimal displacement compatible with the constraints, which in Statics are assumed fixed.

**Theorem 9.8** (Principle of virtual work) *In a mechanical system with ideal constraints, a configuration $\mathscr{C}$ is an equilibrium if and only if the work of the* active *forces is non-positive for all virtual displacements $\delta P$ starting at $\mathscr{C}$:*

$$\delta L^{(\mathrm{a})} \leq 0 \qquad \forall\, \delta P \textit{ from } \mathscr{C}\,. \tag{9.16}$$

*In particular*

$$\delta L^{(\mathrm{a})} = 0 \qquad \forall\, \delta P \textit{ reversible from } \mathscr{C}\,.$$

***Proof*** To prove (9.16) is necessary for the equilibrium we assume $\mathscr{C}$ is an equilibrium and therefore that (9.14) hold, where we have divided the system into the free points $\mathscr{I}_0$ and the constrained points $\mathscr{I}_\varphi$. Then

$$\delta L^{(\mathrm{a})} = \sum_{i=1}^{n} \mathbf{F}_i \cdot \delta P_i = \sum_{P_i \in \mathscr{I}_0} \mathbf{F}_i \cdot \delta P_i + \sum_{P_i \in \mathscr{I}_\varphi} \mathbf{F}_i \cdot \delta P_i = -\sum_{P_i \in \mathscr{I}_\varphi} \mathbf{\Phi}_i \cdot \delta P_i = -\delta L^{(\mathrm{r})}$$

and the ideal constraint hypothesis (8.18) gives us claim (9.16).

Sufficiency: assume (9.16) for all virtual displacements. Consider first only the displacements of the free points. As these are clearly reversible,

$$\sum_{P_i \in \mathscr{I}_0} \mathbf{F}_i \cdot \delta P_i = 0 \quad \forall\, \delta P_i \text{ with } P_i \in \mathscr{I}_0\,. \tag{9.17}$$

In particular (9.17) will hold even if we displace a single $P_k \in \mathscr{I}_0$, and therefore

$$\mathbf{F}_k \cdot \delta P_k = 0 \quad \forall\, \delta P_k\,. \tag{9.18}$$

As $P_k$ is free, its virtual displacements are arbitrary, and so (9.18) implies $\mathbf{F}_k = 0$. Repeating for all points $P_i \in \mathscr{I}_0$ we find that $(9.14)_1$ holds:

$$\mathbf{F}_i = 0 \quad \forall\, P_i \in \mathscr{I}_0\,. \tag{9.19}$$

So there remains to prove $(9.14)_2$. Now the virtual work of the active forces only involves constrained points, by (9.19), and so by assumption

$$\delta L^{(a)} = \sum_{P_i \in \mathscr{I}_\varphi} \mathbf{F}_i \cdot \delta P_i \leq 0 \qquad \forall \delta P_i \in \mathscr{I}_\varphi \,. \tag{9.20}$$

Consider, for all points $P_i \in \mathscr{I}_\varphi$, a hypothetical constraint reaction that is equal and opposite to the active force:

$$\mathbf{\Phi}_i^* = -\mathbf{F}_i \quad \forall P_i \in \mathscr{I}_\varphi \,. \tag{9.21}$$

Clearly, for (9.21) to be equivalent to $(9.14)_2$ we need to prove $\mathbf{\Phi}_i^*$ can be a true constraint reaction. This is easy, since (9.21) and (9.20) tell that the possible constraint reactions do non-negative work:

$$\delta L^{(*)} = \sum_{P_i \in \mathscr{I}_\varphi} \mathbf{\Phi}_i^* \cdot \delta P_i = - \sum_{P_i \in \mathscr{I}_\varphi} \mathbf{F}_i \cdot \delta P_i \geq 0 \,.$$

Due to the "if and only if" characterizing ideal constraints (more precisely, the "if" part) we conclude that the $\mathbf{\Phi}_i^*$, as they belong to the class of reactions that the ideal constraints can in principle generate, will truly be produced, thus warranting equilibrium. Therefore (9.21) reduce to $(9.14)_2$ and the claim is proven. □

**Remark 9.9** It might surprise the reader to call "principle" what should more appropriately be labelled *Theorem of virtual work*, given we have just proved it. In reality, while (9.16) is obviously necessary for a system under ideal constraints to be in equilibrium, that condition is not automatically sufficient. If we look carefully at the proof we see that the crux of the matter is the equivalence between *equilibrium* and *rest*, hence Cauchy's theorem. If we wish to take into account systems of forces that do not satisfy Cauchy's theorem, equilibrium and rest are not the same, and the *Theorem of virtual work* reverts to being a *Principle*. For this reason we chose to preserve the historically sanctioned terminology. □

**Definition 9.10** We call *free equilibrium relation* any condition (equation or inequality) that characterizes the equilibrium configurations and *only involves active forces*. A similar definition holds for the free motion equations.

Evidently the *Principle of virtual work* provides free equilibrium relations: in particular, free equations if the displacements are reversible. This feature is one of its major advantages. As we will see, it allows to deduce and analyze the equilibrium conditions irrespective of any knowledge of the constraint reactions and their computation.

**Example 9.11** Consider a homogeneous rod of weight $p$ whose end $A$ is pinned to a fixed point and is free to rotate around it on a vertical plane, as in the first picture of Fig. 9.7. Suppose that two horizontal barriers are present at $H$ and $K$ that prevent the rod from leaving the upper half-space.

Using the Principle of virtual work we can find the equilibrium configurations and check they agree with our intuition.

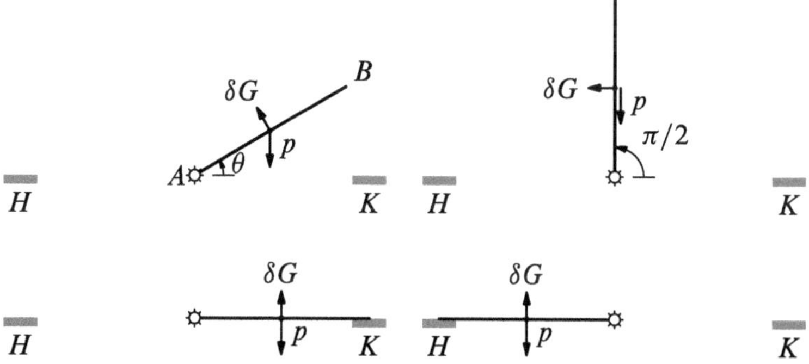


**Fig. 9.7** A rod pinned at one end and subject to unilateral constraints

Note first that because of the constraint at $A$, any virtual displacement of the center of gravity $G$, the rod's midpoint, is necessarily orthogonal to the rod itself. There are two boundary configurations, for which the possible virtual displacements are not reversible, corresponding to $\theta = 0$ and $\theta = \pi$, where $\theta$ is the angle indicated in the first drawing.

In the configuration where $B = K$, i.e. for $\theta = 0$, the unilateral constraint says the only virtual displacements are those with $\delta\theta \geq 0$. When $B = H$, i.e. $\theta = \pi$, we must have $\delta\theta \leq 0$.

The virtual work of the only active force acting on the system, the weight $p$, is

$$\delta L = \mathbf{p} \cdot \delta G = -\tfrac{1}{2} pl \cos\theta \, \delta\theta$$

where $l$ is the rod's length.

Let us focus on the configurations with $0 < \theta < \pi$, where $\delta\theta$ can have any sign. Then $\delta L = 0$ for all virtual displacements if and only if $\cos\theta = 0$, when the rod is in vertical position. In this configuration in fact the weight is *orthogonal* to all virtual displacements $\delta G$, as shown in the second picture.

Consider now the boundary configuration $\theta = 0$, so $\cos\theta = 1$. Here

$$\delta L = \mathbf{p} \cdot \delta G = -\tfrac{1}{2} pl \, \delta\theta$$

and since only variations $\delta\theta \geq 0$ are allowed,

$$\delta L^{(a)} \leq 0 \quad \text{for all virtual displacement} \tag{9.22}$$

holds. Hence the system's equilibrium condition is guaranteed (which is evident from an intuitive viewpoint).

In the position where $\theta = \pi$, i.e. $\cos\theta = -1$, the weight's virtual work is

$$\delta L = \mathbf{p} \cdot \delta G = \tfrac{1}{2} pl \, \delta\theta \, .$$

Due to the unilateral constraint, only variations $\delta\theta \leq 0$ are permitted and so (9.22) holds, meaning that we have equilibrium. □

## 9.5 Equilibrium of a Rigid Body

Through the Principle of virtual work it is possible to show that the equilibrium balance equations, which are a necessary condition for generic material systems, become sufficient to have equilibrium of a rigid body, whether free or constrained.

**Theorem 9.12** (Characterization of equilibrium for rigid bodies) *A rigid system is in equilibrium in a configuration $\mathscr{C}$ if and only if the equilibrium balance equations hold, that is to say, the collection of external forces is balanced.*

***Proof*** Evidently the balance equations are necessary for the equilibrium, of either a rigid body or any material system.

To show they are sufficient as well, we shall consider the worst possible case (from the point of view of equilibria) for simplicity. This is the case where the rigid body is subject to ideal constraints, though we warn the reader that the theorem holds for generic constraints. This agrees with the experience: if a position is an equilibrium for ideal constraints, it will be even more so under real constraints.

By assumption the body is rigid, and occupies a configuration where the equilibrium balance equations (9.15) hold. We recast them separating active forces from constraint reactions:

$$\mathbf{R}^{(\mathrm{e,a})} + \mathbf{R}^{(\mathrm{e,r})} = \mathbf{0}, \qquad \mathbf{M}_O^{(\mathrm{e,a})} + \mathbf{M}_O^{(\mathrm{e,r})} = \mathbf{0}\,. \tag{9.23}$$

We saw in (7.39) the virtual work of forces acting on a rigid body has a special form, in which the forces only appear through the resultant and the resultant's moment:

$$\delta L = \mathbf{R} \cdot \delta O + \mathbf{M}_O \cdot \boldsymbol{\varepsilon}'\,. \tag{9.24}$$

Hence we can multiply $(9.23)_1$ by the virtual displacement $\delta O$ of $O$ and $(9.23)_2$ by $\boldsymbol{\varepsilon}'$, and find for (9.24):

$$\delta L^{(\mathrm{e,a})} + \delta L^{(\mathrm{e,r})} = 0\,. \tag{9.25}$$

On the other hand (9.24) implies the virtual work of the internal forces $\delta L^{(\mathrm{i})}$ is identically zero in a rigid body, since both $\mathbf{R}^{(\mathrm{i})}$ and $\mathbf{M}_O^{(\mathrm{i})}$ vanish (we proved this in Theorem 8.7). So we can add $\delta L^{(\mathrm{i})}$ to the work (9.25) so that $\delta L^{(\mathrm{e,a})} + \delta L^{(\mathrm{e,r})} + \delta L^{(\mathrm{i})} = 0$, i.e.

$$\delta L^{(\mathrm{e,a})} + \delta L^{(\mathrm{e,r})} + \delta L^{(\mathrm{i,a})} + \delta L^{(\mathrm{i,r})} = 0\,.$$

Rearranging terms (gathering together the first and third, and the other two):

$$\delta L^{(\mathrm{a})} + \delta L^{(\mathrm{r})} = 0\,. \tag{9.26}$$

As we have supposed the constraints are ideal, $\delta L^{(r)} \geq 0$ for all virtual displacements, and therefore (9.26) implies $\delta L^{(a)} \leq 0$ for all virtual displacements from $\mathscr{C}$. Since the Principle of virtual work is a sufficient condition, $\mathscr{C}$ is an equilibrium. □

### 9.5.1 Reduction of Forces Applied to a Rigid Body

In Sect. 7.2 we defined as *equivalent* two systems of forces that have the same resultant and the same resultant moment. We also proved (Theorem 7.15) that given any system of forces it is always possibile to find an equivalent system made of one vector and one couple at most. The case of planar systems is especially simple, for there the reduction allows to find an equivalent system with only one vector or only one couple.

We have also just proved that a system of forces allows for the equilibrium of a rigid body if and only if it is balanced. But since that latter condition depends exclusively on the characteristic vectors (resultant and resultant's moment), we have the following consequence.

**Proposition 9.13** *A system of forces acting on* a single rigid body *can be replaced by an equivalent system without altering the equilibrium configurations*

Clearly, the fact the equilibrium balance equations are not enough for non-rigid bodies in general *does not* allow to replace a system of forces applied to a system of several rigid bodies with a equivalent system.

**Example 9.14** Suppose we want to put in equilibrium a massive rigid body (say, a painting). The study of reducible systems of parallel forces (see Sect. 7.4.2) shows we can replace the system of infinitely many weights with their resultant, applied at the center of gravity. It will therefore be enough to apply an equal and opposite force to the total weight at any point on the resultant's line of action (the vertical line through the center of gravity) to attain equilibrium. □

**Example 9.15** Consider a homogeneous rod of length $l$ and weight $p$ in a vertical plane, whose ends $A$ and $B$ are joined to the Cartesian origin $O$ by springs with the same elastic constant $k$, as in Fig. 9.8. The elastic forces $\mathbf{f}_1$ and $\mathbf{f}_2$ meet at $O$ and therefore, imposing $\mathbf{M}_O = \mathbf{0}$, we immediately deduce that in an equilibrium condition the rod's center of gravity $G$ must be along the $y$-axis, as in the figure on the right.

By definition of elastic force $\mathbf{f}_1 = -k(OA)$, $\mathbf{f}_2 = -k(OB)$, so the resultant of the forces acting on the rod is

$$\mathbf{R} = -k(OA) - k(OB) + p\mathbf{j},$$

where $\mathbf{j}$ is the unit vector of the downward $y$-axis Fig. 9.9. Setting equal to zero this resultant is the same as asking $p\mathbf{j} = k(OA + OB)$, which we write

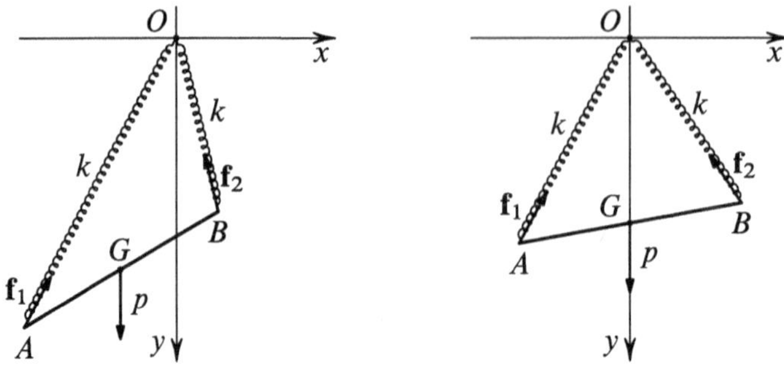


**Fig. 9.8** A rod with weight $p$ subject to the action of two springs

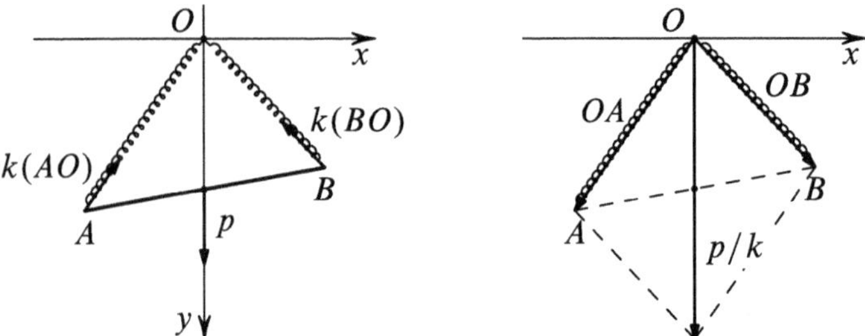


**Fig. 9.9** Deduction of the equilibrium condition

$$(p/k)\mathbf{j} = OA + OB\,.$$

We conclude that the vector $(p/k)\mathbf{j}$ is the diagonal of the parallelogram of edges $OA$ and $OB$, as in Fig. 9.9 right.

Since the diagonals meet at their midpoint, the $y$ coordinate of $G$ is one half of $p/k$. Hence $y_G = p/2k$, while the angle of rotation of the rod is arbitrary. In other words, as long as $G = (0, p/2k)$ the rod can assume infinitely many equilibrium configurations in which $A$ and $B$ are the ends of any diameter in the circle of radius $l/2$ and center $G$. □

## 9.6 Equilibrium of a Constrained Rigid Body

In this section we will examine remarkable cases of equilibrium of rigid bodies admitting different ideal constraints.

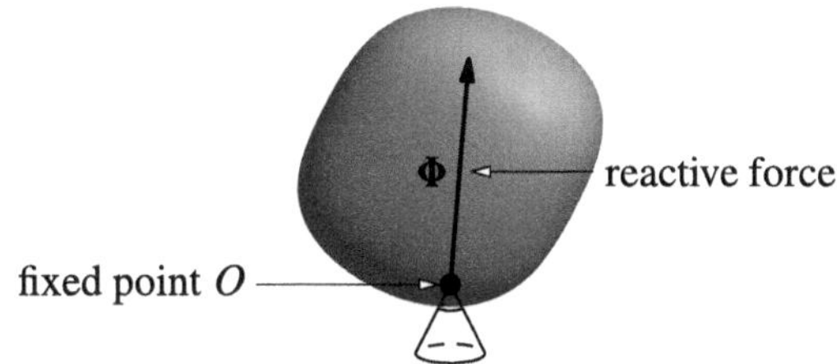


**Fig. 9.10** A rigid body with a fixed point

### 9.6.1 Rigid Body with a Fixed Point

We introduced in Sect. 4.1.2 the constraints represented by a pin, fixed or mobile. In the first case we assume a point $O$ in a rigid body is constrained to a fixed position Fig. 9.10. In the second, two points $Q_1$, $Q_2$, belonging to different rigid bodies, are constrained to be in the same position at any instant. Let us analyze the constraint reactions produced by such constraints.

The constraint reaction $\mathbf{\Phi}$ applied to a fixed pin $O$ does not do any work since by definition of constraint $\delta O = \mathbf{0}$, so

$$\delta L^{(\mathrm{r})} = \mathbf{\Phi} \cdot \delta O = 0\,. \tag{9.27}$$

In this case the virtual displacements are all reversible. We also point out that if the constraint were real and not ideal, there would be a region of points around $O$ subject to constraints. Calling $\mathbf{R}^{(\mathrm{r})}$ and $\mathbf{M}^{(\mathrm{r})}$ the resultant and moment of the constraint reactions, the virtual work would become (see (7.39))

$$\delta L^{(\mathrm{r})} = \mathbf{R}^{(\mathrm{r})} \cdot \delta O + \mathbf{M}_O^{(\mathrm{r})} \cdot \boldsymbol{\varepsilon}' = \mathbf{M}_O^{(\mathrm{r})} \cdot \boldsymbol{\varepsilon}' \neq 0\,,$$

and (9.27) could be violated. But if the radius of the neighbourhood tends to zero then $\mathbf{M}_O^{(\mathrm{r})}$ vanishes, and we fall back to the ideal case.

Let us now find the free equations that characterize the equilibrium of a rigid body constrained by an ideal pin at the fixed point $O$. There is a unique constraint reaction $\mathbf{\Phi}$, applied to $O$, and the balance equations (9.23) become

$$\mathbf{R}^{(\mathrm{e,a})} + \mathbf{\Phi} = \mathbf{0}\,, \qquad \mathbf{M}_O^{(\mathrm{e,a})} = \mathbf{0}\,. \tag{9.28}$$

The second balance equation $(9.28)_2$ represents three free equations in the three unknowns $\theta$, $\phi$, $\psi$ (Euler angles), which in turn give the equilibrium positions. At the same time the first balance equation $(9.28)_1$ allows to determine the value of the constraint reaction $\mathbf{\Phi}$ at the equilibria. The problem is therefore statically determined.

Note the free equilibrium condition $(9.28)_2$ could have been found using the Principle of virtual work. In fact (9.24) implies $\delta L^{(\mathrm{a})} = \delta L^{(\mathrm{e,a})} = \mathbf{M}_O^{(\mathrm{e,a})} \cdot \boldsymbol{\varepsilon}' = 0$ for all $\boldsymbol{\varepsilon}'$, from which $\mathbf{M}_O^{(\mathrm{e,a})} = \mathbf{0}$.

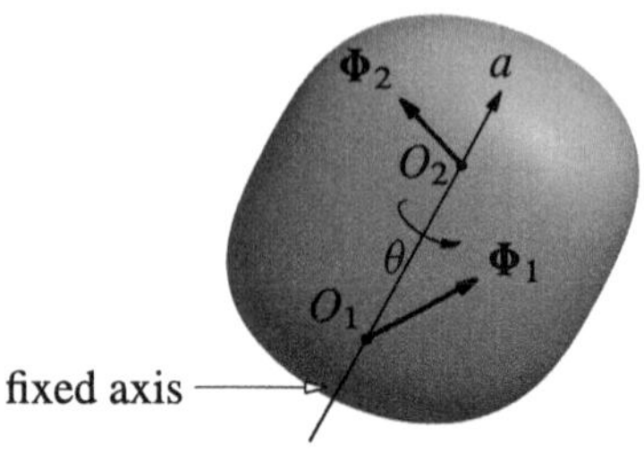


**Fig. 9.11** Equilibrium of a rigid body with a fixed axis

## 9.6.2 *Rigid Body with a Fixed Axis*

Consider a rigid body with a fixed axis. This constraint can be realized in practice in several ways, but usually consists of a *cylindrical pin*, i.e. an appendix in the shape of a round cylinder, comoving with the body, placed in a fixed cylindrical cavity of diameter slightly bigger than the diameter of the appendix. Both appendix and cavity have ridges that prevent the two cylindrical surfaces from sliding on one another along their axis. Common examples are the hinges in doors and windows, which usually have just one ridge since the cylinders' axis is vertical. The usefulness of a single ridge is to make it easier to unhook the door or window. It is assumed in these devices that the two cylinders are ideally coaxial.

To characterize the constraint reactions in a cylindrical pin let us imagine to realize the constraint by two fixed pins, at $O_1$ and $O_2$, as in Fig. 9.11, and set $O_1O_2 = d\mathbf{e}$, where $\mathbf{e}$ is the unit vector parallel to the fixed axis.

Choosing $O_1$ as pole for the moments and remembering that $\mathbf{M}_{O_1}^{(\mathrm{e,r})} = O_1O_2 \times \boldsymbol{\Phi}_2$, equations (9.23) read

$$\mathbf{R}^{(\mathrm{e,a})} + \boldsymbol{\Phi}_1 + \boldsymbol{\Phi}_2 = \mathbf{0}\,, \qquad \mathbf{M}_{O_1}^{(\mathrm{e,a})} + d\mathbf{e} \times \boldsymbol{\Phi}_2 = \mathbf{0}\,. \tag{9.29}$$

In particular, if we take the dot product of the second of (9.29) by $\mathbf{e}$ we obtain that the axial torque of the active external forces must vanish:

$$M_a^{(\mathrm{e,a})} = \mathbf{M}_{O_1}^{(\mathrm{e,a})} \cdot \mathbf{e} = 0\,. \tag{9.30}$$

This scalar equation does not contain the constraint reactions, so it is free and allows to determine the generalized coordinate $\theta$ (angle of rotation about $\mathbf{e}$) for which we have equilibrium. The remaining five scalar equations can be used to find the constraint reactions, which have six unknowns: three components for $\boldsymbol{\Phi}_1$ and three for $\boldsymbol{\Phi}_2$. The problem is statically underdetermined.

The underdetermination can be overcome for example by replacing the spherical pin $O_2$ with a smooth ring (without the ridges that prevent it from gliding on the axis), whose constraint reaction is orthogonal to the fixed axis Fig. 9.12. Equations (9.29) are solvable, and give

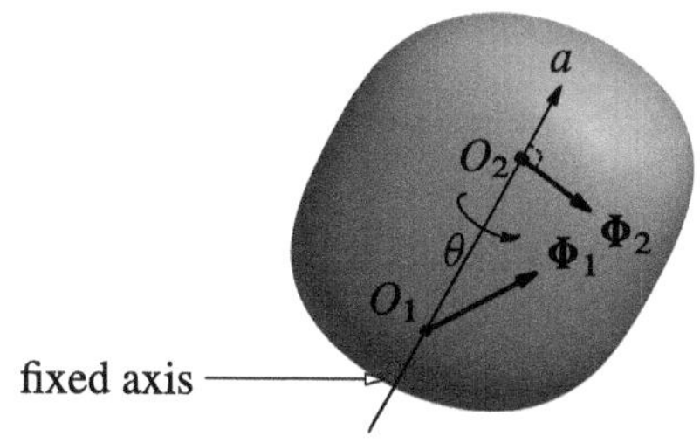


**Fig. 9.12** A statically determined system

$$\mathbf{\Phi}_1 = -\mathbf{R}^{(e,a)} - \frac{1}{d}\,\mathbf{e} \times \mathbf{M}^{(e,a)}_{O_1}\,, \qquad \mathbf{\Phi}_2 = \frac{1}{d}\,\mathbf{e} \times \mathbf{M}^{(e,a)}_{O_1}\,. \tag{9.31}$$

Also in this case it is possible to find the free equilibrium condition (9.30) from the Principle of virtual work. In fact, the constrained rigid body only admits rotational virtual displacements along the constrained axis Fig. 9.11, so

$$\delta L^{(a)} = \mathbf{M}^{(e,a)}_{O_1} \cdot \boldsymbol{\varepsilon}'\,, \qquad \text{with } \; \boldsymbol{\varepsilon}' = \delta\theta\,\mathbf{e}\,.$$

Hence $\delta L^{(a)} = (\mathbf{M}^{(e,a)}_{O_1} \cdot \mathbf{e})\,\delta\theta = M^{(e,a)}_{a}\,\delta\theta = 0$ for all $\delta\theta$, and (9.30) follows.

### 9.6.3 *Rigid Body Rotating and Sliding on a Fixed Axis*

A *cylindrical collar* is a cylindrical pin with ridges removed. It therefore leaves two degrees of freedom because the axis of the cylindrical appendix comoving with the body, despite having fixed direction, is no longer made of fixed elements. This constraint is built with two rings (placed at points $O_1$, $O_2$ on the constrained axis, with $O_1O_2 = d\mathbf{e}$), similar to the ring acting on $O_2$ in the previous example. Choosing once again $O_1$ as pole for the moments, the equilibrium balance equations give

$$\mathbf{R}^{(e,a)} + \mathbf{\Phi}_1 + \mathbf{\Phi}_2 = \mathbf{0}\,, \qquad \mathbf{M}^{(e,a)}_{O_1} + d\mathbf{e} \times \mathbf{\Phi}_2 = \mathbf{0}\,.$$

Take the dot product with the unit vector $\mathbf{e}$ and find the pure equilibrium conditions

$$\mathbf{R}^{(e,a)} \cdot \mathbf{e} = 0, \qquad \mathbf{M}^{(e,a)}_{O_1} \cdot \mathbf{e} = 0\,, \tag{9.32}$$

while the constraint reactions will be given by expressions formally identical to (9.31).

Conditions (9.32) can be found directly from the Principle of virtual work. The constrained rigid body can rotate around the collar's axis and translate along it. Therefore its virtual displacements will be helicoidal:

$$\delta P = \delta O_1 + \boldsymbol{\varepsilon}' \times O_1P\,, \quad \text{with } \; \delta O_1 = \delta z\,\mathbf{e} \; \text{ and } \; \boldsymbol{\varepsilon}' = \delta\theta\,\mathbf{e}\,.$$

The Principle of virtual work then gives the equilibrium condition

$$\delta L^{(a)} = \mathbf{R}^{(e,a)} \cdot \delta O_1 + \mathbf{M}_{O_1}^{(e,a)} \cdot \varepsilon' = (\mathbf{R}^{(e,a)} \cdot \mathbf{e})\, \delta z + (\mathbf{M}_{O_1}^{(e,a)} \cdot \mathbf{e})\, \delta\theta = 0$$

which is equivalent to (9.32) since it must hold for all $\delta z$ and $\delta\theta$.

### Cylindrical Collar

A *cylindrical guide* is a cylindrical collar whose cylindrical appendix has non-circular section. This prevents any rotation for the constrained rigid body, as it only permits translations along the constrained axis. Calling $O$ one of the constrained points, $\delta P = \delta O = \delta z\, \mathbf{e}$ for all points.

Realizing a cylindrical guide requires an axial torque, i.e. a component of the constraint moment parallel to the constraint axis. Therefore it is not possible to build a cylindrical guide combining spherical pins along the constrained axis. The free equilibrium condition is the first in (9.32), while the equilibrium balance equations allow to characterize the constraint force by their resultant (orthogonal to the guide's axis) and resultant's moment (of generic direction).

### Equilibrium of a Rigid Body Subject to a Screw Constraint

As seen in Sect. 8.8.3, the screw constraint is a special cylindrical collar in which the motion along the axis is constrained (by suitable ridges in the cylindrical appendix and the fixed cavity, see Fig. 8.6 and 8.7) to be a multiple of the angle of rotation about the axis: $\delta z = h\delta\theta$. The constraint is characterized by its *pitch* $p$, corresponding to the distance travelled by the rigid body along the axis of rotation after a complete turn.

In an ideal screw asking the constraint reaction to do zero virtual work forces the axial components $\Phi_a$, $M_a$ of the *constraint's* resultant and moment to obey (8.22), i.e. $M_a = -h\Phi_a$ (shown in Fig. 8.6 right, and proven in Sect. 8.8.3).

To find the free equilibrium equations we have to go through the Principle of virtual work, since the constraint is better characterized by the virtual displacement it allows, rather than the constraint reaction it produces. Choose a point $O$ on the screw's axis and a parallel unit vector $\mathbf{e}$, so

$$\delta P = \delta O + \varepsilon' \times OP = \delta z\, \mathbf{e} + \delta\theta \mathbf{e} \times OP\,,$$

for all points $P$ of the rigid body, and with $\delta z = h\, \delta\theta$.

In presence of an ideal screw the Principle of virtual work allows to characterize the equilibrium configurations of the constrained rigid body. In fact,

$$\delta L^{(a)} = \mathbf{R}^{(e,a)} \cdot \delta O + \mathbf{M}_O^{(e,a)} \cdot \varepsilon' = \left(h\mathbf{R}^{(e,a)} + \mathbf{M}_O^{(e,a)}\right) \cdot \mathbf{e}\, \delta\theta = 0 \quad \text{for all } \delta\theta\,,$$

giving the following free equilibrium equation

$$hR_a^{(e,a)} + M_a^{(e,a)} = 0 \tag{9.33}$$

involving the axial components of both resultant and moment of the active forces. In other words: the components of the resultant of the active forces and of their momentum along the unit vector **e** must be related by (9.33), which is therefore the free equilibrium equation.

The balance equations will finally give the resultant and resultant moment of the constraint force. This will contain, apart from the components acting on a cylindrical collar (a resultant and a resultant moment, both orthogonal to the screw's axis), an axial component of resultant and resultant moment, which guarantee the correct ratio between linear motion and rotation.

**Remark 9.16** It is clear that two or more constraints can be applied simultaneously to one rigid body. For example, we have already highlighted how a cylindrical pin can be realized by two spherical pins or by a cylindrical collar and a spherical pin. In these cases we should remember it is not always possible to determine the constraint reactions produced by each single constrain, but only their total resultants and moments. □

## 9.7 Rigid Body Supported by a Smooth Horizontal Plane

**Definition 9.17** (*Support polygon*) Consider a heavy rigid body $\mathscr{C}$ resting on a smooth horizontal plane $\pi$, and assume it touches the plane at a finite number of *support points* $\{P_i,\ i = 1, \ldots, n\}$. We call *support polygon* the unique polygon with these properties:

1. the vertices are support points.
2. The polygon is convex.
3. The support points, when different from vertices, are internal to the polygon.

Finding the support polygon allows to characterize in a simple manner the equilibrium of the supported rigid body Fig. 9.13.

**Theorem 9.18** *Consider a rigid body $\mathscr{C}$ supported by a horizontal plane $\pi$. $\mathscr{C}$ is in equilibrium if and only if the center of pressure is not external to the support polygon, where the* center of pressure *$G^*$ is the projection of the center of gravity of $\mathscr{C}$ onto the support plane $\pi$.*

***Proof*** The proof is straightforward once we recall that rigid bodies satisfy the reduction theorems (Sect. 9.5.1). In fact, the weights can be reduced to the resultant **R** applied at $G^*$, while the constraint reactions, forming a system of parallel vectors with the same orientation, can be reduced to a resultant applied to their center $G^{(r)}$ (see Sect. 7.4.2). The property of the center of coherent parallel vectors applied at

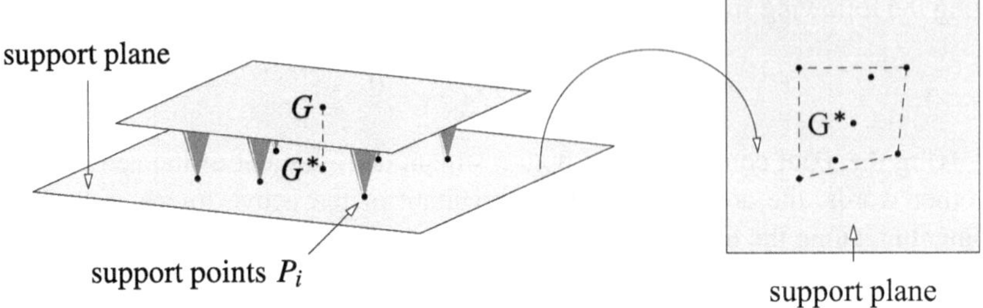


**Fig. 9.13** Equilibrium of a rigid body supported by a horizontal plane

points not all external to a closed convex curve implies $G^{(r)}$ can occupy all positions not external to the support polygon, and only those. In particular, if $G^*$ fell outside the support polygon there would be no system of constraint reactions giving a couple with zero arm and weight applied at $G^*$. □

### *9.7.1 Computing the Reactive Forces*

The issue remains of computing the moduli of the constraint reactions $\boldsymbol{\Phi}_j = \Phi_j \, \mathbf{k}$, which are vertical because of the smooth constraint assumption. Choose a pole $O$ on the support plane $\pi$, and two orthogonal directions $\{\mathbf{i}, \mathbf{j}\}$ in $\pi$, so that $OP_j = x_j \, \mathbf{i} + y_j \, \mathbf{j}$ is the $j$-th support point, and $OG^* = x_G \, \mathbf{i} + y_G \, \mathbf{j}$ the center of pressure. Call $p$ the rigid body's weight. The equilibrium balance equations (9.15) give

$$\sum_{j=1}^{n} \Phi_j = p \,, \qquad \sum_{j=1}^{n} x_j \Phi_j = x_G \, p \,, \qquad \sum_{j=1}^{n} y_j \Phi_j = y_G \, p \,. \tag{9.34}$$

System (9.34) is a linear system of three equations in $n$ unknowns $\{\Phi_j\}$, so it will admit a unique (statically determined) solution only if $n \leq 3$. We exclude the trivial and inconsequential cases $n = 1, 2$ and consider $n = 3$ and $n > 3$ separately. The first case is the so-called "tripod" while the second will lead to discuss an underdetermined system.

### *9.7.2 Tripod*

Consider three non-aligned points $P_1$, $P_2$, $P_3$. Call $\Delta$ the area of the triangle they form, $\Delta_1$ the area of $G^* P_2 P_3$, $\Delta_2$ that of $G^* P_1 P_3$, and $\Delta_3$ the area of $G^* P_1 P_2$ (see Fig. 9.14). Solving system (9.34) with Cramer's rule we find

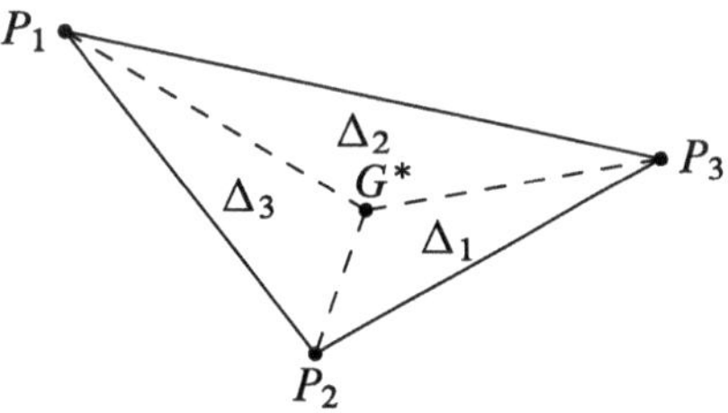


**Fig. 9.14** Equilibrium of a tripod

$$\Phi_j = \frac{\Delta_j}{\Delta} p \qquad \text{for} \quad j = 1, 2, 3.$$

In particular, the three reactions coincide when $G^*$ is the intersection point of the medians.

### Soil Subsidence

Clearly, the physically most interesting case is when there are more than three support points. The problem is then statically underdetermined and it is not possible to determine all constraint reactions from the sole equilibrium balance equations (9.14). *Yet we must remember that if a mathematical model does not answer a question, it often depends on having taken an oversimplification of a real situation. Many times it suffices to refine the modelling to extract an answer from it.*

This is what happens at present. We can check that if we consider not a rigid plane but an elastic one, one that allows for the support positions to lower, we can determine the constraint reactions for any number $n$ of supports. Let $z_j$ denote the height of $P_j$, negative if the points go below the height of the support plane $\pi$. In case of small deformations we can invoke *Hooke's law*, whereby the stress and deformation are proportional. In our problem the stress is given by the unknown constraint reaction $\Phi_j$, so

$$\Phi_j = -kz_j \,, \tag{9.35}$$

where the positive constant $k$ is the elastic coefficient of the support plane. (The previous rigid case, where $z_j = 0$, is a limit as $k \to \infty$ in this model.)

As $\mathscr{C}$ is now rigid, the supporting points after the deformation must still belong to a plane, so

$$z_j = ax_j + by_j + c \,. \tag{9.36}$$

Putting (9.35) and (9.36) together gives

$$\Phi_j = \lambda x_j + \alpha y_j + \beta \tag{9.37}$$

with $\lambda = -ka$, $\alpha = -kb$, $\beta = -kc$. From (9.37) therefore, the $\Phi_j$ depend on three parameters $\lambda$, $\alpha$, $\beta$ only, which become the unknowns of system (9.34). Substituting

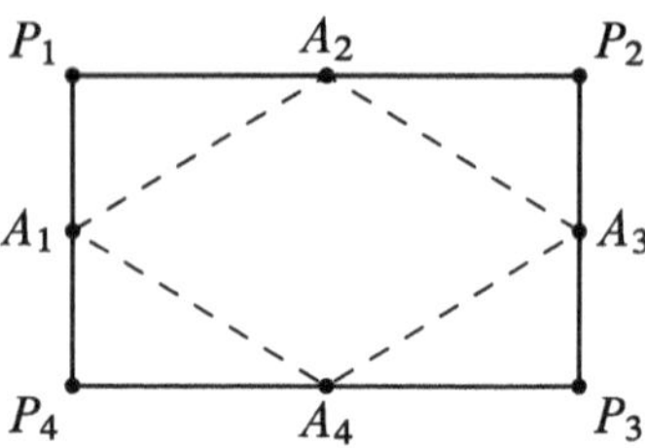


**Fig. 9.15** Supporting polygon and detachment regions for a rectangular table

(9.37) in (9.34) gives the linear system

$$\begin{pmatrix} \sum_{j=1}^{n} x_j & \sum_{j=1}^{n} y_j & n \\ \sum_{j=1}^{n} x_j^2 & \sum_{j=1}^{n} x_j y_j & \sum_{j=1}^{n} x_j \\ \sum_{j=1}^{n} x_j y_j & \sum_{j=1}^{n} y_j^2 & \sum_{j=1}^{n} y_j \end{pmatrix} \begin{pmatrix} \lambda \\ \alpha \\ \beta \end{pmatrix} = p \begin{pmatrix} 1 \\ x_G \\ y_G \end{pmatrix}. \tag{9.38}$$

From (9.38), once we know $\{x_j, y_j\}, x_G, y_G, p$, we can determine $\lambda, \alpha, \beta$, and from (9.37) we recover the $\Phi_j$. Note that system (9.38) is singular only in the particular case in which all support points are collinear. In fact, the determinant on the left is expressible, through the variance and covariance of the $x$- and $y$-coordinates, as $\sigma_x^2 \sigma_y^2 - \sigma_{xy}^2 = \sigma_x^2 \sigma_y^2 (1 - \rho_{xy}^2)$, where $\rho_{xy}$ is the Pearson correlation coefficient between the variables $x$ and $y$. It reaches its extrema $\pm 1$ when the relative variables are linearly related, i.e. the points are collinear.

When system (9.38) is non-singular, it may still happen that when one of the $\Phi_j$ is negative. We should interpret this as the support point $P_j$ detaching itself. In fact if $P_j$ is on the plane, necessarily $\Phi_j \geq 0$ since the constrain is unilateral. When we lose support at some point we should exclude such point(s), and compute the new $\lambda, \alpha, \beta$ with the corresponding $\Phi_j$. If the equilibrium condition of Theorem 9.18 holds, there will be one point among the aforementioned steps after which no point will detach, and all $\Phi_j$ will be non-negative.

**Remark 9.19** The parameters $\lambda, \alpha, \beta$ solving system (9.38) do not depend on the structural constant $k$, and the same goes, by (9.37), for the support constraint reactions. Therefore it is possible to determine uniquely these reactions simply by supposing that the ground is not perfectly rigid, but a tad elastic. Knowing the elastic constant $k$, instead, is necessary if we want to find the plane on which, even after the deformation, the non-detached support points lie. □

System (9.38) can be solved explicitly if we pick suitable Cartesian axes. Fix axes $Oxy$ with origin at the geometric center of the feet, oriented as the" principal axes of inertia" of the system of particles with unit mass associated with the support points. In this frame:

$$\sum_{j=1}^{n} x_j = 0, \qquad \sum_{j=1}^{n} y_j = 0, \qquad \sum_{j=1}^{n} x_j y_j = 0.$$

Now call $I_x = \sum y_j^2$ and $I_y = \sum x_j^2$ the moments of inertia of this hypothetical material system. The solution to (9.38) then reads

$$\lambda = \frac{p x_G}{I_y}, \qquad \alpha = \frac{p y_G}{I_x}, \qquad \beta = \frac{p}{n}$$

and therefore

$$\Phi_j = p\left(\frac{x_G x_j}{I_y} + \frac{y_G y_j}{I_x} + \frac{1}{n}\right). \tag{9.39}$$

For instance, for a rectangular table with four feet $P_1$, $P_2$, $P_3$, $P_4$ coinciding with its vertices, the frame system has origin at the center of the rectangle and axes parallel to the edges. It is immediate to see from (9.39) that if the center of pressure falls inside the rhombus $A_1A_2A_3A_4$ of Fig. 9.15, then $\Phi_j > 0$ for all $j$. But if the center of pressure falls in the triangle $A_1P_1A_2$ then $P_3$ detaches, as $\Phi_3 < 0$. Similarly, in the other external triangles the opposite support point will detach.

## 9.8 Rigid Body Subject to Rough Constraints

In presence of rough constraints it is no longer possible to use the Principle of virtual work, which now is only sufficient in view of an equilibrium. At any rate the equilibrium balance equations, provided we accompany them with a suitable dynamical characterization of the constraint, remain the equilibrium criterion we can always resort to if we wish to determine the equilibrium configurations and corresponding constraint reactions.

Friction tends to *favour* equilibrium, and so from the point of view of *safety* it is preferable to think of constraints as ideal. At the same time it is not always possible to ignore friction, as the following remarkable examples explain.

### *9.8.1 Equilibrium of a Ladder*

An important example where friction is necessary to the equilibrium is the ladder. If the load is symmetrical, and wall and floor are homogeneous, the reduction theorems simplify the problem to a planar problem as in Fig. 9.16. Drawing the respective cones of friction at $A$ and $B$ we immediately see that the possible constraint reactions reduce to a single vector applied at a point not external to the quadrilateral $LMNP$. To have

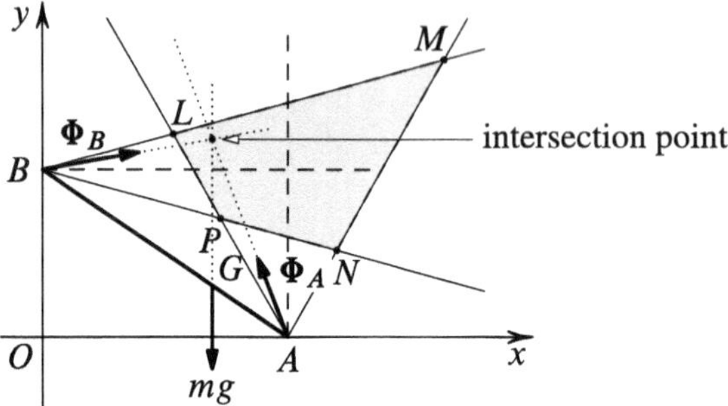


**Fig. 9.16** Equilibrium of a ladder: the intersection point of the lines of action

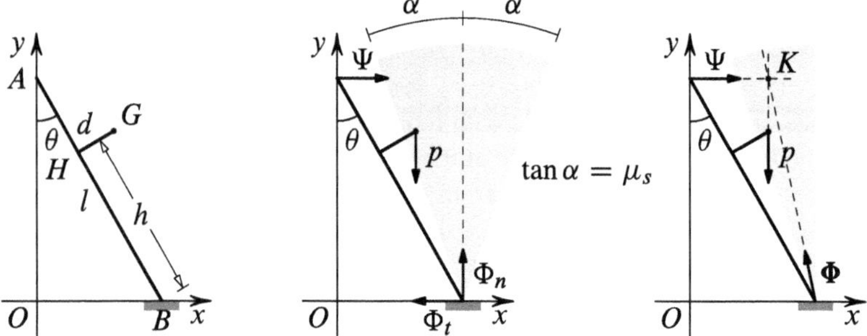


**Fig. 9.17** A ladder supporting a load

equilibrium it will be necessary for the vertical line through the center of gravity (of ladder and person) to contain non-external points. Note how essential the floor's friction is, whereas the wall's friction is irrelevant. If there was no friction at $A$ the only equilibrium configuration of the ladder would be the vertical one!

**Example 9.20** The segment $AB$ in Fig. 9.17 represents a ladder with end $A$ on a smooth vertical wall and end $B$ on a horizontal floor, with coefficient of static friction $\mu_s$. On the ladder, assumed of negligible weight, there is a load $p$ at $G$. This point has distance $d$ from $H$, and $H$ has distance $h$ from $B$. More concretely we could think of $G$ as the center of gravity of the system of the ladder $AB$ and a comoving person.

We wish to determine the range of the angle $\theta$, as per the first picture in Fig. 9.17, that guarantees equilibrium.

The forces acting on the system are represented in the second picture: the load $p$, the wall's horizontal reaction at $A$ and the two components $\Phi_t$ and $\Phi_n$ of the constraint reaction $\mathbf{\Phi}$ at $B$, which obey the Coulomb-Morin inequality $|\Phi_t| \leq \mu_s |\Phi_n|$. Hence $\mathbf{\Phi}$ must fall inside the friction cone, drawn in grey (obviously in a planar problem the cone is an angle). Furthermore, the reaction $\Psi$ must be bigger than or equal to zero (pointing to the right), as the constraint at $A$ is unilateral (the ladder might detach from the wall).

The forces $p$ and $\Psi$ have lines of action meeting at $K$, as in the third picture. Since the total moment with respect to this pole must vanish we conclude $\boldsymbol{\Phi}$ goes from $B$ to $K$, and therefore to have equilibrium $K$ must fall inside the cone of friction. Moreover, as the first balance equation along the horizontal direction says $\Psi - \Phi_t = 0$, we know that $\Psi = \Phi_t$ and hence $\Phi_t \geq 0$ (note that $\Phi_t$ is the component of $\boldsymbol{\Phi}$ oriented by the arrow, i.e. pointing leftwards). From this, the part of cone containing $\boldsymbol{\Phi}$ is the one to the *left* of its axis.

To sum up, an equilibrium is possible if and only if $K$ is inside or on the boundary of the grey left half-cone in the third picture. If $K$ goes beyond the right margin of that region, the ladder will detach from the wall at $A$, while if $K$ is too far on the left, the Coulomb-Morin inequality will not hold and the ladder will start to slip on the floor.

From this analysis such a ladder cannot be too vertical (or there is a risk of the person falling off) nor too horizontal (otherwise the ladder slips and falls on the floor).

To compute precisely the extremal values of $\theta$ it suffices to set to zero the total moment with respect to $B$, together with the vertical and horizontal resultant. Basic trigonometry gives

$$\Psi = \Phi_t, \quad \Phi_n = p, \quad -\Psi l \cos\theta + p(h \sin\theta - d\cos\theta) = 0$$

and so

$$\Phi_t = \Psi = \frac{p}{l}(h \tan\theta - d), \quad \Phi_n = p\,.$$

Condition $\Psi \geq 0$ and the Coulomb-Morin inequality are therefore equivalent to

$$\frac{d}{h} \leq \tan\theta \leq \frac{d + \mu_s l}{h}\,,$$

which provides the range of $\theta$. □

## 9.9 Equilibrium of Holonomic Systems

We know from Sect. 7.5.2 that the virtual work of the active forces of a holonomic system is

$$\delta L^{(\mathrm{a})} = \sum_{h=1}^{N} Q_h \, \delta q_h\,.$$

Recalling (7.44), and as already remarked in (7.45), it is possible to write the above as a dot product in $\mathbb{R}^N$: $\delta L^{(\mathrm{a})} = \mathsf{Q} \cdot \delta \mathsf{q}$.

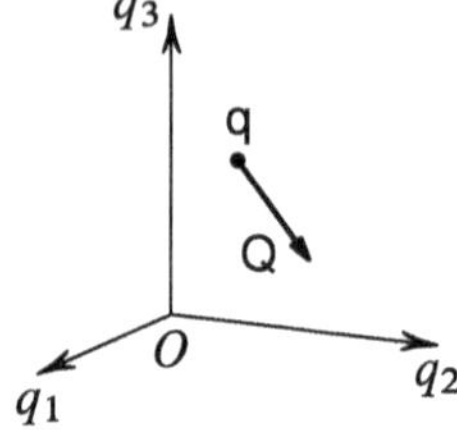


**Fig. 9.18** Configuration space and generalized force for $N = 3$

### 9.9.1 Bilateral Constraints

We begin with looking at systems only subject to bilateral constraints, for which *all* displacements are reversible. The Principle of virtual work tells that there is equilibrium if and only if

$$\delta L^{(a)} = \mathsf{Q} \cdot \delta \mathsf{q} = 0 \qquad \forall\, \delta \mathsf{q}\,.$$

Since the *variations* of the generalized coordinates are independent, which is true only in absence of nonholonomic constraints, we can choose $\delta q_h \neq 0$ and the others zero, and repeat for different $h$. Thus every $Q_h$ must vanish, which in turn implies the equilibrium positions are those for which

$$\mathsf{Q}\,(\mathsf{q}) = \mathsf{0}\,, \tag{9.40}$$

i.e. those for which the generalized force $\mathsf{Q}$ is zero.

Observe the perfect *duality* between the statics of holonomic systems and the statics of the free particle. For a point the equilibrium positions are those where $\mathbf{F}(P) = \mathbf{0}$, while for holonomic systems $\mathsf{Q}(\mathsf{q}) = \mathsf{0}$. We may imagine the point in the $N$-dimensional space of configurations as a real point subject to force $\mathsf{Q}$ Fig. 9.18. Equation (9.40) then is a system of $N$ equations in $N$ unknowns $\mathsf{q} = (q_1, q_2, \ldots, q_N)$

$$\begin{cases} Q_1(q_1, q_2, \ldots, q_N) = 0 \\ Q_2(q_1, q_2, \ldots, q_N) = 0 \\ \ldots \qquad \ldots \qquad \ldots \\ Q_N(q_1, q_2, \ldots, q_N) = 0 \end{cases}$$

### 9.9.2 Unilateral Constraints

In presence of unilateral constraints it becomes necessary to distinguish between *ordinary configurations* and *boundary configurations*.

**Definition 9.21** (*Ordinary configurations*) We call ordinary configurations of a holonomic system those for which the system is described by a point $\mathsf{q}$ of the interior of the set $\Omega$ of allowed configurations.

We recall that a point belongs in the interior of a set if it admits at least one neighbourhood entirely contained in the set. The interior of a set $\Omega$ is denoted by $\overset{\circ}{\Omega}$.

**Definition 9.22** (*Boundary configurations*) A boundary configuration of a holonomic system with unilateral constraints is one where the system is described by a point $\mathsf{q}$ on the frontier of the set $\Omega$ of allowed configurations.

Recall that a point lies on the frontier of a set if each one of its neighbourhoods contains both points of $\Omega$ and points not in $\Omega$. The frontier of a set is indicated by $\partial\Omega$.

The virtual displacements that start from an ordinary configuration are all reversible, and therefore for these configurations the same result (9.40) holds that we saw for bilateral constraints. Therefore ordinary equilibrium configurations are precisely those where $\mathsf{Q}(\mathsf{q}) = 0$, with $\mathsf{q} \in \overset{\circ}{\Omega}$.

The equilibrium in boundary configurations depends instead on the geometry of the domain. Consider first the simpler case of a rectangular domain $\Omega$ (Fig. 9.19), i.e. a domain in the configuration space given by

$$a_h \leq q_h \leq b_h \qquad (h = 1, \ldots, N)\,, \tag{9.41}$$

where $\{a_h < b_h,\ h = 1, \ldots, N\}$ are constants. (We allow that some $a_h$ might be equal to $-\infty$ and some $b_h$ to $+\infty$, in which case the corresponding inequality in (9.41) is strict.)

In a boundary position at least one of (9.41) will be an equality. For example, suppose the first $j$ $(1 \leq j < N)$ generalized coordinates are equal to their smallest value and the remaining $N - j$ in (9.41) are strict inequalities:

$$\begin{aligned} &a_h = q_h < b_h \qquad && (h = 1, \ldots, j < N) \\ &a_k < q_k < b_k && (k = j+1, \ldots, N) \end{aligned} \tag{9.42}$$

From this configuration the admissible virtual displacements are characterized by:

$$\big\{\delta q_h \geq 0,\ h = 1, \ldots, j < N\big\} \quad \text{and} \quad \big\{\delta q_k \gtreqless 0,\ k = j+1, \ldots, N\big\}\,.$$

The system therefore admits virtual displacements of two types, reversible and irreversible, which we now discuss.

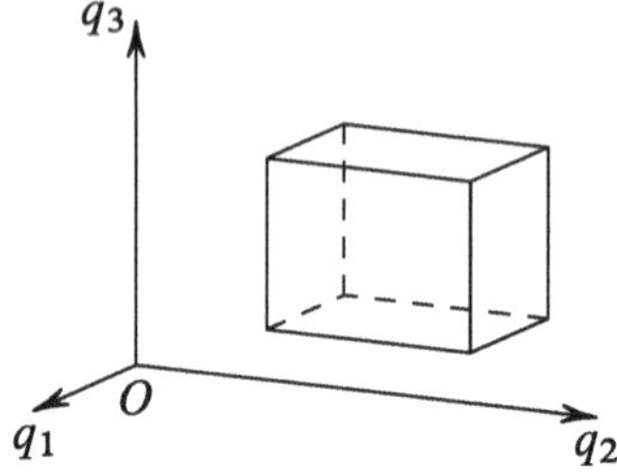


**Fig. 9.19** Unilateral constraints: a rectangular domain for $N = 3$

### Reversible Virtual Displacements

Reversible displacements are those where $\delta q_1 = \delta q_2 = \ldots = \delta q_j = 0$. For this class:

$$\delta L^{(a)} = \sum_{h=j+1}^{N} Q_h \delta q_h = Q_{j+1}\delta q_{j+1} + \ldots + Q_N \delta q_N = 0$$

and since $\delta q_{j+1}, \ldots, \delta q_N$ are arbitrary,

$$\begin{cases} Q_{j+1}(a_1, a_2, \ldots, a_j, q_{j+1}, \ldots q_N) = 0 \\ \quad \vdots \qquad\qquad \vdots \qquad\qquad \vdots \\ Q_N(a_1, a_2, \ldots, a_j, q_{j+1}, \ldots q_N) = 0\,. \end{cases} \tag{9.43}$$

This system has $N - j$ equations in $N - j$ unknowns $\{q_{j+1}, \ldots, q_N\}$.

### Irreversible Virtual Displacements

Irreversible displacements are instead characterized by at least one among the first $j$ $\{\delta q_h\}$ being positive. If, for instance, $\delta q_1 > 0$, and $\delta q_h = 0$ for all $h > 1$, then

$$\delta L^{(a)} = Q_1 \delta q_1 \le 0 \qquad \forall\, \delta q_1 > 0\,,$$

and hence $Q_1 \le 0$. Repeating for the first $j$ virtual displacements we arrive at

$$\begin{cases} Q_1(a_1, a_2, \ldots, a_j, q_{j+1}, \ldots q_N) \le 0 \\ \quad \vdots \qquad\qquad \vdots \qquad\qquad \vdots \\ Q_j(a_1, a_2, \ldots, a_j, q_{j+1}, \ldots q_N) \le 0 \end{cases} \tag{9.44}$$

In other words, system (9.43) allows to determine the values of the "ordinary" coordinates $(q^*_{j+1}, \ldots q^*_N)$. The configuration given by the $N$-tuple

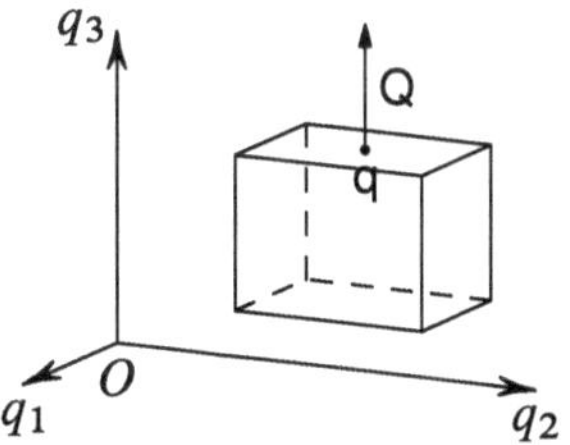


**Fig. 9.20** Equilibrium at a boundary configuration

$$(a_1, \ldots, a_j, q^*_{j+1}, \ldots q^*_N)$$

will be an equilibrium if also (9.44) hold in $\mathscr{C}$. Examining more generally (9.42), (9.43) and (9.44), one deduces the following proposition, which characterizes boundary equilibrium configurations.

**Proposition 9.23** *Boundary configurations are characterized by having at least one generalized coordinate producing virtual displacements of sign determined by the constraint. Such configurations are equilibria if the generalized components of the active forces satisfy:*

$$\begin{aligned} Q_h &= 0 \quad \text{if} \quad \delta q_h \gtrless 0 \quad \text{(reversible)} \\ Q_h &\le 0 \quad \text{if} \quad \delta q_h \ge 0 \quad \text{(irreversible)} \\ Q_h &\ge 0 \quad \text{if} \quad \delta q_h \le 0 \quad \text{(irreversible)}. \end{aligned} \tag{9.45}$$

If only one $Q_h$ is non-zero, the presence of the unilateral constraint is essential for the equilibrium. In fact, there is equilibrium precisely because of a constraint reaction with opposite sign. Conversely, if all $Q_h$ vanish, the boundary equilibrium position would exist even if the corresponding unilateral constraints were absent.

**Example 9.24** To understand better the physical meaning of the previous proposition consider a particular case of boundary equilibrium position. Take a rectangular constraint like (9.41), with $N = 3$, and let us study an equilibrium configuration $(q_1, q_2, q_3)$ in which the system occupies the upper face of the box (see Fig. 9.20), with $a_1 < q_1 < b_1$, $a_2 < q_2 < b_2$, while $q_3 = b_3$. The equilibrium conditions (9.45) impose

$$Q_1(q_1, q_2, b_3) = 0, \qquad Q_2(q_1, q_2, b_3) = 0, \qquad Q_3(q_1, q_2, b_3) \ge 0.$$

In other words, the generalized force **Q** must be orthogonal to the face $q_3 = b_3$ and pointing outwards Fig. 9.20. Here, too, there is a perfect match with the statics of a point. In fact, it is as if the walls of the box were a real smooth surface: the point will be in equilibrium on the upper face if the active force pushes it upwards (i.e., outside the allowed region).

The reader may as an exercise consider the configurations

$$
\begin{array}{lllll}
(1)\ q_1 = a_1 & q_2 = b_2 & a_3 < q_3 < b_3 & \text{(edge)} \\
(2)\ q_1 = a_1 & q_2 = b_2 & q_3 = b_3 & \text{(vertex)}
\end{array}
$$

and find which generalized forces realize boundary equilibrium configurations. □

### *9.9.3 Critical Points of the Potential*

In case the active forces are conservative there is a potential $U = U(\mathbf{q})$, and the generalized components can be simply related to it by (7.46):

$$Q_h(\mathbf{q}) = \frac{\partial U}{\partial q_h}.$$

In this case, therefore, seeking equilibria reduces to studying the potential. In particular, (9.40) immediately proves the following theorem.

**Theorem 9.25** (Critical points of the potential) *The ordinary equilibrium configurations of a holonomic system are precisely those where the derivatives of the potential with respect to all generalized coordinates are zero. Therefore they coincide with the critical points of the potential.*

**Example 9.26** Whenever the active forces are conservative the sole knowledge of the potential enables us to find all equilibria, both ordinary and boundary. Consider, for example, a system with one freedom degree, whose generalized coordinate is constrained to the interval $a \leq q \leq b$. Suppose the potential $U(q)$ behaves as in Fig. 9.21.

The previous theorem guarantees the stationary configurations $\{q^*_{(1)}, q^*_{(2)}, q^*_{(3)}\}$ will be equilibria, irrespective of the fact they may be maxima, minima or horizontal inflections of the potential $U$.

Now let us look at the possible boundary equilibria. When $q = a$ (or $q = b$) the system can only have positive (respectively, negative) virtual displacements. Hence the position is an equilibrium if the generalized component $Q$ (coinciding with the potential's derivative) is zero or negative (positive). Figure 9.21 shows the derivative of the potential is positive at both ends $q = a$ and $q = b$. Therefore the latter is a boundary equilibrium position, while $q = a$ is not an equilibrium. □

## 9.10 Statically Stable Equilibrium

Figure 9.22 clearly show that the nature of equilibrium positions is very diverse. In either case a particle subject to its weight lies on a curve on a vertical plane, but in Fig. 9.22a the point can move in the interior, while in Fig. 9.22b it stays outside.

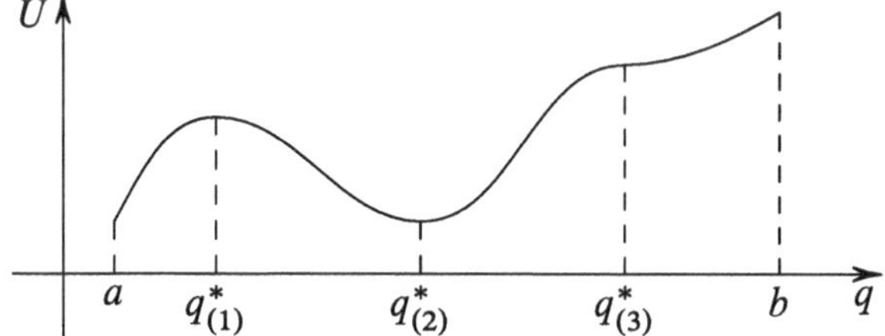


**Fig. 9.21** Equilibrium configurations under conservative forces

**Fig. 9.22** Stable (a) and unstable (b) equilibrium configurations

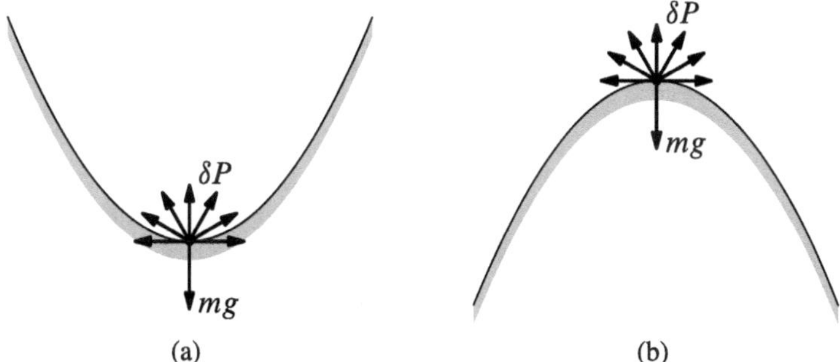


**Fig. 9.23** *Virtual* work of the active forces from the equilibrium configurations: (a) $\delta L^{(a)} \leq 0$; (b) $\delta L^{(a)} \leq 0$

If we put the point, which is initially at rest, close to the equilibrium position, it is easy to see that in case (a) the point will always stay in the proximity of $P^*$, while in case (b) it will tend to move away from $P^*$. In the first case one speaks of *stable equilibrium* and in the second of *unstable equilibrium.*

The detailed analysis of the stability of the equilibrium requires arguments that are typical of dynamics and will be carefully addressed in Sect. 14.7. Here we will simply give a definition of stability, called *stability in statical, or energetic sense*,

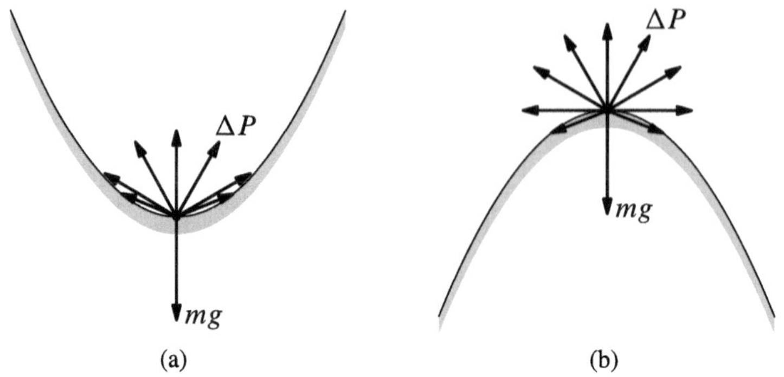


**Fig. 9.24** *Effettive* work of the active forces from the equilibrium configurations: (a) $\Delta L^{(a)} < 0$; (b) $\Delta L^{(a)} \gtreqless 0$

that relies on the concept of work, and is based on intuitive physical considerations regarding the fact that *positive work is a synonym for the ability of a force to produce a given displacement, whereas negative work indicates instead the incapability to do so.*

For example, if we dangle an object outside a window, the weight to make it fall does positive work. The weight cannot make the object levitate mid-air, since to do so it would have to do negative work. We can therefore expect that in a stable equilibrium situation the active force does negative work to move the body from that position.

But what type of work should we consider? If we used the virtual work $\delta L^{(a)}$ we would not be able to distinguish case (a) from (b) since both are equilibria, and by the Principle of virtual work $\delta L^{(a)} \leq 0$ (see Fig. 9.23).

If, instead, we consider the effective work $\Delta L^{(a)}$, built with effective displacements, we have a situation like the one of Fig. 9.24. In case (a) the angle between the weight and the effective displacements is always obtuse, while in (b) we have all possibilities (acute, right and obtuse). This discussion suggests the following definition.

**Definition 9.27** (*Statically stable equilibrium*) An equilibrium configuration $\mathsf{q}^*$ is called statically stable (or stable in statical, or energetic, sense) if the effective work of the active forces to move the system from $\mathsf{q}^*$ to $\tilde{\mathsf{q}}$ is strictly negative,

$$\Delta L^{(a)}_{\mathsf{q}^* \to \tilde{\mathsf{q}}} < 0, \tag{9.46}$$

for all $\tilde{\mathsf{q}} \in I(\mathsf{q}^*, \varepsilon) \setminus \{\mathsf{q}^*\}$, where $I(\mathsf{q}^*, \varepsilon)$ is a suitable neighbourhood of $\mathsf{q}^*$ of radius $\varepsilon$.

Alas, the effective work is a work along a finite path and so to compute (9.46) we need to know the motion $\mathsf{q}(t)$ from $\mathsf{q}^*$ to the final position $\tilde{\mathsf{q}}$. The only case where (9.46) is immediate is that of conservative active forces (see Sect. 7.1), for in that

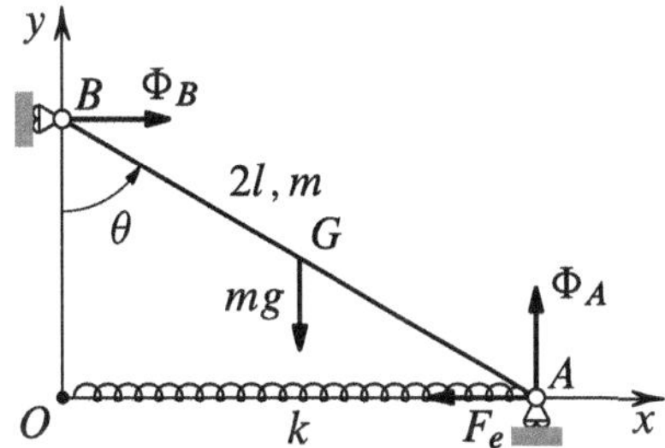


**Fig. 9.25** Example of application of the Principle of virtual work

case $\Delta L^{(a)}_{\mathrm{q}^*\to\tilde{\mathrm{q}}} = U(\tilde{\mathrm{q}}) - U(\mathrm{q}^*)$, and therefore (9.46) becomes

$$U(\tilde{\mathrm{q}}) < U(\mathrm{q}^*) \qquad \forall\, \tilde{\mathrm{q}} \in I(\mathrm{q}^*, \varepsilon) \quad (\tilde{\mathrm{q}} \neq \mathrm{q}^*). \tag{9.47}$$

Inequality (9.47) is precisely the condition that the equilibrium position coincides with an isolated relative maximum of the potential function.

**Proposition 9.28** (Statical stability of conservative systems) *An equilibrium configuration of a conservative holonomic system is statically stable if it corresponds to an isolated relative maximum of the potential.*

Let us emphasize that the proposition holds for both ordinary and boundary equilibrium positions. For example, the potential in Fig. 9.21 admits three ordinary equilibrium positions, only one of which ($q = q^*_{(1)}$) is statically stable. If we apply the definition to the boundary equilibrium position $q = b$ we discover that it too is statically stable, since the potential has an isolated relative maximum at this point as well.

**Remark 9.29** Recall that the potential energy $V$ is by definition minus the potential $V = -U$. Then the stable equilibria are precisely the isolated relative minima of the potential energy. □

**Example 9.30** Let us determine the ordinary and boundary equilibria of the system in Fig. 9.25, analyzing the statical stability. A heavy rigid rod of mass $m$ and length $2l$ is placed on a vertical plane. The ends $A$, $B$ are constrained to a pair of axes $(x, y)$, horizontal and vertical. Two unilateral constraints force the rollers $A$ and $B$ to stay on the positive axes, so that the rod must stay in the first quadrant. A spring of elastic constant $k$ joins $A$ to the intersection $O$ of the axes $(x, y)$.

The system has one freedom degree. Choosing as generalized coordinate the angle $\theta$ in Fig. 9.25, the presence of the unilateral constraints implies $0 \leq \theta \leq \pi/2$. Setting

$$\lambda = \frac{mg}{4kl} > 0,$$

the potential of the active forces equals

$$U(\theta) = -4kl^2 \left( \lambda \cos\theta + \frac{1}{2} \sin^2\theta \right)$$

and the corresponding generalized force is

$$Q_\theta(\theta) = \frac{dU}{d\theta} = 4kl^2 \sin\theta(\lambda - \cos\theta)$$

(for simplicity of notation, and because here there is only one generalized coordinate, in the sequel we will drop the subscript $\theta$ in the component $Q$ of the active force).

*Ordinary equilibrium positions.* For $0 < \theta < \pi/2$ we have $Q(\theta) = 0$ and therefore there exists, when $\lambda < 1$, the ordinary equilibrium position $\theta_1^* = \arccos\lambda$. Instead if $\lambda \geq 1$ there are no ordinary equilibria.

*Boundary equilibria.* For $\theta_2^* = 0$ we have $Q(0) \leq 0$, which holds since $Q(0) = 0$. Consequently $\theta_2^* = 0$ is the boundary equilibrium. Actually, the corresponding unilateral constraint is not essential since this position would exist even if we took away the vertical wall. When $\theta_3^* = \pi/2$ we have $Q\,(\pi/2) \geq 0$. In our case $Q\,(\pi/2) = 4kl^2\lambda = mgl > 0$. Hence $\theta_3^* = \pi/2$ is a boundary equilibrium position with essential unilateral constraint.

*Stability in statical sense.* The ordinary configuration $\theta_1^* = \arccos\lambda$ with $\lambda < 1$ is unstable. In fact it is a minimum of the potential because

$$U''(\theta_1^*) = mgl\frac{\sqrt{1-\lambda^2}}{\lambda} > 0.$$

The equilibrium position $\theta_3^* = \pi/2$ is instead certainly statically stable, as the positivity of $Q\,(\pi/2) = U'\,(\pi/2)$ ensures the point is an isolated relative maximum for the potential.

More delicate is instead the stability analysis of equilibrium configuration $\theta_2^* = 0$. It is easy to verify that $U''(0) = 4kl^2(\lambda - 1)$, so $\theta_2^*$ is an isolated relative maximum (hence stable) of the potential if $\lambda < 1$, while it becomes an isolated relative minimum (unstable) if $\lambda > 1$. In the intermediate case $\lambda = 1$ the study of higher derivatives of the potential allows to show $\theta_2^*$ is an isolated relative minimum, hence again unstable. □

## 9.11 Constraint Release Technique

The method we wish to describe in this section allows to find the constraint reactions with the Principle of virtual work and using in a suitable way the balance equations for non-rigid systems, but systems made of rigid parts. In either case the method relies on the so-called constraint release.

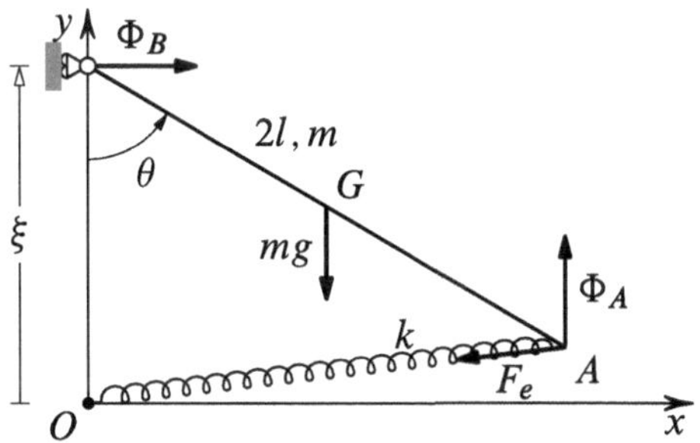


**Fig. 9.26** Constraint release and the computation of the reactive forces

### *9.11.1 Computing the Reactive Forces Through the Principle of Virtual Work*

One of the undisputed advantages of using the Principle of virtual work is working only with free equilibrium equations. Notwithstanding, finding the constraint reactions is often crucial, since their values measure the *effort* they exert to keep the body in equilibrium. Using a simple argument however, it is still possible to use the Principle of virtual work to determine the constraint reactions. To that end consider Example 9.30 again, and assume we wish to find the reaction at $A$ in the ordinary equilibrium position $\theta = \theta_1^*$. Imagine we release $A$, so the system has two degrees of freedom Fig. 9.26. As $A$ is now free *we may view the constraint reaction as the active force that, when applied at A, is able to keep the system in equilibrium in the real position we have already determined*, see Fig. 9.25. In this way the constraint reaction will enter the expression of the work of the active forces and we will be able to compute it.

More precisely, referring to Fig. 9.26, let $\xi$ be the $y$-coordinate of end $B$. After the release, it becomes a new generalized coordinate, independent of $\theta$. So,

$$\delta L^{(a)} = m\mathbf{g} \cdot \delta G + (kAO + \mathbf{\Phi}_A) \cdot \delta A = Q_\xi\, \delta\xi + Q_\theta\, \delta\theta$$

with

$$\begin{cases} Q_\xi = -mg + \Phi_{Ay} + k(2l\cos\theta - \xi) \\ Q_\theta = -l\left(mg + 2k\xi - 2\Phi_{Ay}\right)\sin\theta. \end{cases} \tag{9.48}$$

We want the system in equilibrium when $\xi = 2l\cos\theta$. Imposing that the generalized components of the active force (9.48) vanish, and putting $\xi = 2l\cos\theta$ we obtain a system of two equations in two unknowns $\theta$ and $\Phi_{Ay}$:

$$\begin{cases} -mg + \Phi_{Ay} = 0 \\ \sin\theta\,(\lambda - \cos\theta) = 0 \end{cases}$$

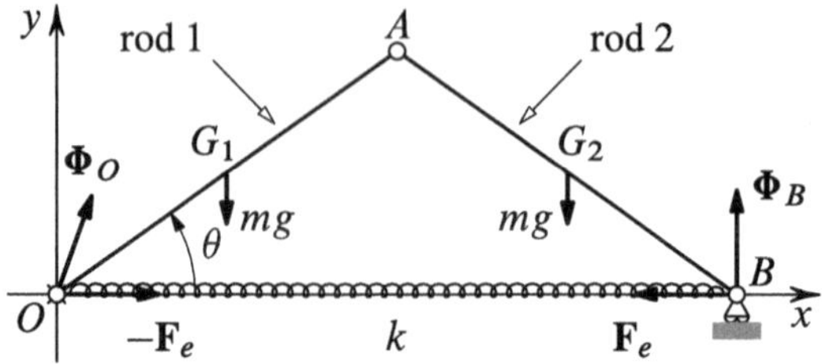


**Fig. 9.27** Equilibrium for a system made by two rigid rods

from which we can extract the constraint reaction and also the equilibrium value of $\theta$. Clearly, the latter coincides with the result obtained earlier.

### *9.11.2 Constraint Release and Balance Equations*

We will now use an explicit example to show how the balance equations can be extremely useful in the study of equilibrium configurations for systems composed by several rigid bodies, although they only guarantee equilibrium when a system is rigid.

Consider the system in Fig. 9.27. Two rigid rods, connected by a moving pin at $A$, move on the vertical plane $Oxy$ so that $O$ is fixed for $OA$ (rod 1), and the end $B$ of $AB$ (rod 2) stays on the horizontal$x$-axis. The rods have both mass $m$ and length $2l$, and there is an elastic force acting on $B$ of elastic constant $k$. The constraints are assumed ideal, and therefore $\mathbf{\Phi}_B = (0, \Phi_{By})$, while $\mathbf{\Phi}_O = (\Phi_{Ox}, \Phi_{Oy})$. Calling $\theta$ the angle in Fig. 9.27, we wish to find the equilibria in the interval $0 < \theta < \pi/2$ and the corresponding constraint reactions.

Clearly the overall system is not rigid, and therefore the balance equations are necessary but not sufficient. On one hand, for any planar system the balance equations projected on the Cartesian axes give three non-trivial scalar equations:

$$R_x^{(e)} = 0, \qquad R_y^{(e)} = 0, \qquad M_{O,z}^{(e)} = 0.$$

On the other hand, there are four unknowns at present: the angle $\theta$ (generalized coordinate) and the components $\Phi_{By}$, $\Phi_{Ox}$, $\Phi_{Oy}$ of the constraint reactions. Moreover (see Sect. 4.1.2 as well) in presence of a smooth moving pin it is possible to replace the constraint with two constraint reactions $\mathbf{\Phi}_{21}$, $\mathbf{\Phi}_{12}$, acting respectively on rod 1 and 2 at their common point $A$. By the action-reaction principle these reactions must be equal and opposite: $\mathbf{\Phi}_{12} + \mathbf{\Phi}_{21} = \mathbf{0}$.

Thus we can reduce the problem to *two* problems on the statics of rigid bodies, shown in Fig. 9.28, for each of which we may invoke the equilibrium balance equations. This will give six scalar equations in the six unknowns

$$\theta, \Phi_{By}, \Phi_{Ox}, \Phi_{Oy}, \Phi_{21x}, \Phi_{21y}.$$

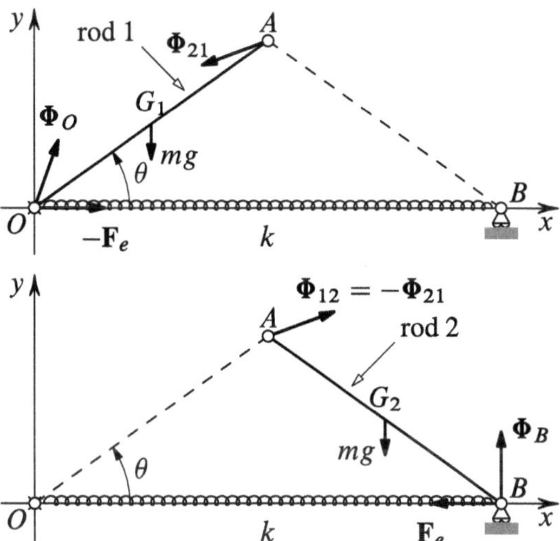


**Fig. 9.28** Constraint release technique

The problem can be solved completely. In fact the balance equations for the rods become

$$\begin{aligned} &\text{rod 1}: \begin{cases} m\mathbf{g} + kOB + \mathbf{\Phi}_O + \mathbf{\Phi}_{21} = \mathbf{0} \\ OG_1 \times m\mathbf{g} + OA \times \mathbf{\Phi}_{21} = \mathbf{0} \end{cases} \\ &\text{rod 2}: \begin{cases} m\mathbf{g} + kBO + \mathbf{\Phi}_B - \mathbf{\Phi}_{21} = \mathbf{0} \\ BG_2 \times m\mathbf{g} - BA \times \mathbf{\Phi}_{21} = \mathbf{0}. \end{cases} \end{aligned} \tag{9.49}$$

Let $G_1, G_2$ be the rods' centers of gravity. Keeping into account that

$$\begin{aligned} OA &= (2l\cos\theta, 2l\sin\theta), & OG_1 &= (l\cos\theta, l\sin\theta), \\ OB &= (4l\cos\theta, 0), & OG_2 &= (3l\cos\theta, l\sin\theta) \end{aligned}$$

we have the following six scalar equations:

$$\begin{cases} 4kl\cos\theta + \Phi_{Ox} + \Phi_{21x} = 0 \\ -mg + \Phi_{Oy} + \Phi_{21y} = 0 \\ -mg\cos\theta + 2\left(\Phi_{21y}\cos\theta - \Phi_{21x}\sin\theta\right) = 0 \\ 4kl\cos\theta + \Phi_{21x} = 0 \\ -mg + \Phi_{By} - \Phi_{21y} = 0 \\ mg\cos\theta + 2\left(\Phi_{21y}\cos\theta + \Phi_{21x}\sin\theta\right) = 0. \end{cases} \tag{9.50}$$

Consider only the significant interval $\theta \in (0, \pi/2)$ and suppose

$$\lambda = \frac{mg}{8kl} < 1.$$

Then there is a unique solution:

$$\begin{aligned} &\sin\theta^* = \lambda, \qquad \boldsymbol{\Phi}_{21} = \left(-4kl\sqrt{1-\lambda^2}, 0\right), \\ &\boldsymbol{\Phi}_B = (0, mg), \quad \boldsymbol{\Phi}_O = (0, mg). \end{aligned} \tag{9.51}$$

Let us emphasize that, as the constraints are ideal, we could have used the Principle of virtual work and immediately recovered the equilibrium equation $Q_\theta(\theta) = 0$, using which system (9.50) would have become linear in the constraint reactions (one of the equations would be a trivial identity). In fact, as the potential of the active forces is

$$U = -\frac{k}{2}|OB|^2 - mgy_{G_1} - mgy_{G_2} = -8kl^2\left(\cos^2\theta + 2\lambda\sin\theta\right) \tag{9.52}$$

we have

$$Q_\theta(\theta) = -16kl^2\cos\theta(\lambda - \sin\theta),$$

and so in $0 < \theta < \pi/2$ the only solution to $Q_\theta(\theta) = 0$ is $\theta^* = \arcsin\lambda$. Substituting in (9.50) immediately gives (9.51).

**Remark 9.31** From this last computation we see that if the system has ideal constraints it is more convenient to find first the equilibria via the Principle of virtual work. In this case the balance equations (using the constraint release technique) reduce to solving a linear system in the unknown constraint reactions. □

**Example 9.32** (Equilibrium in presence of non-ideal constraints) Let us look again at the system in Fig. 9.27, where now we assume friction at $B$ with static coefficient $\mu_s$. The Principle of virtual work does not apply, but we can still use the balance equations (9.49) for the rods, adding to them the component $\Phi_{Bx}$. The unknowns become seven: $\theta$ and the six components of the constraint reactions $\boldsymbol{\Phi}_O$, $\boldsymbol{\Phi}_B$, $\boldsymbol{\Phi}_{21}$. Equations (9.50) change only in line four, which now contains the term $\Phi_{Bx}$. Now though we can use the Coulomb-Morin law

$$\psi = |\Phi_{Bx}| - \mu_s|\Phi_{By}| \le 0. \tag{9.53}$$

The balance equations

$$\begin{cases} 4kl\cos\theta + \Phi_{Ox} + \Phi_{21x} = 0 \\ -mg + \Phi_{Oy} + \Phi_{21y} = 0 \\ -mg\cos\theta + 2\left(\Phi_{21y}\cos\theta - \Phi_{21x}\sin\theta\right) = 0 \\ 4kl\cos\theta + \Phi_{21x} + \Phi_{Bx} = 0 \\ -mg + \Phi_{By} - \Phi_{21y} = 0 \\ mg\cos\theta + 2\left(\Phi_{21y}\cos\theta + \Phi_{21x}\sin\theta\right) = 0. \end{cases}$$

can be solved (Cramer system) for the constraint reactions' components, which will be functions of $\theta$. Easy calculation furnish, in the interval $0 < \theta < \pi/2$:

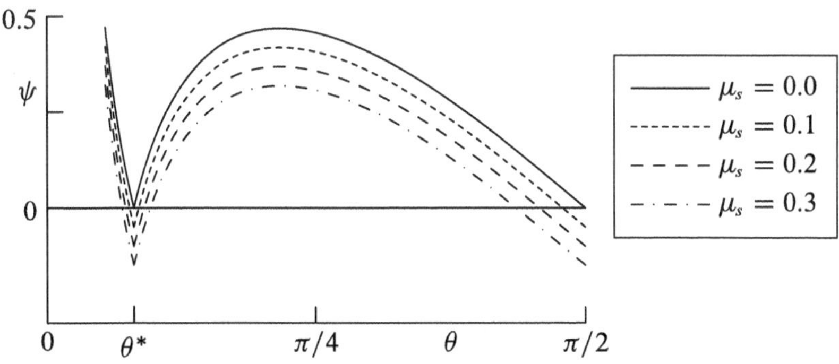


**Fig. 9.29** Equilibrium configurations: all values of $\theta$ such that $\psi(\theta) \leq 0$ ($\lambda = 1/4$)

$$\mathbf{\Phi}_O = \left(4kl\,\frac{\cos\theta}{\sin\theta}\,(\lambda - \sin\theta)\;,\; mg\right)$$

$$\mathbf{\Phi}_{21} = \left(-\frac{mg}{2}\,\frac{\cos\theta}{\sin\theta}\;,\; 0\right)$$

$$\mathbf{\Phi}_B = \left(-4kl\,\frac{\cos\theta}{\sin\theta}\,(\lambda - \sin\theta)\;,\; mg\right)$$

Substituting the components of $\mathbf{\Phi}_B$ in (9.53) we see the equilibrium configurations correspond to those $\theta \in (0, \pi/2]$ such that

$$\psi(\theta) = \frac{\cos\theta}{\sin\theta}\,|\sin\theta - \lambda| - 2\mu_s\lambda \leq 0$$

Figure 9.29 shows that for $\mu_s = 0$ there is a unique solution $\sin\theta^* = \lambda$, while for $\mu_s \neq 0$ we have entire intervals of equilibrium configurations. Note that the equilibrium solution found in the ideal case persists in presence of friction as well. □

## 9.12 Bifurcation Diagrams

If we only look at problems with one degree of freedom we observe that the potential often depends, apart from the generalized coordinate $q$, on a *structural parameter* $\lambda$ related to the system's characteristic physical quantities (for instance masses, lengths, elastic constants and so on): $U = U(q, \lambda)$. Ordinary equilibrium configurations, where

$$Q(q, \lambda) = \frac{\partial}{\partial q} U(q, \lambda) = 0\,,$$

depend on the structural parameter as well, in the sense that the system admits equilibrium configurations that will vary, together with stability, if $\lambda$ varies. Similarly,

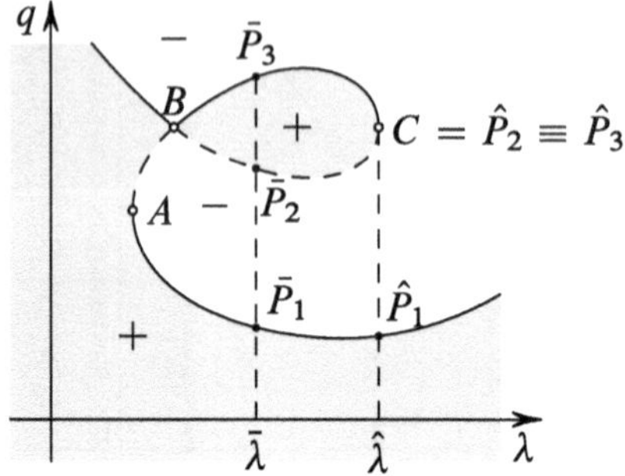


**Fig. 9.30** Bifurcation diagram

the possible boundary positions may or not be equilibrium configurations (stable or unstable) depending on the values of $\lambda$.

**Definition 9.33** The *bifurcation diagram* is constructed by drawing on the $(\lambda, q)$-plane the locus $\mathscr{C}$ of points $(\lambda^*, q^*)$ such that $\lambda^*$ is a physically admissible value for the parameter $\lambda$, and $q^*$ is an equilibrium configuration for $\lambda = \lambda^*$.

Assuming $a \leq q \leq b$, the bifurcation diagram then will contain all ordinary configurations such that $Q(q, \lambda) = 0$, together with the boundary configurations satisfying $Q(a, \lambda) \leq 0$ and $Q(b, \lambda) \geq 0$.

The determination of the bifurcation diagram can simplify an equilibrium's stability analysis. Suppose in fact we find the regions on the plane $(\lambda, q)$ where $Q(q, \lambda) > 0$ (positive regions) and $Q(q, \lambda) < 0$ (negative regions). One proves that there is a 1–1 correspondence between statically stable equilibrium configurations and points of $\mathscr{C}$ with a positive region above and a negative region below. This holds for ordinary and boundary equilibria, although for the latter one should check only a sign, the sign of the unique permitted region adjacent to the equilibrium configuration.

To prove this statement we should recall that only isolated relative maxima of the potential are statically stable. In turn, for isolated relative maxima $U' = Q(q, \lambda)$ changes from being positive to negative, as $q$ passes through an equilibrium configuration when moving along the positive $Oq$ axis.

Consider for example the bifurcation diagram of Fig. 9.30. Solid lines represent stable equilibrium configurations, while dashed lines are unstable positions.

At $A$, $B$, $C$ (*bifurcation points*, those with multiplicity higher than one or where the tangent line is vertical), we have $U''(q) = 0$. The line $\lambda = \bar{\lambda}$, say, meets the curve at $(\bar{P}_1, \bar{P}_2, \bar{P}_3)$ and the system admits, for $\lambda = \bar{\lambda}$, three equilibria. Two are stable $(\bar{P}_1, \bar{P}_3)$ and one is unstable $(\bar{P}_2)$. In case $\lambda = \hat{\lambda}$ we have two equilibria $(\hat{P}_1, \hat{P}_2 \equiv C)$, of which one stable $(\hat{P}_1)$ and the other unstable $(\hat{P}_2)$. As $\lambda$ varies, passing through the bifurcations, the number and nature of the configurations change; the *bifurcation* could occur once or at several points.

Figure 9.31 shows the bifurcation diagram corresponding to Example 9.30.

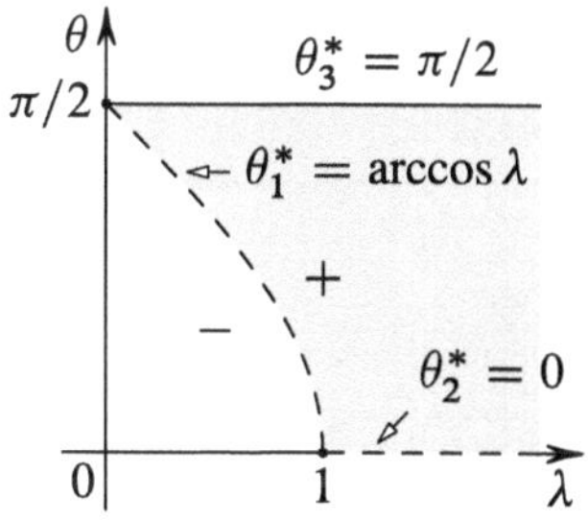

Fig. 9.31 Bifurcation diagram for Example 9.30

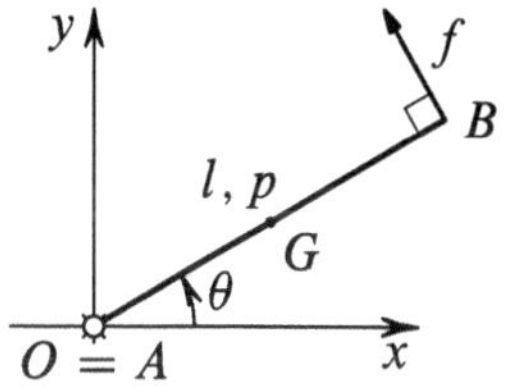

Fig. 9.32 A rod pinned at $A$ and subject to a perpendicular force

## 9.13 Examples in Statics

**Example 9.34** A homogeneous rod of length $l$ and weight $p$ is constrained at end $A$ to a fixed point $O$ by an ideal pin, and is subject to a perpendicular force $f$, as in Fig. 9.32. Setting to zero the forces' moment with respect to the pole $O$, where the pin's constraint reaction is placed, we easily find the free equilibrium equation

$$pl/2 \cos\theta - lf = 0\,.$$

It is important to note the equation can be solved for $\theta$ only if $2f/p \leq 1$. If the inequality is violated ($f$ is "too large") there is no equilibrium configuration. Necessarily therefore $2f/p \leq 1$, i.e. $0 \leq f \leq p/2$, otherwise it is not possible to achieve equilibrium (we limit the discussion to angles $\theta$ between 0 and $\pi/2$). That means every value of $f$ satisfying the limitation corresponds to an angle $\theta$ that can make the moment vanish. With a small $f$ there corresponds an "almost vertical" rod, while near $p/2$ the rod is close to being horizontal (the extreme case $f = p/2$ is that for which the rod is exactly horizontal).

To verify the equilibrium conditions hold it is not necessary to impose $\mathbf{R} = \mathbf{0}$, as the constraint reaction of the pin at $O$ "automatically" yields the necessary force for the resultant to vanish. We shall then write this equation (two scalar equations in the plane) only if we are interested in finding the components of the constraint reaction. □

**Example 9.35** Let us modify the constraint of the previous example, and suppose $A$ rests on a rough guide, with static friction coefficient $\mu_s$ (here, too, we limit ourselves to angles $\theta$ from fra 0 to $\pi/2$). We examine the equilibrium condition and see what

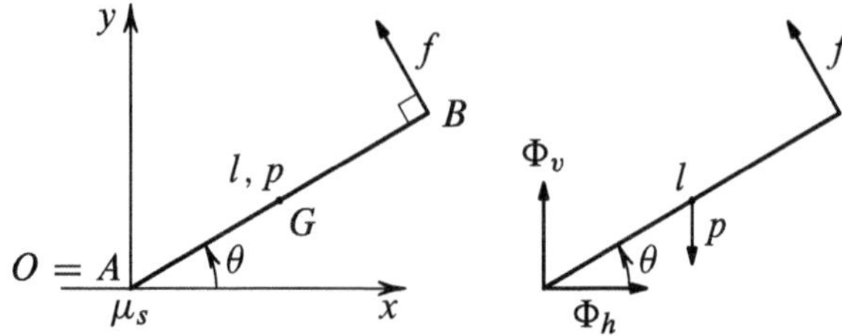


**Fig. 9.33** A rod supported at $A$ by a rough horizontal guide

friction is necessary to satisfy Coulomb's relation. In particular we will find the smallest $\mu_s$ able to guarantee equilibrium for all admissible forces $f$.

The moment, with respect to the pole $O$, will give the equation

$$\mathbf{M}_O = \mathbf{0} \quad \Leftrightarrow \quad fl - pl/2 \cos\theta = 0 \quad \Leftrightarrow \quad f = p/2 \cos\theta. \tag{9.54}$$

As said in the previous example, only the value of $f$ with $0 \le f \le p/2$ allows for an equilibrium position, given by the $\theta$ satisfying (9.54).

In this problem, though, it is further necessary to check the given friction coefficient is sufficient. In fact the rod is no longer pinned at $O$, and might "slip" on the horizontal guide.

Let us pass to the first balance equation: $\mathbf{R} = \mathbf{0}$. Calling $\Phi_h$ and $\Phi_v$ the horizontal and vertical components of the reaction at $A$, see Fig. 9.33 right, the vanishing of the resultant gives

$$\Phi_h = f \sin\theta\,, \qquad \Phi_v = p - f\cos\theta\,.$$

Using the relationship between $f$ and $\theta$ found earlier we can recast the above as

$$\Phi_h = p/2 \cos\theta \sin\theta\,, \qquad \Phi_v = p - p/2\cos^2\theta\,,$$

and deduce from Coulomb's relation that equilibrium is possible if

$$|\Phi_h| \le \mu_s|\Phi_v| \quad \Leftrightarrow \quad |\sin\theta\cos\theta| \le \mu_s|2 - \cos^2\theta|\,. \tag{9.55}$$

For given $\mu_s$ therefore, the $f$ that correspond to angles $\theta$ satisfying inequality (9.55) are precisely those for which equilibrium is possible.

As we only consider $\theta$ in the interval $[0, \pi/2]$, we can eliminate the absolute values are rewrite the condition as

$$\frac{\sin\theta\cos\theta}{2 - \cos^2\theta} \le \mu_s\,. \tag{9.56}$$

It is interesting to find the smallest $\mu_s$ for which the inequality holds. Note the function on the left is positive on $(0, \pi/2)$ and vanishes only at the interval's ends. Computing the derivative we obtain a maximum at $\sin\theta = \sqrt{3}/3$, so $\cos\theta = \sqrt{6}/3$. The function's value at this maximum is $\sqrt{2}/4$. Hence $\mu_s \ge \sqrt{2}/4$ guarantees an

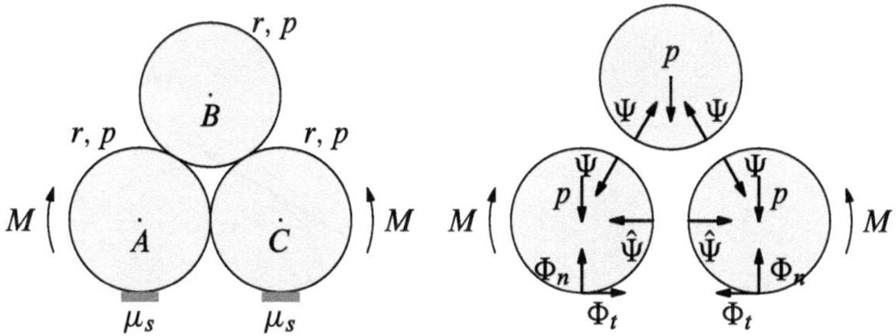


**Fig. 9.34** The equilibrium of three touching disks supported by a rough horizontal guide with coefficient of static friction $\mu_s$

equilibrium is possible (for a suitable slope $\theta$) for any admissible value of the force $f$. In particular, $f = p\sqrt{6}/6$ corresponds to the angle $\theta$ for which the function in (9.56) is largest. This is therefore the intensity of the force requiring the maximum friction to ensure equilibrium. □

**Example 9.36** In a vertical plane the homogeneous disks of equal radius $r$ and weight $p$ of Fig. 9.34 touch each other smoothly. Disks $A$ and $C$ (labelled by their centers) are supported by a horizontal guide with coefficient of friction $\mu_s$. On these two disk act two couples with equal and opposite torques $M$. We ask which are the minimum and maximum values of $M$ and the minimum of $\mu_s$ that guarantee equilibrium.

It is convenient to exploit the material symmetry with respect to the vertical line through $B$, so we can set equal the components of the constraint reactions at the contact points with the guide, as in the right picture, and the forces acting between disks.

The vertical resultant of the forces acting on the three disks is

$$R_v = 0 \quad \Leftrightarrow \quad 2\Phi_n - 3p = 0 \quad \Leftrightarrow \quad \Phi_n = 3p/2\,.$$

For disk $B$, moreover, for the vertical resultant we have

$$R_v = 0 \quad \Leftrightarrow \quad 2\Psi\cos(\pi/6) = p \quad \Leftrightarrow \quad \Psi = p\sqrt{3}/3$$

and for disk $A$

$$R_h = 0 \quad \Leftrightarrow \quad \Phi_t - \hat{\Psi} - \Psi\cos(\pi/3) = 0 \quad \Leftrightarrow \quad \hat{\Psi} = \Phi_t - \Psi/2$$
$$M_A = 0 \quad \Leftrightarrow \quad M = r\Phi_t\,.$$

Applying the Coulomb-Morin inequality,

$$|\Phi_t| \le \mu_s|\Phi_n| \quad \Leftrightarrow \quad M/r \le \mu_s 3p/2 \quad \Leftrightarrow \quad M \le \mu_s 3pr/2\,.$$

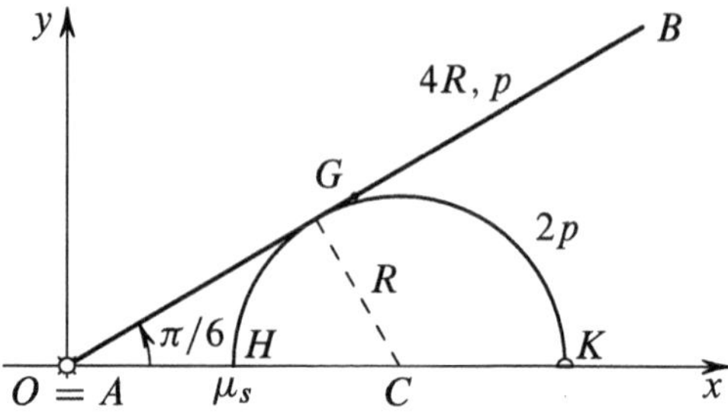


**Fig. 9.35** A rod supported by a semi-circle which is constrained to move on a horizontal guide, with friction at $H$

We must observe that disks $A$ and $C$ touch only if $\hat{\Psi} \geq 0$. Therefore

$$\hat{\Psi} = \Phi_t - \Psi/2 \geq 0 \quad \Leftrightarrow \quad M/r \geq p\sqrt{3}/6 \quad \Leftrightarrow \quad M \geq pr\sqrt{3}/6$$

and so

$$pr\sqrt{3}/6 \leq M \leq \mu_s 3rp/2\,.$$

Consequently $\mu_s \geq \sqrt{3}/9$, giving the minimum value of the coefficient of friction. □

**Example 9.37** In a vertical plane a homogeneous rod of length $4R$ and weight $p$ is pinned at end $A$ to the coordinates' origin and leans without friction on a semi-circle of radius $R$ and weight $2p$. The latter's ends $H$, $K$ move along a horizontal guide through $O$, as in Fig. 9.35. At $H$ there is static friction with coefficient $\mu_s$, while at $K$ the motion is smooth. We seek the *smallest* value of $\mu_s$ necessary to have equilibrium when the angle between rod and $x$-axis is $\pi/6$.

We can separate the two parts of the system to highlight the forces $\mathbf{\Phi}$ and $-\mathbf{\Phi}$ the rods exchange at the contact point, see Fig. 9.36. In the picture the letter $\Phi$ is not in boldface, meaning it *does not* denote the force but the component along the unit vector next to it. By the action-reaction principle the forces exchanged between rod and semi-circle are equal and opposite, so their components along opposite unit directions are the same. We call the common value $\Phi$.

The moment acting on the rod with respect to $O$, where the pin's constraint reaction acts, is zero, so

$$\Phi R\sqrt{3} - pR\sqrt{3} = 0 \quad \Leftrightarrow \quad \Phi = p\,.$$

Let us focus on the semi-circle, for which we impose the moment with respect to the center $C$ is zero. Then $\Psi_v = \Upsilon$, where $\Psi_v$ is the vertical component of the constraint reaction at $H$; at $H$ there will also be a horizontal component due to the friction. The vanishing of the resultant's horizontal and verticale components for the semi-circle gives

$$\Phi/2 + \Psi_h = 0\,, \qquad \Psi_v + \Upsilon - 2p - \Phi\sqrt{3}/2 = 0\,.$$

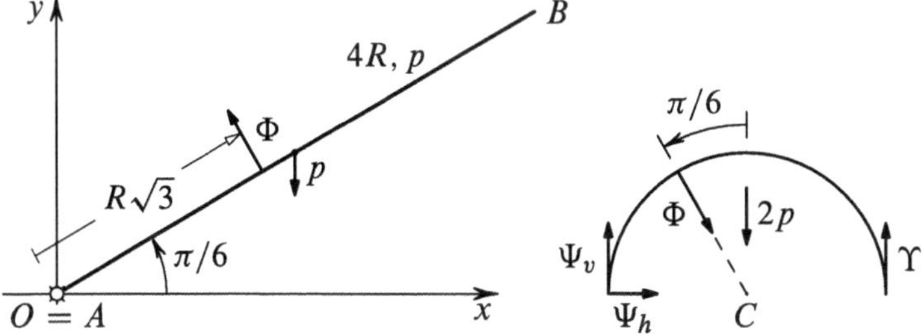


**Fig. 9.36** Equilibrium of the rod and the semi-circle

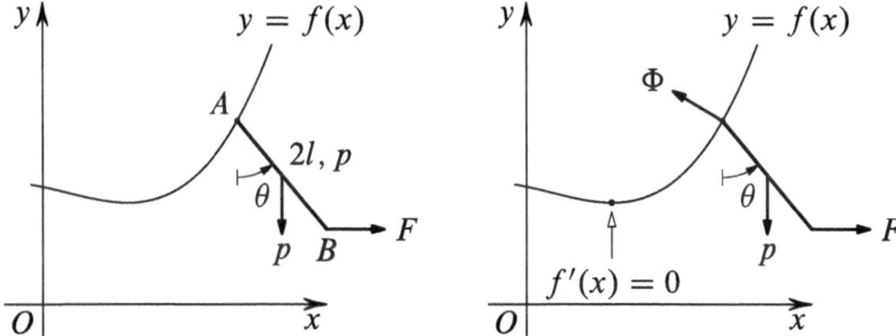


**Fig. 9.37** A rod with weight $p$ constrained to the graph of $y = f(x)$

With the value $\Phi$ already computed, using $\Psi_v = \Upsilon$ a few computations produce

$$\Psi_v = p(1 + \sqrt{3}/4)\,, \qquad \Psi_h = -p/2\,.$$

The minus sign of $\Psi_h$ was to be expected: the component must be negative since it counterbalances the horizontal push to the right of the rod leaning on the semi-circle. Coulomb's relation becomes

$$|\Psi_h| \le \mu_s |\Psi_v| \quad \Leftrightarrow \quad \mu_s \ge \frac{8 - 2\sqrt{3}}{13}$$

and the inequality provides us with the minimum for $\mu_s$. □

**Example 9.38** As easy application of the Principle of virtual work consider a homogeneous rod $AB$ of length $2l$ and weight $p$, with $A$ moving on a vertical plane along a smooth guide described by the graph of a regular map $y = f(x)$, as per Fig. 9.37, and subject to a horizontal force $F$ at $B$.

Call $x$ the abscissa of $A$ and $\theta$ the angle between rod and vertical direction, as in the figure. These are the generalized coordinates we will use to write the virtual work of the active forces. Call $G$ the center of gravity. As

$$y_G = f(x) - l\cos\theta \quad \Rightarrow \quad \delta y_G = f'(x)\delta x + l\sin\theta\delta\theta$$

and

$$x_B = x + 2l \sin\theta \quad \Rightarrow \quad \delta x_B = \delta x + 2l \cos\theta \delta\theta\,,$$

the virtual work of the active forces takes the form

$$\delta L = -p\delta y_G + F\delta x_B = [F - pf'(x)]\delta x + [2lF \cos\theta - pl \sin\theta]\delta\theta\,.$$

The terms in brackets are $Q_x$ and $Q_\theta$, the components of the active force with respect to the generalized coordinates $x$ and $\theta$:

$$Q_x = F - pf'(x)\,, \qquad Q_\theta = 2lF \cos\theta - pl \sin\theta\,. \tag{9.57}$$

The equilibrium condition prescribes that $\delta L = 0$ for all virtual displacements, so here for all arbitrary variations $\delta x$ and $\delta\theta$. This happens if and only if $Q_x = 0$ and $Q_\theta = 0$, so from (9.57) we obtain

$$f'(x) = F/p\,, \qquad \tan\theta = 2F/p\,.$$

The same conclusion can be reached via the balance equations, by annihilating the horizontal and vertical resultant and the moment with respect to $A$ for example (though we must keep into account that at $A$ there is an orthogonal constraint reaction $\Phi$ to the guide). □

**Example 9.39** Trusses are made of a certain number of rods pinned together and variously constrained. They represent standard, and at times not simple, devices where one can apply the balance equations. In many situations it becomes necessary to break up the system in a suitable fashion, and then impose the equilibrium conditions on the separate parts.

An example is represented in Fig. 9.38 left: we have five pinned homogeneous rods in a vertical plane. There are further pins at the fixed points $A$ and $D$, while the pin at $E$ rests on a smooth vertical wall. $AB$ and $BC$ have length $l$, weight $p$ and form a right angle at $B$. $CE$ and $ED$ have negligible weight while $CD$ has weight $q$. The points $A$, $E$, $C$ are aligned vertically, while $C$ and $D$ stay at the same height.

We want to find the constraint reaction $\Phi$ at $E$. We can proceed by separating the system at $C$ into two parts: $ABC$ and $CDE$, and highlighting the forces $\Psi_v$, $\Psi_h$ the two parts exchange through the pin $C$, as in Fig. 9.38 right.

Setting to zero the moment with respect to $A$ for $ABC$ gives

$$+\,p(l/2)\sqrt{2}/2 + p(l/2)\sqrt{2}/2 - \Psi_h l\sqrt{2} = 0 \quad \Rightarrow \quad \Psi_h = p/2\,.$$

Setting also equal to zero the moment acting on $BC$ with respect to $B$ (where a force is applied, due to the rod $AB$ which right now does not contribute) produces

$$\Psi_v l\sqrt{2}/2 - p(l/2)\sqrt{2}/2 - \Psi_h l\sqrt{2}/2 = 0 \quad \Rightarrow \quad \Psi_v = p\,.$$

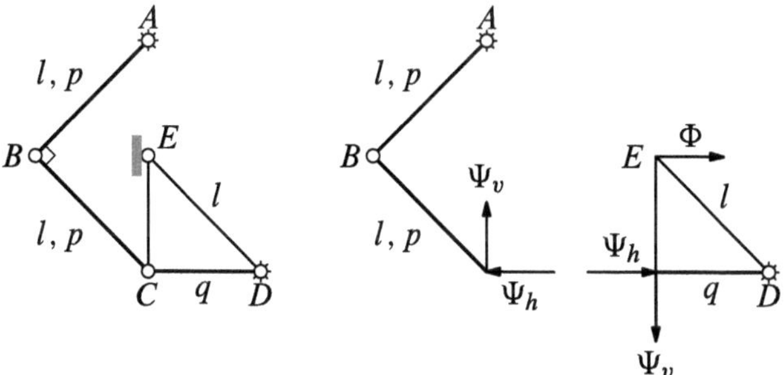


**Fig. 9.38** The equilibrium of a truss made by 5 rods

Finally, imposing that the moment with respect to $D$ of the forces acting on $CDE$ is zero gives

$$+\Psi_v l\sqrt{2}/2 + q(l/2)\sqrt{2}/2 - \Phi\sqrt{2}/2 = 0 \quad \Rightarrow \quad \Phi = p + q/2\,.$$

There are several other approaches to the problem: one could for example break the system into all of its constituents and write the balance equations for each one. In this way we would end up with many equations and we would find unwanted unknowns. In every situation it is important to select the most effective strategy. □

# Chapter 10
# Particle Dynamics

In this chapter we apply the Laws of Mechanics stated in Chap. 8 to analyze the motion of particles, both free and constrained, subject to various kinds of active forces.

Newton's law is a second-order differential equation in the time variable. The principle of mechanical determinism (see Sect. 8.2) guarantees that this ordinary differential equation admits precisely one solution once we know the initial values of the position and the velocity. Starting from the knowledge of the active forces and the constraints acting on the particle, together with the initial conditions, our objective will be to extract the most information possible about the ensuing motions. We will also use qualitative methods that avoid resorting to explicit solutions of the equations of motion, which are not always easy to find.

We shall distinguish the problems as *direct* and *inverse*. We will speak of a direct problem when the active forces and the constraints acting on the system are known, and we seek the motion. A problem is called inverse when the motion we wish to impart on the system is known beforehand, and we want to compute the forces necessary to realize it. There are *semi-inverse* problems too, i.e. problems where one or more forces (to be found) act so that some degrees of freedom evolve in a predetermined way while the remaining degrees have the *status* of unknowns.

## 10.1 Dynamics of an Unconstrained Point

Newton's law postulates that the motion of a free particle is governed by the equation

$$m\,\mathbf{a} = \mathbf{F}(P, \mathbf{v}, t)\,, \tag{10.1}$$

where $m$ is the mass of particle $P$, $\mathbf{v}$, $\mathbf{a}$ are its velocity and acceleration, and $\mathbf{F}$ is the resultant of the active forces applied to $P$. Equation (10.1) must be solved together



P. Biscari et al., *Rational Mechanics*, UNITEXT 177,
https://doi.org/10.1007/978-3-032-07462-1_10

with the initial conditions

$$P(t_0) = P_0\,, \qquad \mathbf{v}(t_0) = \mathbf{v}_0\,. \tag{10.2}$$

### 10.1.1 Energy Integral

When the active forces are conservative equation (10.1) admits a first integral, i.e. it can be transformed in an equation of order one. To find the first integral, suppose $U$ is a potential of the active forces ($\mathbf{F} = \nabla U$), write (10.1) as $\mathbf{0} = m\mathbf{a} - \mathbf{F}$ and take the dot product with the velocity of $P$:

$$0 = \mathbf{v}\cdot(m\mathbf{a} - \mathbf{F}) = m\frac{d}{dt}\left(\frac{1}{2}\mathbf{v}\cdot\mathbf{v}\right) - \nabla U\cdot\mathbf{v} = \frac{d}{dt}\left(\frac{1}{2}mv^2 - U\right)\,. \tag{10.3}$$

Then (10.3) can be simply integrated with respect to time to give

$$\frac{1}{2}mv^2 - U \equiv \text{constant} = E. \tag{10.4}$$

The quantity $E$ is called *mechanical energy* of the particle, while the first summand

$$T = \frac{1}{2}mv^2$$

is the *kinetic energy* of the particle. The mechanical energy $E = T - U$ stays constant during the entire motion and its value is therefore determined by the initial conditions. We emphasize that in terms of the potential energy $V = -U$ (see Sect. 7.1) the mechanical energy is defined as $E = T + V$.

## 10.2 Dynamics of a Constrained Particle

The presence of one or more constraints restricts the set of positions, or more generally of velocity distributions, accessible to the particle. As discussed in Sect. 8.6, the constraint acts through a constraint reaction $\mathbf{\Phi}$, to be added to the possible active forces already acting on the particle. In presence of constraints, Newton's law then becomes

$$m\,\mathbf{a} = \mathbf{F} + \mathbf{\Phi}\,. \tag{10.5}$$

The differential problem associated with (10.5) is profoundly different in spirit to the one characterizing the dynamics of the free particle. In fact, our study needs to account for the fact the constraint reactions are not known beforehand, and also that

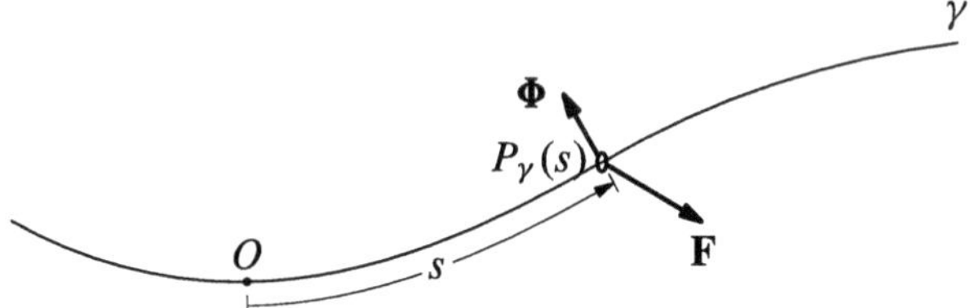


**Fig. 10.1** A particle moving on a prescribed trajectory

not every solution is admissible. Finally, even though the constraint reactions are not completely known a priori, neither are they completely arbitrary. The nature (smooth or rough) of the device that realizes the constraint imposes constitutive restrictions on the type of constraint reactions that can occur. In other words in presence of constraints we have to address the following problem.

- Within the set of constraint reactions the constraint can produce, determine the constraint reaction $\mathbf{\Phi}$ such that, when inserted in (10.5), the motion ensuing from initial conditions (10.2) satisfies the restriction given by the constraint itself.
- Determine, once the constraint reaction is known, the particle's motion.

We will see that in several situations it will be simpler to swap the above problems. Using the constitutive information on the constraint reaction, it will be often possible to project (10.5) in a suitable manner, to find a *free* equation of motion, i.e. a differential equation where only active forces appear. Such equation will enable us to determine the particle's motion, knowing which the remaining components of (10.5) will give the unknown constraint reactions. In presence of holonomic constraints only, Lagrangian mechanics (see Chap. 14) makes this procedure systematic. At any rate, determining the motion and constraint reactions in a constrained system is a *semi-inverse* problem from the point of view of the classification given at the beginning of the chapter.

### 10.2.1 Motion Along a Given Trajectory

Consider a particle $P$ constrained to move along a fixed guide representable by a curve $\gamma$. This model can serve to describe, among other things, the motion of a small ring that is free to move along a guide (Fig. 10.1), or a ball inside a pipe. Parametrizing the positions along $\gamma$ by the arclength $s$ (see Appendix A.2), the curve will be described by a map

$$\begin{aligned} P_\gamma : [0, L] &\to \mathscr{E} \\ s &\mapsto P_\gamma(s)\,. \end{aligned}$$

To identify the position of $P$ at a generic instant $t$, it suffices to know the value of the arclength $s(t)$, since

$$P(t) = P_\gamma\big(s(t)\big)\,. \tag{10.6}$$

The particle has then one degree of freedom. As we found in Sect. 1.2, if we differentiate (10.6) with respect to time and use (A.16), we can prove the velocity of $P$ is parallel to the unit tangent vector along $\gamma$

$$\mathbf{v} = \frac{dP}{dt} = \frac{dP_\gamma}{ds}\,\frac{ds}{dt} = \dot{s}\,\mathbf{t}\,.$$

Differentiating again, and using (A.19), we can prove the acceleration of $P$ has a tangential component and a component along the normal to $\gamma$

$$\mathbf{a} = \frac{d(\dot{s}\mathbf{t})}{dt} = \ddot{s}\,\mathbf{t} + \dot{s}\,\frac{d\mathbf{t}}{ds}\,\frac{ds}{dt} = \ddot{s}\,\mathbf{t} + \frac{\dot{s}^2}{\rho}\,\mathbf{n}\,,$$

where $\rho$ is the curvature radius of $\gamma$ computed at the position of $P$.

These kinematical observations on the direction and components of the acceleration suggest to project Newton's law (10.5) along the trajectory's Frenet frame $\{\mathbf{t}, \mathbf{n}, \mathbf{b}\}$. Then

$$\begin{aligned} m\,\ddot{s} &= \big(\mathbf{F} + \mathbf{\Phi}\big)\cdot\mathbf{t} = F_t(s, \dot{s}, t) + \Phi_t \\ m\,\frac{\dot{s}^2}{\rho} &= \big(\mathbf{F} + \mathbf{\Phi}\big)\cdot\mathbf{n} = F_n(s, \dot{s}, t) + \Phi_n \\ 0 &= \big(\mathbf{F} + \mathbf{\Phi}\big)\cdot\mathbf{b} = F_b(s, \dot{s}, t) + \Phi_b\,. \end{aligned} \tag{10.7}$$

The constraint is holonomic and the system has one degree of freedom with generalized coordinate coinciding with the arclength $s$. So we can define the generalized component of the active forces along $s$ (Definition 7.43)

$$Q_s = \mathbf{F}\cdot\frac{dP}{ds} = \mathbf{F}\cdot\mathbf{t} = F_t\,. \tag{10.8}$$

The previous chapter's results (valid for ideal constraints, see Sect. 9.9) then prove the following.

**Proposition 10.1** *The equilibrium positions of a point constrained to a smooth guide are the positions at which the active force's tangential component vanishes.*

In order to extract more information on the motion it becomes necessary to specify the nature, that is, whether smooth or rough, of the constraint.

### 10.2.2 Smooth Guide

In absence of friction, the tangential component of the constraint reaction is zero (see (8.23) with $\mu_s = 0$):

$$\mathbf{\Phi} \cdot \mathbf{t} = \Phi_t = 0 .$$

Then clearly the first of (10.7) gives the free motion equation, which identifies uniquely the particle's motion if the initial data are given. The second and third equations in (10.7) determine completely the constraint reaction $\mathbf{\Phi}$, once the motion is known. Let us analyze in detail $(10.7)_1$:

$$m\ddot{s} = F_t(s, \dot{s}, t) . \tag{10.9}$$

- *Inertial motion*. In absence of active forces (but also if the active force is normal to the guide) the motion is uniform:

$$F_t = 0 \qquad \Longrightarrow \qquad \dot{s}(t) = \dot{s}(t_0) = \text{constant} \quad \forall\, t \geq t_0 .$$

- *Positional forces*. If the tangential component of the active force depends only on the position of $P$ (hence not on the velocity, nor on time explicitly), the pure equation (10.9) admits a first integral, formally similar to the energy integral (10.4). To find it, note (10.9) can be integrated explicitly if multiplied by $\dot{s}$. In fact,

$$\dot{s}\big(m\ddot{s} - F_t(s)\big) = 0 \qquad \Longrightarrow \qquad \frac{d}{dt}\left(\frac{1}{2}m\dot{s}^2 - U\right) = 0 , \tag{10.10}$$

where $U$ is any primitive of $F_t$,

$$U(s) = \int_{s_0}^{s} F_t(x)\, dx \qquad (s_0 \text{ arbitrary}). \tag{10.11}$$

Note (10.8) and (10.11) show that the generalized component of the active force coincides with the derivative of $U$: $Q_s(s) = F_t(s) = U'(s)$. In particular, the equilibrium positions coincide with the stationary points of $U$. Equation (10.10) immediately gives the integral of motion

$$\frac{1}{2}m\dot{s}^2 - U(s) = \text{constant} \tag{10.12}$$

The function $U$ assumes then a role equivalent to that of potential of a conservative force field. Its stationary points give in fact the equilibrium positions, and it allows to find a first integral perfectly similar to the energy. Clearly, if the active force was conservative, $U$ would be a proper potential. But here we have found a more general property: when a point is constrained to a fixed and smooth guide, we can find a first integral analogous to the energy even in presence non-conservative active force fields. The only requirement is that one component (the tangential one) must be positional.

### 10.2.3 Rough Guide

Friction involves the presence of a tangential component in the constraint reaction, thus turning non-free the equation of motion $(10.7)_1$. We will use the model of dynamic friction introduced in Sect. 8.9.2, and in particular Remark 8.14. Expression (8.28) gives the modulus of the tangent part of the constraint reaction.

The direction of $\boldsymbol{\Phi}_t$ is always opposite to the velocity of a constrained point: $\boldsymbol{\Phi}_t \parallel \mathbf{v}$ and $\boldsymbol{\Phi}_t \cdot \mathbf{v} \leq 0$. In the particular case under exam, and assuming $\dot{s} \neq 0$, we can rewrite (8.27) as

$$\boldsymbol{\Phi}_t = -\mu_d \sqrt{\Phi_n^2 + \Phi_b^2}\, \frac{\dot{s}}{|\dot{s}|}\, \mathbf{t}\,.$$

To find a free motion equation again, we employ $(10.7)_2$ and $(10.7)_3$ to express the normal components of the constraint reaction in terms of velocity and active forces

$$\Phi_n = m\, \frac{\dot{s}^2}{\rho} - F_n\,, \qquad \Phi_b = -F_b\,. \tag{10.13}$$

Substituting (10.13) in $(10.7)_1$ finally gives

$$m\, \ddot{s} = F_t - \mu_d \sqrt{\left(m\, \frac{\dot{s}^2}{\rho} - F_n\right)^2 + F_b^2}\, \frac{\dot{s}}{|\dot{s}|}\,. \tag{10.14}$$

Now, (10.14) is not an easy equation to treat in general. Note however it admits a first integral not very different from (10.12) in case the guide is a straight line and the active forces are positional. In fact, if the guide is straight, the curvature radius diverges ($\rho = +\infty$) and $(10.13)_1$ simplifies to $\Phi_n = -F_n$. If we further restrict to trajectories for which $\dot{s}$ has constant sign (for instance, positive), and consider positional active forces only, (10.14) reads

$$m\, \ddot{s} = F_t(s) - \mu_d \sqrt{F_n^2(s) + F_b^2(s)} = \phi(s)\,.$$

Hence we have a first integral like (10.12), with a "potential" $U$ obtained as primitive of $\phi$ rather than just $F_t$. We emphasize that in this case the value of this first integral remains constant only along any portion of trajectory where $\dot{s}$ has constant sign. Any time the point stops ($\dot{s} = 0$) the model of dynamic friction must be replaced by that of static friction. In particular, one needs to assess with care whether the point is able to start moving again after stopping.

## 10.3 Harmonic Oscillator

At the beginning of this section we discuss a simple mechanical system made of a point constrained to a straight horizontal smooth guide and subject to a spring.

The subject is much more important than what might appear at first, because the ensuing motion, of oscillatory type and called *harmonic motion*, appears in physically different systems, which share with this one the same type of equation of motion. In other words harmonic motion, albeit very simple, is a common feature of a great number of mechanical systems, and for this reason it is important to understand its nature and characteristics.

After presenting *small oscillations* in Sect. 14.8, Chap. 14, we will see that under suitable hypotheses any system with one generalized coordinate undergoes a harmonic motion, approximatively, close to a generic stable equilibrium configuration. For this reason the importance of the subject discussed in the sequel widely transcends the simple context offered here.

Consider then Eq. (10.5) in case of a particle of mass $m$ constrained to move on a smooth horizontal straight guide and subject to an elastic force, as in Fig. 10.2.

Choose the frame of reference with origin at the center $O$ of the elastic force $\mathbf{F} = -kOP$ and axes oriented as the guide and the vertical line. We can write the component along the $x$-axis of (10.5) as

$$m\ddot{x} = -k\,x \tag{10.15}$$

while the vertical component will give $\Phi_y = mg$.

As is well known (see A.5.2 in the Appendix) that the general solution to (10.15), written as

$$\ddot{x} + \omega^2 x = 0 \quad (\omega^2 = k/m)\,, \tag{10.16}$$

contains two arbitrary constants $A$ and $B$ and is given by

$$x(t) = A\cos(\omega t) + B\sin(\omega t)\,. \tag{10.17}$$

This motion, called *harmonic*, can be recast as

$$x(t) = r\cos(\omega t + \delta)\,, \tag{10.18}$$

which is equivalent to (10.17) when we choose the *amplitude* $r$ using

$$r^2 = A^2 + B^2$$

and the *phase* $\delta$ defined by

$$\cos\delta = A/r \qquad \sin\delta = -B/r\,,$$

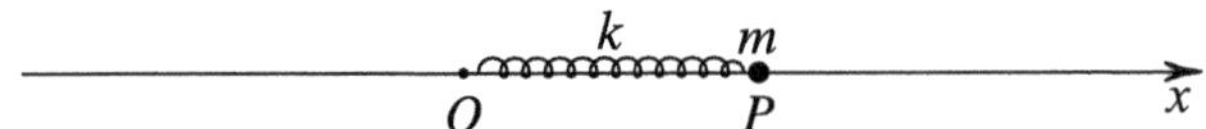


**Fig. 10.2** A particle $P$ constrained to a smooth guide and subject to the force exerted by a spring of elastic constant $k$

so that

$$\tan\delta = -B/A\,.$$

The values of $A$ and $B$ in (10.17) are found using the initial conditions, i.e. the values $x_0$ and $\dot{x}_0$ that $x(t)$ and $\dot{x}(t)$ have at a given instant, for instance $t = 0$. So,

$$x_0 = A\,, \quad \dot{x}_0 = \omega\, B$$

from which we easily deduce the values of $r$ and $\delta$

$$r^2 = x_0^2 + \frac{\dot{x}_0^2}{\omega^2}\,, \qquad \tan\delta = -\frac{\dot{x}_0}{x_0\omega}\,.$$

In conclusion, whichever the initial conditions the particle makes *harmonic* oscillations of *period* $T = 2\pi/\omega$, *amplitude* $r$ and *initial phase* $\delta$. The reciprocal of the period is called the *frequency* of the vibration.

**Remark 10.2** If we place the spring's fixed point in a position other than the origin, the equation of motion (10.16) changes in an inessential way, by an additional constant term on the right. Evidently a simple change of generalized coordinate $x$ takes us back to the present case. Hence in this and the successive sections we shall forsake this possible variant.

**Remark 10.3** There are countless mechanical systems, quite different from that of Fig. 10.2, that are still described by a unique generalized coordinate $q$, for which the equation of motion boils down to

$$\ddot{q} + \omega^2 q = 0\,,$$

where $\omega^2$ is related to the system's physical constants, and whose motion becomes

$$q(t) = A\cos(\omega t) + B\sin(\omega t)$$

or something similar to (10.18). In all of these cases one says that the system moves under *harmonic motion*.

### 10.3.1 Forced Harmonic Oscillator: Beats and Resonance

It is interesting, in view of several possible applications, to study a point subject to, besides the spring, also a force of horizontal component $F(t)$ called *forcing term*, as in Fig. 10.3.

Here, too, it should be said that the discussion goes beyond the present mechanical context, since a qualitatively identical motion holds for systems of other nature than a simple point constrained to a guide.

The equation of motion can easily be recast as

$$\ddot{x} + \omega^2 x = f(t), \tag{10.19}$$

where $f(t) = F(t)/m$. While for Eq. (10.16) one speaks of *free* linear oscillations, here obviously we have *forced* linear oscillations.

The general solution to this equation reads

$$x(t) = A\cos(\omega t) + B\sin(\omega t) + \tilde{x}(t), \tag{10.20}$$

where $\tilde{x}(t)$ is a particular solution of (10.19). A case of particular interest is the sinusoidal forcing term

$$f(t) = \Gamma\cos(\nu t) \qquad (\Gamma > 0).$$

In case $\nu \neq \omega$ a particular solution to (10.19) is

$$\tilde{x}(t) = \frac{\Gamma}{\omega^2 - \nu^2}\cos(\nu t)$$

and (10.20) becomes

$$x(t) = A\cos(\omega t) + B\sin(\omega t) + \frac{\Gamma}{\omega^2 - \nu^2}\cos(\nu t). \tag{10.21}$$

Define now $\epsilon$ to be the difference between $\nu$ and $\omega$, setting $\nu = \epsilon + \omega$. As

$$\cos(\nu t) = \cos(\epsilon t + \omega t) = \cos(\epsilon t)\cos(\omega t) - \sin(\epsilon t)\sin(\omega t)$$

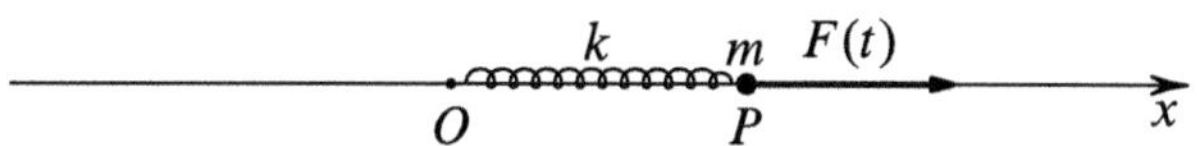


**Fig. 10.3** A particle $P$ constrained to a smooth guide, subject to the force exerted by a spring of elastic constant $k$ and to the action of a *forcing* term with horizontal component $F(t)$

and

$$\omega^2 - \nu^2 = -(2\omega + \epsilon)\epsilon \,,$$

(10.21) can be rewritten as

$$x(t) = \left[\frac{-\Gamma}{(2\omega+\epsilon)\epsilon}\cos(\epsilon t) + A\right]\cos(\omega t) + \left[\frac{\Gamma}{(2\omega+\epsilon)\epsilon}\sin(\epsilon t) + B\right]\sin(\omega t)\,. \tag{10.22}$$

When $\epsilon$ is very small the quantities in square brackets vary slowly in time so the solution $x(t)$ appears as an oscillation with slowly modulated amplitude, known as a *beat*.

The solution for the case $\epsilon = 0$ cannot be obtained by taking the limit in (10.22). In fact

$$\lim_{\epsilon\to 0}\left(-\frac{\Gamma}{(2\omega+\epsilon)\epsilon}\cos(\epsilon t)\right) = -\infty$$

but

$$\lim_{\epsilon\to 0}\left(\frac{\Gamma}{(2\omega+\epsilon)\epsilon}\sin(\epsilon t)\right) = \frac{\Gamma t}{2\omega}\,.$$

In this case, when $\omega^2 = \nu^2$, the particular integral to use in (10.20) is

$$\tilde{x}(t) = \frac{\Gamma}{2\omega^2}\left[\omega t\sin(\omega t) - \cos(\omega t)\right].$$

Notice that $\tilde{x}(0) = 0$ and $\dot{\tilde{x}}(0) = 0$, so if $x_0$ and $\dot{x}_0$ are the initial values the general solution to (10.19) with forcing term $f(t) = \Gamma\cos(\omega t)$ is

$$x(t) = x_0\cos(\omega t) + \frac{\dot{x}_0}{\omega}\sin(\omega t) + \frac{\Gamma}{2\omega^2}\left[\underbrace{\omega t\sin(\omega t)} - \cos(\omega t)\right]. \tag{10.23}$$

Irrespective of the initial conditions, the underscored term makes the oscillations' amplitude grow linearly with time, and we say we are in presence of *resonance*.

### 10.3.2 Damped Oscillations

It is also physically and mathematically interesting to discuss a system involving the resistance of the medium in which the motion occurs. For this one usually assumes the presence of a (viscous) force opposing the motion, with modulus proportional to the velocity of the point or its square–though here we will deal with the former only.

Apart from the spring force we have a viscous resistance of component $-h\dot{x}$, giving equation of motion

$$m\ddot{x} = -h\dot{x} - kx$$

which immediately becomes

$$m\ddot{x} + h\dot{x} + kx = 0 \qquad (h > 0, k > 0). \tag{10.24}$$

Multiplying (10.24) by $\dot{x}$, if not identically null, we obtain

$$\frac{d}{dt}\left(\frac{1}{2}m\dot{x}^2 + \frac{1}{2}kx^2\right) = -h\dot{x}^2$$

after easy operations. In case $h = 0$ (discussed in the previous section), the relation says the sum of kinetic and potential energy of the spring is conserved:

$$\frac{1}{2}m\dot{x}^2 + \frac{1}{2}kx^2 = E_0\,,$$

where $E_0$ is the value of the system's total energy at the initial instant. If $h > 0$ this relation shows instead the mechanical energy is a decreasing function. During the motion there is thus energy dissipation due to the viscosity, as was to be expected.

Let us write (10.24) as

$$\ddot{x} + 2\gamma\dot{x} + \sigma^2 x = 0\,, \qquad (\gamma = h/2m > 0,\ \sigma^2 = k/m > 0) \tag{10.25}$$

(the coefficient $\gamma$ is known as *damping factor*, for reasons that will soon become clear). It is possible to find the general solution to (10.25) with the techniques of Sect. A.5.2 in the appendix. The characteristic equation (A.40) takes the form

$$\lambda^2 + 2\gamma\lambda + \sigma^2 = 0\,. \tag{10.26}$$

The discriminant is $\gamma^2 - \sigma^2$, so with the help of the discussion regarding Eq. A.39 we have three cases, depending on the value of the elastic constant $k$:

(i) $\gamma^2 - \sigma^2 > 0 \Leftrightarrow \gamma > \sigma \Leftrightarrow k < h^2/4m$;
(ii) $\gamma^2 - \sigma^2 = 0 \Leftrightarrow \gamma = \sigma \Leftrightarrow k = h^2/4m$;
(iii) $\gamma^2 - \sigma^2 < 0 \Leftrightarrow \gamma < \sigma \Leftrightarrow k > h^2/4m$;

Let us analyze them one by one:

(i) $\gamma > \sigma$: the roots $\lambda_i$ of (10.26) are real and distinct. Defining

$$\omega^2 = \gamma^2 - \sigma^2 > 0 \tag{10.27}$$

we have

$$\lambda_1 = -\gamma - \omega \qquad \lambda_2 = -\gamma + \omega$$

and the general integral to (10.25) is

$$x(t) = \exp(-\gamma t)\left[c_1 \exp(-\omega t) + c_2 \exp(\omega t)\right] \tag{10.28}$$

with $c_1$ and $c_2$ determined by the initial conditions $x_0 = x(0)$, $\dot{x}_0 = \dot{x}(0)$. In this way

$$x_0 = c_1 + c_2 \qquad \dot{x}_0 = -\gamma(c_1 + c_2) + \omega(c_2 - c_1)$$

so

$$c_1 = \frac{x_0}{2} - \frac{\dot{x}_0 + \gamma x_0}{2\omega} \qquad c_2 = \frac{x_0}{2} + \frac{\dot{x}_0 + \gamma x_0}{2\omega}$$

and substituting in (10.28)

$$x(t) = \exp(-\gamma t)\left[x_0 \cosh(\omega t) + \frac{\dot{x}_0 + \gamma x_0}{\omega} \sinh(\omega t)\right]$$

where we used the hyperbolic cosine and sine functions.
Observe that (10.28) implies $\lim_{t\to\infty} x(t) = 0$ (recall the inequality $\gamma > \omega$, an immediate consequence of (10.27)).

(ii) $\gamma = \sigma$: the roots $\lambda_i$ of (10.26) are real and coincide. The general solution to Eq. (10.25) is

$$x(t) = \exp(-\gamma t)\,[c_1 + c_2 t] \ .$$

The initial conditions allow to determine the constants $c_1$ and $c_2$, so that

$$x(t) = \exp(-\gamma t)\,[x_0 + (\dot{x}_0 + \gamma x_0)t] \ .$$

Here $\lim_{t\to\infty} x(t) = 0$.

(iii) $\gamma < \sigma$: the roots $\lambda_i$ of (10.26) are complex-conjugate. Putting $\omega^2 = \sigma^2 - \gamma^2 > 0$ we obtain

$$\lambda_1 = -\gamma - i\omega \qquad \lambda_2 = -\gamma + i\omega$$

where $i$ is obviously the imaginary unit ($i^2 = -1$). The general solution to (10.25) is

$$x(t) = \exp(-\gamma t)\,[c_1 \cos(\omega t) + c_2 \sin(\omega t)] \tag{10.29}$$

with $c_1$ and $c_2$ determined by the initial conditions. Eventually,

$$x(t) = \exp(-\gamma t)\left[x_0 \cos(\omega t) + \frac{\dot{x}_0 + \gamma x_0}{\omega} \sin(\omega t)\right] .$$

Note again $\lim_{t\to\infty} x(t) = 0$.
The derivative of (10.29) is

$$\dot{x}(t) = \exp(-\gamma t)\,[(-\gamma c_1 + \omega c_2)\cos(\omega t) - (\gamma c_2 + \omega c_1)\sin(\omega t)] \ .$$

Setting $T = 2\pi/\omega$, easy computations give

$$x(t+T) = \exp(-\gamma T)x(t), \qquad \dot{x}(t+T) = \exp(-\gamma T)\dot{x}(t)\,.$$

Function (10.29) is not periodic, but its maximum and minimum points are at a distance $T$.
Moreover

$$\dot{x}(t) = 0 \Longleftrightarrow \tan(\omega t) = \frac{\omega c_2 - \gamma c_1}{\gamma c_2 + \omega c_1}$$

so the times at which the derivative vanishes differ by $T/2$.
The quantity $\mu = \gamma T/2$ is called *logarithmic decay* since if we call $x^{(0)}$, $x^{(1)}$, ..., $x^{(n)}$ the sequence of absolute values of the maxima and minima, then $x^{(n)}=x^{(0)}\exp(-n\mu)$.

As we have seen above, in any case and for any initial conditions there is *damping* in the motion, essentially caused by the term $\exp(-\gamma t)$ appearing in all solutions. It is this that justifies the name *damping coefficient* for $\gamma = h/2m$.

### 10.3.3 Forced and Damped Oscillations

In analogy to Sect. 10.3.1, we may add a sinusoidal forcing term to (10.25) to obtain

$$\ddot{x} + 2\gamma\dot{x} + \sigma^2 x = \Gamma\cos(\nu t)\,. \tag{10.30}$$

As we know from the theory of ordinary linear differential equations, we have to add a particular solution of (10.30) to the general integral of the associated homogeneous equation (10.25), already discussed in Sect. 10.3.2 in all its cases.

A common method to find a particular solution to (10.30) is founded on Complex Analysis. Define an unknown complex-valued function $\xi(t)$ and write (10.30) as

$$\ddot{\xi} + 2\gamma\dot{\xi} + \sigma^2\xi = \Gamma\exp(i\nu t)\,, \tag{10.31}$$

so that the forcing term in (10.30) coincides with the real part of what is on the right-hand side of (10.31) (recall that $\exp(i\phi) = \cos\phi + i\sin\phi$).

Consider a solution of the form

$$\xi(t) = A\exp(i\nu t) \tag{10.32}$$

and compute

$$\dot{\xi} = Ai\nu\exp(i\nu t) \qquad \ddot{\xi} = -A\nu^2\exp(i\nu t)$$

so that, substituting in (10.31),

$$A = \frac{\Gamma}{\sigma^2 - \nu^2 + 2i\gamma\nu}\,. \tag{10.33}$$

Next we use the identities (valid for any non-zero complex number $\alpha + i\beta$)

$$\frac{1}{\alpha+i\beta} = \frac{\alpha-i\beta}{\alpha^2+\beta^2} = \frac{1}{\sqrt{\alpha^2+\beta^2}}\underbrace{\frac{\alpha-i\beta}{\sqrt{\alpha^2+\beta^2}}} = \frac{1}{\sqrt{\alpha^2+\beta^2}}\exp(-i\delta) \tag{10.34}$$

where the underscored term has modulus one, and equals $\exp(-i\delta)$ for

$$\cos\delta = \frac{\alpha}{\sqrt{\alpha^2+\beta^2}}, \quad \sin\delta = \frac{\beta}{\sqrt{\alpha^2+\beta^2}}, \quad \tan\delta = \frac{\beta}{\alpha}. \tag{10.35}$$

Putting in (10.33) $\alpha = \sigma^2 - \nu^2$ and $\beta = 2\gamma\nu$, because of (10.34) and (10.35) we obtain

$$A = \frac{\Gamma}{\sqrt{(\sigma^2-\nu^2)^2+4\gamma^2\nu^2}}\exp(-i\delta) \tag{10.36}$$

with $\delta$ defined by

$$\cos\delta = \frac{\sigma^2-\nu^2}{\sqrt{(\sigma^2-\nu^2)^2+4\gamma^2\nu^2}}, \quad \sin\delta = \frac{2\gamma\nu}{\sqrt{(\sigma^2-\nu^2)^2+4\gamma^2\nu^2}},$$

and so

$$\tan\delta = \frac{2\gamma\nu}{\sigma^2-\nu^2}.$$

In the end, by (10.36) the particular solution (10.32) can be rewritten as

$$\xi(t) = r\exp[i(\nu t - \delta)] \tag{10.37}$$

with

$$r = \frac{\Gamma}{\sqrt{(\sigma^2-\nu^2)^2+4\gamma^2\nu^2}}, \qquad \tan\delta = \frac{2\gamma\nu}{\sigma^2-\nu^2}. \tag{10.38}$$

The real part of (10.37) hence gives a particular real integral of (10.30):

$$\tilde{x}(t) = r\cos(\nu t - \delta) \tag{10.39}$$

with $r$ and $\delta$ given by (10.38).

**Remark 10.4** The amplitude of the oscillation of the particular solution (10.39) is always finite, as the ratio defining $r$ in (10.38) has non-vanishing denominator because of the a priori restrictions on problem's constants. In this sense the situation is considerably different from that of Sect. 10.3.1. It is most notably different from solution (10.21), where we had to assume $\nu \neq \omega$ and we had to employ the particular solution (10.23) which exhibits increasing oscillations in time. This is a natural consequence of having introduced in the model a viscous damping force.

## 10.4 Qualitative Description of a Motion

We shall expand in this section a circle of ideas building on the work of Weierstrass, namely a qualitative analysis that enables to deduce from the first integral (10.12) a lot of information on the particle's motion. We will apply the theory to a particle constrained to a guide, although its reach is broader. Once can prove that similar considerations serve to study the motion of any holonomic system with one degree of freedom, subject to bilateral, ideal and fixed constraints, on which act positional active forces. The subsequent sections will emphasize how some of the conclusions also apply to the motion of particles with several degrees of freedom.

Consider then the motion of a particle $P$ of mass $m$, constrained to a smooth fixed guide, with generalized coordinates $s$, subject to positional active forces admitting first integral (10.12), which we write as

$$T - U = \frac{1}{2}m\dot{s}^2 - U(s) \equiv \text{constant} = E\,, \tag{10.40}$$

where $T$ is the kinetic energy and $U$ a primitive of $F_t(s)$. We insist on the fact (10.40) does not imply the active force $\mathbf{F}$ is conservative, nor that $U$ is a potential. To find (10.40) the presence of the constraint is crucial. It is this that decreases by one the number of degrees of freedom, and hence allows to apply the property whereby in one-dimensional systems the conservative forces are precisely the positional forces.

### 10.4.1 *Barriers*

Solving (10.40) for the velocity gives

$$\dot{s}^2 = \frac{2}{m}\big(E + U(s)\big). \tag{10.41}$$

We remind $E$ is a constant that can be computed from the initial data:

$$E = \frac{1}{2}m\dot{s}(t_0)^2 - U\big(s(t_0)\big). \tag{10.42}$$

In particular, (10.42) shows the energy is always at least equal to minus the initial value of $U$: $E \geq -U\big(s(t_0)\big)$. Back to (10.41), $\dot{s}$ may vanish, so the point can stop instantaneously, only in correspondence to the positions $\tilde{s}$ for which the numerator on the right is zero

$$\dot{s} = 0 \quad\Longleftrightarrow\quad s = \tilde{s}\,, \quad \text{with} \quad E + U(\tilde{s}) = 0. \tag{10.43}$$

The positions $s = \tilde{s}$ satisfying (10.43) represent *barriers* for the motion of $P$, which must (and can) stop only at these positions. We will show in the sequel how $P$ can never *go through* a barrier.

### 10.4.2 Bounded and Unbounded Periodic Motions

Consider a motion bounded by two barriers $\tilde{s}_1 < \tilde{s}_2$, such that

$$\tilde{s}_1 \le s(t_0) \le \tilde{s}_2 , \quad \text{with} \quad E + U(\tilde{s}_1) = E + U(\tilde{s}_2) = 0 .$$

Setting $\tilde{U} = U(\tilde{s}_1) = U(\tilde{s}_2) = -E$, we further suppose

$$U(s) > \tilde{U} \qquad \text{for any} \quad \tilde{s}_1 < s < \tilde{s}_2 \tag{10.44}$$

to exclude the presence of further intermediate barriers. We expect the point will *bounce* from one barrier to the other, giving rise to a periodic motion. We will now show this always happens, and that the barriers do not coincide with critical points of the potential.

Equation (10.41) gives the value of the velocity in any configuration $s$:

$$\dot{s} = \frac{ds}{dt} = \sqrt{\frac{2\big(U(s) - \tilde{U}\big)}{m}} . \tag{10.45}$$

Equation (10.45) is separable and of order one (see Sect. A.5.1). Hence it can be integrated to find the time $t_{12}$ needed to go from one barrier to the other

$$t_{12} = \sqrt{\frac{m}{2}} \int_{\tilde{s}_1}^{\tilde{s}_2} \frac{ds}{\sqrt{U(s) - \tilde{U}}} . \tag{10.46}$$

Note (10.46) is an improper integral since the integrand function diverges at the endpoints. Therefore the time needed to reach a barrier can be finite or infinite. Let us analyze the convergence of (10.46) at the lower end. The result will be easy to generalize for the other end. The Taylor expansion of $U$ near $s = \tilde{s}_1$ gives

$$U(s) = \tilde{U} + \frac{1}{n!} U^{(n)}(\tilde{s}_1) \left(s - \tilde{s}_1\right)^n + o\left(\left(s - \tilde{s}_1\right)^n\right) ,$$

where $n \ge 1$ is the order of the first non-zero derivative of $U$ at the point. Observe that (10.44) implies $U^{(n)}(\tilde{s}_1) > 0$, so that $U(s) > \tilde{U}$ when $s > \tilde{s}_1$. Around $\tilde{s}_1$ we have the estimate

$$\frac{1}{\sqrt{U(s)-\tilde{U}}}=\sqrt{\frac{n!}{U^{(n)}(\tilde{s}_1)}}\,\frac{1}{\left(s-\tilde{s}_1\right)^{n/2}}\,\big(1+O(s-\tilde{s}_1)\big).$$

Therefore, integral $t_{12}$ converges near $\tilde{s}_1$ if $n=1$, i.e. if $U'(\tilde{s}_1)\neq 0$. But if $\tilde{s}_1$ is a critical point of $U$, then $t_{12}$ diverges irrespective of $n\geq 2$.

When $P$ approaches the barrier $\tilde{s}$, we have two possibilities.

- The barrier $\tilde{s}$ is not a critical point of $U$: $U'(\tilde{s})\neq 0$. The point $P$ reaches it in finite time, so that:
  - it cannot go beyond the barrier. In fact, as $U'(\tilde{s})\neq 0$, the function $U(s)-U(\tilde{s})$ from positive would become negative, which is forbidden by, for instance, (10.45).
  - The point cannot stop at the barrier itself, since it was already said that equilibria are stationary points of $U$.
  - In absence of other possibilities, the point is forced to *bounce* off the barrier and go back. For this reason a barrier that is not a critical point of $U$ is called a *reflection point*.
- The barrier $\tilde{s}$ is critical for $U$. Then $P$ reaches it in an infinite time. In practice it remains trapped close to the barrier, towards which it tends with decreasing speed. Hence a barrier that is a critical point of $U$ is called an *asymptotic point*.

To summarize the previous information, we can finally establish the conditions that determine if and when a motion is bounded (periodic or not) or unbounded.

- A motion without barriers will certainly be unbounded.
- Suppose there is one barrier $\tilde{s}_1$ only.
  If $P$ is initially moving away from it, the motion will certainly be unbounded.
  If, on the contrary, initially the point moves towards the barrier, the motion is bounded if $U'(\tilde{s}_1)=0$ and $P$ is trapped near the asymptotic point. Otherwise, if $U'(\tilde{s}_1)\neq 0$, the point hits the reflection point and bounces away never to return.
  Finally, if the point was initially at rest at $\tilde{s}_1$ it can either stay in equilibrium if $U'(\tilde{s}_1)=0$, or start an unbounded motion if $U'(\tilde{s}_1)\neq 0$.
- A motion bounded by two barriers, one to each side of the initial position, will certainly be bounded. Moreover, if both barriers are reflection points the motion is periodic, and the period coincides with twice the time $t_{12}$ found in (10.46).

Figure 10.4 depicts several motions under the effects of the same function $U(s)$. The left diagrams (a), (c), (e) show how one can determine the barriers by solving $E+U(s)=0$, i.e. intersecting the graph of $-U(s)$ with horizontal lines at height $E$. Once the barriers are found, it suffices to evaluate at their position the derivative of $U$ and say whether they are reflection or asymptotic points. The graphs on the right illustrate the induced motions. The various cases arise from the same initial position, but with increasing values of the initial velocity.

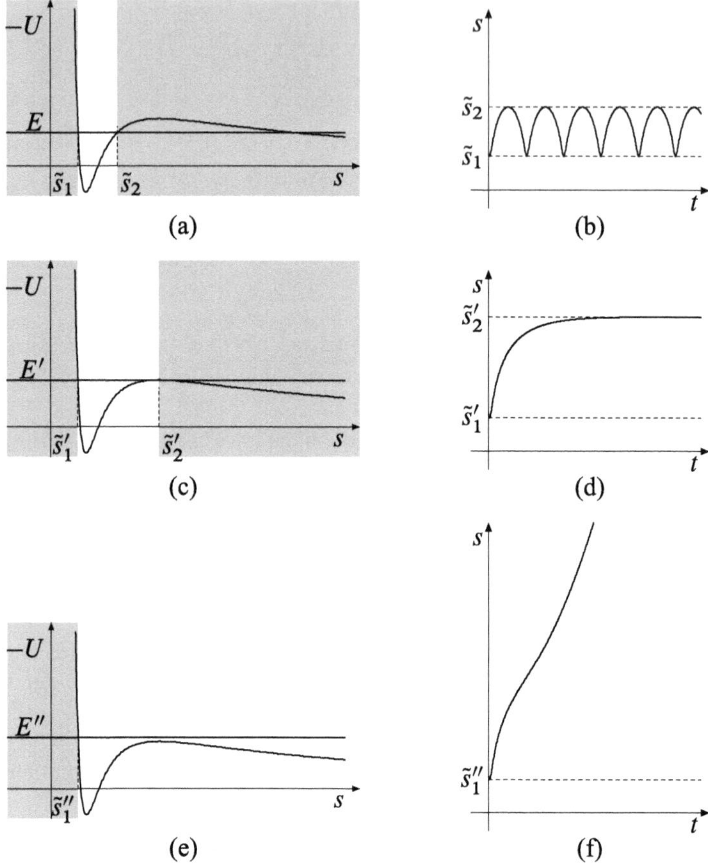


**Fig. 10.4** Different motions under the action of the same active force

In the first case (Fig. 10.4a) the energy is not enough to prevent the existence of two reflection points. The induced motion is periodic, see Fig. 10.4b. The second motion (Fig. 10.4c) happens under higher energy. Here we have a reflection point and an asymptotic point. The motion is not periodic, but remains bounded (Fig. 10.4d). In the last case (Fig. 10.4e) the energy value prevents the presence of barriers on the right of the initial value. The motion (Fig. 10.4f) is unbounded. Recall that if the initial position is the same, the energy increases as the initial speed increases.

## 10.5 Motion Under the Action of Central Forces

We defined in (7.13) central forces as those whose modulus depends only on the distance. We remind that a positional force **F** is called *central* if there exists a fixed point $O$, called *center* of the force, such that

$$OP \times \mathbf{F}(P) = \mathbf{0} \qquad \text{for any } P. \tag{10.47}$$

Among central forces, those whose modulus only depends on the distance to the center are particularly interesting:

$$\mathbf{F}(P) = \Psi(r)\,\mathbf{u}, \qquad \text{where } \; r = |OP| \; \text{ and } \; \mathbf{u} = \frac{1}{r}\,OP\,. \tag{10.48}$$

The scalar function $\Psi(r)$ can be positive or negative, since the central force can respectively be repulsive or attractive. In (7.14) we also found that central forces are conservative, and any primitive of $\Psi$ gives a potential:

$$U(r) = \int^r \Psi(\rho)d\rho + \text{cost} \quad \Longrightarrow \quad \nabla U = \Psi(r)\,\mathbf{u}\,. \tag{10.49}$$

We emphasize that the assumption that $\Psi$ depends only on $r$ is fundamental to obtain a conservative force field. If we take for instance a central force on the plane also depending on the polar angle $\mathbf{G}(P) = \Psi(r, \theta)\,\mathbf{u}$, we obtain

$$\operatorname{rot}\mathbf{G} = -\frac{1}{r}\frac{\partial \Psi}{\partial \theta}\,\mathbf{k}\,,$$

where $\mathbf{k}$ is the unit vector orthogonal to the plane. If a force field admits a potential then its curl must vanish. Hence a central force field whose modulus depends on the orientation is not conservative.

### 10.5.1 Planarity

By definition (10.47) of central motion the cross product of $OP$ and $\mathbf{a} = \mathbf{F}/m$ is zero. Hence

$$\mathbf{0} = OP \times \mathbf{a} = OP \times \frac{d\mathbf{v}}{dt} = \frac{d}{dt}(OP \times \mathbf{v}) - \mathbf{v} \times \mathbf{v} = \frac{d}{dt}(OP \times \mathbf{v})\,.$$

Thus we have proved that the vector

$$\mathbf{c} = OP \times \mathbf{v} \tag{10.50}$$

is constant in time.

If such vector is zero, the position vector is always parallel to the velocity and the motion occurs along a straight line through $O$.

If $\mathbf{c} \neq \mathbf{0}$, the motion is confined to the plane through $O$ orthogonal to $\mathbf{c}$, as this plane contains all $P$ such that $OP$ is perpendicular to $\mathbf{c}$, as prescribed by (10.50).

### The Area Constant

Once we have proved the motion is planar, it becomes convenient to employ polar coordinates $(r, \theta)$, defined with respect to the center of motion and shown in Fig. 10.5. So, introduce the polar unit vectors $\mathbf{e}_r$, $\mathbf{e}_\theta$ defined in (1.6), and recall (1.7), giving velocity and acceleration in polar coordinates:

$$\mathbf{v} = \dot{r}\,\mathbf{e}_r + r\dot{\theta}\,\mathbf{e}_\theta, \qquad \mathbf{a} = \left(\ddot{r} - r\dot{\theta}^2\right)\mathbf{e}_r + \left(r\ddot{\theta} + 2\dot{r}\dot{\theta}\right)\mathbf{e}_\theta. \tag{10.51}$$

From $(10.51)_1$ we have $\mathbf{c} = OP \times \mathbf{v} = r^2\dot{\theta}\,\mathbf{k}$. As $\mathbf{c}$ is a constant vector we obtain the first integral

$$c = r^2(t)\,\dot{\theta}(t) = \text{constant}. \tag{10.52}$$

We call the scalar $c$ the *area constant*, for reasons we will explain in a moment. A central motion is *degenerate* if $c = 0$ and *non-degenerate* if $c \neq 0$. As observed earlier and as confirmed by (10.52), the orbits of degenerate motions are straight lines, since along them $\theta =$ constant. The orbits of non-degenerate central motions not only are planar, but have the additional feature that the time derivative $\dot{\theta}$ cannot change sign. This implies that a clockwise, or counter-clockwise, rotation of a point about the centre of motion cannot reverse its orientation during a central motion.

### Conservation of Areal Speed

Consider on the plane of motion the curve described by the point's orbit, and let $\mathscr{A}(t)$ denote the area swept by the position vector from an initial instant $t_0$ to $t$ (Fig. 10.5).

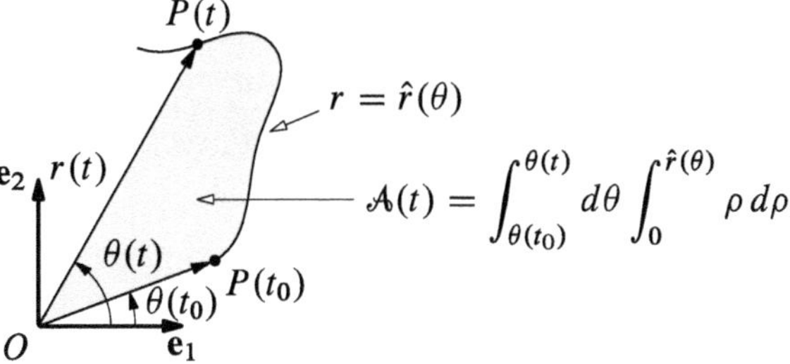


**Fig. 10.5** Polar coordinates and the area swept by $OP(t)$ during the planar motion of $P$

Using polar coordinates again we define the map $\hat{r}(\theta)$ giving, for any $\theta$, the distance of $P$ to $O$.[1] The area $\mathscr{A}(t)$ can then be written in polar coordinates simply as

$$\mathscr{A}(t) = \int_{\theta(t_0)}^{\theta(t)} d\theta \int_0^{\hat{r}(\theta)} \rho\, d\rho = \frac{1}{2} \int_{\theta(t_0)}^{\theta(t)} \hat{r}^2(\theta)\, d\theta,$$

which implies

$$\dot{\mathscr{A}} = \frac{d\mathscr{A}}{dt} = \frac{1}{2} r^2 \dot{\theta} = \frac{c}{2} = \text{constant}. \tag{10.53}$$

The conserved quantity $\dot{\mathscr{A}}$ is called *areal speed*, and (10.53) justifies having called *area constant* the number $c$.

## Binet Formula

Let us take a closer look at $(10.51)_2$. By definition, during a central motion the tangential acceleration must be zero. In fact

$$a_\theta = \mathbf{a} \cdot \mathbf{e}_\theta = r\ddot{\theta} + 2\dot{r}\dot{\theta} = \frac{1}{r} \frac{d}{dt} \left( r^2 \dot{\theta} \right) = \frac{\dot{c}}{r} = 0\,.$$

The above also proves that $c$ being constant is not only necessary, but sufficient as well to ensure a motion on the plane is central.

As for the radial acceleration, from (10.52) we have

$$\begin{aligned} a_r = \mathbf{a} \cdot \mathbf{e}_r = \ddot{r} - r\dot{\theta}^2 &= \frac{d}{dt} \left( \frac{dr}{d\theta} \dot{\theta} \right) - r\dot{\theta}^2 = \frac{d}{d\theta} \left( \frac{dr}{d\theta} \frac{c}{r^2} \right) \dot{\theta} - r \frac{c^2}{r^4} \\ &= -\frac{c^2}{r^2} \frac{d^2}{d\theta^2} \left( \frac{1}{r} \right) - \frac{c^2}{r^3} = -\frac{c^2}{r^2} \left[ \frac{d^2}{d\theta^2} \left( \frac{1}{r} \right) + \frac{1}{r} \right]. \end{aligned} \tag{10.54}$$

This is known as *Binet's formula*. It allows to compute the radial acceleration (hence the radial component of the applied force) once we know the equation $r = \hat{r}(\theta)$ of the trajectory of the point in motion. To know $a_r$ it is enough to know the points through which $P$ passes, without knowing *when* $P$ has actually passed though each of them.

## Effective Potential

Consider the motion of a particle $P$ of mass $m$ and only subject to a radial force field (10.48), with potential (10.49). Clearly, Newton's law implies the motion will be

[1] With more mathematical rigour, observe first that $\theta(t)$ is monotone hence invertible (its derivative $\dot{\theta}$ never changes sign). Calling $\hat{t}(\theta)$ the inverse, set $\hat{r}(\theta) = r\big(\hat{t}(\theta)\big)$.

central. In particular, both the mechanical energy and the area constant are conserved. In terms of polar coordinates on the orbit's plane, these first integrals are

$$E = \frac{1}{2}mv^2 - U(r) = \frac{1}{2}m\big(\dot{r}^2 + r^2\dot{\theta}^2\big) - U(r) = \frac{1}{2}m\dot{r}^2 - U_{\text{eff}}(r)\,, \tag{10.55}$$

$$c = r^2\dot{\theta}\,, \tag{10.56}$$

where we have introduced the *effective potential*

$$U_{\text{eff}}(r) = U(r) - \frac{mc^2}{2r^2}\,. \tag{10.57}$$

Although $P$ is a free point, hence has three degrees of freedom, the central motion allows to find the first integral (10.55); this is formally identical to the conservation law (10.40), which originates all the qualitative analysis of motions with one freedom degree. The distance $r$ plays now the role of "unique" degree of freedom, but when carrying out the qualitative analysis we will need to not forget the generalized coordinate $\theta$, whose evolution can be controlled through (10.56).

The effective potential (10.57) depends on the initial conditions via the area constant $c$. Therefore to gather qualitative information on a central orbit it is not enough to know the initial data for the radial coordinate ($r(t_0)$ and $\dot{r}(t_0)$), but we must also know the tangential velocity, from which we can calculate $\dot{\theta}(t_0)$.

We list below how the qualitative analysis of Sect. 10.4 should be interpreted in the case of a central motion.

- *Apsidal circles, limit circles.* Let us write (10.55) as

$$\dot{r}^2 = \frac{2}{m}\big(E + U_{\text{eff}}(r)\big).$$

Evidently $\dot{r}$ vanishes for those $\tilde{r}$ such that

$$E + U_{\text{eff}}(\tilde{r}) = 0\,. \tag{10.58}$$

Yet to fix the distance to $O$ does not mean to fix the position of $P$, since there is an entire circle corresponding to that distance.

The equivalent concept of reflection points seen in Sect. 10.4 are circles on which the radial velocity is zero and the velocity vector is parallel to $\mathbf{e}_\theta$. Moreover, $\dot{r}$ changes sign when $r$ passes through $\tilde{r}$. In correspondence to distances that satisfy (10.58) therefore, the orbit is tangent to the circle of radius $\tilde{r}$, and the point passes from approaching the center of forces to moving away from it, or conversely. The circle touched by the orbit is called *apsidal circle*, and the tangent point is the orbit's *apsidal point*.

On the other hand, if the distance satisfying (10.58) is also critical for $U_{\text{eff}}$, we obtain the analogous idea of an asymptotic point. Now we have a circle, called

*limit circle*, towards which the orbit tends asymptotically, without ever reaching it.
- *Bounded and unbounded orbits.* As the radial coordinate is never negative ($r \geq 0$), studying whether a central orbit is bounded or not boils down to detecting the presence of an external apsidal circle enclosing the point's orbit.
- *Periodic orbits.* To finish, we stress that not even two apsidal circles, one internal and one external, guarantee the orbit is periodic. In fact, in a central motion bounded by two apsidal circles, the radial coordinate will oscillate between two limit values. But the angular coordinate might not return to the initial value (modulo full $2\pi$-rotations) when the radial coordinate assumes its initial value again.

## 10.6 Kepler and the Law of Gravitation

Between the 16th and 17th centuries, the German mathematician and astronomer Johannes Kepler (Weil, Württemberg 1571—Regensburg 1630) collected a vast amount of astronomical data on the relative positions of the planets of the Solar system. After moving to Prague in 1600, where he became the astronomer at the court of Emperor Rudolf II, he published in 1609 the important treatise *Astronomia Nova*, in which he stated the first two of three laws we will discuss below. In 1619 he published the opus *Harmonice Mundi*, containing his third law of planetary motion. In this section we will examine how from these laws, arising in a purely kinematical context since based on the study of the shape of the planets' orbits around the Sun, one can already infer the analytical expression of the Law of universal gravitation, which Newton formalized in 1687. The analysis and the presentation of the experimental data, together with the possibility of describing the data using three simple laws, make Kepler the founding father of celestial mechanics.

The well-known Kepler laws are the following.

(i) *Planets move on elliptical orbits with the Sun at one of the two foci.*
(ii) *The segment joining each planet to the Sun sweeps out equal areas during equal intervals of time.*
(iii) *The square of the time a planet takes to complete its orbit (orbital period) is proportional to the cube of the length of the ellipse's semi-major axis.*

The first two laws enable us to prove the motion of a planet around the Sun is central. In fact, the first law implies in particular the motion is planar, while the second law forces the areal speed to be constant.

Kepler's first law predicts the explicit shape of planetary orbits, so it actually allows us to know the radial acceleration precisely, by Binet's formula (10.54). In fact, it is possible to prove, and is a well-known geometric fact, that the ellipse of semi-axes $a > b$, in polar coordinates centered at one focus, has equation

$$\hat{r}(\theta) = \frac{p}{1 + e\cos\theta}, \tag{10.59}$$

where

$$p = \frac{b^2}{a} \qquad \text{and} \qquad e = \sqrt{1 - \frac{b^2}{a^2}} \tag{10.60}$$

are a characteristic parameter and the eccentricity, respectively. Substituting (10.59) in (10.54) gives

$$a_r = -\frac{c^2}{r^2}\left[\frac{d^2}{d\theta^2}\left(\frac{1}{r}\right) + \frac{1}{r}\right] = -\frac{c^2}{p\,r^2}\,. \tag{10.61}$$

By the conservation of the areal speed, the area constant can be expressed in terms of the period and the semi-axes:

$$c = 2\dot{\mathscr{A}} = 2\,\frac{\mathscr{A}_{\text{TOT}}}{T} = 2\,\frac{\pi ab}{T}\,. \tag{10.62}$$

Inserting (10.62) and the first of (10.60) in (10.61) we obtain

$$a_r = -4\pi^2\left(\frac{a^3}{T^2}\right)\frac{1}{r^2}\,.$$

The third law, which postulates $T^2$ and $a^3$ are proportional, then enables us to write the radial acceleration

$$a_r = -\frac{K}{r^2}\,, \tag{10.63}$$

with $K$ independent of the specific planet considered. Calling $O$ the Sun's position and $P$ the position of any planet, say of mass $m$, the force the Sun exerts on the planet is

$$\mathbf{F} = -\frac{K\,m}{r^2}\,\mathbf{u}\,, \qquad \text{with} \quad \mathbf{u} = \frac{1}{r}\,OP\,.$$

Kepler's observations stop at this level. They describe the attractive force the Sun generates on each planet, but they do not attempt to generalize the interaction to every pair of bodies.

Starting from Kepler's laws and the action-reaction principle (see Sect. 8.1.3), Newton achieved the Law of universal gravitation.

- Assuming all planets are attracted by the Sun with modulus $F_{\text{PS}} = K\,m/r^2$, where $m$ is the planet's mass and $r$ the distance to the Sun, the first step is to observe that, in turn, each planet will attract every satellite and body orbiting around it, under the equation

$$F' = \frac{k\,m'}{r'^2}\,, \tag{10.64}$$

  where now $m'$ denotes the satellite's mass, $r'$ its distance to the planet, and $k$ is a constant that may a priori depend on the chosen planet.

- The second fundamental observation follows from the action-reaction principle. This says each planet will not only attract the satellites orbiting it, but will respond to the action of the Sun by a force of same modulus (and opposite). Calling $M$ the mass of the Sun and $r$ the Sun-planet distance, (10.64) prescribes for this force

$$F_{SP} = \frac{k\,M}{r^2}\,.$$

Demanding $F_{SP} = F_{PS}$ then implies $K\,m = k\,M$, or in other words that

$$\frac{K}{M} = \frac{k}{m} = h\,.$$

Hence the constant $h$ does not depend on the attracted body nor on the attracting one. It goes by the name of *gravitational constant*, and its approximate numerical value is

$$h = 6.672 \times 10^{-11}\mathrm{N\,m^2\,kg^{-2}} = 6.672 \times 10^{-8}\mathrm{cm^3\,g^{-1}\,s^{-2}}.$$

The experimental value of $h$ is tiny. To understand how tiny, think that the modulus of the gravitational force between two electrons is about $10^{43}$ times smaller than the modulus of the electrostatic forces at play. This explains why in almost all technological applications the gravitational force between moving bodies is negligible, obviously without excluding the gravitational attraction of Earth.
- Newton's final step was to understand that not only planets and satellites, but any pair of points $P_1$, $P_2$, of masses $m_1, m_2$, attract one another with the same Law of universal gravitation

$$\mathbf{F}_{12} = -\mathbf{F}_{21} = -\frac{hm_1m_2}{r_{12}^2}\frac{P_2P_1}{r_{12}}\,, \quad \text{where } r_{12} = |P_1P_2|\,.$$

# Chapter 11
# Dynamics of Systems

This chapter is devoted to finding the equations providing global information on the dynamical behaviour of mechanical systems. Starting from the fundamental system governing the dynamics of material systems (8.7), deduced in Chap. 8, we will obtain relations satisfied by *every* motion of *any material system*. In particular, the next section will deal with:

- the first and second balance equation of Dynamics;
- the Kinetic energy theorem and the Work theorem.

Later we will introduce the important notion of *integral of motion* and see under which conditions there is *conservation of the mechanical energy*.

## 11.1 Balance Equations

We saw in sections Sects. 9.5–9.6 that the balance equations of Statics are very useful tools to study the equilibrium of systems of particles and rigid bodies. We recall that a system of forces acting on a system satisfies the static balance equations if it is balanced, i.e. if its characteristic vectors (resultant and resultant moment) vanish. Moreover, since internal forces are always balanced, the static balance equations hold if and only if the external forces are balanced. These equations are necessary and sufficient conditions for the equilibrium of a single rigid body.

Now we will study what information we can deduce in Dynamics from the resultant and resultant moment of the external forces. Clearly in general such a system will not be balanced, and indeed we will see how the characteristic vectors provide a lot of information on the system's motion. We will prove that in Dynamics, as well, the two balance equations of Dynamics together give necessary and sufficient conditions to determine the motion of one rigid body.



P. Biscari et al., *Rational Mechanics*, UNITEXT 177,
https://doi.org/10.1007/978-3-032-07462-1_11

**Theorem 11.1** (First balance equation of Dynamics) *Let* $\mathbf{R}^{(e)}$ *be the resultant of the external forces, both active and reactive, acting on a generic system with linear momentum* $\mathbf{Q}$. *Then*

$$\mathbf{R}^{(e)} = \dot{\mathbf{Q}} . \tag{11.1}$$

***Proof*** Consider Newton's law of every point, separating the acting forces in internal and external as per (8.7):

$$\mathbf{F}_i^{(i)} + \mathbf{F}_i^{(e)} = m_i \mathbf{a}_i . \tag{11.2}$$

The first balance equation of Dynamics is obtained simply by adding the equations of motion of all points and remembering that, as proved in Theorem 8.5, the resultant of the internal forces is always null:

$$\overbrace{\sum_{i=1}^{n} \mathbf{F}_i^{(i)}}^{\mathbf{R}^{(i)}=\mathbf{0}} + \overbrace{\sum_{i=1}^{n} \mathbf{F}_i^{(e)}}^{\mathbf{R}^{(e)}} = \sum_{i=1}^{n} m_i \mathbf{a}_i , \quad \mathbf{R}^{(e)} = \sum_{i=1}^{n} m_i \frac{d}{dt} \mathbf{v}_i = \frac{d}{dt} \sum_{i=1}^{n} m_i \mathbf{v}_i = \dot{\mathbf{Q}} .$$

□

**Theorem 11.2** (Second balance equation of Dynamics) *Let* $\mathbf{M}_A^{(e)}$ *denote the resultant moment of the external forces acting on an arbitrary system, with respect to the pole* $A$. *Call* $\dot{A}$ *the pole's velocity (see Sect. 6.2.3), and* $\mathbf{K}_A$ *the system's angular momentum with respect to* $A$. *Then*

$$\mathbf{M}_A^{(e)} = \dot{\mathbf{K}}_A + \dot{A} \times \mathbf{Q} . \tag{11.3}$$

***Proof*** Let us start again from the equations of motion of the single points, as in (11.2). Choosing a pole $A$ (fixed or not, coinciding or not with one of the points), we compute the moments of the forces with respect to it:

$$AP_i \times \left( \mathbf{F}_i^{(i)} + \mathbf{F}_i^{(e)} \right) = AP_i \times m_i \mathbf{a}_i . \tag{11.4}$$

Now add equations (11.4), and recalling from Theorem 8.5 that the internal forces' moment is always zero, we obtain

$$\overbrace{\sum_{i=1}^{n} \mathbf{M}_{Ai}^{(i)}}^{\mathbf{M}_A^{(i)}=\mathbf{0}} + \overbrace{\sum_{i=1}^{n} \mathbf{M}_{Ai}^{(e)}}^{\mathbf{M}_A^{(e)}} = \sum_{i=1}^{n} AP_i \times m_i \mathbf{a}_i$$

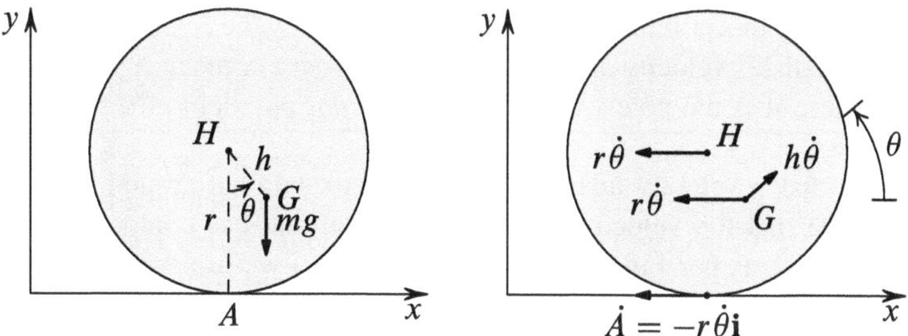


**Fig. 11.1** A *non*-homogeneous disk rolling on a horizontal guide

$$\begin{aligned}\mathbf{M}_A^{(e)} &= \sum_{i=1}^n AP_i \times m_i \frac{d}{dt}\mathbf{v}_i = \sum_{i=1}^n \Big[\frac{d}{dt}\Big(AP_i \times m_i\mathbf{v}_i\Big) - \frac{dAP_i}{dt} \times m_i\mathbf{v}_i\Big] \\ &= \frac{d}{dt}\sum_{i=1}^n AP_i \times m_i\mathbf{v}_i - \sum_{i=1}^n \Big(\mathbf{v}_i - \dot A\Big) \times m_i\mathbf{v}_i \qquad \Big(\mathbf{v}_i \times m_i\mathbf{v}_i = \mathbf{0}\Big) \\ &= \dot{\mathbf{K}}_A + \dot A \times \sum_{i=1}^n m_i\mathbf{v}_i = \dot{\mathbf{K}}_A + \dot A \times \mathbf{Q}\,.\end{aligned}$$

□

Newton's third law Sect. 8.1.3 postulates that the only condition the internal forces obey is to form couples with zero arm. Therefore the only property we can state about such system of forces is that it is balanced. So the balance equations (11.1)–(11.3) are the only consequences of the fundamental system of dynamics (8.7) that are always independent of the internal forces.

**Remark 11.3** The second balance equation (11.3) simplifies if the choice of pole $A$ allows to kill the term $\dot A \times \mathbf{Q} = \dot A \times m\mathbf{v}_G$. These choices are: fixed pole $A$, or $A$ coinciding with the center of mass $G$, or more generally an $A$ with velocity parallel to that of the center of mass. In all cases (11.3) becomes $\mathbf{M}_A^{(e)} = \dot{\mathbf{K}}_A$. □

**Example 11.4** As just noticed, the additional term $\dot A \times \mathbf{Q}$ on the right of (11.3) is zero, provided the pole is chosen to be fixed or with velocity parallel to that of the center of mass. But it is useful to address the case when that does not happen, because the instance has a practical interest of its own and also since it serves to avoid mistakes in similar situations.

In Fig. 11.1 left we see a disk of radius $r$ and mass $m$ in a vertical plane, constrained to roll without slipping on a fixed guide given by the $x$-axis. The disk is not homogeneous and so its center of mass $G$ is at distance $h$ from the geometric center $H$. Suppose $\hat I$, the moment of inertia of the disk with respect to the perpendicular to the plane through $H$, is given. A convenient generalized coordinate is the angle of rotation $\theta$ between the vertical direction and the segment $HG$.

We want to write down the second balance equation of Dynamics using the contact point $A$ with the $x$-axis as pole (the advantage is that we shall eliminate the constraint reaction). Clearly this point is occupied at each instant by the centre of instantaneous rotation of the disk's velocity distribution, but this *does not* mean $\dot{A} \times \mathbf{Q}$ in (11.3) is zero, since here $\dot{A}$ is the pole's velocity, which *is not* parallel to the velocity of the center of mass.

From the disk's velocity distribution, which is rotational around $A$, we immediately deduce that the velocity of $H$ is horizontal and has component $r\dot{\theta}$ opposite to the $x$-axis, as per Fig. 11.1 right. To find $\mathbf{v}(G)$ we use the relation $\mathbf{v}(G) = \mathbf{v}(H) + \boldsymbol{\omega} \times HG$: we move to $G$ the vector $\mathbf{v}(H)$ and add $\boldsymbol{\omega} \times HG$, which is perpendicular to $HG$ and has component $h\dot{\theta}$, as in Fig. 11.1. Projecting onto the axes, the two components of $\mathbf{v}(G)$ eventually are

$$v_x(G) = -r\dot{\theta} + h\dot{\theta}\cos\theta\,, \qquad v_y(G) = h\dot{\theta}\sin\theta\,.$$

The velocity of $A$ is obviously equal to that of $H$, namely

$$\dot{A} = -r\dot{\theta}\mathbf{i}\,.$$

The term $\dot{A} \times \mathbf{Q} = \dot{A} \times m\mathbf{v}(G)$ then equals

$$\dot{A} \times m\mathbf{v}(G) = m(-r\dot{\theta}\mathbf{i}) \times [(-r\dot{\theta} + h\dot{\theta}\cos\theta)\mathbf{i} + (h\dot{\theta}\sin\theta)\mathbf{j}] = -mrh\dot{\theta}^2\sin\theta\mathbf{k}\,,$$

where $\mathbf{k}$ is the outgoing unit vector perpendicular to the plane.

The moment of the forces $\mathbf{M}_A^{(e)}$ is simple, and consists of the weight $mg$ only because the constraint reaction is applied at the same point $A$:

$$\mathbf{M}_A^{(e)} = -mgh\sin\theta\mathbf{k}\,.$$

The angular momentum $\mathbf{K}_A$ reduces to $I_A\dot{\theta}\mathbf{k}$, since the velocity distribution is rotational around $A$ itself. To find the moment of inertia $I_A$ we use Huygens' theorem, and move from $H$ to $G$ by subtracting from $\hat{I}$ the quantity $mh^2$, and then adding $m|AG|^2$. The cosine rule for triangle $AHG$ says $|AG|^2 = r^2 + h^2 - 2rh\cos\theta$, so eventually

$$I_A = \hat{I} - mh^2 + m(r^2 + h^2 - 2rh\cos\theta) = \hat{I} + mr^2 - 2mrh\cos\theta\,.$$

The derivative of the angular momentum has therefore a single component

$$\begin{aligned}\dot{\mathbf{K}}_A t_z &= \frac{d}{dt}[(\hat{I} + mr^2 - 2mrh\cos\theta)\dot{\theta}] \\ &= 2mrh\sin\theta\dot{\theta}^2 + (\hat{I} + mr^2 - 2mrh\cos\theta)\ddot{\theta}\,.\end{aligned}$$

The second balance equation now gives the pure equation of motion

$$-mgh\sin\theta = 2mrh\dot{\theta}^2\sin\theta + (\hat{I} + mr^2 - 2mrh\cos\theta)\ddot{\theta} - mrh\dot{\theta}^2\sin\theta$$

which reduces to the (still complicated) equation

$$(\hat{I} + mr^2 - 2mrh\cos\theta)\ddot{\theta} + mrh\dot{\theta}^2 \sin\theta + mgh\sin\theta = 0\,.$$

It is useful to note (as we expect) that putting $h = 0$ and $\hat{I} = mr^2/2$, as for the *homogeneous* disk, gives the trivial differential equation $\ddot{\theta} = 0$, corresponding to uniform motion with $\dot{\theta}$ constant. □

## 11.2 First Integrals of a Motion

We saw in Chap. 10 how finding one or more first integrals (energy, area constant) allows to deduce a lot of qualitative information regarding the motion of a particle. We revert to the definition of first integral, and observe that balance equations sometimes permit to determine new first integrals.

**Definition 11.5** (*First integral*) Consider a generic system, with fixed or moving constraints, with $N$ generalized coordinates $\mathsf{q} = (q_1, \ldots, q_N)$. A function $F(\mathsf{q}, \dot{\mathsf{q}}, t)$ is a *first integral of motion* if its value remains constant with time:

$$F(\mathsf{q}, \dot{\mathsf{q}}, t) = \text{const.}$$

or, equivalently but more explicitly,

$$F(\mathsf{q}, \dot{\mathsf{q}}, t) = F(\mathsf{q}(t_0), \dot{\mathsf{q}}(t_0), t_0)\,.$$

That means

$$\sum_{k=1}^{N} \frac{\partial F}{\partial q_k}\dot{q}_k + \sum_{k=1}^{N} \frac{\partial F}{\partial \dot{q}_k}\ddot{q}_k + \frac{\partial F}{\partial t} = 0$$

where the functions $q_k(t)$ are solutions to the equations of motion.

We emphasize that the value of a first integral depends on the initial conditions and changes with them. Once these are given, though, $F$ stays constant even if during the motion the points' positions and velocities, hence the $q_k$ and $\dot{q}_k$, vary, besides time $t$. In presence of a first integral $F$ we will say *$F$ is conserved*, or that it is a *conserved quantity*.

In general it is not easy to determine the possible integrals of motion of a mechanical system. As we will see straightaway though, the balance equations make suggestions in this respect, at least in particular situations.

We discover two kinds of first integrals that relate to the balance equation of Dynamics. The proofs of the ensuing propositions are immediate and easily follow by integrating (11.1)–(11.3).

**Proposition 11.6** (Conservation of linear momentum) *If the resultant of the external forces has zero component along a fixed unit vector* $\mathbf{u}$, *the corresponding component of the linear momentum is conserved:*

$$\mathbf{R}^{(e)} \cdot \mathbf{u} = 0 \quad \Longrightarrow \quad \mathbf{Q} \cdot \mathbf{u} = m\mathbf{v}_G \cdot \mathbf{u} = m v_{G,u} \equiv constant.$$

**Proposition 11.7** (Conservation of angular momentum) *If the resultant momentum of the external forces has zero component along a given unit vector* $\mathbf{u}$, *with respect to a pole* $A$ *for which* $\dot{A} \times \mathbf{Q} = \mathbf{0}$ *(see Remark 11.3), the corresponding component of angular momentum with respect to* $A$ *is conserved:*

$$\mathbf{M}_A^{(e)} \cdot \mathbf{u} = 0 \quad \Longrightarrow \quad \mathbf{K}_A \cdot \mathbf{u} = K_{A,u} \equiv constant \qquad \left(if\, \dot{A} \times \mathbf{Q} = \mathbf{0}\right).$$

(In several concrete situations the unit vector $\mathbf{u}$ is one of the vectors in the frame's Cartesian basis.)

The presence of this integral of motion is often evident, and can be deduced from a glance at the external forces acting on the system.

**Example 11.8** In a vertical plane consider a system made of an arbitrary number of interacting points and rigid bodies, resting on the smooth $x$-axis, and subject to one external force, the weight. In presence of a smooth constraint, the constraint reactions will be vertical, so the horizontal component of the external resultant will vanish. Therefore the $x$-component of the velocity distribution is conserved, i.e.: $v_{Gx} \equiv$ constant. □

**Example 11.9** On a horizontal plane $(x, y)$ consider a generic system, again consisting of one or several rigid bodies, possibly interacting (or constrained), but not subject to active external forces. Suppose the only external constraint is a pin fixing a point of one rigid body to a fixed pole $A$. In this case the external forces, represented by the constraint reaction at $A$, have zero moment with respect to the fixed pole, so $\mathbf{K}_A \equiv$ constant. □

**Example 11.10** Consider a system contained in a vertical plane, constrained to rotate around the vertical $y$-axis. Suppose the system is subject to weights, and constrained to the plane by pins or rollers. In these conditions the vertical component of the moment of the external forces with respect to any $A$ on the (fixed) vertical axis is zero, and so $K_{Ay} \equiv$ constant. □

## 11.3 Motion of the Center of Mass

The first balance equation of Dynamics enables us to know the motion of the system's center of mass.

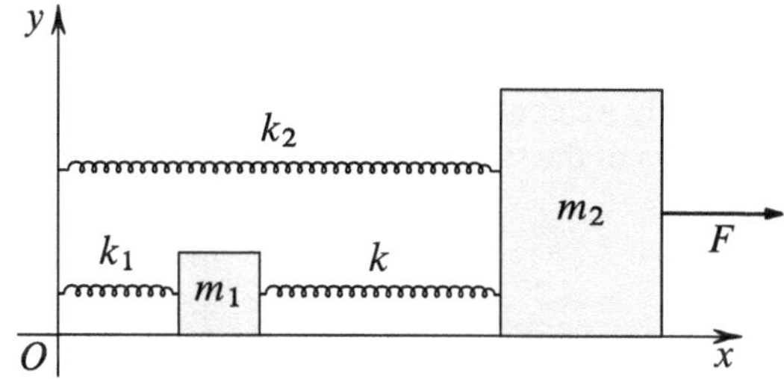


**Fig. 11.2** A model of a system made by two particles and three springs

**Theorem 11.11** (Motion of the center of mass) *The center of mass of a system moves as a point of mass equal to the system's total mass, at which the resultant of the external forces is ideally applied:*

$$m\,\mathbf{a}_G = \mathbf{R}^{(e)}\,. \tag{11.5}$$

***Proof*** Identity (6.1) relates the linear momentum to the velocity of the center of mass: $\mathbf{Q} = m\,\mathbf{v}_G$. Differentiating with respect to time we find $\dot{\mathbf{Q}} = m\,\mathbf{a}_G$, and eventually (11.5) using the first balance equation of Dynamics (11.1). □

The above result implies we cannot affect the motion of the center of mass without acting on the resultant of the external forces $\mathbf{R}^{(e)}$. But it is important to pause and ponder on the explicit independence of the balance equations from the internal forces. In particular, we must pay attention to how the first balance equation relates to the motion of the center of mass. To better understand this issue it is convenient to examine a simple but specific example.

**Example 11.12** Consider two particles $P_1$, $P_2$ of masses $m_1, m_2$, constrained to a smooth horizontal $x$-axis (Fig. 11.2; for better visualization the points are drawn as rectangular blocks). Suppose there are *external* elastic forces with constants $k_1, k_2$ joining the respective point to the origin $O$. On $P_2$ there is also an external horizontal force $\mathbf{F}$. Finally, assume the points exchange an *internal* elastic force, of constant $k$. Let $x_1, x_2$ be the abscissae, $\mathbf{\Phi}_1$, $\mathbf{\Phi}_2$ the constraint reactions. The previous theorem can be written as

$$m\mathbf{a}_G = -k_1 OP_1 - k_2 OP_2 - m\mathbf{g} + \mathbf{F} + \mathbf{\Phi}_1 + \mathbf{\Phi}_2\,, \tag{11.6}$$

where $m = m_1 + m_2$. In particular, the projection of (11.6) on the $x$-axis gives

$$m\ddot{x}_G = -k_1x_1 - k_2x_2 + F\,,$$

or

$$m\ddot{x}_G = -\frac{mk_1}{m_1}x_G + \frac{m_2k_1 - m_1k_2}{m_1}x_2 + F\,, \tag{11.7}$$

keeping into account $mx_G = m_1x_1 + m_2x_2$. This is a differential equation in the unknown $x_G(t)$ *if and only if* the external forces' parameters satisfy a strong condition: $k_1 m_2 = k_2 m_1$. If the latter *does not* hold then (11.7), alone, *does not allow* to determine the motion of the center of mass.

Clearly we could instead determine the system's motion by writing the equations of motion of the two particles

$$\begin{aligned} m_1 \mathbf{a}_1 &= -k_1 OP_1 - kP_2P_1 - m_1 \mathbf{g} + \mathbf{\Phi}_1 \,, \\ m_2 \mathbf{a}_2 &= -k_2 OP_2 + kP_2P_1 - m_2 \mathbf{g} + \mathbf{F} + \mathbf{\Phi}_2 \,. \end{aligned}$$

which, projected onto the horizontal and vertical axes, would give $\mathbf{\Phi}_1 = m_1 \mathbf{g}$, $\mathbf{\Phi}_2 = m_2 \mathbf{g}$ and the system of differential equations

$$\begin{aligned} m_1 \ddot{x}_1 &= -k_1 x_1 + k\,(x_2 - x_1) \\ m_2 \ddot{x}_2 &= -k_2 x_2 - k\,(x_2 - x_1) + F \end{aligned} \tag{11.8}$$

in the unknown functions $x_1(t)$, $x_2(t)$.

To see hands-on how the position of the center of mass depends on the internal elastic force we could compute the equilibrium configuration. In the static case (11.8) becomes

$$\begin{aligned} -\,k_1 x_1 + k\,(x_2 - x_1) &= 0 \\ -k_2 x_2 - k\,(x_2 - x_1) + F &= 0 \end{aligned}$$

An elementary computation gives

$$x_1 = F \frac{k}{kk_1 + kk_2 + k_1 k_2}, \qquad x_2 = F \frac{k + k_1}{kk_1 + kk_2 + k_1 k_2},$$

so

$$x_G = \frac{F}{m} \frac{k(m_1 + m_2) + m_2 k_1}{kk_1 + kk_2 + k_1 k_2}.$$

We clearly see that the equilibrium position of the center of mass depends on the elastic constant $k$, i.e. on the internal elastic force.

This example shows in a direct and easy manner that, in general, the first balance equation of Dynamics *does not* determine the motion of the center of mass, and hence the motion of the system's center of mass *may depend* on the internal forces: the resultant of the external forces might depend on the positions and velocities of its constituents, which in turn are directly affected by the internal forces. □

**Example 11.13** Consider the motion of a skydiver jumping off a plane. The external forces that steer the motion of its center of mass are the weight and air drag. The weight does not depend on any of the system's specific features, except for the total mass. The air drag, on the contrary, depends on the shape of the system. In particular, the resistance acting on the skydiver during free-fall is much smaller than the resistance when the parachute is open. For this reason the skydiver, which deploys the parachute only using internal forces to the system, is able to indirectly modify the motion of the center of mass. □

Having said that, there are special situations, like Example 11.12 in case $k_1m_2 = k_2m_1$, in which the global elements of the external force depend on the positions of the single points of the system and their velocities only through the position and velocity of the center of mass: $m\mathbf{a}_G = \mathbf{R}^{(e)}(G, \mathbf{v}_G, t)$. If so (and only in these cases) the first balance equation is indeed enough to determine the motion of the center of mass, since it becomes completely similar to the equation of a particle with an active force exclusively depending on its velocity distribution. This kind of system is said to be *G-determined*, and the motion of the center of mass is fully independent of the internal forces. Further examples of $G$-determined systems are *isolated systems* (i.e. systems with no external force) and systems whose only external force is the weight. A system that is $G$-determined in one frame will be so in any other frame, as we will see later when talking about relative mechanics.

All in all, we conclude that although the internal forces might not appear explicitly in the balance equations, in general the latter are affected by their influence. and only in special situations the balance equations are truly independent of the internal forces.

## 11.4 Balance of Power

The balance equation of Dynamics have the advantage of relating the system's motion to the external active and reactive forces solely. But there is another equation of motion that often relates the motion to the sole active forces, thus bypassing the discussion of constraint reactions. The Kinetic energy theorem, which we will study in this section, provides in such cases a *free* equation for Dynamics, i.e. an equation linking the motion only to the active forces.

**Theorem 11.14** (Kinetic energy theorem) *Let $T$ denote the kinetic energy of a system, and $\Pi$ the power generated by* all *forces acting on it. Then*

$$\frac{dT}{dt} = \Pi\,. \tag{11.9}$$

***Proof*** To simplify the discussion, but without loss of generality, consider a discrete system of particles. For each element consider Newton's law (11.2), recast succinctly as

$$m_i\mathbf{a}_i = \mathbf{F}_i$$

where $\mathbf{F}_i$ is the resultant of *all* forces acting on $P_i$, both *internal* and *external* to the system. Taking the dot product of the equation with the velocity of the corresponding point we obtain

$$m_i\mathbf{a}_i \cdot \mathbf{v}_i = \mathbf{F}_i \cdot \mathbf{v}_i\,.$$

The quantity on the right coincides by definition with the power of the forces on $P_i$ (see (7.5)). The quantity on the left is the derivative of the kinetic energy of $P_i$, since

$$m_i \mathbf{a}_i \cdot \mathbf{v}_i = m_i \frac{d}{dt}\mathbf{v}_i \cdot \mathbf{v}_i = \frac{d}{dt}\Big(\frac{1}{2} m_i \mathbf{v}_i \cdot \mathbf{v}_i\Big) = \frac{d}{dt}\Big(\frac{1}{2} m_i v_i^2\Big).$$

Summing over all points we obtain (11.9). □

The Kinetic energy theorem 11.9 can be formulated equivalently as the so-called work theorem, or kinetic energy theorem *in integral form.*

**Theorem 11.15** (Work theorem) *The variation of the kinetic energy of a material system, over any time interval and during a given motion, equals the work done by* all *forces acting on the system:*

$$\Delta T = L\,. \tag{11.10}$$

***Proof*** This is a direct consequence of the relationship between work and power. Applying (7.6) to all forces in the interval $[t_1, t_2]$ says that

$$L = \int_{t_1}^{t_2} \Pi\, dt\,. \tag{11.11}$$

If we integrate (11.9) in time, we obtain

$$\int_{t_1}^{t_2} \frac{dT}{dt}\, dt = \int_{t_1}^{t_2} \Pi\, dt$$

so

$$T(t_2) - T(t_1) = L$$

by (11.11) and after a simple integration. Putting $\Delta T = T(t_2) - T(t_1)$ to denote the *variation of the kinetic energy* over the interval $[t_1, t_2]$ we obtain (11.10) □

We must not forget that in general (11.9) and (11.10) should include the power and (respectively) the work of all forces: external and internal, active and reactive.

With reference to the Kinetic energy theorem we now recognize two important situations, where we know a priori that the power of some forces is zero. We will see that:

- for systems subject to *ideal*, *bilateral* and *fixed* constraints the power of the constraint reactions is zero.
- For *rigid* systems the power of *internal* forces is zero.

Each of the above evidently has an equivalent version formulated within the Work theorem, which follows as immediate consequence without the need of another proof:

- For systems subject to ideal, bilateral and fixed constraints the work of the constraint reactions is zero.

- For rigid systems the work of the internal forces is zero.

**Proposition 11.16** (Power of reactive forces) *The power of the constraint reactions acting on a system subject to* ideal, bilateral *and* fixed *constraints is zero. In this case* (11.9) *can be written*

$$\frac{dT}{dt} = \Pi^{\text{active}}$$

***Proof*** *Ideal* constraints are identified by (see the characterization in Sect. 8.7) their virtual work (or, respectively, their virtual power) being bigger than or equal to zero for any virtual displacement (virtual velocity). In presence of ideal *bilateral* constraints, all virtual displacements (and all virtual velocities) are reversible, so inequality (8.17) becomes an equality in the sense that the virtual power is zero for any admissible virtual velocity

$$\Pi'_{\text{reactive}} = \sum_{i=1}^{n} \mathbf{\Phi}_i \cdot \mathbf{v}'_i = 0\,.$$

When we further restrict to *fixed* constraints, among the infinitely many virtual displacements there is the effective one, exactly as among the infinitely many virtual velocities $\mathbf{v}'_i$ there is the effective velocity $\mathbf{v}_i$ (see Sect. 4.1.4). In this case, then, among the infinitely many vanishing virtual powers there will be the effective power, and therefore the power of the constraint reactions is zero

$$\Pi_{\text{reactive}} = \sum_{i=1}^{n} \mathbf{\Phi}_i \cdot \mathbf{v}_i = 0\,.$$

□

It is interesting to note that in order to ensure the reactive forces have zero power, it is indispensable to demand not just that the constraints be ideal and bilateral, but also that they be *fixed*. For a counterexample it suffices to take a small ring threaded onto a smooth guide that rotates with given angular velocity, as in Fig. 11.3. For the ring the guide is an ideal, bilateral but *moving* constraint, with constraint reaction $\mathbf{\Phi}$ orthogonal to the guide. As the virtual velocities $\mathbf{v}'$ must be compatible with the constraint "as if it were fixed", each velocity is parallel to the guide, and consequently the virtual power $\Pi'$ will certainly vanish. The effective velocity $\mathbf{v}$ on the other hand, as we see in Fig. 11.3 right, has a component perpendicular to the rod as well, because the ring during its effective motion must follow the guide, and therefore $\Pi \neq 0$.

The previous proposition is of great importance, since it allows us to say the majority of constraint reactions of this book (for instance: point belonging to a fixed smooth surface or curve, fixed pin, internal moving pin, roller, pure rolling) do not produce any power. Hence in presence of such constraints the Kinetic energy theorem is a free motion equation.

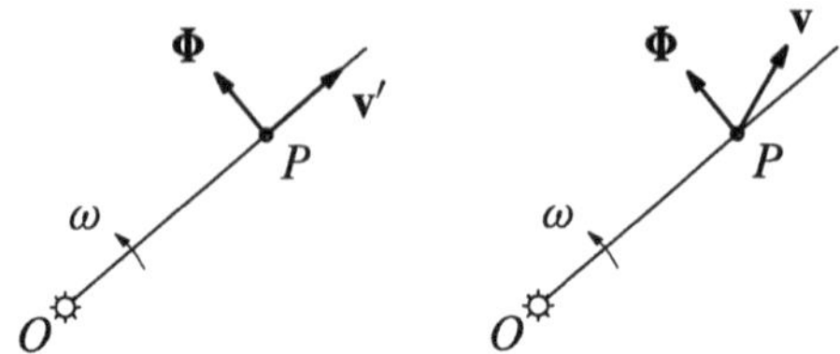


**Fig. 11.3** The virtual and effective power of the reactive force for a constraint which is ideal, bilateral and *moving*: $\Pi' = 0$, $\Pi \neq 0$

**Proposition 11.17** *For rigid bodies the Kinetic energy theorem's formula* (11.9) *can be written as*

$$\frac{dT}{dt} = \Pi^{\text{ext}}. \tag{11.12}$$

***Proof*** We proved in Theorem 8.7 and again in Proposition 8.10 that the power of the internal forces of a *rigid body* is always zero. Hence (11.9) here reduces to (11.12). □

**Remark 11.18** An immediate consequence of Theorem 8.7, in particular of $(8.9)_1$, is that the Work theorem (11.10), in case of a *rigid body*, takes the form

$$\Delta T = L^{\text{ext}}.$$

□

A further important property is that while the Kinetic energy theorem provides us in general with an equation of motion that is *independent* of the balance equations, since only internal forces crop up, we will see later (Chap. 12) that for *rigid bodies* the theorem arises by taking a linear combination of the balance equations.

**Proposition 11.19** (Power of conservative active forces) *If a system of conservative active forces acting on an arbitrary system has potential $U$, their power is*

$$\Pi = \frac{dU}{dt}. \tag{11.13}$$

***Proof*** This property is a direct consequence of Sect. 7.7. It coincides with relation (7.49), when applied to the sole conservative active forces acting on the system. □

## 11.5 Conservation of Mechanical Energy

The Work theorem detects another first integral of motion, this time coming from the Kinetic energy theorem. We will prove that under certain hypotheses on constraints and forces, the quantity $E = T - U$, which we call *mechanical energy*, remains constant during the system's motion.

**Proposition 11.20** (Conservation of mechanical energy) *Consider a system subject to ideal, bilateral and fixed constraints. Suppose the active forces are conservative, with potential $U$. Then the mechanical energy is conserved:*

$$E = T - U \equiv \textit{constant}. \tag{11.14}$$

***Proof*** From Proposition 11.16, under the assumptions made on the constraints,

$$\frac{dT}{dt} = \Pi^{\text{active}}.$$

By Proposition 11.19 furthermore, for *active conservative* forces (11.13) holds:

$$\Pi^{\text{active}} = \frac{dU}{dt}.$$

Comparing the two yields easily

$$\frac{d}{dt}(T - U) = 0$$

i.e.

$$E = T - U = \text{const.}$$

This says the difference $E$ of the kinetic energy $T$ and the potential $U$ remains constant during motion. □

We remind that the kinetic energy is a function of the variables $(\mathsf{q}, \dot{\mathsf{q}}, t)$ in general, and the potential $U$ a function of $(\mathsf{q}, t)$. For this reason the mechanical energy takes the form $E(\mathsf{q}, \dot{\mathsf{q}}, t)$. The conservation of the mechanical energy is then a particular integral of motion, as per Definition 11.5.

It is useful to note that Theorem 11.20 takes a perhaps more "natural" form if we use the *potential energy* $V$, defined as the opposite of the potential $U$: $V = -U$. In this way (11.14) reads

$$E = T + V = \text{const.}$$

This might be the main reason why several texts privilege the *potential energy* $V$ over the conservative active forces' potential $U$. It seems in fact more natural to say the *sum* of two quantities is conserved: when one decreases the other increases, and conversely.

When browsing books on mechanics one should take heed of the notation, since it is not rare to see $V$ indicate what we called $U$, or the opposite.

**Remark 11.21** The mechanical energy may be conserved also in presence of non-conservative active forces, as long as these do not have power. We will see an example in Chap. 13 regarding the Coriolis force. Other examples were presented in

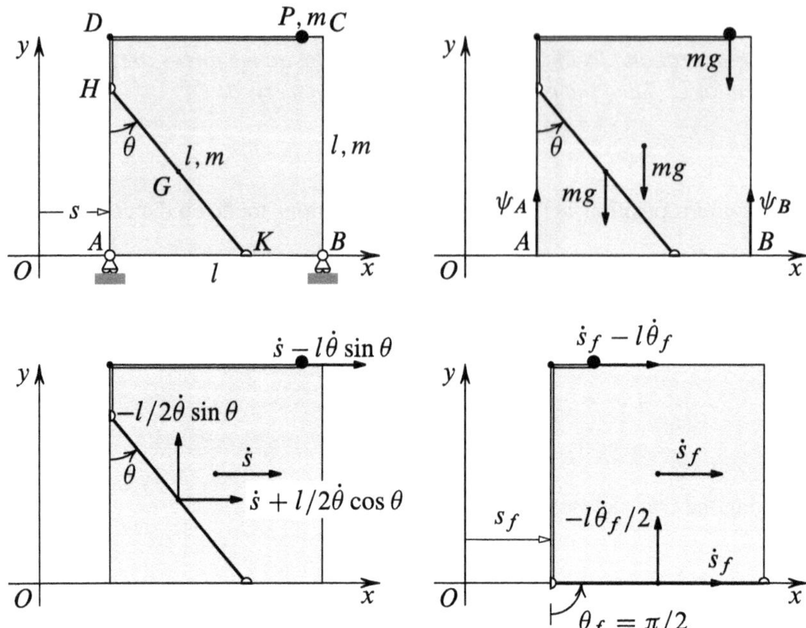


**Fig. 11.4** Conservation of the energy and of the horizontal component of the linear momentum

Sect. 10.2.1, where we proved that an integral similar to the energy arises when a particle moves on a prescribed trajectory with positional active forces. □

**Example 11.22** The conservation of mechanical quantities, that is, the existence of "integrals of motion", is a relevant property, one that should be observed straightaway, since at times it allows to answer questions we would not know how to address otherwise.

Let us provide an example to support this observation. Consider the system in Fig. 11.4. On a vertical plane a thin homogeneous square plate of edge $l$ and mass $m$ has the vertices $A$ and $B$ move without friction on a horizontal guide. The ends $H$, $K$ of a homogeneous rod of length $l$ and mass $m$ move along two of the square's edges (bilateral constraints). A string of negligible mass, passing through a smooth pin at the vertex $D$, connects the end $H$ to a point $P$ of mass $m$ that moves without friction along the upper edge $DC$ (that the three parts have equal masses is only practical for the ensuing calculations). It is convenient to introduce coordinate axes as in the figure.

Suppose at the initial instant the system is at rest and the angle $\theta$ between the vertical direction and the rod is $\pi/6$. Due to its weight the rod will clearly slide downwards and the square will move with it. We want to find out the velocity distribution when $\theta = \pi/2$, i.e. when the rod reaches a horizontal position. The configuration is given Fig. 11.4 bottom right, where obviously we should think of an ideal situation where $H$ and $K$ can occupy the square's vertices.

The *external* forces are the three weights (the plate's, the rod's and the point's) and the two vertical constraint reactions the guide exerts at $A$ and $B$, called $\psi_A$, $\psi_B$ in Fig. 11.4 top right. There are internal forces, naturally (between the rod, the point and the plate), that cannot appear in the first balance equation, when we apply the latter to the overall system. Hence $R_x^{\text{ext}} = 0$, and so the horizontal component of the linear momentum remains constant from the *initial instant* to the instant when $\theta = \pi/2$ (henceforth, the *final instant*)

$$R_x^{\text{ext}} = 0 \quad \Rightarrow \quad Q_x = \text{const.} \quad \Rightarrow \quad Q_x^0 = Q_x^f \,. \tag{11.15}$$

Using the Linear momentum theorem we deduce from (11.15) the velocity of the center of mass is zero, so its position does not change from the initial time to the final time.

Since the constraints are ideal and fixed, and the active forces conservative (weights), the total energy is another conserved mechanical quantity

$$T - U = \text{const.} \quad \Rightarrow \quad T_0 - U_0 = T_f - U_f \,. \tag{11.16}$$

Convenient generalized coordinates are the variables $s$ and $\theta$ in the figure. In view of computing mechanical quantities it is important to express the velocities of certain important points in function of them. The coordinates of $G$ are $x_G = s + l/2 \sin\theta$, $y_G = l/2 \cos\theta$, so

$$\dot{x}_G = \dot{s} + (l/2)\dot{\theta}\cos\theta \,, \qquad \dot{y}_G = -(l/2)\dot{\theta}\sin\theta \,.$$

The velocity of the translating plate equals $\dot{s}$, and so is the velocity of its center of mass. For $P$ it is convenient to use the frame comoving (translating) with the plate. As $y_H = l\cos\theta$ and the string from $H$ to $P$ is inextensible, the velocity of $P$ relative to the plate is $-l\dot{\theta}\sin\theta$ and oriented rightwards. To that we must add the drag velocity $\dot{s}$. Overall, the relevant velocities are represented in Fig. 11.4 bottom left.

By the Linear momentum theorem we can write its total horizontal component as

$$\begin{aligned} Q_x &= m\dot{s} + m(\dot{s} + (l/2)\dot{\theta}\cos\theta) + m(\dot{s} - l\dot{\theta}\sin\theta)\,, \\ &= 3m\dot{s} + m(l/2)\dot{\theta}\cos\theta - ml\dot{\theta}\sin\theta \,. \end{aligned}$$

Condition $Q_x = \text{const.}$, and the fact the system is initially at rest, imply that at each instant $Q_x = 0$, i.e.

$$3\dot{s} - l\dot{\theta}\sin\theta + l/2\dot{\theta}\cos\theta = 0$$

which we integrate to find

$$3\,s + l\cos\theta + l/2\sin\theta = \text{const.}$$

Evaluating at $\theta = \pi/6$ and $\theta = \pi/2$ and comparing, we obtain

$$3s_0 + l\sqrt{3}/2 + l/4 = 3s_f + l/2$$

so in the given time interval the plate is displaced by

$$\Delta s = s_f - s_0 = l(2\sqrt{3} - 1)/12\,.$$

The kinetic energies of plate and point are

$$T^{\text{plate}} = \frac{1}{2}m\dot{s}^2\,, \quad T^{\text{point}} = \frac{1}{2}\,m(\dot{s} - l\dot{\theta}\sin\theta)^2\,,$$

while for the rod $HK$ we have to invoke König's theorem:

$$\begin{aligned} T^{\text{rod}} &= \frac{1}{2}mv_G^2 + \frac{1}{2}I_{z,G}\omega^2 \\ &= \frac{1}{2}m[(\dot{s} + (l/2)\dot{\theta}\cos\theta)^2 + (-(l/2)\dot{\theta}\sin\theta)^2] + \frac{1}{2}(ml^2/12)\dot{\theta}^2 \\ &= \frac{1}{2}m[\dot{s}^2 + l^2\dot{\theta}^2/3 + l\dot{s}\dot{\theta}\cos\theta]\,. \end{aligned}$$

We conclude the total kinetic energy equals

$$T^{\text{tot}} = \frac{1}{2}m(3\dot{s}^2 + l^2\dot{\theta}^2/3 + l^2\dot{\theta}^2\sin^2\theta - 2l\dot{s}\dot{\theta}\sin\theta + l\dot{s}\dot{\theta}\cos\theta)\,.$$

The potential of the active forces is

$$U = -mg\frac{l}{2}\cos\theta + \text{const.}$$

Next we make the dependence of $Q_x$, $T$, $U$ upon the generalized coordinates and their derivatives

$$Q_x(\dot{s}, \theta, \dot{\theta})\,, \quad T(\dot{s}, \theta, \dot{\theta})\,, \quad U(\theta)$$

explicit. We know the values of $\dot{s}$, $\theta$, $\dot{\theta}$ at the initial instant

$$\dot{s}_0 = 0, \quad \theta_0 = \pi/6, \quad \dot{\theta}_0 = 0\,,$$

and we want to find the velocity distribution when $\theta = \pi/2$ using $\dot{s}_f, \dot{\theta}_f$. The conservation laws (11.15) and (11.16) can be rewritten to give a system

$$\begin{aligned} Q_x(\dot{s}_0, \theta_0, \dot{\theta}_0) &= Q_x(\dot{s}_f, \theta_f, \dot{\theta}_f) \\ T(\dot{s}_0, \theta_0, \dot{\theta}_0) - U(\theta_0) &= T(\dot{s}_f, \theta_f, \dot{\theta}_f) - U(\theta_f) \end{aligned} \tag{11.17}$$

where $\dot{s}_0 = 0$, $\theta_0 = \pi/6$, $\dot{\theta}_0 = 0$, $\theta_f = \pi/2$, while $\dot{s}_f, \dot{\theta}_f$ are the two unknowns to be found.

After substituting the above known values, the conservation equations (11.17) read

$$0 = 3\dot{s}_f - l\dot{\theta}_f\,, \qquad mgl\frac{\sqrt{3}}{4} = \frac{1}{2}m(3\dot{s}_f^2 + \frac{4}{3}l^2\dot{\theta}_f^2 - 2l\dot{s}_f\dot{\theta}_f)\,.$$

From the first one we obtain $\dot{s}_f = l\dot{\theta}/3$, and substituting in the second a few elementary manipulations yield

$$\dot{\theta}_f^2 = \frac{g\sqrt{3}}{2l}\,.$$

In this phase we have to choose the positive value of $\dot{\theta}_f$, while in the next (same configuration) we will take the negative root.

It is extremely important to notice that only due to the two conservation laws we were able to find the velocity distribution in the final configuration, without computing the system's motion (which would have been impossible in an explicit elementary way, due to the complicated differential equations). This is a typical "start-end" problem, where two states must be compared without finding out what happens inbetween. □

# Chapter 12
# Dynamics of Rigid Bodies

The balance equations are necessarily true during the motion of any material system, but in case of a free rigid body they become sufficient to determine the motion caused by some external force. Before we analyze particular problems related to the dynamics of variously constrained rigid bodies, we specialize the balance equations of Dynamics to the case of a single rigid body.

**Proposition 12.1** *Let $Q$ be a point coinciding with the center of mass of a rigid body, or a fixed point belonging at each instant to the instantaneous axis of rotation of the rigid body. The balance equation of Dynamics imply*

$$m\mathbf{a}_G = \mathbf{R}^{(e)} \qquad \textit{and} \qquad \mathbf{I}_Q\dot{\boldsymbol{\omega}} + \boldsymbol{\omega} \times \mathbf{I}_Q\boldsymbol{\omega} = \mathbf{M}_Q^{(e)} . \tag{12.1}$$

*Projecting* $(12.1)_2$ *along the basis* $\{\mathbf{e}_1, \mathbf{e}_2, \mathbf{e}_3\}$ *of principal axes at $Q$, then*

$$\begin{cases} I_1\dot{\omega}_1 - (I_2 - I_3)\omega_2\omega_3 = M_1^{(e)} \\ I_2\dot{\omega}_2 - (I_3 - I_1)\omega_3\omega_1 = M_2^{(e)} \\ I_3\dot{\omega}_3 - (I_1 - I_2)\omega_1\omega_2 = M_3^{(e)} \end{cases} \tag{12.2}$$

*where $I_1, I_2, I_3$ are the principal moments of inertia at $Q$, and $\{\omega_1, \omega_2, \omega_3\}$ are the components of the angular velocity $\boldsymbol{\omega}$ in the basis of principal axes. Equations* (12.2) *are called* Euler equations.

***Proof*** Equation $(12.1)_1$ coincides with the equation of motion of the center of mass (11.5). Moreover, the choice of pole allows to write the second balance equation (11.3) as $\dot{\mathbf{K}}_Q = \mathbf{M}_Q^{(e)}$. Considering that the system is a single rigid body, (6.20) in Proposition (6.11) plus Remark 6.12 give the two expressions of $\dot{\mathbf{K}}_Q$ that lead to $(12.1)_2$–(12.2). □

**Remark 12.2** The second balance equation of Dynamics of a rigid body reduces to the Euler equations even when it refers to a pole $Q$, still on the instantaneous axis of rotation, but in motion with $\dot{Q} \parallel \mathbf{v}_G$ (so that $\dot{\mathbf{K}}_Q = \mathbf{M}_Q^{(e)}$). □



P. Biscari et al., *Rational Mechanics*, UNITEXT 177,
https://doi.org/10.1007/978-3-032-07462-1_12

We emphasize that it is not always possible to split problem (12.1) into two independent equations, one for studying the motion of the center of mass and the other for the behaviour of the Euler angles. In fact, the resultant of the external forces may depend on the rigid body's orientation, just like their resultant moment may depend on the motion of the center of mass. In the ensuing two sections we will see particularly significant cases (free rigid body, rigid body with fixed point, rigid body with fixed axis) where the Euler equations truly differ from the first balance equation, and alone enable to determine the Euler angles' evolution.

## 12.1 Free Rigid Body

We have already observed that in general the balance equations do not enable us to complete the study of the dynamics of an arbitrary material system. Already in the static setting (see the example of Fig. 9.6) they are necessary conditions, usually not sufficient, to study the equilibrium of a non-rigid system. The situation is completely different for a single rigid body. Exactly as we did in Statics (Theorem 9.12) we will prove that in dynamics too the balance equations provide also sufficient conditions to determine the dynamics of a rigid body.

For free rigid bodies in space there are six degrees of freedom (irrespective of the number of points forming the body). Usually the suggestion is to choose as free parameters the coordinates of the center of mass $(x_G, y_G, z_G)$ with respect to a suitable fixed frame, and, locally, the three Euler angles $\Theta = \{\theta, \phi, \psi\}$, which allow to find the orientation, with respect to a fixed basis, of a frame comoving with the body (Sect. 2.2).

**Theorem 12.3** (Dynamical characterization of a free rigid body) *The balance equations of Dynamics give two vector equations, or six scalar equations, that are both necessary and sufficient to determine the dynamics of a single free rigid body.*

***Proof*** In absence of constraint reactions, and taking the center of mass $G$ as pole for the moments, (12.1) and (12.2) give the system

$$\begin{cases} m\ddot{x}_G = R_x^{(e)}(G, \mathbf{v}_G, \Theta, \dot{\Theta}, t) \\ m\ddot{y}_G = R_y^{(e)}(G, \mathbf{v}_G, \Theta, \dot{\Theta}, t) \\ m\ddot{z}_G = R_z^{(e)}(G, \mathbf{v}_G, \Theta, \dot{\Theta}, t) \\ I_1\dot{\omega}_1 - (I_2 - I_3)\omega_2\omega_3 = M_1^{(e)}(G, \mathbf{v}_G, \Theta, \dot{\Theta}, t) \\ I_2\dot{\omega}_2 - (I_3 - I_1)\omega_3\omega_1 = M_2^{(e)}(G, \mathbf{v}_G, \Theta, \dot{\Theta}, t) \\ I_3\dot{\omega}_3 - (I_1 - I_2)\omega_1\omega_2 = M_3^{(e)}(G, \mathbf{v}_G, \Theta, \dot{\Theta}, t), \end{cases} \tag{12.3}$$

where we have emphasized the external forces are now active. The three components $(\omega_1, \omega_2, \omega_3)$ of the angular velocity can be written in terms of the Euler angles and their first derivatives (see Sect. 3.5). In particular, (3.14) proves that the relation linking the components of $\boldsymbol{\omega}$ to the derivatives $\dot{\Theta}$ is linear, with coefficients depending on the angles $\Theta$. Taking a further derivative shows that $(\dot{\omega}_1, \dot{\omega}_2, \dot{\omega}_3)$ depend linearly on the *second* derivatives $\ddot{\Theta}$. Therefore it is possible to write the Euler equations exclusively in terms of the Euler angles and their first and second derivatives (besides obviously the coordinates and velocity of the center of mass). More algebraic manipulations (see Sect. 12.1.3) enable to solve the last three equations in (12.3) for the second derivatives $\ddot{\Theta}$.

In other words, system (12.3) is made by *six* scalar ordinary differential equations of order two in the *six* unknowns $\{x_G(t), y_G(t), z_G(t); \theta(t), \phi(t), \psi(t)\}$ and can be put in *normal form*: the second derivatives of the unknown functions are expressible by the values of the unknown functions, the first derivatives and possibly time. Assuming the functions on the right-hand side of (12.3), which represent the characteristic vectors of the external force, are regular enough in their variables, the Cauchy theorem holds (see Sect. 8.2). Hence system (12.3), *equivalent to the balance equations*, admits a unique solution once we have initial conditions $G(t_0) = G_0$, $\mathbf{v}_G(t_0) = \mathbf{v}_0$, $\Theta(t_0) = \Theta_0$, $\dot{\Theta}(t_0) = \dot{\Theta}_0$. The latter are by the way the conditions allowing to determine the position and velocity distribution of all points of the rigid body at the given instant $t = t_0$.

In conclusion, the balance equations of Dynamics of a free rigid body (just like the static ones) are not only necessary but also sufficient to know the motions compatible with the external force. □

**Remark 12.4** Note that the balance equations (12.3) are written in an inertial frame. Now, while the first equation is also projected along the axes of the same reference frame, the second equation is projected along the axes of the comoving (non-inertial) frame, whose basis vectors are central principal axes. This allows to use the diagonal form and the fact the inertia matrix does not depend on time. □

In the sequel we examine a few consequences of the balance equation of Dynamics being sufficient for a single free rigid body.

### 12.1.1 Equivalent Loads

In equations (12.3) the system of external forces only shows up through the characteristic vectors (resultant and resultant moment). Hence two equivalent external forces determine the same dynamics of a rigid body (the same motions). As a matter of fact this observation is the origin of the usual definition of equivalence between systems of forces. The definition implicitly says the two forces are understood to be equivalent when studying the dynamics of a rigid body. For example, if we only look at a rigid body it is possible to replace the weight (which is spread across all points of the body) with the relative resultant applied at the center of gravity. For a generic material system this operation would change the ensuing motions.

### The Role of the Internal Forces

A straightforward consequence of the previous remark is that any balanced system of forces can be ignored, meaning we may replace it with the zero system. In particular, *the dynamics and statics of rigid bodies are not affected in any way by the internal forces*. In fact, these form a balanced system by virtue of Newton's third law. The internal forces do not appear in the balance equation of Dynamics of any system, rigid or not. The peculiarity of the single rigid body is that the balance equations contain all of the information needed to determine the motion, while in generic systems we must accompany them with other equations that, in general, will involve the internal forces.

Before we continue we think it is useful to clarify the meaning of the previous statement on ignoring the internal forces when the objective is the mechanics of a single rigid body. We should more precisely say that *the dynamics and statics of rigid bodies are not affected at all by the internal forces, so long as the rigidity constraint is in place*. This explanation is necessary lest we think, for instance, that we can replace the system of internal constraint reactions guaranteeing the rigidity constraint with the zero system. If we did that, the system would cease to be rigid and the aforementioned property on equivalent applied forces would collapse! On the contrary, a correct way to interpret the property is the following: it makes no sense to ask ourselves which particular constraints (rods, pins) fix the distances of the points of a rigid body, thus ensuring the rigidity constraint. Any two systems will be balanced, hence equivalent.

### Equivalent Rigid Bodies

The balance equations (12.3) give us another significant equivalence relation, this time between rigid bodies. Consider two rigid bodies of equal mass and equal central principal moments of inertia. If we apply identical systems of forces (or rather, equivalent systems), both will have the same equations of motion, so they will undergo identical motions, given the same initial conditions. In fact, mass and central principal moments of inertia are the only constitutive elements of a rigid body that enter (12.3).

The above equivalence properties (for systems of forces and for rigid bodies) have a fundamental applicative interest. Suppose in fact we wish to simulate the behaviour of a rigid body, of any shape and composition, subject to a given system of forces. Few easy tests allow to measure mass and central principal moments of inertia. At this point it is possible to replace the system with a simple but equivalent test system (say, a parallelepiped whose sides are determined by the inertia moments) and subject it to the original system of forces (or an equivalent one if this simplifies the study further). Equations (12.3) confirm the dynamical results of the experiment will perfectly simulate the behaviour of the original rigid body.

### 12.1.2 Balance of Power

As the balance equations are necessary and sufficient to determine the motion of a rigid body, evidently we should be able to deduce from them *any* statement regarding its dynamics. Consider for instance the Kinetic energy theorem.

**Proposition 12.5** *For a rigid body the Kinetic energy theorem is a direct consequence of the balance equations.*

***Proof*** Take the dot product of equation (11.1) by $\mathbf{v}_G$ and of (11.3) (the center of mass is the pole) by $\boldsymbol{\omega}$. Then add:

$$m\mathbf{a}_G \cdot \mathbf{v}_G = \mathbf{R}^{(e)} \cdot \mathbf{v}_G$$
$$\dot{\mathbf{K}}_G \cdot \boldsymbol{\omega} = \mathbf{M}_G^{(e)} \cdot \boldsymbol{\omega}$$

$$m\mathbf{a}_G \cdot \mathbf{v}_G + \dot{\mathbf{K}}_G \cdot \boldsymbol{\omega} = \mathbf{R}^{(e)} \cdot \mathbf{v}_G + \mathbf{M}_G^{(e)} \cdot \boldsymbol{\omega}\,. \tag{12.4}$$

The right-hand side of (12.4) coincides exactly with the power of the external forces, by (7.31). If we recall that the power of the internal forces is zero in a rigid velocity distribution (see $(8.9)_2$ in Theorem 8.7), the right-hand side of (12.4) coincides exactly with the power of all forces acting on the body.

To prove the left-hand side of (12.4) is $\dot{T}$, we first observe that from (6.20) we have

$$\dot{\mathbf{K}}_G \cdot \boldsymbol{\omega} = \underbrace{\mathbf{I}_G\dot{\boldsymbol{\omega}} \cdot \boldsymbol{\omega}}_{\mathbf{I}_G = \mathbf{I}_G^T} + \underbrace{\boldsymbol{\omega} \times \mathbf{I}_G\boldsymbol{\omega} \cdot \boldsymbol{\omega}}_{0} = \boldsymbol{\omega} \cdot \mathbf{I}_G\boldsymbol{\omega} = \boldsymbol{\omega} \cdot \mathbf{K}_G\,, \tag{12.5}$$

where we have used property (A.24), guaranteeing each symmetric map (such as $\mathbf{I}_G$) satisfies $\mathbf{I}_G\mathbf{u} \cdot \mathbf{v} = \mathbf{u} \cdot \mathbf{I}_G\mathbf{v}$ for any $\mathbf{u}$, $\mathbf{v}$. Then (12.5) implies

$$\dot{\mathbf{K}}_G \cdot \boldsymbol{\omega} = \frac{1}{2}\Big(\dot{\mathbf{K}}_G \cdot \boldsymbol{\omega} + \dot{\mathbf{K}}_G \cdot \boldsymbol{\omega}\Big) = \frac{1}{2}\Big(\dot{\mathbf{K}}_G \cdot \boldsymbol{\omega} + \mathbf{K}_G \cdot \dot{\boldsymbol{\omega}}\Big) = \frac{d}{dt}\Big(\frac{1}{2}\mathbf{K}_G \cdot \boldsymbol{\omega}\Big).$$

At the same time

$$m\mathbf{a}_G \cdot \mathbf{v}_G = m\frac{d\mathbf{v}_G}{dt} \cdot \mathbf{v}_G = \frac{d}{dt}\Big(\frac{1}{2}m\mathbf{v}_G \cdot \mathbf{v}_G\Big),$$

and so by (6.26) the left-hand side of (12.4) equals

$$m\mathbf{a}_G \cdot \mathbf{v}_G + \dot{\mathbf{K}}_G \cdot \boldsymbol{\omega} = \frac{d}{dt}\Big(\frac{1}{2}mv_G^2 + \frac{1}{2}\mathbf{K}_G \cdot \boldsymbol{\omega}\Big) = \dot{T}\,.$$

This completes the proof. □

### 12.1.3 Euler Equations

The result whereby the balance equations (12.3) are necessary and sufficient requires we write the Euler equations (12.2) in normal form for the Euler angles, i.e. solved for the second derivatives. The relationship between angular velocity and Euler angles follows from (3.14):

$$\omega_1 = \dot{\theta}\cos\phi + \dot{\psi}\sin\phi\sin\theta\,, \qquad \omega_2 = -\dot{\theta}\sin\phi + \dot{\psi}\cos\phi\sin\theta\,,$$
$$\omega_3 = \dot{\psi}\cos\theta + \dot{\phi}\,. \tag{12.6}$$

Differentiating in time we obtain

$$\begin{aligned}
\dot{\omega}_1 &= \ddot{\theta}\cos\phi + \ddot{\psi}\sin\phi\sin\theta - \dot{\theta}\dot{\phi}\sin\phi + \dot{\psi}\left(\dot{\phi}\cos\phi\sin\theta + \dot{\theta}\sin\phi\cos\theta\right)\\
\dot{\omega}_2 &= -\ddot{\theta}\sin\phi + \ddot{\psi}\cos\phi\sin\theta - \dot{\theta}\dot{\phi}\cos\phi + \dot{\psi}\left(\dot{\theta}\cos\phi\cos\theta - \dot{\phi}\sin\phi\sin\theta\right)\\
\dot{\omega}_3 &= \ddot{\psi}\cos\theta + \ddot{\phi} - \dot{\psi}\dot{\theta}\sin\theta
\end{aligned} \tag{12.7}$$

When we substitute (12.6)–(12.7) in (12.2), the Euler angles' second derivatives only crop up in (12.7). So we can rewrite (12.2) as

$$\mathbf{A}\begin{pmatrix}\ddot{\theta}\\ \ddot{\psi}\\ \ddot{\phi}\end{pmatrix} = \mathbf{b}\,, \tag{12.8}$$

where

$$\mathbf{A} = \begin{pmatrix}\cos\phi & \sin\phi\sin\theta & 0\\ -\sin\phi & \cos\phi\sin\theta & 0\\ 0 & \cos\theta & 1\end{pmatrix}. \tag{12.9}$$

The vector on the right in (12.8), in the comoving frame, is

$$\mathbf{b} = \begin{pmatrix}\dfrac{M_1^{(e)}}{I_1} + \dfrac{I_2 - I_3}{I_1}\omega_2\omega_3 + \dot{\theta}\dot{\phi}\sin\phi - \dot{\psi}\left(\dot{\phi}\cos\phi\sin\theta + \dot{\theta}\sin\phi\cos\theta\right)\\ \dfrac{M_2^{(e)}}{I_2} + \dfrac{I_3 - I_1}{I_2}\omega_1\omega_3 + \dot{\theta}\dot{\phi}\cos\phi - \dot{\psi}\left(\dot{\theta}\cos\phi\cos\theta - \dot{\phi}\sin\phi\sin\theta\right)\\ \dfrac{M_3^{(e)}}{I_3} + \dfrac{I_1 - I_2}{I_3}\omega_1\omega_2 + \dot{\psi}\dot{\theta}\sin\theta\end{pmatrix}$$

where the components $M_i^{(e)}$ of the moment of the external forces are functions of $(\Theta, \dot{\Theta}, t)$

The determinant of the matrix $\mathbf{A}$ defined in (12.9) is

$$\det\mathbf{A} = \cos^2\phi\sin\theta + \sin^2\phi\sin\theta = \sin\theta\,.$$

Since the nutation $\theta$ is strictly between 0 and $\pi$ by definition, we infer (in the spatial region where the Euler angles exist) $\det \mathbf{A} \neq 0$, so the inverse $\mathbf{A}^{-1}$ exists. This guarantees the possibility of putting the Euler equations in *normal form*

$$\ddot{\Theta} = \mathbf{A}^{-1}(\Theta, \dot{\Theta}, t)\mathbf{b}(\Theta, \dot{\Theta}, t)$$

## 12.2 Poinsot Motion

The balance equation of Dynamics of the rigid body can be studied in detail in a particular case, which subsumes the motion of an isolated rigid body.

**Definition 12.6** (*Poinsot motion*) The motion of a rigid body is called *Poinsot* if there exists a point $Q$, either *fixed* or *coinciding with the center of mass*, such that the moment of the external forces with respect to $Q$ is zero at all times: $\mathbf{M}_Q^{(e)} = \mathbf{0}$.

A Poinsot motion can occur in several ways. The first and most obvious is the motion of an isolated rigid body, i.e. free and not subject to any active force (in this case $Q$ coincides with the center of mass $G$). Alternatively, we have Poinsot motions for non-isolated rigid bodies, but the system of external forces must be equivalent to its resultant applied at a point, either fixed or coinciding with the center of mass. Two special cases are given by the weight (and again $Q \equiv G$) or by the inertial motion (i.e. without active forces) of a rigid body with a fixed point $Q$.

In Poinsot motions the Euler equations are sufficient to determine how the rigid body's orientation evolves. The motion of the center of mass can then be deduced from the first balance equation. For example, if the rigid body is isolated, the center of mass will move under uniform linear motion, while for a free rigid body subject to its weight the center of mass will obey $m\mathbf{a}_G = m\mathbf{g}$, i.e. it will be uniformly accelerated (with straight or parabolic trajectory, depending on the initial conditions).

What makes Poinsot motions interesting is the possibility of gathering a lot of information on the evolution of the angular velocity from the Euler equations.

**Proposition 12.7** (Integral of the angular momentum) *The Poinsot motion of a rigid body preserves the moment of the velocity distribution with respect to the pole $Q$: $\mathbf{K}_Q = \mathbf{I}_Q\boldsymbol{\omega} \equiv constant = \mathbf{K}_0$.*

***Proof*** The property follows trivially from the second balance equation of Dynamics, referred to $Q$:

$$\dot{\mathbf{K}}_Q = \mathbf{M}_Q^{(e)} = \mathbf{0} \qquad \Longrightarrow \qquad \mathbf{K}_Q \equiv \text{constant}.$$

□

**Remark 12.8** That the angular momentum is constant does not imply in general the angular velocity is, albeit $\mathbf{K}_Q = \mathbf{I}_Q\boldsymbol{\omega}$. Poinsot motions where the angular velocity is conserved as well will be discussed shortly (see Sect. 12.2.1). □

We also emphasize that the conservation of $\mathbf{K}_Q$ does not imply the conservation of $\mathbf{K}_A$ with respect to a generic pole $A$. If for instance we consider the case $Q \equiv G$, (6.15) and the assumption that $\dot{\mathbf{K}}_G = \mathbf{0}$ imply

$$\dot{\mathbf{K}}_A = \frac{d}{dt}\big(m\, AG \times \mathbf{v}_G + \mathbf{K}_G\big) = \frac{d}{dt}\big(m\, AG \times \mathbf{v}_G\big).$$

**Proposition 12.9** (Integral of the kinetic energy) *The Poinsot motion of a rigid body preserves the quantity* $T_0 = \frac{1}{2}\mathbf{I}_Q\boldsymbol{\omega} \cdot \boldsymbol{\omega} = \frac{1}{2}\mathbf{K}_Q \cdot \boldsymbol{\omega}$. *If* $Q$ *is a fixed point, this quantity coincides with the kinetic energy of the rigid body, while if* $Q \equiv G$ *it represents the kinetic energy in the motion relative to the center of mass* $T_{\mathrm{rel}}^{(G)}$.

***Proof*** The conservation of $T_0$ follows from the Kinetic energy theorem. If in fact $Q$ is fixed, then $T_0 = T$, and using (7.31) we obtain

$$\dot{T} = \Pi = \underbrace{\mathbf{v}_Q}_{\mathbf{0}} \cdot \mathbf{R} + \boldsymbol{\omega} \cdot \underbrace{\mathbf{M}_Q}_{\mathbf{0}} = 0\,,$$

since the moment of all forces $\mathbf{M}_Q$ coincides with the moment of the external forces, which is zero since the motion is Poinsot. If $Q \equiv G$, consider the frame relative to the center of mass, where $T_0$ coincides with $T^{(G)}$, and then

$$\dot{T}^{(G)} = \Pi^{(G)} = \mathbf{R} \cdot \underbrace{\mathbf{v}_G^{(G)}}_{=\mathbf{0}} + \underbrace{\mathbf{M}_G}_{=\mathbf{0}} \cdot \boldsymbol{\omega}^{(G)} = 0\,,$$

since $\mathbf{v}_G^{(G)}$ is the velocity of the center of mass with respect to itself. □

It is useful to express the two first integrals of Propositions 12.7 and 12.9 in terms of the components of the angular velocity in the basis $\{\mathbf{e}_1, \mathbf{e}_2, \mathbf{e}_3\}$ of principal axes of inertia with respect to $Q$. Call $\{I_1, I_2, I_3\}$ the respective principal moments of inertia, still with respect to $Q$, and set $\boldsymbol{\omega} = \omega_1\mathbf{e}_1 + \omega_2\mathbf{e}_2 + \omega_3\mathbf{e}_3$. Then:

$$I_1^2\omega_1^2 + I_2^2\omega_2^2 + I_3^2\omega_3^2 = K_0^2 \tag{12.10}$$

$$I_1\omega_1^2 + I_2\omega_2^2 + I_3\omega_3^2 = 2T_0 \tag{12.11}$$

These first integrals enable us to prove the component of the angular velocity $\boldsymbol{\omega}$ along the (fixed) direction $\mathbf{K}_0$ is constant:

$$\omega_{K_0} = \frac{\boldsymbol{\omega} \cdot \mathbf{K}_0}{K_0} = \frac{2T_0}{K_0}\,.$$

Yet we cannot say in general that the *angle* determined by $\boldsymbol{\omega}$ with the moment's fixed direction is constant, since the modulus of the angular velocity might not be constant.

The existence of the independent first integrals (12.10)–(12.11) allows to determine analytically the general integral of the Euler equations (12.2), because in a

Poinsot motion the latter become homogeneous differential equations *of order one* in the unknowns $(\omega_1(t), \omega_2(t), \omega_3(t))$. The explicit integration requires, though, special functions in the general case where the principal moments of inertia are all different. Therefore, instead of attempting to determine the general integral, we will focus in the sequel on special classes of solutions, that anyhow possess a precise mechanical meaning.

### *12.2.1 Permanent Rotations*

It has already been stressed that the first integrals coming from a Poinsot motion do not warrant the conservation of the angular velocity. In particular, we will now prove that not all motions of an isolated rigid body have constant angular velocity.

**Definition 12.10** (*Permanent rotation*) We will call *permanent rotation* a Poinsot motion with constant angular velocity.

**Theorem 12.11** (Characterization of permanent rotations) *A Poinsot motion is a permanent rotation if and only if the initial angular velocity is parallel to a principal inertia axis with respect to Q.*

***Proof*** The result follows from the derivative (6.20): $\dot{\mathbf{K}}_Q = \mathbf{I}_Q\dot{\boldsymbol{\omega}} + \boldsymbol{\omega} \times \mathbf{I}_Q\boldsymbol{\omega}$. Since now $\dot{\mathbf{K}}_Q = \mathbf{0}$, we have

$$\dot{\boldsymbol{\omega}} = \mathbf{0} \iff \boldsymbol{\omega} \times \mathbf{I}_Q\boldsymbol{\omega} = \mathbf{0} \iff \mathbf{I}_Q\boldsymbol{\omega} \parallel \boldsymbol{\omega}\,.$$

Asking $\mathbf{I}_Q\boldsymbol{\omega} \parallel \boldsymbol{\omega}$ forces $\boldsymbol{\omega}$ to be an eigenvector of $\mathbf{I}_Q$. But we know (Sect. 5.4.3) the eigenvectors of $\mathbf{I}_Q$ are the principal directions, and the claim follows. □

Therefore in a motion of this type the angular velocity remains constant if and only if it is parallel to a principal inertia axis (Fig. 12.1). We emphasize that the request on

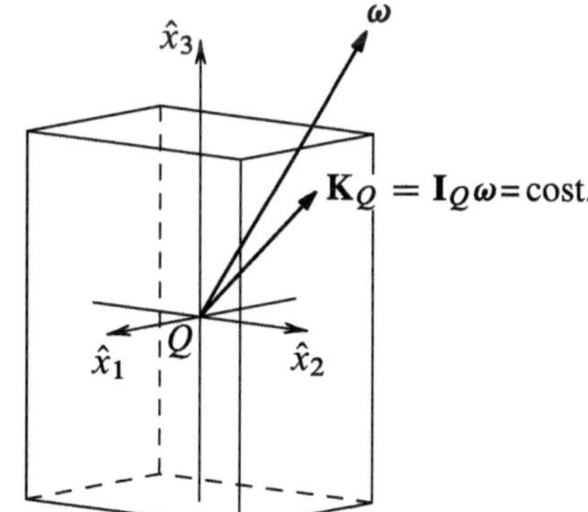


**Fig. 12.1** In a Poinsot motion the angular momentum $\mathbf{K}_Q$ is constant in time, but in general $\boldsymbol{\omega}$ is not. The angular velocity is constant if and only if it is parallel to one of the principal axes of inertia through $Q$ (permanent rotations)

the initial condition that characterizes permanent rotations regards exclusively the direction of $\boldsymbol{\omega}$. Put otherwise, in these rotations the angular velocity can have any modulus (obviously equal to the initial modulus).

**Example 12.12** (Poinsot motions of spherical rigid bodies) Consider a rigid body with the property that the three principal moments of inertia with respect to $Q$ are equal:

$$I_1 = I_2 = I_3 = I\,. \tag{12.12}$$

If $Q \equiv G$, the body could be a sphere or a homogeneous cube, or a round cone of height equal to the base diameter (see Example 5.15). At any rate, asking (12.12) forces the inertia ellipsoid to be a sphere, and the corresponding inertia matrix a multiple of the identity. In particular, all directions are principal since $\mathbf{I}_Q\boldsymbol{\omega} = I\boldsymbol{\omega}$, whatever the direction of $\boldsymbol{\omega}$. Therefore *all Poinsot motions of a body satisfying* (12.12) *are permanent rotations.* □

## 12.2.2 Stability of Permanent Rotations

It is patently very rare that the initial angular velocity is *exactly* aligned with a central principal axis of inertia. In reality we can only expect the initial angular velocity to be *approximately* parallel to one such axis. So it is only legitimate and reasonable to ask the following: if at the initial instant the angular velocity is almost parallel to a principal inertia axis, can we expect it to stay approximately constant during the motion? This is to all effects a stability problem, which we briefly discuss below.

**Definition 12.13** (*Stable permanent rotation*) Let $\tilde{\boldsymbol{\omega}}$ be the permanent rotation of a Poinsot motion of a rigid body (hence such that the initial condition $\boldsymbol{\omega}(t_0) = \tilde{\boldsymbol{\omega}}$ implies $\boldsymbol{\omega}(t) = \tilde{\boldsymbol{\omega}}$ for any $t \geq t_0$). We call it stable if for any $\varepsilon > 0$ we can find a $\delta_\varepsilon > 0$ such that

$$\left|\boldsymbol{\omega}(t_0) - \tilde{\boldsymbol{\omega}}\right| < \delta_\varepsilon \quad \Longrightarrow \quad \left|\boldsymbol{\omega}(t) - \tilde{\boldsymbol{\omega}}\right| < \varepsilon \quad \text{for any } t \geq t_0\,.$$

In other words, a permanent rotation is stable when, given some tolerance $\varepsilon$, we are able to find a precision $\delta_\varepsilon$ (obviously depending on the tolerance) such that all motions starting with $\boldsymbol{\omega}(t_0)$ sufficiently precise (i.e. less than $\delta_\varepsilon$ away from the permanent rotation) possess, from that moment onwards, angular velocity that differs from $\tilde{\boldsymbol{\omega}}$ not more than $\varepsilon$.

**Theorem 12.14** (Stability of permanent rotations) *Let* $\{\mathbf{e}_1, \mathbf{e}_2, \mathbf{e}_3\}$ *be a basis of principal axes of inertia with respect to $Q$ for a rigid body, and $I_1 \leq I_2 \leq I_3$ the corresponding principal moments of inertia. In the particular case $I_1 = I_2 = I_3$ all permanent rotations are stable. In all other cases the permanent rotations with angular velocity parallel to* $\mathbf{e}_2$ *are unstable. Permanent rotations with angular velocity parallel to* $\mathbf{e}_1$ *(or* $\mathbf{e}_3$*) are stable if and only if $I_1 < I_2$ (respectively, $I_2 < I_3$).*

The above theorem says there are two directions around which we have permanent rotations, in the general case where the principal moments of inertia are all different. If two moments coincide, only the axis with different moment is stable, while for a spherical ellipsoid every permanent rotation is stable.

***Proof*** We postpone the full analysis of the degenerate case (spherical and gyroscopic ellipsoids) to Examples 12.15, 12.19. Now we prove the stability of rotations around axes parallel to $\mathbf{e}_3$ when $I_1 \leq I_2 < I_3$.

Set $\boldsymbol{\omega} = \omega_1\mathbf{e}_1 + \omega_2\mathbf{e}_2 + \omega_3\mathbf{e}_3$ and $\mathbf{K}_Q = I_1\omega_1\mathbf{e}_1 + I_2\omega_2\mathbf{e}_2 + I_3\omega_3\mathbf{e}_3$. We expect that $\boldsymbol{\omega} \approx \omega_3\mathbf{e}_3$ and $\mathbf{K}_Q \approx I_3\omega_3\mathbf{e}_3 \approx I_3\boldsymbol{\omega}$. So let us compute the difference

$$\left|\boldsymbol{\omega} - \frac{\mathbf{K}_Q}{I_3}\right| = \frac{1}{I_3}\sqrt{(I_3 - I_1)^2\omega_1^2 + (I_3 - I_2)^2\omega_2^2} \leq \frac{I_3 - I_1}{I_3}\sqrt{\omega_1^2 + \omega_2^2}\,, \tag{12.13}$$

where we have used that $|I_3 - I_2| \leq |I_3 - I_1|$. Note that in (12.13) we are looking at how different $\boldsymbol{\omega}$ is from $\mathbf{K}_Q/I_3$ (constant in a Poinsot motion). If we manage to show the right-hand side remains small at all times, the distance of $\boldsymbol{\omega}(t)$ from a constant will stay small, and we will have proved the stability of the given permanent rotation. Our goal is therefore to show that for any given $\varepsilon > 0$, if the initial values $|\omega_1(t_0)|$, $|\omega_2(t_0)|$ are small enough (less than $\delta_\varepsilon$), then $|\omega_1(t)|$, $|\omega_2(t)|$ will remain smaller than $\varepsilon$. In this way from (12.13) we will deduce that the angular velocity $\boldsymbol{\omega}(t)$, as it stays close to $\mathbf{K}_Q/I_3$, must be *almost* constant.

Consider the first integrals (12.10)-(12.11). Multiply (12.11) by $I_3$ and subtract from it (12.10). Then

$$I_1(I_3 - I_1)\omega_1^2 + I_2(I_3 - I_2)\omega_2^2 = 2I_3T_0 - K_0^2\,. \tag{12.14}$$

The right-hand side $2I_3T_0 - K_0^2$ cannot be negative, as the left side is a sum of non-negative terms. Moreover, this term can only vanish in the trivial case $\omega_1 = \omega_2 = 0$, corresponding to the permanent rotation we are perturbing. Define constants

$$a^2 = \frac{2I_3T_0 - K_0^2}{I_1(I_3 - I_1)} \quad \text{and} \quad b^2 = \frac{2I_3T_0 - K_0^2}{I_2(I_3 - I_2)}\,,$$

so that we can rewrite (12.14) in the simpler way

$$\frac{\omega_1^2}{a^2} + \frac{\omega_2^2}{b^2} = 1\,. \tag{12.15}$$

Now, (12.15) says that on the plane of $\{\mathbf{e}_1, \mathbf{e}_2\}$ the point of Cartesian coordinates $\big(\omega_1(t), \omega_2(t)\big)$ describes an *ellipse* centered at the origin, with semi-axes $a$, $b$ depending on the principal moments of inertia and the initial data of the motion (via the first integrals $T_0$, $K_0$). To ensure $|\omega_1(t)|$, $|\omega_2(t)|$ are less than $\varepsilon$ for any $t$ it suffices to choose the initial data so that $a$, $b$ are smaller than $\varepsilon$. This is possible since $2I_3T_0 - K_0^2$ can be arbitrarily small (by virtue of (12.14)) if we choose $|\omega_1(t_0)|$, $|\omega_2(t_0)|$ small enough. Hence the permanent rotations around the axis of the unit vector $\mathbf{e}_3$ are stable: if initially the angular velocity $\boldsymbol{\omega}$ has sufficiently small orthogonal components to $\mathbf{e}_3$, it will stay approximately constant.

An essentially similar argument proves, when $I_1 < I_2 \leq I_3$, the permanent rotations with axis parallel to $\mathbf{e}_1$ are stable.

The situation changes radically when we study the stability of rotations around axes parallel to $\mathbf{e}_2$. In this case the equivalent equation to (12.15) has a minus sign on the left (caused by the differences of the inertia moments). Therefore it is possible to show that $(\omega_1, \omega_3)$ (the two components of the angular velocity we would like to be small for the rotation about $\mathbf{e}_2$ to be stable) describes a *hyperbola*, not an ellipse. Moreover, a suitable argument we do not replicate in full proves that this point describes a finite portion of the hyperbola, and moves the angular velocity away from the permanent rotation around $\mathbf{e}_2$. Such rotation is thus *unstable*: if the initial angular velocity is not *exactly* collinear with $\mathbf{e}_2$ it will end up differing a lot from the initial value and will vary in a significant manner. □

The above phenomenon can be verified experimentally by tossing a rigid body, for example a right parallelepiped (a box or the like). By prescribing an initial velocity distribution with angular velocity clearly aligned with a principal axis with maximal or minimal moment of inertia, we produce a motion with almost constant angular velocity. In the other case the motion is completely different (seeing is believing …).

**Example 12.15** (Stability of permanent rotations of spherical rigid bodies) We saw in Example 12.12 all Poinsot motions of a rigid body obeying $I_1 = I_2 = I_3 = I$ with respect to the point $Q$ are permanent rotations. Proving that these rotations are stable is trivial. In fact all Poinsot motions are characterized by $\boldsymbol{\omega}(t) \equiv$ constant. Therefore $\boldsymbol{\omega}(t) \equiv \boldsymbol{\omega}(t_0)$, and in particular if $\boldsymbol{\omega}(t_0)$ is close to the permanent rotation $\tilde{\boldsymbol{\omega}}$, so will $\boldsymbol{\omega}(t)$ for any $t \geq t_0$. □

### 12.2.3 *Poinsot Motion of a Gyroscope*

Consider the case where the central ellipsoid has a gyroscopic structure, say $I_{G1} = I_{G2}$, and let us add the assumption the point $Q$ characterizing the Poinsot motion belongs to the spinning axis (see Definition 5.14). Evidently this last hypothesis will hold trivially if $Q \equiv G$. If $Q$ is instead on the spinning axis, parallel to $\mathbf{e}_3$ and through $G$, the inertia ellipsoid at $Q$ will be rotationally symmetric around $\mathbf{e}_3$. In fact, the Huygens-Steiner Theorem (Sect. 5.3) implies $I_1 = I_2$. Assume at last the third central principal moment $I_3$ is different from the other two, so we do not fall back to the spherical case of Example 12.12. The Euler equations (12.2), specialized to the Poinsot motion of a gyroscope, give

$$\dot{\omega}_1 + \frac{I_3 - I_1}{I_1}\omega_3\,\omega_2 = 0\,, \qquad \dot{\omega}_2 - \frac{I_3 - I_1}{I_1}\omega_3\,\omega_1 = 0\,, \qquad \dot{\omega}_3 = 0\,. \tag{12.16}$$

The third equation can be integrated directly $\omega_3 \equiv \omega_{30}$, with $\omega_{30}$ arbitrary but *constant*. Denoting

$$\alpha = \frac{I_3 - I_1}{I_1}, \tag{12.17}$$

we rewrite $(12.16)_{1,2}$ as system of two linear ordinary differential equations of order one and constant coefficients

$$\begin{cases} \dot{\omega}_1 + \alpha\omega_{30}\omega_2 = 0 \\ \dot{\omega}_2 - \alpha\omega_{30}\omega_1 = 0 \end{cases} \implies \begin{cases} \ddot{\omega}_1 + \alpha^2\omega_{30}^2\omega_1 = 0 \\ \ddot{\omega}_2 + \alpha^2\omega_{30}^2\omega_2 = 0 \end{cases}. \tag{12.18}$$

System (12.18) can be integrated elementarily to give the solution to (12.16):

$$\omega_1(t) = \omega_0 \cos(\alpha\omega_{30}t + \phi), \qquad \omega_2(t) = \omega_0 \sin(\alpha\omega_{30}t + \phi), \qquad \omega_3(t) = \omega_{30}, \tag{12.19}$$

where the integration constants $\omega_0$, $\phi$ and $\omega_{30}$ must be determined via the initial conditions, i.e. assigning the velocity distribution at the initial instant.

**Proposition 12.16** (Constants of motion) *The angular velocity of a gyroscope in a Poinsot motion can be written*

$$\boldsymbol{\omega} = \frac{\mathbf{K}_Q}{I_1} - \alpha\omega_{30}\mathbf{e}_3, \tag{12.20}$$

*using notation* (12.17). *Therefore, beside the first integrals* $\mathbf{K}_Q$ *and* $T_0$, *common to all Poinsot motions, gyroscopes possess the following conserved quantities:*

- *the component of* $\boldsymbol{\omega}$ *along the (comoving) axis* $\mathbf{e}_3$;
- *the modulus of the angular velocity* $|\boldsymbol{\omega}|$;
- *the angles between* $\boldsymbol{\omega}$ *and the (fixed) axis of the moment* $\mathbf{K}_Q$, *and between* $\boldsymbol{\omega}$ *and the (comoving) axis* $\mathbf{e}_3$.

***Proof*** Set $\boldsymbol{\omega} = \omega_1\mathbf{e}_1 + \omega_2\mathbf{e}_2 + \omega_3\mathbf{e}_3$, and use simple manipulations of the angular momentum's first integral to find

$$\mathbf{K}_Q = \underline{I_1\omega_1\mathbf{e}_1} + \underline{I_1\omega_2\mathbf{e}_2} + I_3\omega_3\mathbf{e}_3 + (\underline{I_1\omega_3\mathbf{e}_3} - I_1\omega_3\mathbf{e}_3) = \underline{I_1\boldsymbol{\omega}} + (I_3 - I_1)\omega_3\mathbf{e}_3 .$$

Recalling Definition (12.17), this implies (12.20).

The component $\omega_3 = \boldsymbol{\omega} \cdot \mathbf{e}_3$ is constant, as is clear from $(12.16)_3$.

To prove the modulus of $\boldsymbol{\omega}$ is constant we use (12.11):

$$\omega_1^2 + \omega_2^2 = \frac{2T_0 - I_3\omega_3^2}{I_1} \equiv \text{cost.} \implies |\boldsymbol{\omega}|^2 = \left(\omega_1^2 + \omega_2^2\right) + \omega_3^2 \equiv \text{cost.}$$

At this point we have constant components of $\boldsymbol{\omega}$ along the gyroscopic axis $\mathbf{e}_3$ and constant moment $\mathbf{K}_Q$ (true in all Poinsot motions). Hence the angles formed by the angular velocity with those directions is constant. □

**Proposition 12.17** (Regular precessions) *Every Poinsot motion of a gyroscope, referred to the fixed point $Q$ or in the barycentric frame (if $Q \equiv G$), is a regular precessions (see Sect. 2.5.2) with precession axis parallel to $\mathbf{K}_Q$ and axis of intrinsic rotation coinciding with the spinning axis.*

***Proof*** It is enough to compare (12.20) to (2.33), which characterizes precessional motions. The precession is regular because the moduli of the components $K_Q/I_1$ and $\alpha\omega_{30}$ are both constant. □

**Example 12.18** (Earth's precession) The motions determined in this section are of particular interest since the Earth can be roughly considered a rigid body. It is slightly squashed at the poles, so its central ellipsoid of inertia is not spherical, but has $I_1 = I_2 \approx 0.9966 I_3$, whence $\alpha \approx 3.4 \times 10^{-3}$. Moreover, if we suppose the gravitational attractions acting on it can be replaced by their resultant at the center of mass, the motion can be approximated by the Poinsot motion of a gyroscope. It is a regular precession where the term $\lambda \mathbf{e}_3$ of (2.33) represents the Earth's rotation around its axis (lasting 24 hours), so with

$$\lambda = \frac{2\pi}{24 \cdot 60 \cdot 60} \approx 7.2722 \times 10^{-5} \mathrm{s}^{-1} .$$

The term $\mu \mathbf{i}_3$ in (2.33) represents instead the precession the Earth and its axis undergoes during a *Platonic year* (26,000 years)

$$\mu = \frac{\lambda}{365.25 \times 26,000} \approx 7.6578 \times 10^{-12} \mathrm{s}^{-1} .$$

As $\mu$ is so much smaller than $\lambda$, it is easy to see how the Earth's motion with respect to a barycentric frame (with origin always at the center of the Earth) is treated as a uniform rotational motion having angular velocity equal to the daily rotation. However, if we want to perceive certain special phenomena, such as the *precession of the equinoxes* (which every 13,000 years swaps summer with winter, spring with fall), we cannot neglect the discrepancy between the precessional and intrinsic rotational velocities. □

**Example 12.19** (Stability of permanent rotations of a gyroscope) Consider a gyroscope such that $I_1 = I_2 < I_3$ (the case $I_1 < I_2 = I_3$ is dealt with similarly, while the spherical central ellipsoid was discussed in Examples 12.12 and 12.15). The gyroscope admits permanent rotations around $\mathbf{e}_3$, or around any axis $\tilde{\mathbf{e}}$ on the plane $\{\mathbf{e}_1, \mathbf{e}_2\}$ (see Property i). The proof of Theorem 12.14 on stable permanent rotations shows the rotations around the gyroscopic axis are stable. Consider the permanent rotation $\tilde{\boldsymbol{\omega}} = \tilde{\omega}\mathbf{e}_1$. The result we will find will also apply, by symmetry, to any permanent rotation $\tilde{\boldsymbol{\omega}} = \tilde{\omega}\tilde{\mathbf{e}}$. Formulas (12.19) enable to determine the exact evolution of $\boldsymbol{\omega}$. Consider in particular the motion starting with initial condition $\boldsymbol{\omega}(0) = \tilde{\omega}\mathbf{e}_1 + \delta\mathbf{e}_3$. When $\delta$ is small, this initial condition is arbitrarily close to the permanent rotation

$\tilde{\omega}$. Examining (12.19), it is easy to convince ourselves the integration constants corresponding to the given initial condition are $\omega_0 = \tilde{\omega}$, $\phi = 0$, $\omega_{30} = \delta$. Hence we have the motion

$$\omega_1(t) = \tilde{\omega}\cos(\alpha\delta t), \qquad \omega_2(t) = \tilde{\omega}\sin(\alpha\delta t), \qquad \omega_3(t) = \delta\,.$$

In particular, after a time $t = \pi/(2\alpha\delta)$, the angular velocity will completely change direction: $\boldsymbol{\omega}\big(\pi/(2\alpha\delta)\big) = \tilde{\omega}\mathbf{e}_2 + \delta\mathbf{e}_3$. Rotations around axes on the plane $\{\mathbf{e}_1, \mathbf{e}_2\}$ are therefore unstable. □

**Example 12.20** Let us study the motion of a cylindrical homogeneous rigid body of mass $m$, represented in Fig. 12.2, tossed under the action of its weight alone. Suppose at the initial instant the cylinder's axis is tilted by $\beta$ with respect to the vertical direction, and the initial angular velocity is vertical: $\boldsymbol{\omega}(0) = \omega_0\mathbf{u}$. Choose a fixed frame $\{\mathbf{i}_1, \mathbf{i}_2, \mathbf{i}_3\}$ with origin $O$ and vertical $z$-axis ($\mathbf{i}_3 = \mathbf{u}$). Suppose the initial position of the center of mass is $OG(0) = \delta\mathbf{i}_2 + \delta\tan\beta\mathbf{i}_3$ (Fig. 12.2) and the initial velocity is $\mathbf{v}_G(0) = v_0\mathbf{i}_2$.

The cylinder is a gyroscope in Poinsot motion, since we can replace the weight with its resultant applied at the body's center of gravity. The balance equations can be written as

$$m\mathbf{a}_G = m\mathbf{g}\,, \qquad \text{and} \qquad \dot{\mathbf{K}}_G = \mathbf{0}\,.$$

The first balance equation allows to determine the motion of the center of mass, which is that of a particle with initial velocity $\mathbf{v}_G(0)$ moving under its weight, and that describes a concave parabola on the plane $x = 0$.

The second balance equation allows to determine the motion around the center of mass. The integral of the angular momentum is $\mathbf{K}_G = \mathbf{I}_G\boldsymbol{\omega}_0$, where $\mathbf{I}_G$ denotes the barycentric inertia matrix. Excluding the trivial case of a spherical central ellipsoid (we assume $I_1 = I_2 \neq I_3$), the moment $\mathbf{K}_G$ is parallel to $\boldsymbol{\omega}_0$ (vertical) if and only if the initial angular velocity is parallel to a principal axis, i.e. if $\beta = 0$ or $\beta = \pi/2$. To determine the direction of $\mathbf{K}_G$ we recall (see (12.16)) that among the first integrals of a gyroscope's Poinsot motion there is the component of $\boldsymbol{\omega}$ along the spinning axis:

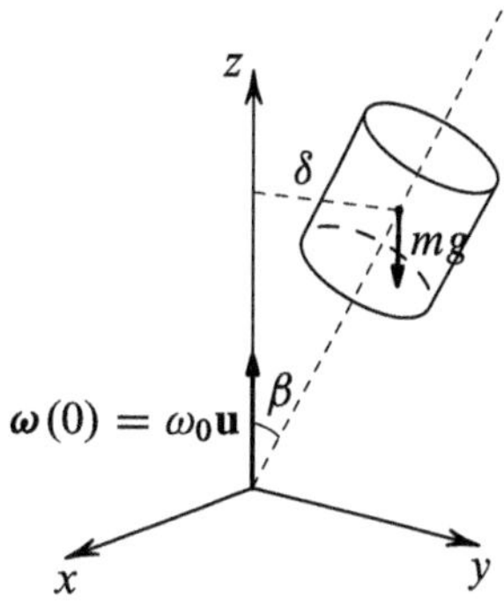


**Fig. 12.2** Initial configuration of a cylinder only subject to its own weight

$$\omega_3 = \boldsymbol{\omega}(0) \cdot \mathbf{e}_3 = \omega_0 \mathbf{u} \cdot \mathbf{e}_3 = \omega_0 \cos\beta \,.$$

Call $\mathbf{e}_{30} = \sin\beta \mathbf{i}_2 + \cos\beta \mathbf{i}_3$ the unit vector parallel to the spinning axis at the initial instant. Then (12.20) gives

$$\begin{aligned} \mathbf{K}_G &= I_1 \boldsymbol{\omega} + (I_3 - I_1)\omega_3 \mathbf{e}_3 = I_1 \omega_0 \mathbf{u} + (I_3 - I_1)\omega_0 \cos\beta \mathbf{e}_{30} \\ &= (I_3 - I_1)\omega_0 \sin\beta \cos\beta \,\mathbf{i}_2 + \left(I_3 \cos^2\beta + I_1 \sin^2\beta\right)\omega_0 \,\mathbf{i}_3 \,, \end{aligned}$$

from which $K_G = \left(I_3^2 \cos^2\beta + I_1^2 \sin^2\beta\right)^{1/2} \omega_0$. Now set $\mathbf{p} = \mathbf{K}_G / K_G$, so the motion around the center of mass is a *regular precession*, with precession axis $\mathbf{p}$ and axis of intrinsic rotation coinciding with the cylinder's axis. □

## 12.3 Constrained Rigid Body

The ensuing sections are dedicated to the detailed study of the dynamics of variously constrained rigid bodies. In most examples we consider *ideal* constraints, and refer to the characterization of Sect. 8.7 which will hold in Dynamics as it did in Statics. We emphasize the procedure used for the analysis of these simple systems: the procedure explained in the next sections is the *concrete* and *rational* method to be used when examining any mechanical system, even if complex. Such a procedure can be broken down into the following steps.

- First of all we must determine the number of degrees of freedom each constraint bestows on the body, and hence the number of *kinematic unknowns* (generalized coordinates) needed to find the possible motions (or equilibria) of the system.
- Using the dynamic characterization of ideal constraint (Definition 8.12), step two consists in identifying the *dynamical unknowns* (constraint reactions) introduced by the constraint.
- Using the balance equations of Dynamics write the system of second-order scalar differential equations in *normal form*, so that it is *complete* and *free* in terms of the unknowns. That involves suitable combinations of the scalar components of the balance equations, and should lead to enough consequences that do not contain the constraint reactions, so that they suffice to determine uniquely the solution to the initial-value problem associated with the unknowns.
- Finally, the remaining consequences of the balance equations, i.e. the scalar components that have not been used to detect the unknowns, allow to find the dynamical unknowns. It is fundamental that these consequences are always sufficient to determine the characteristic vectors (resultant and resultant moment) of the constraint reactions.

In this way one determines all compatible motions with force and constraint for the rigid body in question. Fixing position and velocity distribution at a given instant will allow to determine the particular motion given by the specific initial conditions chosen.

## 12.4 Rigid Body with a Fixed Point

To begin with we examine the motion of a rigid body constrained to a fixed pin. Introduce a fixed frame in the reference space and a comoving frame, both with origin at the pinned point $Q$. The rigid body's configuration can be described by the Euler angles. The unknowns are the functions $\{\theta(t), \psi(t), \phi(t)\}$. In general these parameters can be used to describe the kinematics of the rigid body only locally.

The constraint reactions' analysis in Sect. 9.6.1 shows that the pin is ideal if and only if the resultant moment of the constraint with respect to the center $Q$ is zero

$$\mathbf{M}_Q^{(r)} = \mathbf{0}\,. \tag{12.21}$$

The constraint force is therefore equivalent to the only force $\mathbf{R}^{(r)} = \mathbf{\Phi}_Q$, applied at $Q$. Hence the constraint introduces three dynamical unknowns, the scalar components of $\mathbf{\Phi}_Q$. Choosing $Q$ as pole for the second balance equation of Dynamics, and by (12.21), we obtain a free motion equation. The Euler equations, referred to the fixed point, are then necessary and sufficient to characterize the motion of a rigid body with a fixed point. The first balance equation allows to determine the components of the unknown constraint reaction $\mathbf{\Phi}_Q$.

### 12.4.1 *Motion of a Rigid Body Under Its Weight*

Consider a rigid body with a fixed point $Q$ where the only active force is the weight (Fig. 12.3). The second balance equation of Dynamics gives $\dot{\mathbf{K}}_Q = QG \times m\mathbf{g}$. Let $\{\mathbf{e}_1, \mathbf{e}_2, \mathbf{e}_3\}$ be a triple of comoving axes, principal at $Q$, and $\{I_1, I_2, I_3\}$ the corresponding principal moments of inertia. Call $\{\omega_1, \omega_2, \omega_3\}$ the components of the angular velocity in the chosen basis, $\{x_1, x_2, x_3\}$ the coordinates (fixed in the comoving system) of the center of gravity, and $\{u_1, u_2, u_3\}$ the components (variable in the chosen frame) of the vertical unit vector $\mathbf{u} = \mathbf{i}_3$ such that $\mathbf{g} = -g\mathbf{u}$. The Euler equations (12.2) give

$$\begin{cases} I_1\dot{\omega}_1 - (I_2 - I_3)\omega_2\omega_3 = mg(u_2x_3 - u_3x_2) \\ I_2\dot{\omega}_2 - (I_3 - I_1)\omega_3\omega_1 = mg(u_3x_1 - u_1x_3) \\ I_3\dot{\omega}_3 - (I_1 - I_2)\omega_1\omega_2 = mg(u_1x_2 - u_2x_1) \end{cases} \tag{12.22}$$

To define the Euler angles we pick the fixed basis with vertical third axis, so that

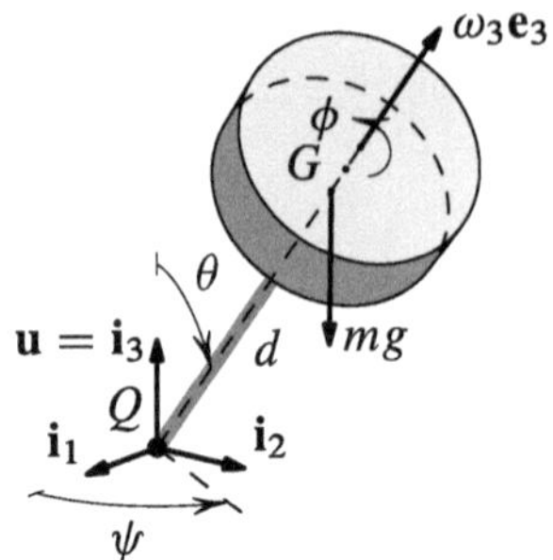


**Fig. 12.3** A rigid body constrained at a fixed point $Q$

$$u_1 = \sin\phi \sin\theta, \quad u_2 = \cos\phi \sin\theta\,, \quad u_3 = \cos\theta\,, \tag{12.23}$$

where $\phi$ is the intrinsic rotation angle. Clearly to have a second-order system in normal form we must not only substitute (12.23) into (12.22), but also the components of the angular velocity in terms of the Euler angles and their derivatives (see (3.14)). Then system (12.22) is no longer algebraic, indeed it contains trigonometric functions in the unknowns. But this can be overcome using formula (3.1) to express that the vertical direction stays constant, i.e.

$$\underbrace{\dot{\mathbf{u}}}_{\text{fixed system}} = \underbrace{\mathbf{u}'}_{\text{comoving system}} + \boldsymbol{\omega} \times \mathbf{u} = \mathbf{0}\,.$$

Projecting the latter onto the comoving basis we obtain

$$\dot{u}_1 + u_3\omega_2 - u_2\omega_3 = 0\,, \quad \dot{u}_2 + u_1\omega_3 - u_3\omega_1 = 0\,, \quad \dot{u}_3 + u_2\omega_1 - u_1\omega_2 = 0\,. \tag{12.24}$$

Now we can solve the first-order differential system in the six unknowns $u_1$, $u_2$, $u_3$, $\omega_1$, $\omega_2$, $\omega_3$ made by the six equations (12.22)–(12.24). It is in normal form, so the initial-value problem is well defined.

Once solved, we can recover the Euler angles from the usual components of the angular velocity depending on them, using exclusively integrations. Clearly the general integral is much more complicated, except for in some special cases. For example if $Q \equiv G$ the resultant moment of the active force is zero. The possible motions are just Poinsot motions and the general solution can be reduced to the cases seen in Sect. 12.2.

**Example 12.21** (Uniform rotations of a gyroscope under its weight) Consider a gyroscope of mass $m$, constrained by a spherical pin $Q$ on the spinning axis at distance $d$ from the center of gravity, that moves under its weight. We wish to determine whether the rigid body can rotate uniformly around the vertical line through $Q$.

Consider two frames with common origin at the fixed point $Q$. Suppose the first frame is fixed, with axis $\mathbf{i}_3$ coinciding with the vertical, while the second frame comoves with the gyroscope and has axis $\mathbf{e}_3$ parallel to the spinning axis. Hence $QG = d\mathbf{e}_3$. We complete the comoving basis (principal at $Q$) with a horizontal axis

$\mathbf{e}_1$ and the corresponding $\mathbf{e}_2$. This implies $\mathbf{e}_1$ is the line of nodes of the transformation between frames (Fig. 2.2), and therefore the intrinsic rotation angle $\phi$ is zero. Let at last $\{I_1 = I_2; I_3\}$ be the principal moments of inertia at $Q$.

In the required uniform rotation the nutation angle $\theta$, between the spinning axis and the vertical direction, is constant, so the rotation will be defined by the precession angle only, or its complement $\psi$, determined by the horizontal projection of $\mathbf{e}_3$ and $\mathbf{i}_1$ (Fig. 12.3). Because of the choice made for $\mathbf{e}_1$ we can decompose the vertical unit vector $\mathbf{u} = \mathbf{i}_3$ along the comoving basis as $\mathbf{u} = \sin\theta\,\mathbf{e}_2 + \cos\theta\,\mathbf{e}_3$. So,

$$\boldsymbol{\omega} = \dot{\psi}\,\mathbf{u} = \dot{\psi}\sin\theta\,\mathbf{e}_2 + \dot{\psi}\cos\theta\,\mathbf{e}_3$$

where, for uniform rotations, $\dot{\psi}$ is constant. The only non-trivial Euler equation is

$$(I_3 - I_2)\dot{\psi}^2 \sin\theta\cos\theta = mgd\sin\theta\,. \tag{12.25}$$

A careful analysis of the solutions to (12.25) gives the following result.

- If $d = 0$ ($Q \equiv G$) we recover known results (see uniform rotations in Poinsot motions). In the sequel we assume $d > 0$.
- Equation (12.25) always admits the solutions $\theta = 0$ and $\theta = \pi$. In either case the center of gravity lies on the vertical through $Q$ (below or above the fixed point respectively), and can rotate uniformly with arbitrary angular velocity $\dot{\psi}$.
- If $I_2 = I_3$ (spherical inertia ellipsoid with respect to $Q$) the only solutions to (12.25) are the previous ones.
- If $I_2 > I_3$ (elongated gyroscope, stretched along the spinning axis), equation (12.25) admits solutions for any $\theta \in (\pi/2, \pi)$ (studying the sign shows that the cosine must be negative). Given the value of $\theta$, the angular velocity of the uniform rotation is
$$\dot{\psi}^2 = \frac{mgd}{(I_2 - I_3)\cos\theta}\,. \tag{12.26}$$
The condition $\theta \in (0, \pi/2)$ implies the center of gravity's height is *less* than that of $Q$, as $QG \cdot \mathbf{i}_3 = d\cos\theta$.
- Similar arguments lead to conclude that for $I_2 < I_3$ (the gyroscope is squashed along its axis), (12.25) admits solutions for any $\theta \in (0, \pi/2)$. Once $\theta$ has been found, the angular velocity of the uniform rotation satisfies (12.26). The center of gravity is *above* $Q$.

The first balance equation eventually allows to find the constraint reaction $\mathbf{\Phi}_Q = m\mathbf{a}_G - m\mathbf{g}$. As (in the fixed frame)

$$QG = d\sin\theta(\cos\psi\mathbf{i}_1 + \sin\psi\mathbf{i}_2) + \cos\theta\mathbf{i}_3\,,$$

we obtain $\mathbf{a}_G = -d\dot{\psi}^2\sin\theta(\cos\psi\mathbf{i}_1 + \sin\psi\mathbf{i}_2)$ and so

$$\mathbf{\Phi}_Q = -md\dot{\psi}^2\sin\theta(\cos\psi\mathbf{i}_1 + \sin\psi\mathbf{i}_2) - mg\,\mathbf{i}_3\,.$$

The constraint reaction therefore is periodic, because it depends on the inertia forces. □

**Example 12.22** (Gyroscopic effect) Let us analyze further how a gyroscope, suspended at a point of the spinning axis, moves under its weight. Now we wish to study the motion with initial condition $\boldsymbol{\omega}(0) = \omega_{30}\mathbf{e}_3$, i.e. when the angular velocity is initially parallel to the device's axis.

Under the previous assumptions $(12.22)_3$ gives $I_3\dot{\omega}_3 = 0$, so the component of $\boldsymbol{\omega}$ along the spinning axis remains constant: $\omega_3 \equiv \omega_{30}$. Moreover, the weight is a conservative force, so the system admits energy integral (11.14). Calling $z_G = d\cos\theta$ the center of gravity's height (along $\mathbf{i}_3$) we obtain

$$T - U = \frac{1}{2}\left(I_1\omega_1^2 + I_1\omega_2^2 + I_3\omega_3^2\right) + mgd\cos\theta \equiv \text{cost.}$$

$$\Rightarrow \omega_1^2 + \omega_2^2 = \frac{2mgd}{I_1}(\cos\theta_0 - \cos\theta).$$

But $|\cos\theta_0 - \cos\theta| \leq 2$ so we find the estimate

$$I_1^2\omega_i^2 \leq 4mgdI_1 \qquad (i = 1, 2). \tag{12.27}$$

Consider now the angular momentum with respect to the fixed point: $\mathbf{K}_Q = I_1\omega_1\mathbf{e}_1 + I_1\omega_2\mathbf{e}_2 + I_3\omega_3\mathbf{e}_3$. The first two components of $\mathbf{K}_Q$ are bounded by (12.27), while the third is constant. Hence if at the initial instant the third component of $\boldsymbol{\omega}$ obeys

$$I_3^2\omega_{30}^2\mathbf{g}4mgdI_1 \iff \omega_{30}^2\mathbf{g}\frac{4mgdI_1}{I_3^2},$$

*during the entire motion* the component $I_3\omega_3$ will be dominant over the others. More precisely, the components' ratio satisfies

$$\frac{I_1^2\omega_i^2(t)}{I_3^2\omega_3^2} \leq \frac{4mgdI_1}{I_3^2\omega_{30}^2} \quad \text{for } i = 1, 2, \quad \forall t \geq 0,$$

and so it can be made as small as we want by choosing the third component's initial value suitably large.

We can therefore ignore the other two components of the angular momentum (by deciding the approximation error) and suppose

$$\mathbf{K}_Q(t) \approx I_3\omega_{30}\mathbf{e}_3(t). \tag{12.28}$$

The above is sometimes called *gyroscopic approximation*. Some properties typical of the motion of gyroscopes follow from it.

- Inserting (12.28) into the second balance law for the pole $Q$ we obtain

$$\mathbf{M}_Q = \dot{\mathbf{K}}_Q \approx I_3\omega_{30}\dot{\mathbf{e}}_3 \quad \Rightarrow \quad \dot{\mathbf{e}}_3 \approx \frac{\mathbf{M}_Q}{I_3\omega_{30}}. \tag{12.29}$$

This shows that, for the same applied moment, the spinning axis' variation is proportional to the component's inverse. This effect is known as *tenacity* of the spinning axis, and is the basis for the use of gyroscopes as motion stabilizers.

- Formula (12.29) also shows the variation of the spinning axis is not parallel to the external force, but to its moment. Hence the weight, rather than *laying down* the axis (as one would be tempted to think), makes it precess around the vertical. Moreover, the spinning axis stays constant only when it is parallel to the applied external force, and explains the phenomenon whereby the axis *tends to be parallel*.

□

## 12.5 Rigid Body with a Fixed Axis

Consider a rigid body constrained by a cylindrical pin, which exclusively allows rotation around its axis. We saw in Sect. 9.6.2 how the two-fixed-pin constraint produces a statically indeterminate system, but the indeterminacy is resolved once we replace one pin with a smooth ring, whose constraint reaction is thus orthogonal to the fixed axis (Fig. 9.12).

In the particular case of a planar rigid body, only moving on its plane, the present constraint is necessarily realized by an orthogonal axis to the plane of motion, and is equivalent to fixing with a pin the position of a point of the body.

To study the dynamics of a body thus constrained it is convenient to introduce in space two half-planes $\pi_0$, $\pi$, both bounded by the fixed axis, but one fixed and one comoving with the rigid body. The angle between the half-planes is sufficient to determine the position in time of any element of the rigid body (see Sect. 2.2.2). In fact, referring to the Euler angles, the fixed axis can be taken to be $\mathbf{i}_3$ and $\mathbf{e}_3$ (zero nutation angle). In this case the half-planes $\pi_0$, $\pi$ are identified with those containing $\{\mathbf{i}_1, \mathbf{i}_3\}$ and $\{\mathbf{e}_1, \mathbf{e}_3\}$ respectively, so the only angle needed for the orientation is the precession angle $\psi$ (Fig. 2.3), where $\{\tilde{\mathbf{e}}_1, \tilde{\mathbf{e}}_3\}$ coincides with $\{\mathbf{e}_1, \mathbf{e}_3\}$). All in all the motions of a rigid body with a fixed axis are rotational around the axis, and given by the unique generalized coordinate $\psi = \psi(t)$ and the angular velocity $\boldsymbol{\omega} = \dot{\psi}\mathbf{i}_3$.

Analyzing the constraint force in the ideal case proceeds exactly as in the static case, since the characterization of ideal constraints in Sect. 8.7 holds either way. Introducing a reference frame with origin $Q$ at a point on the fixed axis $\mathbf{i}_3$, it follows that $\delta Q = \mathbf{0}$, while the constraint gives us virtual infinitesimal rotation $\boldsymbol{\varepsilon}' = \delta\psi\, \mathbf{i}_3$. Using (7.39) we can then write the virtual work of the constraint force as

$$\delta L^{(\mathrm{r})} = \mathbf{R}^{(\mathrm{r})} \cdot \delta Q + \mathbf{M}_Q^{(\mathrm{r})} \cdot \boldsymbol{\varepsilon}' = (\mathbf{M}_Q^{(\mathrm{r})} \cdot \mathbf{i}_3)\delta\psi\,.$$

Asking ideal and bilateral constraints, $\delta L^{(\mathrm{r})} = 0$, forces

$$\mathbf{M}_Q^{(\mathrm{r})} \cdot \mathbf{i}_3 = 0\,, \tag{12.30}$$

i.e. that the axial component of the moment of the constraint force is zero. This scalar relation alone characterizes the constraint reactions of an ideal cylindrical pin. In general, therefore, there are five dynamical unknowns introduced with the constraint: three components of the resultant of the constraint force and two component of the constraint force's momentum orthogonal to the fixed axis.

We reached the same conclusion in Statics, Sect. 9.6.2, but through another argument. We observed that realizing the constraint by two spherical pins introduced two unknowns constraint reactions $\mathbf{\Phi}_1$, $\mathbf{\Phi}_2$, applied at points $Q_1$, $Q_2$ on the fixed axis. We also said the balance equations enable to determine only five of the six unknown components of the reactions.

**Remark 12.23** In a planar case, the component of the constraint momentum that vanishes by (12.30) is the only one that is not identically zero due to the system being planar. Therefore the total momentum of the constraint reaction is necessarily zero. □

### 12.5.1 Free Motion Equation

Now we would like to pin down which consequences of the balance equations enable us to determine the kinematic unknown. Consider the second balance equation with respect to fixed pole $Q$

$$\frac{d\mathbf{K}_Q}{dt} = \mathbf{M}_Q^{(\mathrm{a})} + \mathbf{M}_Q^{(\mathrm{r})} .$$

The only information we have about the constraint reactions is contained in (12.30). In order to use it we can project the second balance equation along the fixed axis. This produces the free equation

$$\frac{d\mathbf{K}_Q}{dt} \cdot \mathbf{i}_3 = \mathbf{M}_Q^{(\mathrm{a})} \cdot \mathbf{i}_3 .$$

The unit vector $\mathbf{i}_3 = \mathbf{e}_3$ is fixed both in space and on the body, so

$$\frac{d\mathbf{K}_Q}{dt} \cdot \mathbf{i}_3 = \frac{d}{dt}\left(\mathbf{K}_Q \cdot \mathbf{i}_3\right) = \frac{d}{dt}\left(I_3\dot{\psi}\right) = I_3\ddot{\psi} ,$$

where we have used (6.12) to express the axial component of $\mathbf{K}_Q$. The free motion equation then is

$$I_3\ddot{\psi} = M_{Q3}^{(\mathrm{a})}\left(\psi, \dot{\psi}, t\right) . \tag{12.31}$$

It is worth emphasizing that for a rigid body with fixed axis the notion of *equivalent force* can be generalized. In fact the possible motions of a rigid body constrained by a cylindrical pin do not change when we replace a given force with any other force with the same total torque with respect to the fixed axis.

**Remark 12.24** (Positional active forces) The free motion equation (12.31) admits a first integral that is formally similar to the energy, under the assumption that the active forces are positional. Suppose in fact the axial torque is a function of the position only: $M^{(a)}_{Q3}(\psi)$. If we introduce a primitive $U(\psi)$ such that $M^{(a)}_{Q3}(\psi) = U'(\psi)$, and multiply both sides of (12.31) by $\dot{\psi}$, we obtain

$$I_3\dot{\psi}\ddot{\psi} = U'(\psi)\dot{\psi} \implies \frac{d}{dt}\left(\frac{1}{2}I_3\dot{\psi}^2 - U(\psi)\right) = 0$$
$$\implies \frac{1}{2}I_3\dot{\psi}^2 - U(\psi) = T - U \equiv \text{constant} = E. \qquad (12.32)$$

The first integral thus found is perfectly similar to the one studied in detail in Sect. 10.4. In fact we have already observed that such a first integral arises every time a system with ideal constraints has one degree of freedom and the active forces are positional (possibly not conservative as well). □

**Example 12.25** (Compound pendulum [1/3]) As application of the above results it is interesting to consider the case where the external force reduces to the weight. Suppose the fixed axis $\mathbf{i}_3 = \mathbf{e}_3$ is horizontal, and pick $\mathbf{i}_1$ vertical and oriented downwards. It is also convenient to choose the comoving half-plane $\pi$ so to contain the body's center of gravity, and the pole $Q$ coinciding with the center of gravity's projection on the fixed axis, i.e. $QG = \delta\mathbf{e}_1$. In this way $\psi$ is the angle between $OG$ and the vertical $\mathbf{i}_1$, and so $\mathbf{e}_1 = \cos\psi\mathbf{i}_1 + \sin\psi\mathbf{i}_2$. We then have

$$\mathbf{M}^{(a)}_Q = QG \times m\mathbf{g} = \delta\mathbf{e}_1 \times mg\mathbf{i}_1 = -mg\delta\sin\psi\,\mathbf{i}_3\,.$$

The free motion equation (12.31) then gives $I_3\ddot{\psi} = -mg\delta\sin\psi$, i.e.

$$\ddot{\psi} + \frac{g\delta}{\rho_3^2}\sin\psi = 0\,, \qquad (12.33)$$

where $\rho_3$ is the gyration radius of the rigid body with respect to the fixed axis (see (5.10)). Equation (12.33) is equivalent to the pendulum's equation of motion, and for this reason it is called *equation of the compound pendulum*. In case the center of gravity lies on the axis of rotation we have $\delta = 0$ and (12.33) simplifies to $\ddot{\psi} = 0$. Then the possible motions are exclusively uniform rotations.

Coherently with (12.32), also (12.33) admits first integral

$$\frac{1}{2}I_3\dot{\psi}^2 - mg\delta\cos\psi \equiv \text{constant} = E\,.$$

### 12.5.2 Computing the Reactive Forces

The resultant of the constraint reactions is found using the first balance equation of Dynamics: $\mathbf{R}^{(r)} = m\mathbf{a}_G - \mathbf{R}^{(a)}$. Moreover, the projections of the second balance equation along the axes orthogonal to the fixed axis give the non-zero components of the resultant moment of the constraint force: $\mathbf{M}_Q^{(r)} = \dot{\mathbf{K}}_Q - \mathbf{M}_Q^{(a)}$.

Let us call again $\delta$ the distance of the center of gravity $G$ to the fixed axis, and choose the pole $Q$ and the axis $\mathbf{e}_1$ as in the previous example of the compound pendulum. Then $QG = \delta\mathbf{e}_1$ and $\mathbf{v}_G = \dot{\psi}\mathbf{i}_3 \times QG = \delta\dot{\psi}\mathbf{e}_2$, while $\mathbf{K}_Q = (I_{13}\mathbf{e}_1 + I_{23}\mathbf{e}_2 + I_3\mathbf{e}_3)\dot{\psi}$. Differentiating in time, and recalling that $\dot{\mathbf{e}}_i = \boldsymbol{\omega} \times \mathbf{e}_i$, we obtain

$$\begin{aligned}
\mathbf{a}_G &= \delta\ddot{\psi}\mathbf{e}_2 + \delta\dot{\psi}(\dot{\psi}\mathbf{e}_3 \times \mathbf{e}_2) = -\delta\dot{\psi}^2\mathbf{e}_1 + \delta\ddot{\psi}\mathbf{e}_2 \qquad \text{and} \\
\dot{\mathbf{K}}_Q &= (I_{13}\mathbf{e}_1 + I_{23}\mathbf{e}_2 + I_3\mathbf{e}_3)\ddot{\psi} + (I_{13}\dot{\psi}\mathbf{i}_3 \times \mathbf{e}_1 + I_{23}\dot{\psi}\mathbf{i}_3 \times \mathbf{e}_2)\dot{\psi} \\
&= (I_{13}\ddot{\psi} - I_{23}\dot{\psi}^2)\mathbf{e}_1 + (I_{23}\ddot{\psi} + I_{13}\dot{\psi}^2)\mathbf{e}_2 + I_3\ddot{\psi}\mathbf{e}_3\,.
\end{aligned} \tag{12.34}$$

The modulus of the center of gravity's acceleration then is $\delta\sqrt{\ddot{\psi}^2 + \dot{\psi}^4}$. Similarly, the component orthogonal to $\mathbf{e}_3 = \mathbf{i}_3$ of $\dot{\mathbf{K}}_Q$ has modulus

$$\sqrt{(I_{13}^2 + I_{23}^2)(\ddot{\psi}^2 + \dot{\psi}^4)}\,.$$

**Example 12.26** (Compound pendulum [2/3]) Let us return to the case where the only active force is the weight, orthogonal to the fixed axis $\mathbf{i}_3$. Keeping the earlier notation, the balance equation of Dynamics give us the characteristic vectors of the constraint force. The resultant equals

$$\begin{aligned}
\mathbf{R}^{(r)} &= m\mathbf{a}_G - \mathbf{R}^{(a)} \\
&= \big(-mg - m\delta(\ddot{\psi}\sin\psi + \dot{\psi}^2\cos\psi)\big)\mathbf{i}_1 + m\delta(\ddot{\psi}\cos\psi - \dot{\psi}^2\sin\psi)\mathbf{i}_2\,,
\end{aligned} \tag{12.35}$$

where we used the expressions $\mathbf{e}_1 = \cos\psi\,\mathbf{i}_1 + \sin\psi\,\mathbf{i}_2$, $\mathbf{e}_2 = -\sin\psi\,\mathbf{i}_1 + \cos\psi\,\mathbf{i}_2$. Similarly,

$$\begin{aligned}
\mathbf{M}_Q^{(r)} = \dot{\mathbf{K}}_Q - \mathbf{M}_Q^{(a)} &= (I_{13}\ddot{\psi} - I_{23}\dot{\psi}^2)\mathbf{e}_1 + (I_{23}\ddot{\psi} + I_{13}\dot{\psi}^2)\mathbf{e}_2 \\
&= \big((I_{13}\ddot{\psi} - I_{23}\dot{\psi}^2)\cos\psi - (I_{23}\ddot{\psi} + I_{13}\dot{\psi}^2)\sin\psi\big)\mathbf{i}_1 \\
&+ \big((I_{13}\ddot{\psi} - I_{23}\dot{\psi}^2)\sin\psi - (I_{23}\ddot{\psi} + I_{13}\dot{\psi}^2)\cos\psi\big)\mathbf{i}_2\,.
\end{aligned} \tag{12.36}$$

□

### 12.5.3 Static and Dynamical Balancing

Expressions (12.34) show that the characteristic vectors of the constraint force may depend on the system's motion. When, in particular, we think of fast-rotating systems, these forces can be so intense that they break the constraint at some of its points, and the resulting motion might be far from the desired rotation. Moreover, the explicit dependence of the constraint reactions on time subjects the system to resonance phenomena. Note in fact that even in presence of uniform rotations the modulus of the characteristic vectors of the constraint force is proportional to the square of the angular velocity's modulus, i.e. $m\delta\omega^2$ and $\omega^2\sqrt{(I_{13}^2 + I_{23}^2)}$ respectively.

This fact renders extremely important, when manufacturing the rigid body, that one try to minimize the components (12.34). There are two ways to attempt to reach the desired motion: decrease the distance $\delta$ of the center of gravity to the fixed axis, and lower the inertia products $I_{13}$, $I_{23}$, making the fixed axis a principal axis of inertia. The cases where these objectives are attained can be classified as follows.

**Definition 12.27** (*Static balancing*) A rigid body with fixed axis is called *statically balanced* when its center of gravity lies on the fixed axis.

The term originates in the fact that the center of gravity of a statically balanced body is fixed. Therefore $\mathbf{a}_G = \mathbf{0}$, and the constraint force's resultant we obtain from the first balance equation is $\mathbf{R}^{(\mathrm{r})} = -\mathbf{R}^{(\mathrm{a})}$. Hence the resultant of the constraint force *during the motion of the rigid body* is identical to that of a system at rest.

**Definition 12.28** (*Dynamic balancing*) A rigid body with fixed axis is called *dynamically balanced* when $\mathbf{K}_Q$ is parallel to $\boldsymbol{\omega}$, i.e. when the fixed axis is a principal axis of inertia with respect to its points.

When a rigid body is dynamically balanced, equation (12.34) gives $\dot{\mathbf{K}}_Q = I_3\ddot{\psi}\mathbf{e}_3$, so also the resultant moment of the constraint force coincides with the equilibrium's moment.

**Example 12.29** (Compound pendulum [3/3]) For a compound pendulum that is both statically and dynamically balanced, equations (12.35)–(12.36) give $\mathbf{R}^{(\mathrm{r})} = -m\mathbf{g}$ and $\mathbf{M}_Q^{(\mathrm{r})} = \mathbf{0}$. □

**Example 12.30** (Balancing through two particles) Any rigid body rotating around a fixed axis can be made both statically and dynamically balanced in a very easy way, by applying two suitable masses. This is the core idea at the heart of automotive tire balancing. There, small lead weights are applied to the rim of a wheel to compensate for the tire's non-uniform distribution of mass. Let us find explicitly how this balancing should be done.

Consider a slightly deformed wheel, modelled as a rigid body of mass $m$ with a fixed axis (the axis of rotation through the center and orthogonal to it). Choose a comoving frame of reference where the $z$-axis is fixed, the origin $O \in z$ is the geometric center of the wheel and the $x_1$-axis is such that the wheel's center of gravity lies on the $(x, z)$-plane: $OG = \delta\mathbf{i} + \beta\mathbf{k}$. Let us fix two external circles centered at

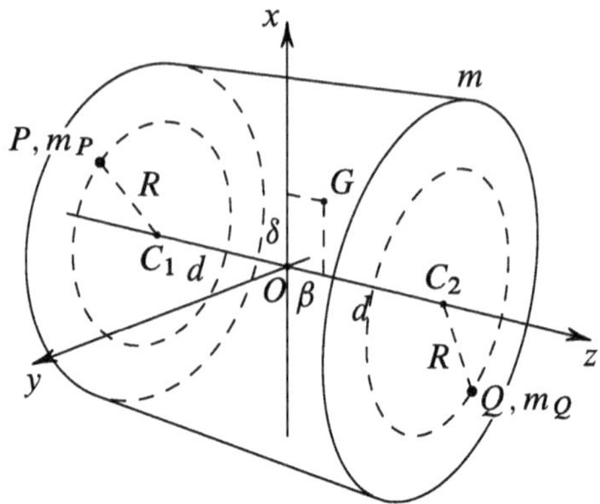


**Fig. 12.4** Balancing of a rotating rigid body by means of two particles $P$ and $Q$

$C_1, C_2$, on which we will apply the pointwise masses. These circles, both of radius $R$, will have origin on the $z$-axis and will belong to the planes $z = \pm d$ Fig. 12.4 respectively. Call $I_{13}, I_{23}$ the inertia products of the deformed wheel relative to the $z$-axis. We shall prove that it is possible to balance, statically and dynamically, the wheel with respect to the $z$-axis, by attaching two particles $P$, $Q$ of suitable masses $m_P, m_Q$ on the two circles Fig. 12.4.

Let

$$OP = x_P\mathbf{i} + y_P\mathbf{j} - d\mathbf{k} \qquad \text{and} \qquad OQ = x_Q\mathbf{i} + y_Q\mathbf{j} + d\mathbf{k}$$

denote the coordinates of the points to be applied. Statical balancing requires the body's center of gravity to belong to the $z$-axis. For that,

$$\begin{cases} (mOG + m_P OP + m_Q OQ) \cdot \mathbf{i} = 0 \\ (mOG + m_P OP + m_Q OQ) \cdot \mathbf{j} = 0 \end{cases} \Rightarrow \begin{cases} m\delta + m_P x_P + m_Q x_Q = 0 \\ m_P y_P + m_Q y_Q = 0 \end{cases} \tag{12.37}$$

To balance dynamically we must impose that $z$ be a principal axis of inertia, i.e. that the relative inertia products vanish:

$$I_{13} + m_P x_P d - m_Q x_Q d = 0\,, \qquad I_{23} + m_P y_P d - m_Q y_Q d = 0\,. \tag{12.38}$$

Finally, asking that $P$, $Q$ belong to the circles of centers $C_1, C_2$ implies

$$x_P^2 + y_P^2 = R^2\,, \qquad x_Q^2 + y_Q^2 = R^2\,. \tag{12.39}$$

Conditions (12.37)–(12.38)–(12.39) form a system of six equations in the six unknowns $\{m_P, m_Q, x_P, y_P, x_Q, y_Q\}$. The subsystem formed by (12.37)–(12.38) can be solved for $\{x_P, y_P, x_Q, y_Q\}$, giving

$$
\begin{aligned}
x_P &= -\frac{m\delta d + I_{13}}{2dm_P}, \qquad & y_P &= -\frac{I_{23}}{2dm_P}, \\
x_Q &= -\frac{m\delta d - I_{13}}{2dm_Q}, \qquad & y_Q &= \frac{I_{23}}{2dm_Q}.
\end{aligned} \tag{12.40}
$$

Substituting (12.40) in (12.39) produces the relations prescribing which masses should be applied

$$
\begin{aligned}
m_P^2 &= \frac{I_{23}^2 + (m\delta d + I_{13})^2}{4R^2d^2} \\
m_Q^2 &= \frac{I_{23}^2 + (m\delta d - I_{13})^2}{4R^2d^2}
\end{aligned} \tag{12.41}
$$

Each one of the quadratic equations (12.41) has one positive and one negative root. Indeed, in some cases it is more convenient to remove masses rather than adding them. Although our car mechanic prefers to apply masses on the external circles, one can check that in several domestic appliances, like kitchen blenders, a small quantity of mass is actually removed. □

## 12.6 Rigid Body on a Surface

In presence of unilateral constraints the constraint reactions cannot take arbitrary values. As we showed in Chapter 9, the components orthogonal to the unilateral constraint have sign prescribed by the touching condition (see, for instance, the discussion of Sect. 8.8.1 and equations (9.5)). Hence it becomes imperative to distinguish between boundary positions and ordinary positions, as was already the case in Statics (see Sect. 9.9.2). Consider for instance a rigid body that cannot penetrate a fixed surface.

- In ordinary configurations, occurring when no point of the rigid body touches the surface, the system obviously has six degrees of freedom, since it is a free rigid body.
- When a point $P_1$ comes into contact with the surface, and for as long as there is contact, the number of degrees of freedom is five: two coordinates give the position of $P_1$ on the surface, and three angles describe the rotations around $P_1$.
- Whenever a second point $P_2$ comes into contact with the surface, the number of generalized coordinates needed to describe the configurations decreases to four. In fact, there are only two independent angles, since the rigid body can only do two rotations without $P_1$, $P_2$ detaching from the surface: it can rotate around the axis parallel to $P_1P_2$, or around an axis orthogonal to the surface.
- Finally, when three or more points must touch the surface, the number of degrees of freedom becomes three: two coordinates still give the position of $P_1$ on the surface, and the only allowed rotation is the one around axes orthogonal to the surface.

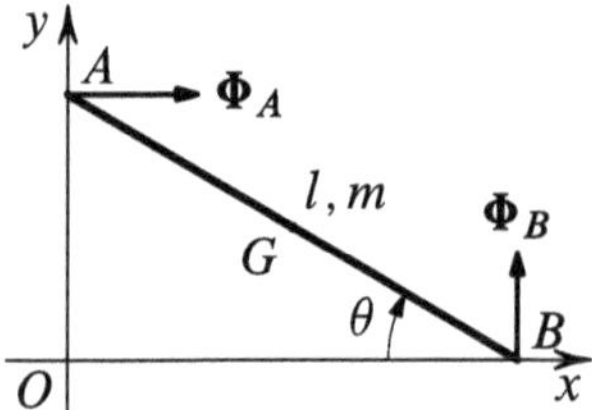


**Fig. 12.5** A model of a ladder leaning against a wall and supported by a horizontal floor

Despite the above remarks, the number of degrees of freedom of a rigid body on a surface is always six, since the body might at any time detach itself. This fact complicates considerably the study of the statics, and therefore that of the dynamics, of systems with unilateral constraints. One way to bypass the hurdle is the following: every time the initial conditions are compatible with a boundary motion we might assume the motion keeps of boundary type at all successive instants. Once we have found the required motion, we compute the constraint reactions, and check in particular that they satisfy suitable conditions, such as (9.5), on the signs of the components orthogonal to the constraints. As long as the latter are fulfilled, the characterization of ideal constraints (see Sect. 8.7) guarantees the reactions will occur. The uniqueness of the solutions to the equations of motion required by mechanical determinism—which we have assumed in Sect. 8.2—guarantees that the motion is the actual motion of the system. If, at some time $\bar{t}$, the orthogonal component of one or more constraint reactions changes sign, the corresponding contact disappears. In such a case the number of degrees of freedom increases, and we calculate the solutions starting from $\bar{t}$ using as initial data the solutions just found, which do hold until that instant.

**Example 12.31** (Falling ladder) Consider a homogeneous ladder of length $l$ and mass $m$, modelled as a rod $AB$, with vertices lying on two smooth axes (Fig. 12.5). Let us study its motion assuming that at the initial instant it is at rest, with angle at the base $\theta_0$. At the initial instant the ladder leans on the wall at one end, while the other end touches the floor, so it only has one degree of freedom. Call $\theta$ the angle it forms with the horizontal, and let the origin $O$ be the intersection of the axes. The position of the center of gravity is $OG = \frac{l}{2}\cos\theta\,\mathbf{i} + \frac{l}{2}\sin\theta\,\mathbf{j}$, so

$$\begin{aligned} \mathbf{v}_G &= \frac{l\dot{\theta}}{2}(-\sin\theta\,\mathbf{i} + \cos\theta\,\mathbf{j}), \\ \mathbf{a}_G &= \frac{l}{2}(-\ddot{\theta}\sin\theta - \dot{\theta}^2\cos\theta)\,\mathbf{i} + \frac{l}{2}(\ddot{\theta}\cos\theta - \dot{\theta}^2\sin\theta)\,\mathbf{j}. \end{aligned} \tag{12.42}$$

We can study the motion using the energy integral. As $T = \frac{1}{6}ml^2\dot{\theta}^2$ and $U = -\frac{1}{2}mgl\sin\theta$, we have

$$\dot{\theta}^2 = 3\frac{g}{l}(\sin\theta_0 - \sin\theta) \qquad \Longrightarrow \qquad \ddot{\theta} = -\frac{3}{2}\frac{g}{l}\cos\theta. \tag{12.43}$$

Substituting in (12.42) we find in particular

$$\mathbf{a}_G = \frac{3}{4}g(3\sin\theta - 2\sin\theta_0)\cos\theta\,\mathbf{i} + \frac{3}{4}g\big((3\sin\theta - 2\sin\theta_0)\sin\theta - 1\big)\,\mathbf{j}\,.$$

Let $\mathbf{\Phi}_A = H_A\mathbf{i}$ and $\mathbf{\Phi}_B = V_B\mathbf{j}$ denote the constraint reactions. The first balance equation of Dynamics gives

$$m\mathbf{a}_G = -mg\mathbf{j} + \mathbf{\Phi}_A + \mathbf{\Phi}_B \quad \Rightarrow \quad \begin{cases} H_A = \frac{3}{4}mg(3\sin\theta - 2\sin\theta_0)\cos\theta \\ V_B = \frac{3}{4}mg(3\sin\theta - 2\sin\theta_0)\sin\theta + \frac{1}{4}mg \end{cases}$$

Therefore the detachment will occur at $A$, when $H_A = 0$, since at that instant $V_B$ will still be strictly positive. More precisely, the rod detaches from the vertical axis in correspondence to the angle $\theta_\mathrm{d}$ such that

$$\sin\theta_\mathrm{d} = \frac{2}{3}\sin\theta_0\,. \tag{12.44}$$

Equation $(12.43)_1$ allows to find the angular velocity at the instant of detachment:

$$\boldsymbol{\omega}_\mathrm{d} = -\dot{\theta}_\mathrm{d}\mathbf{k} = \sqrt{\frac{g\sin\theta_0}{l}}\,\mathbf{k}. \tag{12.45}$$

Conditions (12.44)–(12.45) are the initial conditions for the motion after detachment. In this phase the rod has two degrees of freedom: the angle $\theta$ and the abscissa $x$ of $B$, now a generalized coordinate. For this second coordinate the initial conditions following the detachment instant are

$$x_\mathrm{d} = l\cos\theta_\mathrm{d} \qquad \text{and} \qquad \dot{x}_\mathrm{d} = -l\dot{\theta}_\mathrm{d}\sin\theta_\mathrm{d} = \frac{2}{3}\sqrt{gl\sin^3\theta_0}\,.$$

Analyzing the motion in this phase is simpler because there are two first integrals: the energy integral and now also the integral of the linear momentum's horizontal component. As

$$OG = \big(x - \tfrac{1}{2}l\cos\theta\big)\mathbf{i} + \tfrac{1}{2}l\sin\theta\mathbf{j} \quad \Rightarrow \quad \mathbf{v}_G = \big(\dot{x} + \tfrac{1}{2}l\dot{\theta}\sin\theta\big)\mathbf{i} + \tfrac{1}{2}l\dot{\theta}\cos\theta\mathbf{j}\,,$$

we have

$$\begin{aligned} T - U &= \frac{1}{2}m\dot{x}(\dot{x} + l\dot{\theta}\sin\theta) + \frac{1}{6}ml^2\dot{\theta}^2 + \frac{1}{2}mgl\sin\theta = \frac{1}{2}mlg\sin\theta_0 \\ Q_x &= m\Big(\dot{x} + \frac{1}{2}l\dot{\theta}\sin\theta\Big) = \frac{m}{3}\sqrt{gl\sin^3\theta_0}\,. \end{aligned} \tag{12.46}$$

Differentiating with respect to time (12.46) we find the free equations

$$\ddot{x} = \frac{2\cos\theta(3g\sin\theta - 2l\dot{\theta}^2)}{5 + 3\cos 2\theta} \quad \text{and} \quad \ddot{\theta} = \frac{6\cos\theta(l\dot{\theta}^2\sin\theta - 2g)}{(5 + 3\cos 2\theta)l}. \tag{12.47}$$

Now (12.46)–(12.47) enable us to use the vertical component of the first balance equation of Dynamics to write down the remaining constraint reaction $\mathbf{\Phi}_B = V_B\mathbf{j}$ in terms of $\theta$ only. The result is that the vertical component of the constraint reaction stays positive until the ladder falls on the floor ($\theta = 0$). Moreover, (12.46) also allow to determine the velocity distribution at the final instant:

$$\dot{x}_{\mathrm{f}} = \frac{1}{3}\sqrt{gl\sin^3\theta_0}\,, \qquad \dot{\theta}_{\mathrm{f}}^2 = \frac{g}{3l}\sin\theta_0\left(9 - \sin^2\theta_0\right).$$

□

## 12.7 Disk on a Linear Guide

In this section we examine in detail the boundary motion of a disk on a straight guide, possibly inclined. We will study the conditions under which the constraint, assumed not smooth, is capable of forcing the disk to roll along the guide. These considerations mimic the motion of a wheel, viewed as a rigid disk, moving on the surface of a road.

Consider a disk (homogeneous, of mass $m$ and radius $r$) rolling on an axis $x$, tilted by an angle $\alpha$ with respect to the horizontal. Choose a frame of reference with origin on the $x$-axis, and with $y$-axis coinciding with the upward perpendicular. The center $G$ is determined by its abscissa $x_G$, and the orientation by a counter-clockwise angle $\theta$, as in Fig. 12.6. Suppose, in view of the previous section, at each instant the disk touches the plane, i.e. $y_G = r$.

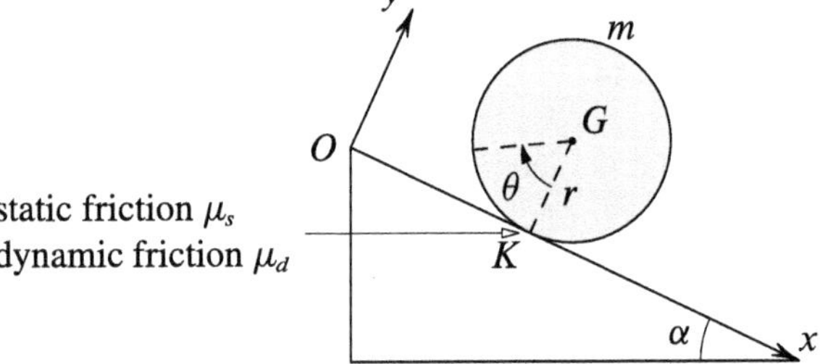


**Fig. 12.6** A disk on a ramp

### 12.7.1 Rolling Without Slipping

We start by supposing that, in presence of friction, the disk rolls without slipping because the plane is rough. Then

$$\dot{x}_G = r\dot{\theta} . \tag{12.48}$$

The forces acting on the disk are the weight, which can be reduced to a force $m\mathbf{g}$ applied at $G$, and the reaction $\mathbf{\Phi}_K$ applied at the contact point $K$.

The power of the constraint reaction is zero since $\Pi^{(\mathrm{r})} = \mathbf{\Phi}_K \cdot \mathbf{v}_K = 0$, due to the pure rolling constraint $\mathbf{v}_K = \mathbf{0}$. Therefore the free equation in the only kinematic unknown left is found directly from the Kinetic energy theorem. As

$$T = \frac{m}{2}\dot{x}_G^2 + \frac{1}{2}I_G\dot{\theta}^2 = \frac{3}{4}m\dot{x}_G^2 \quad \Longrightarrow \quad \dot{T} = \frac{3}{2}m\dot{x}_G\ddot{x}_G$$

and

$$\Pi = \Pi^{(\mathrm{a})} = m\mathbf{g} \cdot \mathbf{v}_G = mg\dot{x}_G \sin\alpha ,$$

we obtain

$$\ddot{x}_G = \frac{2}{3}g\sin\alpha \tag{12.49}$$

The free equation (12.49) could have been deduced using the second balance equation of Dynamics referred to the pole $K$. To that end let us observe that although $K$ moves (recall that $\dot{K} \neq \mathbf{v}_K$), this equation does not require any correction term $\dot{K} \times \mathbf{Q}$ since $\dot{K}$ is parallel to (actually, coinciding with) $\mathbf{v}_G$. Hence (12.49) shows that the disk's center describes a straight trajectory parallel to the $y$-axis, with uniformly accelerated motion. To compute the constraint reaction $\mathbf{\Phi} = \Phi_t\mathbf{i} + \Phi_n\mathbf{j}$ we may use the first balance equation

$$m\ddot{x}_G = mg\sin\alpha + \Phi_t , \qquad m\ddot{y}_G = -mg\cos\alpha + \Phi_n .$$

Keeping in account (12.49) and $y_G(t) = r$, implying $\ddot{y}_G \equiv 0$, we obtain

$$\Phi_t = -\frac{1}{3}mg\sin\alpha , \qquad \Phi_n = mg\cos\alpha . \tag{12.50}$$

Equation $(12.50)_2$ shows that the normal reaction $\Phi_n$ is non-negative for any $\alpha \in [0, \pi/2]$, thus guaranteeing we correctly assumed a boundary motion. Moreover, $(12.50)_1$ says that pure rolling imposes the presence of a precise frictional component. The Coulomb-Morin law (see (8.23)) tells that for a given coefficient of static friction $\mu_s$, the component of the friction must obey $|\Phi_t| \leq \mu_s\,|\Phi_n|$. In our case this reads

$$\mu_s \geq \frac{\tan\alpha}{3} . \tag{12.51}$$

When the coefficient of friction $\mu_s$ is not large enough to warrant (12.51), we have to abandon the hypothesis of pure rolling, and study the motion without resorting to (12.48), as we will do later. We emphasize that, under pure rolling, we have used the static version of the Coulomb-Morin law, not the dynamic counterpart (8.24). The reason is that the contact point does not move during pure rolling, making the contact between disk and guide of static type. When pure rolling disappears, and $\mathbf{v}_K \neq \mathbf{0}$, the friction coefficient to use in the Coulomb-Morin law is the dynamic one.

### 12.7.2 Rolling and Slipping

If (12.51) does not hold the disk must slip, and we then must consider two unknowns: $x_G$ and $\theta$, no longer related by (12.48). The dynamic version (8.24) of the Coulomb-Morin law establishes that $|\Phi_t| = \mu_d\,|\Phi_n|$, and the sign of the frictional component must be the opposite of the velocity's. We emphasize that, assuming (12.51) does not hold, for an even greater reason we have

$$\mu_d \leq \mu_s < \frac{\tan\alpha}{3}. \tag{12.52}$$

If the initial conditions of motion ensure the velocity of $G$ always points downwards (i.e. $\dot{x}_G(0) = \dot{x}_{G0} \geq 0$), the first balance equation of Dynamics gives

$$\begin{cases} m\ddot{x}_G = mg\sin\alpha - \mu_d\,|\Phi_n| \\ 0 = -mg\cos\alpha + \Phi_n \end{cases} \implies \begin{cases} \ddot{x}_G = g(\sin\alpha - \mu_d\cos\alpha) \\ \Phi_n = mg\cos\alpha \end{cases} \tag{12.53}$$

Because of (12.52), equations (12.53) imply $\ddot{x}_G > 0$: the dynamic friction will slow down the disk, but not stop it completely.

The evolution of the rotation angle $\theta$ can be determined using the second balance equation, using as pole the disk's center

$$-\frac{1}{2}mr^2\ddot{\theta} = -\mu_d\,|\Phi_n|\,r \implies \ddot{\theta} = \frac{2\mu_d g\cos\alpha}{r}.$$

We complement the initial conditions by supposing $x_G(0) = 0$, $\theta(0) = 0$, and that the initial angular velocity is $\boldsymbol{\omega}(0) = -\dot{\theta}_0\mathbf{k}$ (note the minus sign, due to $\theta$ being counter-clockwise). Then

$$x_G(t) = \dot{x}_{G0}t + \frac{gt^2}{2}\big(\sin\alpha - \mu_d\cos\alpha\big), \qquad \theta(t) = \dot{\theta}_0 t + \frac{\mu_d g t^2\cos\alpha}{r}. \tag{12.54}$$

Define the *slip velocity* $u$ to be the horizontal component of the velocity of the contact point: $u = \mathbf{v}_K \cdot \mathbf{i}$. We have

$$\mathbf{v}_K = \mathbf{v}_G + \boldsymbol{\omega} \times GK = \dot{x}_G \, \mathbf{i} - \dot{\theta} \, \mathbf{k} \times (-r\mathbf{j}) \quad \Longrightarrow \quad u = \dot{x}_G - r\dot{\theta} \, ,$$

(the vanishing of the slip velocity corresponds exactly to a pure rolling constraint).

Using (12.54), and calling $u_0 = \dot{x}_{G0} - r\dot{\theta}_0$ the initial slip velocity,

$$u(t) = u_0 + gt \, (\sin\alpha - 3\mu_d \cos\alpha) \; .$$

The slip velocity increases with $t$, because of (12.52). Therefore if $u_0 > 0$ the disk will slip at every successive instant. If instead $u_0 \leq 0$, there will be an instant (the initial one if $u_0 = 0$) at which the slip velocity vanishes and the disk rolls without slipping. If (12.51) were to hold, the disk would continue to roll without slipping. Otherwise, the slip velocity will eventually become positive.

**Example 12.32** (Horizontal ground) Let us study carefully the particular case in which the axis along which the disk rolls is horizontal ($\alpha = 0$). Taking the same initial conditions as before, we now assume the initial slip velocity $u_0$ is positive ($\dot{x}_{G0} > r\dot{\theta}_0$). The solutions (12.54) to the equations of motion are

$$x_G(t) = \dot{x}_{G0}t - \frac{\mu_d g t^2}{2} \, , \qquad \theta(t) = \dot{\theta}_0 t + \frac{\mu_d g t^2}{r} \, , \tag{12.55}$$

so $u(t) = u_0 - 3\mu_d g t$. The slip velocity then vanishes at

$$\bar{t} = \frac{u_0}{3\mu_d g} = \frac{\dot{x}_{G0} - r\dot{\theta}_0}{3\mu_d g} > 0,$$

at which instant the contact point between disk and ground has zero velocity.

Consider now the motion after $t = \bar{t}$. Suppose that at some $t > \bar{t}$ the disk starts to slip. The slip velocity would be positive, so by the same reasoning the friction would cancel out the slip velocity. The conclusion is that this is not possible, and that for any $t > \bar{t}$ the disk *must* continue to roll without slipping. The values $V_G$ and $\dot{\Theta}$ taken by $\dot{x}_G$ and $\dot{\theta}$ at time $\bar{t}$ must obviously respect the pure rolling $V_G - r\dot{\theta} = 0$, and indeed (12.55) gives

$$V_G = \dot{x}_G(\bar{t}) = \frac{2}{3}\dot{x}_{G0} + \frac{1}{3} r\dot{\theta}_0 = r\dot{\theta} \, . \tag{12.56}$$

Finally, observe that for a horizontal ground, equation $(12.50)_1$ explains that we do not need a frictional component to maintain the pure rolling. The static Coulomb-Morin law will definitely hold in this regime since (12.51) just requires $\mu_s \geq 0$. During pure rolling, moreover, (12.49) gives $\ddot{x}_G = 0$. Hence at any instant $t > \bar{t}$ the motion is uniform with $\dot{x}_G(t) = V_G$ for any $t \geq \bar{t}$. □

**Remark 12.33** If at the initial instant the disk's center moves with $\dot{x}_{G0} > 0$, but the angular velocity is such that $\dot{\theta}_0 < 0$ (counter-clockwise rotation), with $|\dot{\theta}_0| > 2\dot{x}_{G0}/r$, the center's final velocity $V_G$ will be negative, as per (12.56). The counter-clockwise rotation can therefore cause, at the end, the center of gravity to move

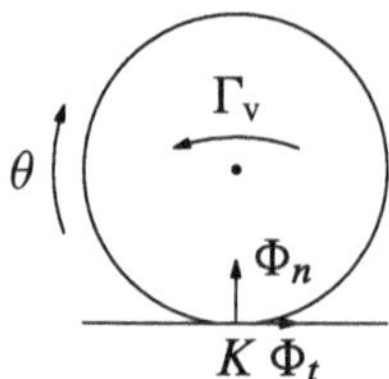

**Fig. 12.7** A model of a wheel with a tire subject to a rolling friction moment $\Gamma_v$

opposite to the initial condition. This effect is clearly visible in a ping-pong ball rolling on a sufficiently rough surface like felt. If at the initial time we are able to impart enough *spin* on the ball, it will slip in some direction for a while, but then it will revert its motion and roll backwards without slipping. □

### *12.7.3 Disk Subject to Rolling Friction*

Let us drop the assumption that the disk is rigid, and consider a real wheel covered by a tire. Now it is no longer reasonable to think the wheel-road contact is a single point due of tire's specific feature of being able to deform, accomodate the irregularities of the tarmac and increase adherence. The contact occurs on an entire region, technically called *contact patch*. Therefore the force of the contact between wheel and surface cannot reduce to a single force (the resultant of the contact constraint reaction) applied at the contact point. We must keep into account a couple with moment $\mathbf{\Gamma}_v$, expressing the presence of *rolling friction*.

As we explained in Sect. 8.9.3, Chap. 8, a more realistic model involves a normal reaction of modulus $\Phi_n$, applied at the geometric contact point, a tangent reaction $\Phi_t$ again at the geometric contact point, which opposes the motion and obeys the Coulomb-Morin law, and finally a couple $\Gamma_v$, orthogonal to the plane of the disk, which opposes the rolling. The couple is called *rolling friction* moment. One assumes (in the simplest case) the existence of a dynamical characterization for the rolling friction moment given by (8.29), similar to the Coulomb-Morin laws of static and dynamic friction (Fig. 12.7).

**Example 12.34** (Setting a real wheel in motion) One problem when we consider the effect of rolling friction is the computation of the minimal force required to make a disk at rest to start rolling. Given the contact between disk and ground, the first equilibrium balance equation implies the center of the disk will be at rest until $\Phi_t = F$ and $\Phi_n = mg$, where $\mathbf{F} = F\mathbf{i}$ is the traction force, assumed applied to the wheel at a distance $h$ from the (horizontal) guide. Furthermore, if we choose the contact point $K$ as pole for the second balance equation, the latter prescribes $Fh - \delta_s \Phi_n = 0$.

The usual Coulomb-Morin law $|\Phi_t| \le \mu_s |\Phi_n|$ allows us to recover from the first balance equation

$$F \le F_{\mathrm{cr},1} = \mu_s mg \,, \qquad (12.57)$$

while from the second balance equation we obtain

$$F \le F_{\mathrm{cr},2} = \frac{\delta_s mg}{h} \,. \tag{12.58}$$

If both (12.57)–(12.58) hold, the wheel stays in equilibrium. If not, it starts to move. The ensuing motion in the second case depends on which of (12.57)–(12.58) fails.

- The first requirement that is breached is (12.58) if the traction is applied high enough: $h > \delta_s/\mu_s$. If so, the critical traction force for setting the wheel in motion equals $F_{\mathrm{cr},2} < F_{\mathrm{cr},1}$, and the disk starts to roll *without slipping*, since the law of static friction is not violated during the initial phase of the motion.
- If instead $h \le \delta_s/\mu_s$, the critical traction force is $F_{\mathrm{cr},1} < F_{\mathrm{cr},2}$, and the disk starts to *slip*.

□

**Remark 12.35** The coefficient of rolling friction can be measured experimentally by taking a disk touching the ground, at whose center of gravity $G$ we apply a vertical force $\mathbf{P}$. Now, shift this force to increasing distances $b$ from the center of gravity, so that it stays parallel to itself on the plane of the disk. On the disk there is a couple with moment $M(b) = Pb$. The limit value $b_{\mathrm{cr}}$ for which the disk starts to move allows to find the static coefficient of rolling friction $\delta_s$, because

$$Pb_{\mathrm{cr}} = \delta_s \Phi_n = \delta_s (P + mg) \quad \Longrightarrow \quad \delta_s = \frac{Pb_{\mathrm{cr}}}{P + mg} \,,$$

where $m$ denotes the mass of the disk. The effects of rolling friction are usually much smaller than those caused by static or dynamic friction. For rubber tires in contact with asphalt, for instance, the value of $\delta_s$ can roughly be considered to be proportional to the wheel's radius: $\delta_s \approx \gamma_s r$, with $\gamma_s \approx 0.0035$. Hence, remembering that the corresponding static friction coefficient is close to $1/2$, it will suffice to apply a traction force bigger than $1/100$ of the radius' height to guarantee $F_{\mathrm{cr},2} < F_{\mathrm{cr},1}$ in the previous example. □

**Example 12.36** (Optimal braking) Let us study a rigid disk that rolls without slipping and suddenly brakes because of rolling friction. Call $\mathbf{M}_{\mathrm{f}} = -M_{\mathrm{f}}\mathbf{k}$ the moment of the braking couple. The balance equations of Dynamics (the second of which is always referred to the contact point $K$) give

$$m\ddot{x}_G = \Phi_t, \qquad \Phi_n = mg\,, \qquad \frac{3}{2}mr^2\ddot{\theta} = -M_{\mathrm{f}} - \delta_s \Phi_n \,. \tag{12.59}$$

The kinematic relation of pure rolling implies $\ddot{x}_G = r\ddot{\theta}$, so $(12.59)_1$ gives $\Phi_t = mr\ddot{\theta}$. The Coulomb-Morin law requires

$$|\Phi_t| \le \mu_s |\Phi_n| \quad \Longrightarrow \quad M_{\mathrm{f}} \le \left(\tfrac{3}{2}\mu_s - \gamma_s\right) mgr \,, \tag{12.60}$$

where we have set $\delta_s = \gamma_s r$. If we want to brake and prevent slipping, inequality (12.60) must be enforced. If we apply the maximum possible braking couple allowed by (12.60), the component $\Phi_t$ will attain the maximum value permitted by the static Coulomb-Morin law: $\Phi_t = -\mu_s mg$. Then, $(12.59)_1$ can be integrated:

$$m\ddot{x}_G + \mu_s mg = 0 \quad \Longrightarrow \quad \dot{x}_G + \mu_s gt = V\,, \tag{12.61}$$

where $V$ is the velocity of the center of gravity at the time $t = 0$ when braking begins. From (12.61) we deduce the instant $T_1$ at which the center of gravity stops, assuming the braking couple is the largest possible under pure rolling:

$$T_1 = \frac{V}{\mu_s g}\,.$$

In case the wheel slips, instead, the law of dynamic friction gives a relation similar to (12.61): $\dot{x}_G + \mu_d gt = V$, from which we can compute the stopping time

$$T_2 = \frac{V}{\mu_d g}\,.$$

As $\mu_d < \mu_s$, necessarily $T_2 > T_1$, so the optimal stopping time (and hence, the optimal braking distance) is achieved by the braking couple $M_{\mathrm{f,ott}} = (3\mu_s/2 - \gamma_s)\, mgr$. If the braking couple exceeds this value the wheel slips, and both the braking time and the braking distance increase. Observe that in presence of slipping, as $x$ and $\theta$ are decoupled, there is no upper bound for the braking moment $M_{\mathrm{f}}$. □

# Chapter 13
# Non-inertial Reference Frames

The Laws of Mechanics, which we stated and discussed in Chap. 8, posit the existence of inertial systems, namely those systems where isolated points stay in their state of rest or of linear uniform motion. Newton's law (8.1) establishes how a particle moves (in an inertial system) under the action of forces. These two principles alone, however, would not be capable of explaining the motion we observe in our day-to-day reality, given that no frame of reference accessible to us behaves as inertial. A frame of reference comoving with Earth is affected by the planet's rotation; a more precise system comoving with the Sun would still be impacted by the latter's motion together with the solar system's. Even the motion with respect to the system that Kepler still called the *sphere of fixed stars* can only approximate the concept of *inertial system*. In Sect. 8.5 we did solve this problem, by observing how Relative Kinematics, introduced in Chap. 3, enables us to study the motion of particles in non-inertial frames. Newton's law (8.1) includes in that case the *drag force* and the *Coriolis force* (see Sect. 8.5 and Definitions 8.12). Both are *apparent forces*, since they do not arise through a physical interaction, but are rather born out of the wish of a non-inertial observer to associate a force with every kind of acceleration measured.

The present chapter intends to examine in detail the effects that apparent forces cause on the statics and dynamics of particles and systems of rigid bodies.

We find it convenient to call Relative Mechanics the part of Classical Mechanics dealing with equilibrium conditions and equations of motion with respect to *non-inertial reference frames*. Accordingly, in this chapter we shall speak about "relative statics", "relative equilibrium", "relative stability" and so on.

Relative Mechanics, in other words, contains all equations, theorems and properties that involve apparent forces, or their power and work, as defined in Sect. 8.5.

We are aware that this terminology is not widely used, but we believe it captures neatly and precisely the context of the ideas presented here. It also respects the Italian tradition of books on Rational Mechanics, where the term "meccanica relativa" (relative mechanics) is used. Indeed, we are convinced it is important to devote an



P. Biscari et al., *Rational Mechanics*, UNITEXT 177,
https://doi.org/10.1007/978-3-032-07462-1_13

entire chapter to Relative Mechanics due to its pronounced impact in the applications, as we will demonstrate with the help of examples.

## 13.1 Apparent Forces

All frames detect the same interactions and the same forces among particles, but each one measures a different acceleration. Despite that, one can use Newton's law in non-inertial frames as well, with the proviso of including the apparent forces among the forces acting on the system.

As we have seen in Sect. 8.5, it is easy to take the composition law of accelerations

$$\mathbf{a} = \mathbf{a}_r + \mathbf{a}_\mathcal{T} + \mathbf{a}_c$$

and the law $\mathbf{F} = m\mathbf{a}$ written for an inertial frame, and deduce

$$\mathbf{F} - m\mathbf{a}_\mathcal{T} - m\mathbf{a}_c = m\mathbf{a}_r \,. \tag{13.1}$$

It is then natural to define the "apparent" forces

$$\mathbf{F}_\mathcal{T} = -m\mathbf{a}_\mathcal{T} \qquad \mathbf{F}_c = -m\mathbf{a}_c \tag{13.2}$$

calling the former *drag force* and the latter *Coriolis force*. In this way the (13.1) reads

$$\mathbf{F} + \mathbf{F}_\mathcal{T} + \mathbf{F}_c = m\mathbf{a}_r \tag{13.3}$$

where $\mathbf{a}_r$ is the acceleration of the point with respect to the non-inertial frame.

Formally speaking, (13.3) says Newton's law holds in non-inertial frames too, as long as the effective force $\mathbf{F}$ is added to the apparent forces $\mathbf{F}_\mathcal{T}$ and $\mathbf{F}_c$. We must remark that in order to compute these forces it is indispensable to know the drag acceleration $\mathbf{a}_\mathcal{T}$ and the Coriolis acceleration $\mathbf{a}_c$, that is to say we must know the motion of the non-inertial frame with respect to the inertial one. All in all, as is easy to perceive, we have simply taken two quantities, made them switch sides, and labelled them "apparent forces". This is a pure manipulation, but as its several applications attest, Relative Mechanics, i.e. the part of Mechanics entirely based upon (13.3), has nevertheless an enormous use, even though it does not bring about any conceptual novelty compared to what we have seen so far.

It is important to add some comments:

- "Apparent" forces have that name because, in some sense, they only "appear" in, and manifest themselves to, non-inertial frames, as opposed to effective forces, which are due to interactions (whether or not by direct contact) between bodies, and are present in the equations and mechanical laws that *every* frame uses.

- Apparent forces, in contrast to effective forces, violate the action-reaction principle since they are not caused by interactions of pairs of bodies.
- Apparent forces must be included, with the effective forces, in the relations and equations we have found and discussed thus far, *only* when we want to study the equilibrium or motion of a system from a non-inertial point of view.

Relative Mechanics is useful mainly because in many situations the description of a system's motion is greatly simplified when we put ourselves in a frame that moves non-inertially. It is easy to convince ourselves of this fact with a few significant examples, as we will see later.

For the considerations and proofs that follow we should keep in mind definitions (3.8), and more generally the picture of Fig. 3.1. The vector $\boldsymbol{\omega}$ will hence denote the angular velocity of a frame in motion with respect to an inertial frame, and $O$ will be its origin.

Based on that, for simplicity and completeness we rewrite more explicitly definitions (13.2) for the apparent forces acting on a point $P$ of mass $m$:

$$\mathbf{F}_{\tau} = -m\mathbf{a}_{\tau}(P) = -m\,(\mathbf{a}_O + \dot{\boldsymbol{\omega}} \times OP + \boldsymbol{\omega} \times (\boldsymbol{\omega} \times OP)) \tag{13.4}$$

$$\mathbf{F}_{\mathrm{c}} = -m\mathbf{a}_{\mathrm{c}}(P) = -2m\boldsymbol{\omega} \times \mathbf{v}_{\mathrm{r}}(P) \tag{13.5}$$

The problems and applications of Relative Mechanics cover the important section normally referred to as *Relative Statics*. The classical problem in this area seeks to determine the equilibrium conditions for a system relative to a non-inertial frame.

It is straightforward and important to observe that in the equations of Relative Statics the Coriolis forces are absent because, as (13.5) shows, they vanish together with the relative velocities of the points they refer to. Let us summarize this and other immediate properties of Coriolis forces in the next proposition.

**Proposition 13.1** *The Coriolis force is always zero when the point to which it refers has zero velocity* $\mathbf{v}_{\mathrm{r}}$, *for which reason it* does not *appear in the problems of relative statics. Similarly, this force vanishes when the moving frame translates with respect to the inertial frame, i.e. when* $\boldsymbol{\omega} = \mathbf{0}$.

*The Coriolis force produces zero power on the point to which it is applied:*

$$\Pi_{\mathrm{c}} = \mathbf{F}_{\mathrm{c}} \cdot \mathbf{v}_{\mathrm{r}} = -2m\boldsymbol{\omega} \times \mathbf{v}_{\mathrm{r}} \cdot \mathbf{v}_{\mathrm{r}} = 0\,.$$

Note finally that the Coriolis force depends on the velocity of the point, and hence *cannot* be positional, whence neither conservative. That said, it is important to stress that the presence of the Coriolis force does not hinder the existence of the energy integral (10.4) in case the other active forces are all conservative. If we look at the passages (10.3) needed to obtain the energy integral of the mechanics of a point, any force with zero power (in particular any force orthogonal to the velocity) vanishes in the first step of (10.3). Similarly, the Coriolis forces do not belong among those to consider when we deduce the theorem of conservation of the energy in more general mechanical systems, as shown by Proposition 11.20.

**Remark 13.2** In the sequel of this chapter we will refer to *discrete distributions* of particles $\{(P_i, m_i),\ i = 1, \dots, n\}$, for simplicity. However, it is evident that every result extends naturally to continuous distributions of masses. □

## 13.2 Resultant of Apparent Forces

There are two important, and similar, properties that allow to compute easily the resultant of the drag forces and the resultant of the Coriolis forces acting on the points of a system.

**Proposition 13.3** (Resultant of the drag forces) *The resultant of the drag forces equals the drag force that would act on the center of mass if the latter had a mass equal to the system's total mass.*

***Proof*** Consider a system of particles $\{(P_i, m_i),\ i = 1, \dots, n\}$ with total mass $m$ and center of mass $G$. Then

$$OG = \frac{1}{m}\sum_{i=1}^{n} m_i OP_i, \qquad \text{where } m = \sum_{i=1}^{n} m_i$$

and then, in the light of (13.4),

$$\begin{aligned}
\mathbf{R}_\tau &= -\sum_{i=1}^{n} m_i \mathbf{a}_\tau(P_i) = -\sum_{i=1}^{n} m_i\big(\mathbf{a}_O + \dot{\omega} \times OP_i + \omega \times [\omega \times OP_i]\big) \\
&= -\Big(\sum_{i=1}^{n} m_i\Big)\mathbf{a}_O - \dot{\omega} \times \Big(\sum_{i=1}^{n} m_i OP_i\Big) - \omega \times \Big[\omega \times \Big(\sum_{i=1}^{n} m_i OP_i\Big)\Big] \\
&= -m\mathbf{a}_O - m\dot{\omega} \times OG - m\omega \times (\omega \times OG) \\
&= -m\mathbf{a}_\tau(G)
\end{aligned}$$

□

**Remark 13.4** This proposition is extremely useful in that it permits to compute with ease the resultant of the drag forces, which is then typically inserted in the first balance equation. However, it *does not* prove the system of the drag forces is equivalent to the resultant applied at the center of mass. In fact, the equivalence of two systems of forces (and in particular the use of equivalent systems towards the second equilibrium balance equation) also requires the moments to coincide.

It is then important to recall that:

- the resultant of the drag forces is computed at the center of mass.
- The drag forces' resultant does not in general possess a line of action.
- Even in cases where the collection of drag forces admits a line of action, nothing guarantees this line passes through the center of mass of system. □

**Proposition 13.5** (Resultant of the Coriolis forces) *The resultant of the Coriolis forces equals the Coriolis force that would act on the center of mass if this had as mass the total mass of the system.*

***Proof*** Let $\{(P_i, m_i),\ i = 1, \dots, n\}$ be a system of particles with velocities $\mathbf{v}_i$ measured in a frame that rotates with angular velocity $\boldsymbol{\omega}$ with respect to an inertial system. By (13.5),

$$\mathbf{R}_{\mathrm{c}} = -\sum_{i=1}^{n} m_i \mathbf{a}_{\mathrm{c}}(P_i) = -2\boldsymbol{\omega} \times \sum_{i=1}^{n} m_i \mathbf{v}_{\mathrm{r}}(P_i) = -2m\boldsymbol{\omega} \times \mathbf{v}_{\mathrm{r}}(G) = -m\mathbf{a}_{\mathrm{c}}(G)\,. \tag{13.6}$$

where we have used the relation $\sum m_i \mathbf{v}_r(P_i) = m\mathbf{v}_r(G)$, i.e. the Linear momentum theorem from the viewpoint of the moving frame. □

**Remark 13.6** Here, too, it is important to remember that while we may compute the value of the resultant of the Coriolis forces at the center of mass, we cannot be certain that this collection of forces reduces to the sole resultant applied at the center of mass, or elsewhere. This must be decided on a case-by-case analysis. □

**Example 13.7** (*Translating frame of reference*) Consider a non-inertial frame whose axes stay parallel to those of an inertial system, but whose origin $O$ has acceleration $\mathbf{a}_O$. As the angular velocity $\boldsymbol{\omega}$ vanishes, the apparent force acting on each point $P_i$ of mass $m_i$ reduces to the drag force $\mathbf{F}_{\tau i} = -m_i \mathbf{a}_O$.

In this case the system of drag forces is equivalent to its resultant $\mathbf{R}$, applied at the center of mass $G$.

Proposition 13.3 proves equivalence as regards the resultant. Moreover, taken any pole $Q$,

$$\mathbf{M}_Q = \sum_{i=1}^{n} QP_i \times (-\, m_i \mathbf{a}_O) = -\Big(\sum_{i=1}^{n} m_i QP_i\Big) \times \mathbf{a}_O = QG \times \mathbf{R}\,,$$

which shows in this case that the sole resultant generates, if applied at $G$, a moment equal to the moment of the entire system of drag forces. □

## 13.3 Centrifugal Force

One of the most common situations where the ideas of Relative Mechanics become useful is that where a non-inertial frame rotates uniformly about a fixed axis.

The drag force (13.4), under the assumption

$$\mathbf{a}_O = \mathbf{0}\,, \qquad \dot{\boldsymbol{\omega}} = \mathbf{0}\,,$$

simplifies considerably and reduces to

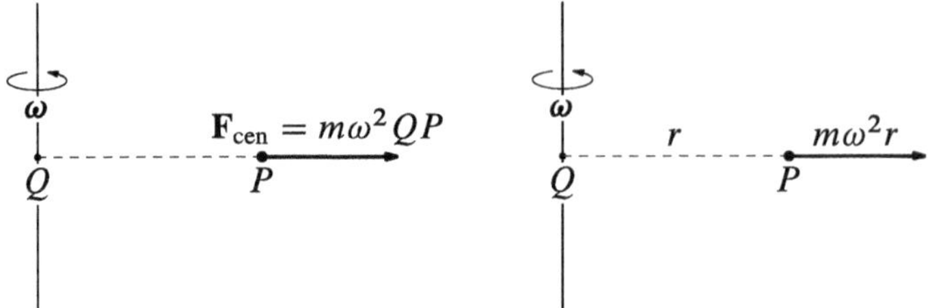


**Fig. 13.1** The centrifugal force acting on a point $P$ with mass $m$ at a distance $r$ from the rotation axis

$$\mathbf{F}_{\text{cen}} = -m\boldsymbol{\omega} \times (\boldsymbol{\omega} \times OP)\,. \tag{13.7}$$

This is called *centrifugal force*, as suggested by the subscript.

The centrifugal force, then, is a particular drag force: it is the name we give the drag force when the moving frame rotates with constant angular velocity around a fixed axis.

To understand better (13.7) let us choose a frame with $z$-axis parallel to $\boldsymbol{\omega}$, so that $\boldsymbol{\omega} = \omega\,\mathbf{k}$, and also set $OP = x\mathbf{i} + y\mathbf{j} + z\mathbf{k}$. Using expression (A.8) to compute the triple product we obtain:

$$\mathbf{F}_{\text{cen}} = m\,\omega^2\big[OP - (OP \cdot \mathbf{k})\mathbf{k}\big] = m\,\omega^2(x\mathbf{i} + y\mathbf{j})\,. \tag{13.8}$$

Referring to Fig. 13.1 let $Q$ denote the orthogonal projection of $P$ onto the axis of rotation (so $OQ = z\mathbf{k}$). If we decompose the position vector $OP$ into its parallel and orthogonal components to the axis, so that $OP = (x\mathbf{i} + y\mathbf{j}) + (z\mathbf{k}) = QP + OQ$ with $QP = x\mathbf{i} + y\mathbf{j}$, Eq. (13.8) simply becomes

$$\mathbf{F}_{\text{cen}} = m\,\omega^2\,QP\,. \tag{13.9}$$

As shown in Fig. 13.1 the centrifugal force is therefore radial from the rotation axis, with modulus proportional to the distance point-axis and to the angular velocity squared.

### *13.3.1 Planar Systems of Centrifugal Forces*

Consider a rigid system contained in a plane $\pi$ that rotates with constant angular velocity $\boldsymbol{\omega}$ around an axis on the plane, which will be our $y$-axis: $\boldsymbol{\omega} = \omega\,\mathbf{j}$. Fix a pole $O$ on the axis, and assume to simplify the calculation that the system consists of $n$ particles $\{(P_i, m_i),\ i = 1, \ldots, n\}$, with $OP_i = x_i\,\mathbf{i} + y_i\,\mathbf{j}$ and $Q_iP_i = x_i\,\mathbf{i}$. We will now show how we can reduce the system of centrifugal forces.

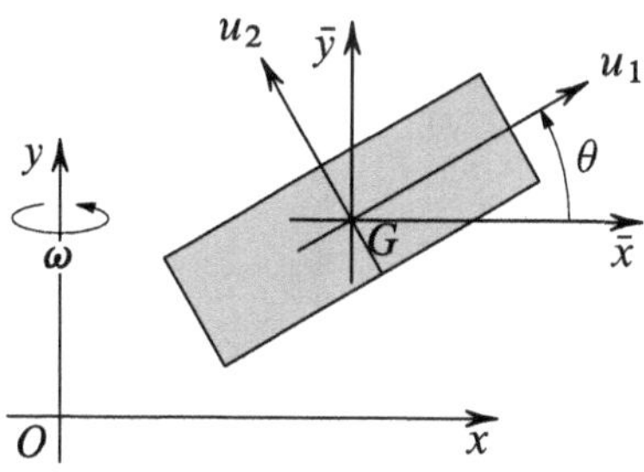


**Fig. 13.2** A rigid plate constrained to a rotating plane

- The moment of the centrifugal forces equals

$$\mathbf{M}_O = \sum_{i=1}^{n} m_i\, OP_i \times \left(\omega^2 Q_i P_i\right) = -\sum_{i=1}^{n} m_i\, x_i y_i\, \omega^2 \mathbf{k} = I_{xy}\, \omega^2 \mathbf{k}\,,$$

where $I_{xy}$ is the inertia product with respect to the rotation axis and to the orthogonal line to it through $O$ lying on the plane. Using property (5.28) we can relate $I_{xy}$ to the inertia product with respect to the pair $(\bar{x}, \bar{y})$ parallel to $(x, y)$ but passing through the center of mass (Fig. 13.2). Then (5.43) allows us to express that inertia product $I_{\bar{x}\bar{y}}$ in function of the central principal moments of inertia $I_1$, $I_2$ and of the angle $\theta$ between the first principal axis and the $\bar{x}$-axis:

$$\mathbf{M}_O = \left(I_{\bar{x}\bar{y}} - m x_G y_G\right) \omega^2 \mathbf{k} = ((I_1 - I_2) \sin\theta \cos\theta - m x_G y_G)\, \omega^2 \mathbf{k}\,.$$

We should remark that in using (5.43) the inertia product changes sign because the angle $\theta$ defined in Fig. 13.2 is the opposite of the one in (5.43).
The resultant moment $\mathbf{M}_O$ is orthogonal to the plane of the system, and in particular to the resultant. This implies that every time the latter is non-zero (provided the center of mass does not belong to the axis of rotation), the system of centrifugal forces will have a resultant line of action, since the scalar invariant $I = \mathbf{R} \cdot \mathbf{M}_O$ is zero (see (7.21) and Proposition 7.8.
- Assuming $x_G \neq 0$, and keeping (7.25) in mind, the system of centrifugal forces is equivalent to the sole resultant applied at a point $P^*$ with second coordinate

$$y^* = -\frac{I_{xy}}{m x_G} = y_G - \frac{(I_1 - I_2)\sin\theta\cos\theta}{m x_G}\,. \tag{13.10}$$

When, in particular, the two central principal moments of inertia coincide ($I_1 = I_2$) we can definitely apply the forces' resultant at the center of mass. This happens for instance if the system is a homogeneous disk or a homogeneous square plate.

**Example 13.8** We shall find the resultant of the centrifugal forces and its line of action in case of a homogeneous rod $AB$, of length $l$ and mass $m$, constrained to move on a plane $(x, y)$ that rotates with constant angular velocity $\boldsymbol{\omega} = \omega\mathbf{j}$, under the

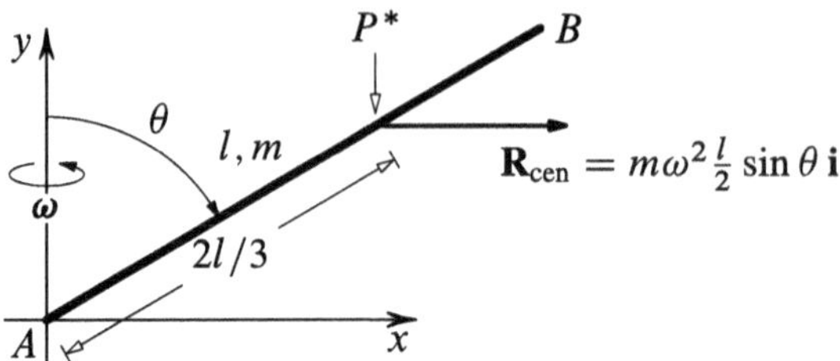


**Fig. 13.3** Reduction of the centrifugal force system acting on a rod

assumption that the end $A$ belongs on the axis of rotation (Fig. 13.3). Call $\theta$ the angle between the rod and the axis, so $x_G = l/2 \sin\theta$, $I_{G1} = 0$, $I_{G2} = ml^2/12$, and hence

$$\mathbf{R} = m\,\omega^2 \tfrac{1}{2} l \sin\theta\, \mathbf{i} \qquad y^* = y_G + \tfrac{1}{6} l \cos\theta \implies AP^* = \tfrac{2}{3} l\,.$$

In other words, the resultant is computed assuming the entire mass is concentrated in the center of mass, but is applied at a point at distance equal to $2/3$ of the rod's length, with respect to a point on the axis. □

### 13.3.2 Generic Systems of Centrifugal Forces

When the system is not planar the scalar invariant $I = \mathbf{R} \cdot \mathbf{M}_O$ might not vanish, and then the system of centrifugal forces might not be reducible to its resultant. Still, it could prove useful to determine in which special cases the centrifugal force is equivalent to its resultant applied at the center of mass.

The resultant of the centrifugal force can always be written as $m\omega^2 QG$, where $Q$ is the projection of $G$ on the axis of rotation. Supposing $QG \neq 0$, the centrifugal force will be equivalent to its resultant (applied at some point to be determined) if and only if its scalar invariant is zero

$$m\omega^2 QG \cdot \mathbf{M}_O = 0\,. \tag{13.11}$$

Assuming (13.11) holds, we can then find the necessary and sufficient condition so that the centrifugal force is equivalent to its resultant *applied at the center of mass* $G$ of the body on which it acts. Clearly, this condition requires that the moment of the centrifugal force with respect to $G$ is zero:

$$\mathbf{M}_G = \sum_{i=1}^{n} GP_i \times m_i\omega^2 Q_i P_i = \mathbf{0}\,, \tag{13.12}$$

where $Q_i$ is the orthogonal projection of $P_i$ onto the axis of rotation.

**Proposition 13.9** *A system of centrifugal forces is equivalent to its resultant applied at the center of mass if and only if the axis of rotation is parallel to a central principal axis of inertia.*

***Proof*** Let us show initially that the expression of $\mathbf{M}_G$ does not change when we replace the projections $Q_i$ on the axis of rotation with the projections $S_i$ on the axis parallel to $\boldsymbol{\omega}$ through $G$. Indeed, calling $Q_G$ the projection of the center of mass on the axis of rotation, we have $Q_iS_i = Q_GG$. Setting $Q_iP_i = Q_iS_i + S_iP_i = Q_GG + S_iP_i$ it follows that

$$\sum_{i=1}^{n} GP_i \times m_i\omega^2 Q_GG = \left(\sum_{i=1}^{n} m_iGP_i\right) \times \omega^2 Q_GG = \mathbf{0}\,.$$

Now we introduce a frame of reference with origin at the center of mass $G$, and $z$-axis parallel to the axis of rotation. Inserting $GP_i = x_i\mathbf{i} + y_i\mathbf{j} + z_i\mathbf{k}$ and $S_iP_i = GP_i - z_i\mathbf{k}$ in (13.12) gives

$$\begin{aligned}
\mathbf{M}_G &= \sum_{i=1}^{n} \left(x_i\mathbf{i} + y_i\mathbf{j} + z_i\mathbf{k}\right) \times m_i\omega^2\left(x_i\mathbf{i} + y_i\mathbf{j}\right) \\
&= \sum_{i=1}^{n} z_i\mathbf{k} \times m_i\omega^2\left(x_i\mathbf{i} + y_i\mathbf{j}\right) \\
&= -\Big(\sum_{i=1}^{n} m_iy_iz_i\Big)\omega^2\mathbf{i} + \Big(\sum_{i=1}^{n} m_ix_iz_i\Big)\omega^2\mathbf{j} \\
&= \left(I_{23}\mathbf{i} - I_{13}\mathbf{j}\right)\omega^2
\end{aligned}$$

where we have used (5.26) for the inertia products of discrete distributions of particles.

The request that $\mathbf{M}_G = \mathbf{0}$ holds if and only if $I_{13} = I_{23} = 0$, i.e. if and only if the $z$-axis, parallel to $\boldsymbol{\omega}$ through the center of mass, is a central principal axis of inertia. □

## 13.4 Conservative Components of the Apparent Forces

The drag force $\mathbf{F}_\tau$ is positional. The natural question is then whether it is conservative, which would allow to define a potential also in non-inertial frames. We will see in the sequel that in general the answer is no: it is not always possible to find a potential whose gradient equals the drag force. That said, we will find some special cases where the potential does indeed exist.

The first term in the expression of $\mathbf{F}_\tau$ in (13.4) is related to the acceleration of the origin of the non-inertial frame, it is independent of the point's position and is proportional to the point's mass. These features enable one to interpret this force as a sort of weight, with acceleration $-\mathbf{a}_O$ taking the place of the acceleration of gravity $\mathbf{g}$. In any case, if we recall that conservative forces depend explicitly neither on the velocities, nor on time, it becomes evident that the force $-m\,\mathbf{a}_O$ will be conservative precisely when the acceleration $\mathbf{a}_O$ is constant in time. When that happens it is easy to show that

$$-\,m\,\mathbf{a}_O = \operatorname{grad} U_O\,, \quad \text{with} \quad U_O = -m\,\mathbf{a}_O \cdot OP = -m\left(a_{Ox}\,x + a_{Oy}\,y + a_{Oz}\,z\right).$$

The second term of the drag force (13.4) does not admit potential. In fact, we recall that having a potential implies that a positional force has zero curl (see (7.7)). It then suffices to differentiate once to prove that

$$\operatorname{rot}\left(-\,m\,\dot{\boldsymbol{\omega}} \times OP\right) = -m\,\dot{\boldsymbol{\omega}}\,.$$

A further necessary condition for the drag force to be conservative is that the angular velocity of the non-inertial frame be constant. In this case the corresponding term vanishes and can be ignored.

The discussion of the last summand in (13.4) deserves special attention. As already remarked, if the angular velocity $\boldsymbol{\omega}$ of the non-inertial frame is constant, this quantity is called centrifugal force, and is expressed by (13.9), as also made clear in Fig. 13.1.

It is important to note that this force is not only positional but conservative as well. Its potential has a compact and remarkable expression.

**Proposition 13.10** *The centrifugal force is conservative and has potential*

$$U_{\text{cen}} = \frac{1}{2}\,I_\omega\,\omega^2\,, \tag{13.13}$$

*where $I_\omega$ denotes the system's moment of inertia with respect to the rotation axis of the frame.*

***Proof*** Choose the axis $\mathbf{k}$ parallel to $\boldsymbol{\omega}$. Setting $OP = x\mathbf{i} + y\mathbf{j} + z\mathbf{k}$ we have $\mathbf{F}_{\text{cen}} = m\,\omega^2(x\,\mathbf{i} + y\,\mathbf{j}) = \operatorname{grad} U_{\text{cen}}$, with

$$U_{\text{cen}} = \frac{1}{2}\,m\,\omega^2\left(x^2 + y^2\right) = \frac{1}{2}\left(m\,r^2\right)\omega^2\,, \tag{13.14}$$

where $r$ denotes the distance of the particle to the axis of rotation, as in Fig. 13.1. Note that the quantity $mr^2$ in brackets in (13.14) is nothing but the moment of inertia of a point with respect to the axis of rotation. Clearly, for systems made of several particles or rigid bodies, or for continuous systems, the potential $U_{\text{cen}}$ of the centrifugal forces is the sum of the single potentials. Hence it can be written in the form (13.13). □

**Example 13.11** Consider the thin homogeneous square plate (of edge $l$ and mass $m$) in Fig. 13.4. It is constrained to a vertical plane, which rotates with constant

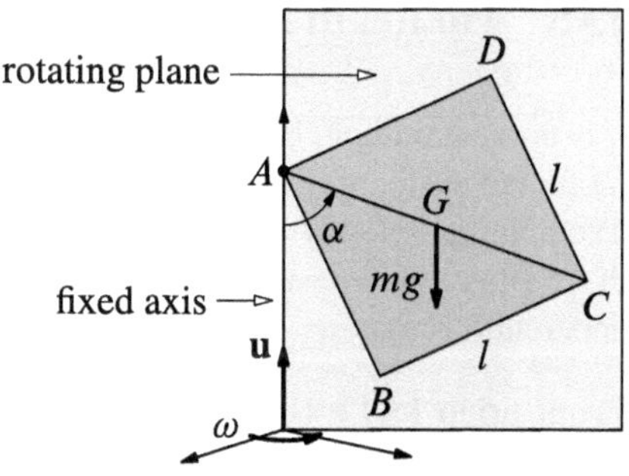


**Fig. 13.4** A square plate constrained to a uniformly rotating plane

angular velocity $\boldsymbol{\omega} = \omega \mathbf{u}$ around the vertical axis through the fixed point $A$. We wish to determine the relative equilibrium positions of the plate on the rotating plane when the only active force is the weight. Based on (13.10) we can reduce the centrifugal force to its resultant at the center of mass, since the principal moments of a square plate coincide. Call $\alpha$ the angle between $AG$ and the vertical direction, and set $\omega_0 = (\sqrt{2}g/l)^{1/2}$. The second balance equation from (Relative) Statics with respect to pole $A$ gives

$$\sin\alpha\big(\omega^2\cos\alpha - \omega_0^2\big) = 0 \implies \begin{cases} \sin\alpha = 0 \\ \cos\alpha = \omega_0^2/\omega^2, \quad \text{if } \omega \geq \omega_0 \end{cases} \tag{13.15}$$

We can find the equilibrium configurations (13.15) using the characterization of the potential's critical points, here applied to the centrifugal potential (13.13). Call $u'$ the vertical axis through $G$. Since the plate has two equal principal moments, $u'$ is principal as well at $G$, with the same moment $I_{u'} = ml^2/12$. The Huygens-Steiner theorem says $I_u = I_{u'} + mx_G^2 = ml^2/12 + ml^2/2\sin^2\alpha$, and the potential equals

$$U(\alpha) = -mgy_G(\alpha) + \frac{1}{2}\,I_\omega(\alpha)\,\omega^2 = \frac{\sqrt{2}}{2}mgl\cos\alpha + \frac{1}{4}m\omega^2 l^2\sin^2\alpha + \text{constant},$$

with obvious stationary points (13.15). Finding the potential also allows to study the statical stability of the equilibria. Applying what we found in Sect. 9.10 we obtain

$$U''(0) = \frac{1}{2}ml^2\big(\omega^2 - \omega_0^2\big) \qquad \begin{cases} \text{stable if } \omega \leq \omega_0 \\ \text{unstable if } \omega > \omega_0 \end{cases}$$

$$U''(\pi) = \frac{1}{2}ml^2\big(\omega^2 + \omega_0^2\big) \qquad \text{unstable}$$

$$U''\Big(\arccos\frac{\omega_0^2}{\omega^2}\Big) = \frac{1}{2}\frac{ml^2}{\omega^2}(\omega_0^4 - \omega^4) \qquad \text{stable if } \omega \geq \omega_0$$

where we have used the fact that in the critical case $\omega = \omega_0$ we have $U'''(0) = 0$ and $U^{(iv)}(0) = -3mgl/\sqrt{2}$. □

## 13.5 Motion in Non-inertial Planar Frames

It is not uncommon, because of the several applications, to study the equilibrium or motion of systems of points and rigid bodies that are constrained to a rotating plane. The angular velocity may be constant or not with respect to an inertial frame. For these situations it is useful to keep in mind a number of special properties of the apparent forces.

**Proposition 13.12** (Planar systems) *The Coriolis forces do not affect the relative dynamics of a planar rigid body, if the latter is in a frame that rotates around an axis lying on the plane.*

***Proof*** Consider a rigid system of particles as per the hypothesis. The resultant moment of the Coriolis forces is

$$\mathbf{M}_O = -2\sum_{i=1}^{n} m_i OP_i \times \big(\boldsymbol{\omega} \times \mathbf{v}_i\big) = -2\Big(\sum_{i=1}^{n} m_i\, OP_i \cdot \mathbf{v}_i\Big)\boldsymbol{\omega} + 2\sum_{i=1}^{n} m_i\,(OP_i \cdot \boldsymbol{\omega})\mathbf{v}_i\,. \tag{13.16}$$

As $\boldsymbol{\omega}$ and $\mathbf{v}_G$ belong on the plane of motion, the resultant (13.6) is orthogonal to the plane. On the other hand (13.16) proves that the resultant moment lies on the plane of motion. Hence the Coriolis forces cannot interfere with the rigid body's motion. The latter is determined by the components of the resultant that belong to the plane and by the component of the moment orthogonal to the plane. □

**Remark 13.13** The above property depends critically on the fact that the axis of rotation lies on the plane of motion. If in fact $\boldsymbol{\omega}$ had an orthogonal component, the resultant (13.6) would have a component on the plane, and the first summand in (13.16) would acquire an orthogonal component. □

### *13.5.1 Uniformly Rotating Planar Systems*

Consider a planar system, whose plane of motion keeps in uniform rotation around some axis lying on the plane itself, so that the constant angular velocity $\boldsymbol{\omega}$ belongs to the plane of motion. Consider two frames, whose common origin $O$ is on the axis. We suppose one frame (fixed) is inertial, while the second, comoving with the plane, rotates with respect to the fist with the same angular velocity as the plane.

Excluding for the time being the constraint of belonging to the plane, let us assume any other constraints acting on the system (pins, rollers or other) are ideal, bilateral and fixed, so that their power is zero. As regards the active forces we suppose they are conservative, with potential $U^{(\mathrm{a})}$.

By studying the motion in the (non-inertial) comoving frame we can conclude the following. All constraints, including that of belonging to the plane, here fixed, are ideal, bilateral and fixed, and hence have zero power. As for the active forces, they are conservative. In fact, besides the forces with potential $U^{(a)}$ we must consider the drag forces, which reduce to the centrifugal force and have potential (see (13.13))

$$U_{\text{cen}} = \frac{1}{2}\, I_\omega\, \omega^2\,, \tag{13.17}$$

where $I_\omega$ is the moment of inertia with respect to the axis of rotation of the frame. Call $\mathbf{v}_{i\text{r}}$ the velocity of the $i$-th particle in the relative frame, and define the *relative* kinetic energy

$$T_{\text{r}} = \frac{1}{2}\sum_{i=1}^{n} m_i v_{i\text{r}}^2.$$

From the Kinetic energy theorem we know that

$$\dot{T}_{\text{r}} = \Pi = \dot{U}^{(a)} + \dot{U}_{\text{cen}} \quad \Longrightarrow \quad T_{\text{r}} - U^{(a)} - U_{\text{cen}} \equiv \text{constant} = E. \tag{13.18}$$

Equation (13.18) is a free motion equation, and if the system has one degree of freedom the equation is enough to determine its motion.

Now we will study the system in the inertial frame. Observe that the constraints are no longer fixed, since the rotating plane represents a moving constraint. In fact, if we wish to maintain the system's rotation at constant angular velocity we need to apply a motion that supports the rotation. The Kinetic energy theorem, in this frame, allows to express the power of this motion in a simple way. Consider the *absolute* kinetic energy $T_{\text{a}}$ measured in the fixed frame. It is built with the squares of the absolute velocities $\mathbf{v}_i = \mathbf{v}_{i\text{r}} + \mathbf{v}_{i\tau}$, where the drag velocities are $\mathbf{v}_{i\tau} = \boldsymbol{\omega} \times OP_i$. Due to the system's symmetry, $OP_i$ and $\boldsymbol{\omega}$ lie on the rotating plane, so $\mathbf{v}_{i\tau}$ will be orthogonal to it, and in particular orthogonal to $\mathbf{v}_{i\text{r}}$. Therefore

$$v_i^2 = v_{i\text{r}}^2 + 2\mathbf{v}_{i\text{r}} \cdot \mathbf{v}_{i\tau} + v_{i\tau}^2 = v_{i\text{r}}^2 + v_{i\tau}^2\,.$$

Therefore the kinetic energy $T_{\text{a}}$ decomposes as a sum of two kinetic energies: the relative one and the drag kinetic energy: $T_{\text{a}} = T_{\text{r}} + T_\tau$. The latter, moreover, has the exact same expression as the potential of the centrifugal forces (13.17):

$$T_\tau = U_{\text{cen}} = \frac{1}{2}\, I_\omega\, \omega^2\,, \tag{13.19}$$

since $\mathbf{v}_{i\tau}$ represents a rigid rotation of angular velocity $\boldsymbol{\omega}$. As for the power of the forces acting on the system, it will need to include the power $\Pi^{(v)}$ of the motion that maintains the plane's uniform rotation and the power $\Pi^{(a)} = \dot{U}^{(a)}$ of the active forces. It will not contain the power of the apparent forces, as the frame is inertial. Consequently

$$\dot{T}_{\text{a}} = \dot{T}_{\text{r}} + \dot{T}_\tau = \Pi^{(v)} + \dot{U}^{(a)}\,. \tag{13.20}$$

Inserting (13.18) and (13.19) in (13.20) produces the following interesting result:

$$\left(\dot{U}^{(a)} + \dot{U}_{cen}\right) + \dot{U}_{cen} = \Pi^{(v)} + \dot{U}^{(a)} \quad \Longrightarrow \quad \Pi^{(v)} = 2\dot{U}_{cen}\,.$$

The power required to maintain a planar system in uniform rotation (around some axis that lies on the plane) is twice the relative power of the centrifugal forces.

## 13.6 The Weight

An interesting problem in Relative Statics is that of a particle $P$ of mass $m$ in equilibrium on the surface of the Earth at latitude $\lambda$ (Fig. 13.5). The absolute force acting on the point is the gravitational force

$$\mathbf{F} = h\frac{mM_T}{r^2}\mathbf{u}, \qquad \text{with} \quad \mathbf{u} = \frac{P\Omega}{|P\Omega|}$$

where $h$ is the Cavendish constant ($h \approx 6.67 \times 10^{-11}$ m$^3$ kg$^{-1}$ s$^{-2}$), $\Omega$ is the Earth's center and $M_T$ is its mass ($M_T \approx 5.98 \times 10^{24}$ kg). Supposing the Earth rotates around its axis with roughly constant angular velocity $\omega_T$, the drag force is the centrifugal force and $\mathbf{F}_\tau = m\omega_T^2 QP$. Denote by $\mathbf{\Phi}$ the reaction the surface of the Earth exerts on the point, so

$$h\frac{mM_T}{r^2}\mathbf{u} + m\omega_T^2 QP + \mathbf{\Phi} = \mathbf{0}. \tag{13.21}$$

By introducing the *acceleration of gravity*

$$\mathbf{g} = h\frac{M_T}{r^2}\mathbf{u} + \omega_T^2 QP\,. \tag{13.22}$$

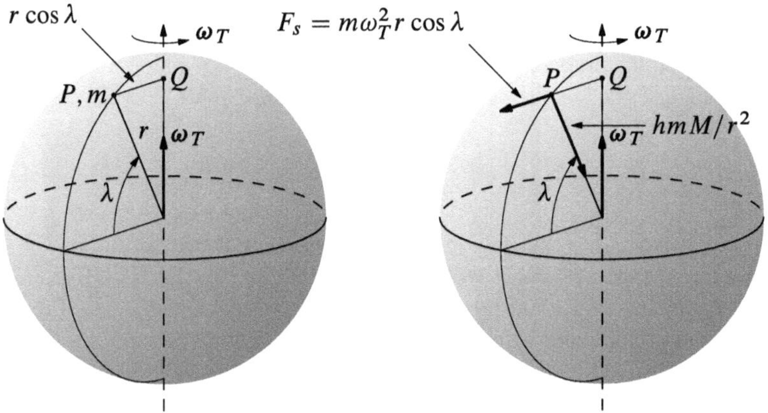


**Fig. 13.5** Equilibrium of a particle on the surface of the Earth

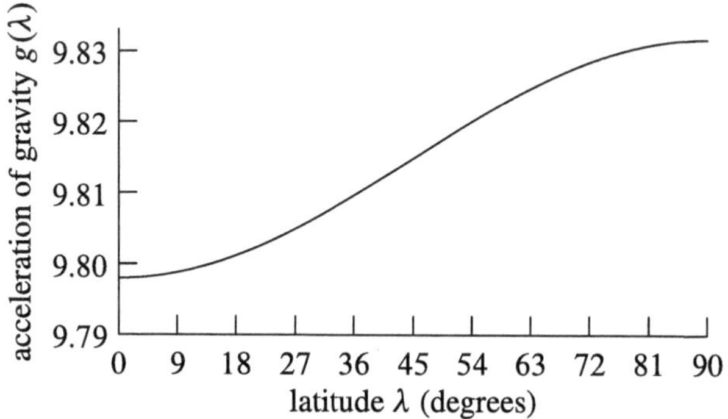


**Fig. 13.6** The intensity of the acceleration of gravity $g$ as a function of the latitude $\lambda$

Equation (13.21) can be formally written as an equation of inertial statics:

$$m\mathbf{g} + \mathbf{\Phi} = \mathbf{0}.$$

The force $m\mathbf{g}$ is the so-called *weight*. What we have shown is that the weight is the vector sum of the gravitational force and the centrifugal force. The acceleration of gravity, hence the weight, depend on the latitude by way of the centrifugal force: a body is heavier at the poles, where the drag force is zero, and becomes increasingly lighter as we approach the equator, where the drag force is maximal and opposite to the gravitational force. In general, from (13.22) we deduce:

$$g(\lambda) = \sqrt{h^2 \frac{M_T^2}{r^4} \sin^2 \lambda + \left( h \frac{M_T}{r^2} - \omega_T^2 r \right)^2 \cos^2 \lambda}\,,$$

where $\lambda$ and $r$ respectively represent the latitude of particle $P$ and its distance from the center of the Earth.

Figure 13.6 shows the variation of the modulus of the acceleration of gravity $g$ with the latitude $\lambda$, measured in degrees, for a particle $P$ on the Earth's surface.

## 13.7 Two-Body Problem

The term *two-body problem* refers to the motion of two particles $P_1$, $P_2$ of masses $m_1, m_2$, that interact between themselves but are otherwise isolated, studied in the translating non-inertial frame with origin at one of the points (we will assume it is $P_1$). The motion of $P_1$ in this frame is trivial, since it occupies the origin and hence is at relative rest. It is far from easy to imagine what the motion of $P_2$ will be. Although Newton's third law says that the force on $P_2$ will always point to the origin $P_1$, we cannot be certain that the considerations of Sect. 10.5 apply to $P_2$, since on $P_2$ there will be apparent forces as well. In this case (translating frame) the apparent force of concern is the drag force $\mathbf{F}_\tau = -m_2\mathbf{a}_1$. The historical interest in the problem is that it permits to study the motion of a celestial body $P_2$ (the Sun, say), using a frame with origin at another celestial body $P_1$ (representing the Earth), under the approximating hypothesis that the bodies' interactions with other planets or satellites are imperceptible.

To study the relative motion of $P_2$ it is anyhow useful to write Newton's equation for the two points with respect to an inertial frame

$$m_1\mathbf{a}_1 = \mathbf{F}_{12}, \qquad m_2\mathbf{a}_2 = \mathbf{F}_{21}\,, \tag{13.23}$$

with

$$\mathbf{F}_{21} = -\mathbf{F}_{12} = \Psi(r)\,\mathbf{u}, \qquad r = \left|\mathbf{r}\right|, \qquad \mathbf{u} = \frac{\mathbf{r}}{r}, \quad \text{and} \quad \mathbf{r} = P_1P_2\,. \tag{13.24}$$

We emphasize that the force acting on each point *is not* central, albeit always oriented towards the other point. In fact both forces are directed towards moving points, while the definition of central force requires the center to be fixed.

### 13.7.1 Motion of the Center of Mass

Formulas (13.24) allow to integrate trivially the motion of the center of mass of the points. Indeed if we add the equations of motion (13.23) we obtain

$$\mathbf{0} = m_1\mathbf{a}_1 + m_2\mathbf{a}_2 = \frac{d}{dt}\big(m_1\mathbf{v}_1 + m_2\mathbf{v}_2\big) \implies \mathbf{v}_G = \text{constant},$$

where

$$OG = \frac{m_1\,OP_1 + m_2\,OP_2}{m_1 + m_2}$$

indicates the position of the center of mass. The latter moves uniformly along a straight line in every inertial frame.

### 13.7.2 Relative Motion

Let us concentrate on the motion of $P_2$ *relative* to $P_1$, where by motion relative to a point we mean a motion with respect to a translating frame with origin at the point in question. The position vector of $P_2$ in this frame is precisely the vector $\mathbf{r}$ defined in (13.24). Taking in account the drag force in Newton's law for $P_2$, and using (13.23), we obtain

$$m_2\ddot{\mathbf{r}} = \mathbf{F}_{21} + \mathbf{F}_\tau = \mathbf{F}_{21} - m_2\mathbf{a}_1 = \mathbf{F}_{21} - \frac{m_2}{m_1}\mathbf{F}_{12} = \frac{m_1 + m_2}{m_1}\mathbf{F}_{21}\,, \tag{13.25}$$

whence

$$\mu\ddot{\mathbf{r}} = \mathbf{F}_{21} = \Psi(r)\mathbf{u}\,, \tag{13.26}$$

where

$$\mu = \left(\frac{1}{m_1} + \frac{1}{m_2}\right)^{-1} = \frac{m_1 m_2}{m_1 + m_2}$$

is the *reduced mass* of the system, which enjoys the following properties.

- The reduced mass is less than either of the two masses that build it:

$$\mu = \frac{m_1 m_2}{m_1 + m_2} = m_1\,\frac{m_2}{m_1 + m_2} < m_1$$

  (similarly, $\mu < m_2$).
- When one mass is much bigger than the other mass ($m_2 \gg m_1$) the reduced mass tends to the smaller one:

$$\mu = \frac{m_1 m_2}{m_1 + m_2} = m_1\left[1 - \frac{m_1}{m_2} + o\left(\frac{m_1}{m_2}\right)\right] \qquad \text{as} \quad \frac{m_1}{m_2} \to 0\,.$$

- If the total mass is fixed, the reduced mass is largest when $m_1 = m_2$. In fact, setting $m_1 + m_2 = M$ and $m_1 = \alpha M$, $m_2 = (1 - \alpha)M$, we have

$$\mu(\alpha) = \alpha(1 - \alpha)M \quad \Longrightarrow \quad \mu_{\max} = \mu(1/2) = \frac{M}{4}\,.$$

Equation (13.26) proves that *the motion of $P_2$ relative to $P_1$ is central.* In fact, in a frame translating with $P_1$ the force exerted on $P_2$ always points towards the origin, and its modulus only depends on the distance. The motion of $P_2$ is then that of a hypothetical point of mass $\mu$ if, in an inertial system, it interacted under (13.24) with a fixed point. It is interesting to note that the only effect of the drag force in (13.25) is to *renormalize* the mass of $P_2$, replacing $m_2$ with the reduced mass $\mu$.

### 13.7.3 A Critique of Kepler's Laws

That Kepler's laws do not hold exactly is to be expected, since the planets are affected not just by the solar attraction but also by the attraction of their own satellites, other planets, and in general every other celestial body. This effect is anyhow truly small and may, in a first approximation, be disregarded.

A more serious mistake would arise if we disregarded that the motion of the Earth-Sun system is not central, even if we study it in an inertial frame, since the forces of attraction are not directed towards fixed points.

The study of the two-body problem shows though that planetary motion, with respect to the Sun, is central, even when we include the Sun's motion (and therefore that the frame with origin at the Sun is not inertial). The first two laws hold, while the third one must be corrected. In fact, it requires (see (10.63)) that the radial acceleration of any planet is

$$a_r = -\frac{K}{r^2},$$

with multiplicative constant $K$ independent of the planet under exam. Calling $m$ and $M$ the masses of planet and Sun, (13.26) shows instead that the radial acceleration equals

$$a_r = \frac{1}{\mu}\left(-\frac{hmM}{r^2}\right) = -\frac{h\big(m+M\big)}{r^2}.$$

Replacing the planet's mass with the reduced mass implies that Kepler's third law holds up to an approximating error, which is as small as the ratio of the planet's mass over the Sun's.

## 13.8 Eastward Deviation of Free-Falling Objects

In this section we will study the effect of terrestrial rotation on simple relative motions of heavy particles. To do that we will suppose Earth is a sphere, in uniform rotation around a fixed axis through the poles, with angular velocity

$$\boldsymbol{\omega}_{\mathrm{T}} = \omega_{\mathrm{T}}\,\mathbf{k}\,, \qquad \text{with} \quad \omega_{\mathrm{T}} = 2\pi/(24\mathrm{h}) \doteq 7.272\times 10^{-5}\mathrm{s}^{-1}\,.$$

Figure 13.7 shows the frame we shall use to project the various position vectors. Assuming we are on the northern hemisphere (positive latitude $\lambda$), we define a unit vector $\mathbf{e}_1$ pointing eastward and another unit vector $\mathbf{e}_2$, still tangent to the surface, pointing north. The radial unit vector $\mathbf{e}_3 = \mathbf{e}_1\times\mathbf{e}_2$ completes the former two to give an orthonormal basis. In particular, $\mathbf{k} = \cos\lambda\,\mathbf{e}_2 + \sin\lambda\,\mathbf{e}_3$.

Consider a non-inertial frame, comoving with the surface, with origin at the center of the Earth $O$. Define the pole $Q$, comoving with Earth, so that $OQ = R_{\mathrm{T}}\,\mathbf{e}_3$. The position of any point $P$ is given by three coordinates $(x, y, z)$ such that $OP = OQ + QP = R_{\mathrm{T}}\,\mathbf{e}_3 + x\,\mathbf{e}_1 + y\,\mathbf{e}_2 + z\,\mathbf{e}_3$. The drag force, which in our hypotheses is just the centrifugal force, is

$$\mathbf{F}_{\mathcal{T}} = \mathbf{F}_{\mathrm{cen}} = -m\,\boldsymbol{\omega}_{\mathrm{T}}\times(\boldsymbol{\omega}_{\mathrm{T}}\times OP) = -m\omega_{\mathrm{T}}^2\mathbf{k}\times\big(\mathbf{k}\times(R_{\mathrm{T}}\mathbf{e}_3 + x\,\mathbf{e}_1 + y\,\mathbf{e}_2 + z\,\mathbf{e}_3)\big). \tag{13.27}$$

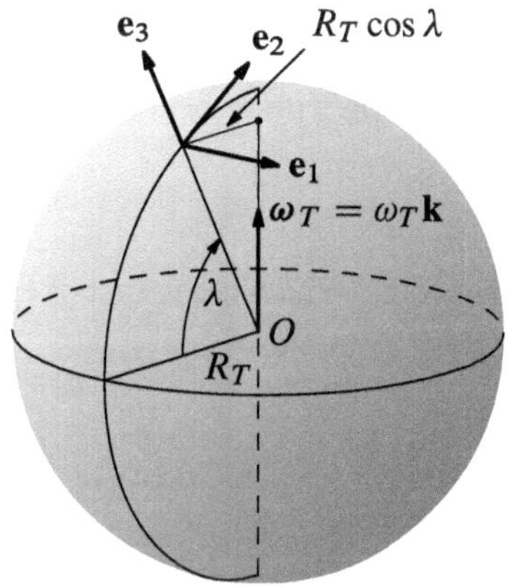


**Fig. 13.7** A model of the Earth as a sphere under uniform rotation. The basis used in the text is the following one: axis $\mathbf{e}_1$ is oriented to the East, axis $\mathbf{e}_2$ is oriented to the North, while axis $\mathbf{e}_3 = \mathbf{e}_1 \times \mathbf{e}_2$ is radial. The unit vector $\mathbf{k}$ points from the south pole to the north pole

As $R_{\mathrm{T}}$ is much larger than any one of the coordinates $(x, y, z)$, we may approximate $OP \approx R_{\mathrm{T}}\, \mathbf{e}_3$ in (13.27), thus obtaining

$$\mathbf{F}_{\tau} \approx m\omega_{\mathrm{T}}^2 R_{\mathrm{T}} \cos\lambda\,(-\sin\lambda\,\mathbf{e}_2 + \cos\lambda\,\mathbf{e}_3).$$

To assess the Coriolis force let us take a point $P$ in motion with relative velocity $\mathbf{v} = \dot{x}\,\mathbf{e}_1 + \dot{y}\,\mathbf{e}_2 + \dot{z}\,\mathbf{e}_3$. Acting on it is a force

$$\mathbf{F}_{\mathrm{c}} = -2m\boldsymbol{\omega}_{\mathrm{T}} \times \mathbf{v} = 2m\omega_{\mathrm{T}}\big[(\sin\lambda\,\dot{y} - \cos\lambda\,\dot{z})\,\mathbf{e}_1 - \sin\lambda\,\dot{x}\,\mathbf{e}_2 + \cos\lambda\,\dot{x}\,\mathbf{e}_3\big]\,. \tag{13.28}$$

The centrifugal force is positional, so its effect is felt irrespective of the state of motion or rest of the point in question. At any rate, this force adds up (vectorially) with the Earth's gravitational attraction, hence it changes in modulus and direction in function of the latitude $\lambda$. We have measured in Sect. 13.6 the order of magnitude of such an effect.

Next we analyze the effect of the terrestrial rotation, more precisely the Coriolis force, on free-falling objects. Since the effect we will find is small, in our discussion we will neglect the centrifugal effect, and identify the vertical direction with that of the radial vector $\mathbf{e}_3$.

Consider the free fall of a particle of mass $m$, initially at rest and subject to the weight $m\mathbf{g} = -mg\,\mathbf{e}_3$ and to the Coriolis force (13.28). Projecting Newton's law along the basis in Fig. 13.7 gives:

$$\begin{aligned} m\ddot{x} &= 2m\omega_{\mathrm{T}}\,(\sin\lambda\,\dot{y} - \cos\lambda\,\dot{z})\,, \\ m\ddot{y} &= -2m\omega_{\mathrm{T}}\dot{x}\,\sin\lambda\,, \\ m\ddot{z} &= -mg + 2m\omega_{\mathrm{T}}\dot{x}\,\cos\lambda\,. \end{aligned} \tag{13.29}$$

To describe how an object falls freely from a height $h$, we solve (13.29) with initial conditions

$$x(0) = y(0) = 0\,, \quad z(0) = h; \qquad \dot{x}(0) = \dot{y}(0) = \dot{z}(0) = 0\,.$$

Integrating with respect to time the second and third equations in (13.29) we obtain

$$\dot{y} = -2\omega_{\rm T} x \sin\lambda \quad \text{and} \quad \dot{z} = -gt + 2\omega_{\rm T} x \cos\lambda\,. \tag{13.30}$$

Substituting now in the first of (13.29) we find:

$$\ddot{x} + 4\omega_{\rm T}^2 x = 2\omega_{\rm T} \cos\lambda\, gt\,. \tag{13.31}$$

Solving (13.31) and then (13.30) we arrive at

$$\begin{aligned} x(t) &= \frac{g\cos\lambda}{4\omega_{\rm T}^2}\,(2\omega_{\rm T}t - \sin 2\omega_{\rm T}t) \\ y(t) &= \frac{g\sin\lambda\cos\lambda}{4\omega_{\rm T}^2}\left(1 - \cos 2\omega_{\rm T}t - 2\omega_{\rm T}^2 t^2\right) \\ z(t) &= h - \frac{1}{2}gt^2 - \frac{g\cos^2\lambda}{4\omega_{\rm T}^2}\left(1 - \cos 2\omega_{\rm T}t - 2\omega_{\rm T}^2 t^2\right). \end{aligned} \tag{13.32}$$

But $\omega_{\rm T} \doteq 7.272 \times 10^{-5}{\rm s}^{-1}$, so evidently the product $\omega_{\rm T}t$ will be small provided we do not consider motions that last hours.[1] We can understand the magnitude of the corrections occurring in solutions (13.32) via a Taylor expansion of the functions in brackets, for small $\omega_{\rm T}t$:

$$\begin{aligned} x(t) &= \frac{g\cos\lambda}{3\omega_{\rm T}^2}\left(\omega_{\rm T}^3 t^3 + O(\omega_{\rm T}^5 t^5)\right) \\ y(t) &= -\frac{g\sin\lambda\cos\lambda}{6\omega_{\rm T}^2}\left(\omega_{\rm T}^4 t^4 + O(\omega_{\rm T}^6 t^6)\right) \\ z(t) &= h - \frac{1}{2}gt^2 + \frac{g\cos^2\lambda}{6\omega_{\rm T}^2}\left(\omega_{\rm T}^4 t^4 + O(\omega_{\rm T}^6 t^6)\right). \end{aligned}$$

[1] The approximation should be considered more than acceptable, because in one hour a particle would travel more than 10Km, in which case one should even reconsider the assumption the acceleration of gravity does not depend on the height.

Considering a fall time $\tau$ such that $h = \frac{1}{2}g\tau^2$, we obtain:

$$\begin{aligned} x(\tau) &= \frac{2h\cos\lambda}{3}\left(\omega_{\mathrm{T}}\tau + O(\omega_{\mathrm{T}}^3\tau^3)\right) \\ y(\tau) &= -\frac{h\sin\lambda\cos\lambda}{3}\left(\omega_{\mathrm{T}}^2\tau^2 + O(\omega_{\mathrm{T}}^4\tau^4)\right) \\ z(\tau) &= \frac{h\cos^2\lambda}{3}\left(\omega_{\mathrm{T}}^2\tau^2 + O(\omega_{\mathrm{T}}^4\tau^4)\right). \end{aligned}$$

The dominant correction term appears in $x(\tau)$: as the sign is positive, it indicates a deflection towards the East. The size of the deflection depends on how long the object falls. For instance, suppose $h \approx 20\,\mathrm{m}$. The fall time will have order $\tau = \sqrt{2h/g} \approx 2\,\mathrm{s}$, so $\omega_{\mathrm{T}}\tau \approx 1.5 \times 10^{-4}$. Putting $\lambda = \pi/4$ (the average approximate latitude of northern Italy), the eastward deflection will have order

$$x(\tau) \approx 7 \times 10^{-5}\,h = 1.4\,\mathrm{mm}.$$

Roughly one millimeter, all in all: small, but definitely detectable.

For completeness, we also point out that the smaller corrections, along $\mathbf{e}_2$ and $\mathbf{e}_3$ have a sign that forces the object to deflect very slightly to the equator, accelerating the fall. At last, we emphasize that one can use a similar discussion to address objects tossed with non-zero initial velocity. The result is that the direction and orientation of the dominant correction term to the object's orbit depend, in this case, on the direction of the initial velocity.

## 13.9 Non-inertial Frames: Examples

**Example 13.14** (*Relative equilibrium*) Consider the problem from Relative Statics depicted in Fig. 13.8. Let $D$ be a rough disk, rotating with constant angular velocity $\boldsymbol{\omega}$ around its vertical symmetry axis, and $P$ a point of mass $m$ on the disk. The coefficient of static friction is $\mu_s$. We want to know if the point can stay in equilibrium with respect to a frame comoving with the disk, and if yes under which conditions.

The drag force coincides with the centrifugal force: $\mathbf{F}_\tau = m\omega^2 OP$, since $\boldsymbol{\omega}$ is constant and $Q \equiv O$. The equilibrium condition becomes

$$m\mathbf{g} + m\omega^2 OP + \boldsymbol{\Phi} = \mathbf{0},$$

which projected along the vertical axis and radial direction decomposes as

$$-mg + \Phi_z = 0 \qquad m\omega^2 r + \Phi_r = 0 \tag{13.33}$$

(refer to Fig. 13.8). The Coulomb-Morin inequality (8.23) in this case reads

$$|\Phi_r| \leq \mu_s|\Phi_z|$$

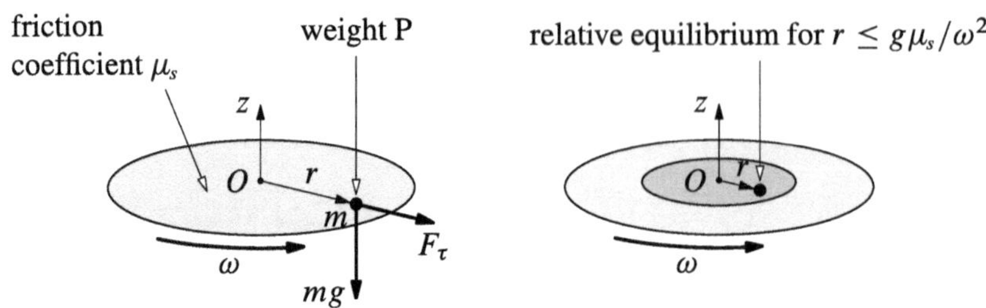


**Fig. 13.8** An example of equilibrium relative to a non-inertial frame: a point on a rough rotating plane

or equivalently

$$\omega^2 r \leq \mu_s g$$

in the light of (13.33). There will be equilibrium at all positions contained inside the circle of radius $R = \mu_s g/\omega^2$ with center $O$. Obviously the smaller we take $\mu_s$, the smaller this radius is. It shrinks to zero if $\mu_s = 0$. This last remark says that only in presence of friction there will be relative equilibrium positions for $P$ other than the origin. □

**Example 13.15** Figure 13.9 shows a system of two perpendicular rods $OD$ and $AB$ of respective lengths $l + r$ and $4r$, rigidly welded at $D$. The system rotates on a *horizontal* plane around the fixed point $O$ with *constant* angular velocity $\omega$ with respect to an inertial frame, under a suitable couple with torque $M$, a function of time, caused by a motor.

A homogeneous disk of radius $r$ and mass $m$ rolls without slipping along $AB$. A spring of elastic constant $k$ parallel to $AB$ attached at $H$ to one rod acts on the center $G$. The disk is initially at relative rest with respect to the rods, with spring length $s$ initially equal to $2r$, so that the contact point $C$ coincides with $A$.

First things first, we wish to write down the equation of motion of the disk, using the angle $\theta$ that measures the rotation *relative* to the frame comoving with the rods, then find the relative angular velocity $\dot{\theta}$ in the configuration where the spring length is zero, $C = D$, and finally verify which condition allows this particular configuration to occur.

It is convenient to take the point of view of the frame with $\omega$ constant. Apart from the effective forces, among which there is the elastic force, this frame will seeÂ the apparent forces (drag and Coriolis). In this case the drag force is just the centrifugal force, which as we know is conservative. The Coriolis forces, on the other hand, do not do work because they are always orthogonal to the velocities of the points at which they are applied, so they do not partake in the Kinetic energy theorem nor in the conservation law of mechanical energy. For the rotating frame, the system is subject to ideal and fixed constraints, and conservative active forces (spring and centrifugal force), so we are entitled to write $T - U = E$. The disk's kinetic energy is due to its rotational velocity distribution around $C$, so

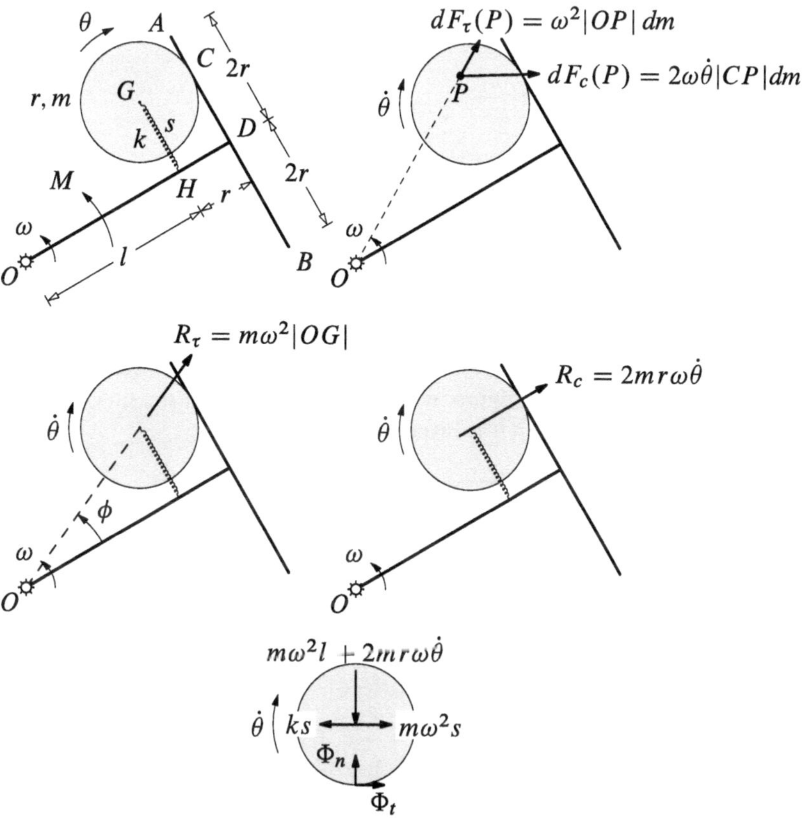


**Fig. 13.9** A disk rolling without slipping on a rod which is uniformly rotating with respect to an inertial frame

$$T^{\mathrm{rel}} = \frac{1}{2}\left(\frac{3mr^2}{2}\right)\dot\theta^2 = \frac{3mr^2}{4}\dot\theta^2 .$$

The elastic potential is $U^{\mathrm{spring}} = -ks^2/2$ while for the centrifugal force it is better to use the relation

$$\begin{aligned} U_\tau &= \frac{1}{2}I_O\omega^2 = \frac{1}{2}\left(\frac{mr^2}{2} + m|OG|^2\right)\omega^2 = \frac{1}{2}\left(\frac{mr^2}{2} + m(s^2 + l^2)\right)\omega^2 \\ &= \frac{1}{2}ms^2\omega^2 + \mathrm{cost}. \end{aligned}$$

Observe that $s = r\theta$. If we put $s = 2r$ at the initial time, the initial conditions will read

$$\theta(0) = 2, \quad \dot\theta(0) = 0 ,$$

and the conservation of the energy $T - U = T_0 - U_0$ becomes

$$\frac{3mr^2}{4}\dot{\theta}^2 + kr^2\frac{\theta^2}{2} - \frac{1}{2}mr^2\theta^2\omega^2 = 2kr^2 - 2mr^2\omega^2 . \tag{13.34}$$

Now, putting $\theta = 0$ in the first term, the squared (relative) angular velocity at the final time $\dot{\theta}_f^2$ will be

$$\dot{\theta}_f^2 = \frac{8}{3m}(k - m\omega^2)$$

from which we deduce that the final configuration can be reached only if $k \geq m\omega^2$ (for otherwise $\dot{\theta}_f^2 < 0$). A more careful analysis, or just intuitive considerations, show that the orientation of $\theta$ in the figure, when the disk passes for the first and second time through the midpoint of $AB$ is respectively determined by

$$\dot{\theta} = -\sqrt{\frac{8}{3m}(k - m\omega^2)}, \quad \dot{\theta} = +\sqrt{\frac{8}{3m}(k - m\omega^2)},$$

($\dot{\theta}$ is negative during the motion from $A$ to $B$ and positive from $B$ to $A$).

It is interesting to find the pure equation of motion, which is of order two, by differentiating (13.34) with respect to time. Easy calculations give

$$\ddot{\theta} + \frac{2(k - m\omega^2)}{3m}\theta = 0, \tag{13.35}$$

a linear homogeneous equation with constant coefficients. Suppose $k \geq m\omega^2$ and set

$$\Omega^2 = \frac{2(k - m\omega^2)}{3m}$$

so that (13.35) turns into an equation for harmonic motion,

$$\ddot{\theta} + \Omega^2\theta = 0,$$

The general integral is

$$\theta(t) = A\sin(\Omega t) + B\cos(\Omega t), \quad \text{where} \quad \Omega = \sqrt{\frac{2(k - m\omega^2)}{3m}}. \tag{13.36}$$

Imposing the initial conditions $\theta(0) = 2, \dot{\theta}(0) = 0$ we find the motion and the angular velocity of the disk:

$$\theta(t) = 2\cos(\Omega t), \quad \dot{\theta}(t) = -2\Omega\sin(\Omega t). \tag{13.37}$$

What would happen if $k < m\omega^2$? Equation (13.35) would become

$$\ddot{\theta} - \Omega^2\theta = 0\,, \quad \text{where} \quad \Omega^2 = \frac{2(m\omega^2 - k)}{3m}$$

(note the different $\Omega^2$).

Now the general integral is

$$\theta(t) = A\exp(\Omega t) + B\exp(-\Omega t), \quad \text{with} \quad \Omega = \sqrt{\frac{2(m\omega^2 - k)}{3m}}\,,$$

and the initial conditions yield the motion

$$\theta(t) = \exp(\Omega t) + \exp(-\Omega t)$$

which is no longer oscillating. The angle $\theta$ would increase from the initial configuration and make the disk move away from $O$, beyond $A$ (if the rod $AB$ were ideally infinitely long...)

Let us explain how to compute the components $\Phi_t$ and $\Phi_n$ of the constraint reaction the rod exerts on the disk. For this we can resort to the balance equations, and we must understand the role played by the apparent forces acting on the disk. Hence consider a generic point $P$ where we imagine to place an infinitesimal mass $dm$, as in the second picture of Fig. 13.9. The infinitesimal drag force $d\mathbf{F}_\tau$ (centrifugal, in this case) is radial, along $OP$, and has component $\omega^2|OP|dm$. The Coriolis force $d\mathbf{F}_c = -2\boldsymbol{\omega} \times \mathbf{v}_r dm$, instead, is on the plane of motion and perpendicular to the relative velocity. Hence it is aligned to the segment through $P$ and $C$, the disk's instantaneous center of rotation. The component of this force, oriented by the arrow, is equal to $2\omega\dot{\theta}|CP|dm$. In the phases where $\dot{\theta} > 0$ the Coriolis force tends to push the disk towards the rod, and tends to move it away if $\dot{\theta} < 0$.

The resultant $\mathbf{R}_\tau$ of the centrifugal forces is computed at the center of mass, so it equals $m\omega^2 OG$. For the line of action we can exploit the fact that, by the previous observations, the total momentum of the centrifugal forces with respect to $O$ is zero, so the line of action of $\mathbf{R}_\tau$ passes through $O$ and $G$, as shown in the third picture of Fig. 13.9. This force decomposes into a part parallel to $AB$ and a part perpendicular to it. To see this observe that the angle $\phi$ in the picture satisfies

$$|OG|\cos\phi = l, \quad |OG|\sin\phi = s\,,$$

so the tangent and normal components of $\mathbf{R}_c$ are equal to $m\omega^2 s$ (from $C$ to $A$) and $m\omega^2 l$ (from $G$ to $C$).

Also the resultant of the Coriolis forces can be computed at the center of mass, to give $\mathbf{R}_c = 2m\omega\dot{\theta}GC$. As the momentum with respect to $C$ of the Coriolis forces is necessarily zero (all pass through $C$), the line of action of $\mathbf{R}_c$ must pass through $G$ and $C$.

To sum up, the components of the forces acting on the disk are represented in the bottom of Fig. 13.9. We may write the first balance equation for the (relative) motion of the disk to get

$$\Phi_t + (m\omega^2 - k)r\theta = mr\ddot{\theta}\,, \quad \Phi_n - m\omega^2 l - 2m\omega r\dot{\theta} = 0\,.$$

Because $\dot{\theta}$ changes sign during the motion there is no guarantee that $\Phi_n \geq 0$, i.e. that the disk and the rod stay in contact (assuming the constraint unilateral). So let us impose $\Phi_n \geq 0$, namely

$$m\omega^2 l + 2m\omega r\dot{\theta} \geq 0\,.$$

Due to (13.37), that condition becomes

$$m\omega^2 l \geq 4m\omega r\Omega\sin(\Omega t)\,.$$

This is always true during the motion if and only if

$$m\omega^2 l \geq 4m\omega r\Omega$$

or, solving for $l$,

$$l \geq \frac{4r\Omega}{\omega}\,. \tag{13.38}$$

This result determines the *minimal* length $l$ that ensures the contact between disk and rod. Recalling the definition of $\Omega$ in (13.36) and substituting it in (13.38), we can then solve the latter for $\omega$. Squaring, and further computations, lead to

$$\omega^2 \geq \frac{k}{m}\frac{32r^2}{32r^2 + 3l^2}\,,$$

now saying what is the *minimal* value of the angular velocity $\omega$ that, for a *given* $l > 0$, warrants contact.

If we employ the condition $k \geq m\omega^2$ (necessary for an oscillating motion) we may encode everything into

$$\frac{k}{m}\frac{32r^2}{32r^2 + 3l^2} \leq \omega^2 \leq \frac{k}{m}\,.$$

Altogether, for *given* values of $r, l, m, k$, the angular velocity $\omega$ of the rods cannot be too large, for otherwise the centrifugal force would prevail over the spring and the motion would not be oscillatory; but it cannot be too small either, because if so the disk would detach from the rod.

We finish by finding the couple $M(t)$ the motor exerts on the rods to maintain their uniform rotation. It is convenient to notice that the power $\Pi^{\text{mot}}$ of the motor is $M(t)\omega$, and compute directly this quantity via the Kinetic energy theorem in the fixed inertial frame.

The absolute kinetic energy of the system consists of a constant part, due to the rods, and of the disk's kinetic energy, which we find using König's theorem. The absolute angular velocity of the disk is equal to $\omega - \dot{\theta}$ (by composition of the angular velocities) while the velocity of the center of mass $G$ has one component parallel to $AB$, equal to $r\dot{\theta} + l\omega$, and one perpendicular, equal to $s\omega = r\theta\omega$ (using Galilei's theorem and the definition of drag velocity). All in all,

$$T = \frac{1}{2}\frac{mr^2}{2}(\omega - \dot{\theta})^2 + \frac{1}{2}m[(r\dot{\theta} + l\omega)^2 + r^2\theta^2\omega^2] + \text{const.}$$

At the same time, the power of the active forces, in the fixed frame, is the sum of the spring's power plus the motor's

$$\Pi = -ks\dot{s} + M(t)\omega = -kr^2\theta\dot{\theta} + \Pi^{\text{mot}}\,.$$

Differentiating in time the kinetic energy and setting the result equal to $\Pi$ produces

$$\Pi^{\text{mot}} = kr^2\theta\dot{\theta} + mr^2\omega^2\theta\dot{\theta} - \frac{mr^2}{2}(\omega - \dot{\theta})\ddot{\theta} + mr(r\dot{\theta} + l\omega)\ddot{\theta}\,.$$

Because we know the system's motion $\theta(t) = 2\cos(\Omega t)$ explicitly, we can substitute $\dot{\theta}(t)$ and $\ddot{\theta}(t)$ and find the motor's power, whence the couple, in function of time.

Note though that even if the equation of motion were not explicitly integrable, we could still find, also via the energy integral, the expressions of both $\ddot{\theta}$ and $\dot{\theta}$ in function of $\theta$ (rather than time), and then write the power of the motor in this coordinate. □

# Chapter 14
# Lagrangian Mechanics

In Newtonian Mechanics, which has been our framework until now, we have studied the equilibrium and motion of systems of points and constrained rigid bodies by making use of the *Constraints postulate* (see Sect. 8.6). It guarantees that the action caused by the constraints on each point in the system is representable by a suitable force, called *constraint reaction*.

Such postulate allows to write Newton's law for every point of a constrained system:

$$m_i\,\mathbf{a}_i = \mathbf{F}_i + \mathbf{\Phi}_i\,, \tag{14.1}$$

where $m_i$ denotes the mass of the $i$th particle and $\mathbf{a}_i$ its acceleration, while $\mathbf{F}_i$ and $\mathbf{\Phi}_i$ are respectively the resultant of the active forces and the constraint reactions acting on $P_i$.

The collection of Newton's equations for every point enables us in turn to analyze the equilibrium and the motions of several types of systems of points and constrained rigid bodies. One of the major difficulties we had to face during our study arose from a special feature of equations (14.1), in which we know in advance only the resultant of the active forces, while both the motion and the constraint reactions are unknown. In the jargon of Chap. 10, the equation of motion (14.1) is a *semi-inverse* problem in Dynamics.

Lagrangian Mechanics flips this point of view. It renounces to introduce (and hence compute) the constraint reactions, and focuses all its efforts in finding the equilibrium configurations and the motions of systems of points and constrained rigid bodies. To do that, the constraints are characterized by the velocity distributions they permit, rather than by the forces they exert.

When using Lagrangian Mechanics the advantage will be to work always and exclusively with *free motion equations* (without, that is, constraint reactions). There is a price to pay for this: Lagrangian Mechanics, at least in its standard formulation, is not capable of dealing with any constraint, since its reach is limited to *ideal*



P. Biscari et al., *Rational Mechanics*, UNITEXT 177,
https://doi.org/10.1007/978-3-032-07462-1_14

holonomic constraints (see Sect. 8.7). We will devote a final section to explain the modifications to our presentation that permit to extend the analytical methods to certain nonholonomic constraints.

## 14.1 D'Alembert Principle

In the chapter dedicated to Statics, more precisely Sect. 9.4, we met a tool that provides free equilibrium equations in constrained systems: the Principle of virtual work. Now we introduce d'Alembert's principle, which allows to transform this principle of Statics into its dynamical analogue.

We start from the observation that any particle obviously stays at rest in the reference frame (in general not inertial) that translates and has origin at the point itself. The *drag force* it experiences,

$$\mathbf{F}_{\tau} = -m\,\mathbf{a}\,,$$

is referred to as *inertial force*, and Newton's second law of motion (14.1), for a single particle, turns into an equation for statics:

$$(\mathbf{F} + \mathbf{F}_{\tau}) + \boldsymbol{\Phi} = (\mathbf{F} - m\mathbf{a}) + \boldsymbol{\Phi} = \mathbf{0}. \tag{14.2}$$

In (14.2), $\mathbf{F}$ denotes the resultant of the active forces, while $\boldsymbol{\Phi}$ is the resultant of the constraint reactions. The term $\mathbf{F} - m\mathbf{a}$, sum of the active and the inertial force, is called *lost force*. The equation of motion (14.2) can then be interpreted by saying that, at every instant, in each particle's motion the resultant of the constraint reactions acting on it balances the lost force. This observations generalizes as follows.

**Theorem 14.1** (D'Alembert principle) *In a system of particles, Newton's equations of motion can be obtained from the equilibrium equations simply by including the inertial forces among the forces acting on the system.*

**Remarks**

- When deploying d'Alembert's principle we must remember that among the inertial forces are those acting on *all* points, not just the points at which the forces are applied. To clarify this fact, consider a discrete system of particles $\{(P_i, m_i),\ i = 1, \dots, n\}$. Suppose there is a system of forces $\{\mathbf{F}_i,\ i = 1, \dots, k < n\}$ acting on some the points (for instance, the first $k < n$). The first equilibrium balance equation requires, as necessary condition for equilibrium,

$$\sum_{i=1}^{k} \mathbf{F}_i = \mathbf{0}. \tag{14.3}$$

Applying d'Alembert's principle to (14.3) *should not* give as dynamical necessary condition

$$\sum_{i=1}^{k}\big(\mathbf{F}_i - m_i\,\mathbf{a}_i\big) = \mathbf{0},$$

but rather

$$\sum_{i=1}^{k}\big(\mathbf{F}_i - m_i\,\mathbf{a}_i\big) - \sum_{i=k+1}^{n} m_i\,\mathbf{a}_i = \mathbf{0}.$$

- Albeit implicit in the statement of d'Alembert's principle, one should be careful about the following. Suppose there are, among the forces effectively acting on the system, forces that depend on the velocity. These will typically give no contribution in the equilibrium equations. But it is important to point out that when we pass to dynamics we obviously need to re-introduce them among the acting forces.
- D'Alembert's principle can be proved along the same lines as the Principle of virtual work (see Sect. 9.4). As for the choice of respecting the traditional name "principle", and not calling it *d'Alembert's theorem*, we recall Remark 9.9, where we stressed how that proof relies in a critical way on having assumed mechanical determinism.

### *14.1.1 Inertial Forces and Balance Equations*

As an application of the d'Alembert principle let us consider a system of inertial forces and express their resultant and resultant moment. We show that, as a consequence, the first and second balance equation of Dynamics can be given a form analogous to the balance equations for Statics. Next, when the system on which the forces act is a rigid body, we show how to find a simpler set of equivalent forces that allows to analyze the dynamics of a rigid body using the equations and techniques developed in Statics.

Reducing a system of forces (see Sect. 7.4) depends on the characteristic vectors of the system. We start by computing the resultant and the resultant moment of the inertial forces. For the resultant we immediately have

$$\mathbf{R}^{(\text{in})} = \sum_{i=1}^{n}\big(-m_i\mathbf{a}_i\big) = -\frac{d}{dt}\sum_{i=1}^{n} m_i\mathbf{v}_i = -\dot{\mathbf{Q}} \tag{14.4}$$

while for the moment

$$\begin{aligned}\mathbf{M}_O^{(\mathrm{in})} &= \sum_{i=1}^{n} OP_i \times (-m_i\mathbf{a}_i) \\ &= -\frac{d}{dt}\left(\sum_{i=1}^{n} OP_i \times m_i\mathbf{v}_i\right) - \sum_{i=1}^{n} \dot{O} \times m_i\mathbf{v}_i \\ &= -\dot{\mathbf{K}}_O - \dot{O} \times \mathbf{Q}.\end{aligned} \tag{14.5}$$

(the reader should see Sect. 6.2.3 for the difference between $\dot{O}$ and $\mathbf{v}_O$).

The balance equations of Dynamics (11.1) and (11.3) can be written as

$$\mathbf{R}^{(\mathrm{e})} = \dot{\mathbf{Q}} \qquad \mathbf{M}_O^{(\mathrm{e})} = \dot{\mathbf{K}}_O + \dot{O} \times \mathbf{Q} \tag{14.6}$$

(denoting the pole with $O$ rather than $A$). If all terms are taken to the left of the equal sign we have

$$\mathbf{R}^{(\mathrm{e})} - \dot{\mathbf{Q}} = 0 \qquad \mathbf{M}_O^{(\mathrm{e})} - \dot{\mathbf{K}}_O - \dot{O} \times \mathbf{Q} = 0$$

In view of (14.4) and (14.5) we conclude that the balance equations of Dynamics (14.6) can be rewritten in this compact and useful form:

$$\mathbf{R}^{(\mathrm{e})} + \mathbf{R}^{(\mathrm{in})} = 0 \qquad \mathbf{M}_O^{(\mathrm{e})} + \mathbf{M}_O^{(\mathrm{in})} = 0\,. \tag{14.7}$$

It is very important to notice that (14.7) can be interpreted as balance equations for Statics. More explicitly, if we just add the inertial forces to the external forces acting on a given system, we are allowed to write the balance equations exactly as when solving a problem of Statics.

This might be seen as just a simple trick, and in a sense it is, but it has much usefulness in many applications. The real issue, of course, is to be able to express conveniently the resultant and moment of the inertial forces.

First, we turn to the resultant: since, in view of (6.1), the linear momentum can be expressed as $\mathbf{Q} = m\mathbf{v}_G$ it is quite easy to deduce that

$$\dot{\mathbf{Q}} = m\mathbf{a}_G \quad \Rightarrow \quad \mathbf{R}^{(\mathrm{in})} = -m\mathbf{a}_G \tag{14.8}$$

Thus, the resultant of the inertial forces can be computed as the opposite of the total mass times the acceleration of the center of mass. In most applications to rigid bodies we are led anyway to the computation of the velocity $\mathbf{v}_G$ as a function of the generalized coordinates and their derivatives and, as a consequence, it is not too difficult to differentiate one more time and obtain $\mathbf{a}_G$, and from this $\mathbf{R}^{(\mathrm{in})}$.

Expression (14.5) for the moment of the inertial forces greatly simplifies if we choose $G$ as a pole, since $\dot{G} = \mathbf{v}_G$ is parallel to $\mathbf{Q}$:

$$\mathbf{M}_G^{(\mathrm{in})} = -\dot{\mathbf{K}}_G \tag{14.9}$$

Suppose now that the balance equations (14.7) are applied to a rigid body, in which case we are allowed to substitute a system of forces with a simpler but equivalent one, according to the discussion presented in Sect. 7.4 of Chap. 7.

In view of (14.8) and (14.9), if we choose to apply the resultant $\mathbf{R}^{(\text{in})} = -m\mathbf{a}_G$ at the center of mass $G$ (as is quite natural), we only need to add a couple with moment $\mathbf{M}_G^{(\text{in})} = -\dot{\mathbf{K}}_G$ in order to have a system which is equivalent to the set of inertial forces acting on the rigid body.

Thus, the system of inertial forces is equivalent to:

$$\begin{cases} \text{An } \textit{inertial resultant} - m\mathbf{a}_G \text{, applied at the center of mass } G \\ \text{An } \textit{inertial torque}\text{, with moment } - \dot{\mathbf{K}}_G \,. \end{cases} \tag{14.10}$$

If we decide to apply the resultant $\mathbf{R}^{(\text{in})}$ at some different point $O$ we need to add a couple (or torque) with moment $\mathbf{M}_O^{(\text{in})}$ given by

$$\mathbf{M}_O^{(\text{in})} = OG \times \mathbf{R}^{(\text{in})} + \mathbf{M}_G^{(\text{in})} = OG \times (-m\mathbf{a}_G) - \dot{\mathbf{K}}_G$$

as we can readily obtain from the rule for changing poles (7.18) in Chap. 7.

This reduction further simplifies when the system is composed by a single rigid body constrained to rotate around its principal axis, as in the important case of planar motions of rigid bodies contained in the plane of motion, for instance. If so, denoting with $\mathbf{u}$ the unit vector perpendicular to the plane of motion, if $\boldsymbol{\omega} = \dot{\theta}\,\mathbf{u}$ is the angular velocity and $I_{Gu}$ the moment of inertia with respect to the axis through the center of mass and parallel to the rotation axis, the inertial torque is simply

$$-\dot{\mathbf{K}}_G = -I_{Gu}\ddot{\theta}\,\mathbf{u}$$

If we want to write the balance equations of a *system* of rigid bodies and particles, using d'Alembert's principle, we must insert an inertial force $-m_i\mathbf{a}_i$ for every particle $P_i$, while for every rigid body (of mass $m_j$ and center of mass $G_j$) we must add an inertial resultant $-m_j\mathbf{a}_{G_j}$, applied at $G_j$, together with an inertial torque $-\dot{\mathbf{K}}_{G_j}$.

**Remark 14.2** In this section we used the definition of inertial forces for writing the balance equations of Dynamics in an equivalent form, which has some usefulness in applications to rigid body dynamics. In a sense, the situation is quite similar to what we saw in Chap. 13, when we introduced apparent forces in Relative Mechanics. As is clear, all problems can be solved in an inertial frame and, in principle, there is no reason to introduce apparent forces and all the concepts related with them. However, as we showed through specific examples, a non-inertial point of view turns out to be occasionally quite natural and useful. The same holds true here: solving a problem by adding resultants and moments of inertial forces to each rigid body and then using the balance equations of Statics (rather than Dynamics) can be very convenient. Indeed, the use of d'Alembert principle by means of its consequences is, mostly for plane systems, quite useful, even if never necessary, strictly speaking.

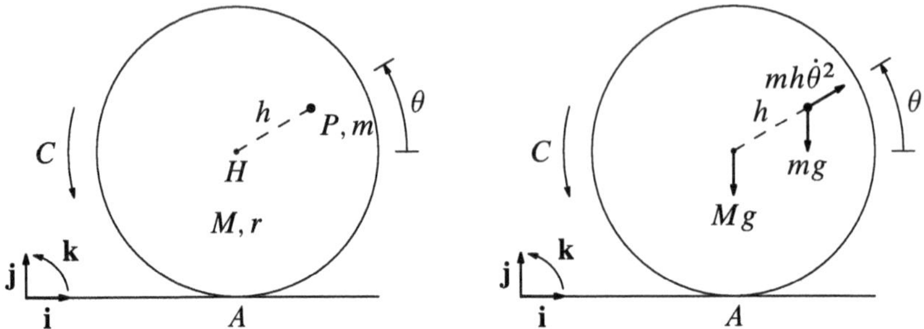


**Fig. 14.1** A disk rolling on a guide, subject to a couple with moment $C$, such that the angular velocity is *constant*. On the right, the forces, inertial and not, acting on the disk and the point are shown (except for the contact force in $A$)

**Example 14.3** We discuss a simple problem to which we apply the method of d'Alembert, for a better understanding of the concepts introduced in Sect. 14.1.1.

A homogeneous disk of mass $M$ and radius $r$ rolls without slipping on a horizontal fixed guide, in a vertical plane. A point $P$ of mass $m$ is fixed on the disk, at a distance $h$ from the center $H$. The disk is subject to a couple with moment $C$, which we are asked to compute, such that the angular velocity is kept constant (see Fig. 14.1).

Since the angular velocity is constant it is clear that both the inertial torque $-\dot{\mathbf{K}}_H = -I_H\ddot{\theta}\,\mathbf{k}$ and the inertial force $-m\mathbf{a}_H$ acting on the disk are zero. The inertial force acting on $P$ is given by $-m\mathbf{a}_P$, where the acceleration can be computed using Coriolis' Theorem 3.4 in Chap. 3 or, more simply, applying the acceleration distribution law in its form (2.39) of Chap. 2 (see also Fig. 2.15).

Under the given assumptions, $\ddot{\theta} = 0$ and as a consequence $\mathbf{a}_H = \mathbf{0}$, the acceleration of $P$ is centripetal towards $H$ with modulus $h\dot{\theta}^2$. We conclude that the inertial force acting on $P$ is centrifugal from $H$ and of modulus $mh\dot{\theta}^2$. Thus, if we want to compute $C$ using d'Alembert's approach, it is sufficient to write the second balance equation of *statics*, as in (14.7), with respect to $A$, the contact point of the disk with the horizontal guide. After computing the moment with respect to $H$ of the weight of $P$ and of the inertial force applied at $P$ we easily get

$$C = mgh\cos\theta + mhr\dot{\theta}^2\cos\theta\,. \tag{14.11}$$

What should we do, if we wanted to use the second balance equation of *Dynamics*, as expressed in (11.3) of Chap. 11? Using $A$ as the pole, the total moment of the weight $mg$ and of the applied couple is easily expressed as

$$\mathbf{M}_A = (C - mgh\cos\theta)\,\mathbf{k}\,.$$

The angular momentum is

$$\mathbf{K}_A = (3Mr^2/2\dot{\theta} + md^2\dot{\theta})\mathbf{k}$$

where

$$d^2 = |AP|^2 = r^2 + h^2 + 2hr \sin\theta \,.$$

Thus, assuming $\ddot{\theta} = 0$,

$$\dot{\mathbf{K}}_A = (2mhr\dot{\theta}^2 \cos\theta)\mathbf{k}$$

We should now remember that the term $\dot{A} \times \mathbf{Q}$ in (11.3) is *not* zero, since the center of mass of the system is *not* $H$. Thus

$$\dot{A} \times \mathbf{Q} = \dot{A} \times M\mathbf{v}_H + \dot{A} \times m\mathbf{v}_P = \dot{A} \times m\mathbf{v}_P \,.$$

Since

$$\dot{A} = -r\dot{\theta}\mathbf{i} \qquad \mathbf{v}_P = (-r\dot{\theta} - h\sin\theta\dot{\theta})\mathbf{i} + h\cos\theta\dot{\theta}\mathbf{j}$$

the conclusion is that

$$\dot{A} \times \mathbf{Q} = (-mhr\dot{\theta}^2 \cos\theta)\mathbf{k}$$

and, finally, the second balance equation of dynamics (11.3) yields

$$C - mgh\cos\theta = 2mhr\dot{\theta}^2 \cos\theta - mhr\dot{\theta}^2 \cos\theta$$

which simplifies to (14.11).

As expected, we reach exactly the same result following different routes. Notice, however, that when using d'Alembert's approach we do not have the inconvenience arising in Dynamics when using a moving pole, for which the extra term $\dot{A} \times \mathbf{Q}$ does not vanish. After inserting the appropriate inertial terms we are free to behave as in statics, where the choice of poles for writing the second balance equation is unhindered by such issues.

The example discussed here is very simple but the advantage of using the approach of d'Alembert is even more evident when discussing complex plane problems.

We might wonder what would be the distribution of the inertial forces acting on the disk if the hypothesis that the angular velocity is constant is abandoned. In this case the resultant and the moment of inertial forces applied to the disk would be easily expressed as

$$\mathbf{R}^{(\text{in})} = -M\mathbf{a}_H = Mr\ddot{\theta}\mathbf{i} \qquad \mathbf{M}_H^{(\text{in})} = -Mr^2/2\ddot{\theta}\mathbf{k}$$

with the resultant applied at $H$. What would be more complex is the expression of the inertial force acting on $P$, since this requires a computation of the acceleration $\mathbf{a}_P$, using again (2.39) (see also Fig. 2.15).

In Fig. 14.2 all the inertial forces acting on the system when the angular velocity is *arbitrary* are shown.

In conclusion, d'Alembert's approach is mostly useful under the following conditions:

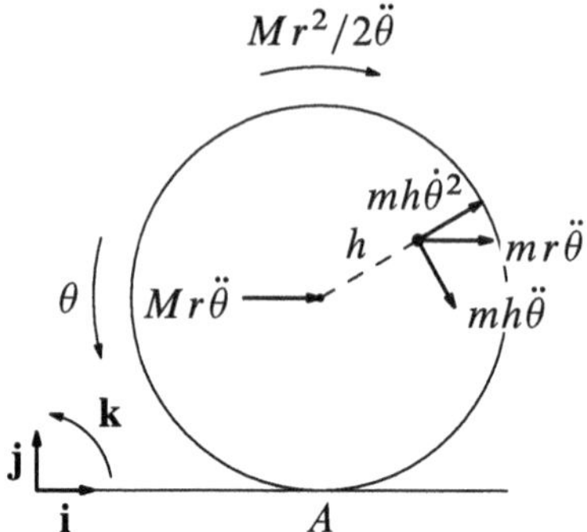


**Fig. 14.2** The distribution of *inertial* forces acting on the disk of Example 14.3 when the angular velocity is *arbitrary* (all other forces are not shown)

- the mechanical problem is described by a *planar* system, in which case it is very easy to express both $\mathbf{R}^{(\mathrm{in})}$ and $\mathbf{M}_G^{(\mathrm{in})}$ for each rigid body;
- we are capable of expressing the acceleration of particles to which inertial forces should be applied. □

## 14.2 D'Alembert's Equation

In this section we shall apply d'Alembert's principle to construct, from the Principle of virtual work of Sect. 9.4, a principle capable of describing the dynamics of a system of particles and rigid bodies subject to ideal constraints. In this way we will find the cornerstone principle of Lagrangian Mechanics: it will allow to find the free conditions of equilibrium and motion for systems subject to ideal constraints. The proof of the following relation comes from putting together the Principle of virtual work and the d'Alembert principle (i.e., adding to the active forces the inertial forces).

**Theorem 14.4** (D'Alembert relation) *Let* $\{(P_i, m_i),\ i = 1, \dots, n\}$ *be a system of free particles, or particles subject to ideal constraints. Denote by* $\{(P_i, \mathbf{F}_i),\ i = 1, \dots, n\}$ *the system of* active *forces acting on the system. Then the collection of accelerations* $\{(P_i, \mathbf{a}_i),\ i = 1, \dots, n\}$ *gives the system's motion if and only if*

$$\sum_{i=1}^{n} (\mathbf{F}_i - m_i\, \mathbf{a}_i) \cdot \delta P_i \leq 0 \tag{14.12}$$

*for any virtual displacements* $\{(P_i, \delta P_i),\ i = 1, \dots, n\}$ *permitted by the constraints.*

**Remarks**
Being a direct consequence of the Principle of virtual work, many observations made in Sect. 9.4 apply to d'Alembert's equation.

- Relation (14.12) is not a scalar inequality, as it might seem: it subsumes infinitely many conditions (one for each possible choice of virtual displacements). Obviously we will see in the sequel that not all of those relations are independent.
- D'Alembert's relation postulates (hence requires) an existence and uniqueness theorem for solutions of the free motion equations. In fact, that the relation is necessary can be proved starting from Newton's second law of motion. That it is sufficient has to do with the solutions being unique.
- In writing (14.12) we should take into account not only points on which active forces are applied, but all points with mass, in order to account for all inertial forces.
- In Sects. 7.5 and 7.6 we proved that equivalent systems of forces on one rigid body generate the same work, hence the same power (be it effective or virtual). Therefore, when using d'Alembert's equation we may replace the inertial forces of a rigid body with the equivalent system (14.10).

When the constraints acting on the system, apart from ideal, are bilateral as well, d'Alembert's inequality simplifies and becomes an equation. Recall that a constraint is called bilateral if it only admits reversible virtual displacements. That is, a constraint is bilateral when in presence of a virtual displacement $\delta P_i$ the opposite displacement $-\delta P_i$ is virtual, i.e. permitted by the constraint.

For ideal bilateral constraints we can write (14.12) for any virtual displacements $\{(P_i, \delta P_i),\ i = 1, \dots, n\}$ and for the opposites $\{(P_i, -\delta P_i),\ i = 1, \dots, n\}$. In all cases the left-hand side of (14.12) must be non-positive:

$$\sum_{i=1}^{n} (\mathbf{F}_i - m_i\, \mathbf{a}_i) \cdot (\ \ \delta P_i\,) \le 0, \quad \sum_{i=1}^{n} (\mathbf{F}_i - m_i\, \mathbf{a}_i) \cdot (-\,\delta P_i) \le 0.$$

Therefore the quantities on the left are both non-negative and non-positive, so zero. This proves the following particular case of d'Alembert's relation.

**Theorem 14.5** (D'Alembert equation) *Let $\{(P_i, m_i),\ i = 1, \dots, n\}$ be a system of particles, free or subject to ideal bilateral constraints. Call $\{(P_i, \mathbf{F}_i),\ i = 1, \dots, n\}$ the system of active forces. Then the accelerations $\{(P_i, \mathbf{a}_i),\ i = 1, \dots, n\}$ give the system's motion if and only if*

$$\sum_{i=1}^{n} (\mathbf{F}_i - m_i\, \mathbf{a}_i) \cdot \delta P_i = 0$$

*for any virtual displacements $\{(P_i, \delta P_i),\ i = 1, \dots, n\}$ permitted by the constraints.*

**Remark 14.6** D'Alembert's inequality and equation are the exact analogue of the Principle of virtual work of Dynamics. In presence of ideal constraints, they allow to completely determine the dynamics of any holonomic system. In particular, they allow to extend to a dynamical setting Theorem 9.12, so that the balance equations are necessary and sufficient to establish the dynamics of a single rigid body. Recall that in Chap. 12 this property was proved for free rigid bodies and rigid bodies with a fixed point or fixed axis. D'Alembert's equation allows to extend that characterizing feature of the balance equations to any rigid body with ideal constraints. □

## 14.3 Lagrange Equations

Consider a system with holonomic constraints, *ideal* and *bilateral*. Since the constraints are holonomic the virtual displacements of all particles are expressible by the (mutually independent) virtual displacements of the generalized coordinates $(q_1, \ldots, q_N)$ (see (4.23)):

$$\delta P_i = \sum_{k=1}^{N} \frac{\partial P_i}{\partial q_k}\, \delta q_k \,.$$

Assuming ideal constraints means we can use d'Alembert's relation, which becomes an equation because the constraints are bilateral. So let $\mathbf{F}_i$ be the resultant of the active forces acting on the point $P_i$ of mass $m_i$, and $\mathbf{a}_i$ its acceleration. Then

$$\begin{aligned}\sum_{i=1}^{n} \big(\mathbf{F}_i - m_i\,\mathbf{a}_i\big)\cdot \delta P_i &= \sum_{i=1}^{n} \big(\mathbf{F}_i - m_i\,\mathbf{a}_i\big)\cdot \sum_{k=1}^{N} \frac{\partial P_i}{\partial q_k}\,\delta q_k \\ &= \sum_{k=1}^{N}\left(\sum_{i=1}^{n} \big(\mathbf{F}_i - m_i\,\mathbf{a}_i\big)\cdot \frac{\partial P_i}{\partial q_k}\right)\delta q_k = \sum_{k=1}^{N} \big(Q_k - \tau_k\big)\,\delta q_k = 0\,,\end{aligned} \tag{14.13}$$

*for any* choice of virtual displacements $(\delta q_1, \ldots \delta q_N)$.

- In the first line of (14.13) we swapped the summation over $i$ (counting the particles) and $k$ (counting generalized coordinates). This is possible since the sums are finite.
- In passing from line one to line two we introduced new symbols. The first, already met in Sects. 7.5.2 and 10.4, is for the Lagrangian components of the active forces:

$$Q_k = \sum_{i=1}^{n} \mathbf{F}_i \cdot \frac{\partial P_i}{\partial q_k}\,. \tag{14.14}$$

  Parallel to that, we now introduce the *Lagrangian components of the opposite of the inertial forces*:

$$\tau_k = \sum_{i=1}^{n} m_i\,\mathbf{a}_i \cdot \frac{\partial P_i}{\partial q_k}\,. \tag{14.15}$$

We emphasize once more the difference between the initial and final form in (14.13). In either expression the components (possibly Lagrangian) of the forces are multiplied by virtual displacements, and the sum of all products vanishes for any virtual displacements. But while the particles' virtual displacements depend on one another due to the constraints (if we move a point the constraint might move other points as a consequence), the virtual displacements of the generalized coordinates are totally independent of one another. In other words, before using the first expression with displacements $\{\delta P_i,\ i = 1, \ldots, n\}$ we ought to verify that such set is allowed by the constraints. On the other hand, any displacement of type $\{\delta q_k,\ k = 1, \ldots, N\}$ is automatically admitted.

**Theorem 14.7** (Free motion equations) *D'Alembert's equation is equivalent to the $N$ independent equations:*

$$Q_k = \tau_k \qquad \textit{for any} \quad k = 1, \ldots, N. \tag{14.16}$$

***Proof*** That (14.16) are sufficient is evident. In fact, if every $Q_k$ equals the corresponding $\tau_k$, evidently all summands in (14.13) vanish for any virtual displacements, so their sum is zero.

For the converse we exploit the freedom to choose the virtual displacements of the generalized coordinates. Pick an arbitrary generalized coordinate (say, the $\bar{k}$-th one), and consider the following virtual displacements:

$$\delta q_{\bar{k}} \neq 0 \qquad \text{and} \qquad \delta q_k = 0 \ \ \text{for any } k \neq \bar{k}.$$

We are thus only moving the $\bar{k}$-th generalized coordinate. As (14.13) holds for all virtual displacements, it holds for our choice. All $\delta q_k$ are zero (except for the $\bar{k}$-th one), so the sum simplifies to

$$\sum_{k=1}^{N} (Q_k - \tau_k)\,\delta q_k = \left(Q_{\bar{k}} - \tau_{\bar{k}}\right)\delta q_{\bar{k}} = 0 \qquad \left(\text{with } \delta q_{\bar{k}} \neq 0\right),$$

and then $Q_{\bar{k}} = \tau_{\bar{k}}$. As $\bar{k}$ was chosen arbitrarily, in reality $Q_k = \tau_k$ for all $k = 1, \ldots, N$. □

Equations (14.16) are *free*, since only the active forces contribute to the $Q_k$. We will soon see that the $\tau_k$ descend from the expression of the kinetic energy written in term of generalized coordinates and derivatives.

**Theorem 14.8** (Lagrangian binomials) *Let $T$ be the kinetic energy of a holonomic system with generalized coordinates $\{q_1, \ldots, q_N\}$. The Lagrangian components of the opposite of the inertial forces satisfy the identities:*

$$\tau_k = \frac{d}{dt}\left(\frac{\partial T}{\partial \dot{q}_k}\right) - \frac{\partial T}{\partial q_k} \qquad \textit{for any} \quad k = 1, \ldots, N. \tag{14.17}$$

*We call the right-hand sides* Lagrangian binomials.

***Proof*** From definition (14.15) we have

$$\begin{aligned}\tau_k &= \sum_{i=1}^{n} m_i\, \mathbf{a}_i \cdot \frac{\partial P_i}{\partial q_k} = \sum_{i=1}^{n} m_i\, \frac{d\mathbf{v}_i}{dt} \cdot \frac{\partial P_i}{\partial q_k} \\ &= \sum_{i=1}^{n} m_i\, \frac{d}{dt}\left(\mathbf{v}_i \cdot \frac{\partial P_i}{\partial q_k}\right) - \sum_{i=1}^{n} m_i\, \mathbf{v}_i \cdot \frac{d}{dt}\frac{\partial P_i}{\partial q_k}\,, \end{aligned} \tag{14.18}$$

where the first equality is simply the definition of acceleration, while in the next we used the identity

$$\frac{d\mathbf{v}_i}{dt} \cdot \frac{\partial P_i}{\partial q_k} = \frac{d}{dt}\left(\mathbf{v}_i \cdot \frac{\partial P_i}{\partial q_k}\right) - \mathbf{v}_i \cdot \frac{d}{dt}\frac{\partial P_i}{\partial q_k}\,,$$

direct consequence of the chain rule.

The position of each particle depends on time both explicitly and through the generalized coordinates: $P_i(t) = P_i\big(q_1(t), \ldots, q_N(t); t\big)$. We then have

$$\mathbf{v}_i = \frac{dP_i}{dt} = \sum_{j=1}^{N} \frac{\partial P_i}{\partial q_j}\, \dot{q}_j + \frac{\partial P_i}{\partial t}\,. \tag{14.19}$$

Apart from the time derivative, the velocity of the $i$th point depends linearly on the Lagrangian velocities $\{\dot{q}_1, \ldots, \dot{q}_N\}$. In particular,

$$\frac{\partial \mathbf{v}_i}{\partial \dot{q}_k} = \frac{\partial P_i}{\partial q_k}\,. \tag{14.20}$$

Moreover, differentiating (14.19) with respect to the $k$th generalized coordinate,

$$\frac{\partial \mathbf{v}_i}{\partial q_k} = \sum_{j=1}^{N} \frac{\partial^2 P_i}{\partial q_k \partial q_j}\, \dot{q}_j + \frac{\partial^2 P_i}{\partial q_k \partial t} = \frac{d}{dt}\frac{\partial P_i}{\partial q_k}\,, \tag{14.21}$$

since derivatives can be swapped.

Consider again expression (14.18) for $\tau_k$. It contains the right-hand-side terms of (14.20) and (14.21). Replacing both with the respective left-hand sides,

$$\begin{aligned}\tau_k &= \sum_{i=1}^{n} m_i\, \frac{d}{dt}\left(\mathbf{v}_i \cdot \frac{\partial \mathbf{v}_i}{\partial \dot{q}_k}\right) - \sum_{i=1}^{n} m_i\, \mathbf{v}_i \cdot \frac{\partial \mathbf{v}_i}{\partial q_k} \\ &= \frac{d}{dt}\left(\frac{1}{2}\sum_{i=1}^{n} m_i\, \frac{\partial}{\partial \dot{q}_k}\left(\mathbf{v}_i \cdot \mathbf{v}_i\right)\right) - \frac{1}{2}\sum_{i=1}^{n} m_i\, \frac{\partial}{\partial q_k}\left(\mathbf{v}_i \cdot \mathbf{v}_i\right) = \frac{d}{dt}\left(\frac{\partial T}{\partial \dot{q}_k}\right) - \frac{\partial T}{\partial q_k}\,,\end{aligned}$$

whereby $\tau_k$ is written using the derivatives of the kinetic energy. □

Substituting now the Lagrangian binomials (14.17) in the free equation (14.16) we obtain *Lagrange's equations*

$$\frac{d}{dt}\left(\frac{\partial T}{\partial \dot{q}_k}\right) - \frac{\partial T}{\partial q_k} = Q_k \quad \text{for any} \quad k = 1, \ldots, N. \tag{14.22}$$

## 14.4 Lagrange's Equations as Projection of Newton's on Constraint Surfaces

The true nature of the Lagrange equations (14.22) becomes clearer through a simple example, which illustrates how they can be naturally derived from Newton's laws, via scalar multiplication with the partial derivatives of the point position with respect to generalized coordinates $q_k$, at least in the case of holonomic systems.

Let $P$ be a particle of mass $m$, constrained to move on a smooth surface $\mathcal{S}$. The position of $P$ is described by two independent parameters $q_1$, $q_2$ (surface coordinates). If the surface $\mathcal{S}$ itself moves according to a prescribed law, time $t$ may also be involved, so the position becomes $P(q_1, q_2, t)$. An active force $\mathbf{F}$ acts on the particle, in addition to the constraint force $\mathbf{\Phi}$, which is orthogonal to the surface (see Fig. 14.3).

From elementary surface geometry, we know that partial derivatives

$$\frac{\partial P}{\partial q_1}, \quad \frac{\partial P}{\partial q_2} \tag{14.23}$$

yield vectors tangent to the surface at each point and at any given instant, as shown in Fig. 14.3 (notice that time $t$ is kept fixed while performing this differentiation). Thus, they are perpendicular to $\mathbf{\Phi}$:

$$\mathbf{\Phi} \cdot \frac{\partial P}{\partial q_k} = 0 \quad (k = 1, 2). \tag{14.24}$$

Newton's law for particle $P$ states that

$$m\mathbf{a} = \mathbf{F} + \mathbf{\Phi}$$

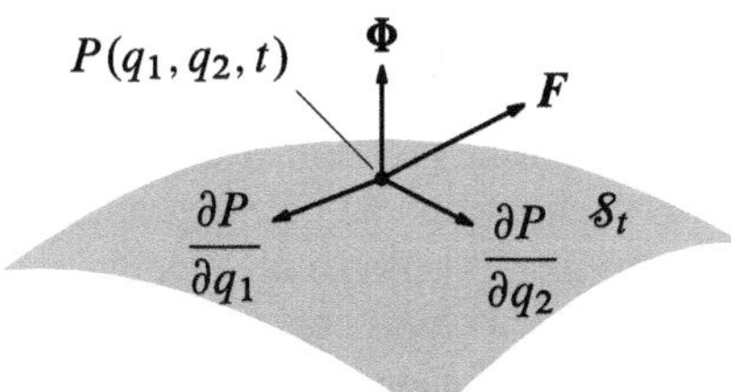


**Fig. 14.3** A particle $P$ with mass $m$ on a smooth moving surface $\mathcal{S}_t$

and, when multiplied by (14.23), gives

$$m\mathbf{a}\cdot\frac{\partial P}{\partial q_k}=(\mathbf{F}+\boldsymbol{\Phi})\cdot\frac{\partial P}{\partial q_k}$$

which, in view of (14.24), simplifies to

$$m\mathbf{a}\cdot\frac{\partial P}{\partial q_k}=\mathbf{F}\cdot\frac{\partial P}{\partial q_k}.$$

On the right-hand side, we easily recognize the quantities $Q_k$, as defined in (14.14), while the term on the left side can be manipulated exactly as in Theorem 14.8. Thus, Newton's equation, when projected on the tangent space of the constraint surface at any given time, is equivalent to

$$\frac{d}{dt}\left(\frac{\partial T}{\partial \dot{q}_k}\right)-\frac{\partial T}{\partial q_k}=Q_k \qquad (k=1,2)$$

which are exactly Lagrange's equations for a system with two degrees of freedom. This example clearly illustrates the connection between Newton and Lagrange equations.

It is worth noting that, by definition, partial derivatives of $P(q_k, t)$ with respect to $q_k$ are taken at fixed time $t$. They are therefore naturally associated with *virtual displacements*, which are also defined with the constraints held fixed in time.

This approach naturally extends to a more general system with $n$ degrees of freedom, subject to ideal, fixed or moving *holonomic* constraints. In such a system, the position of a generic point is expressed as $P_i(q_k, t)$. To proceed, each Eq. (14.1) is multiplied by the partial derivative of $P_i$ with respect to the $q_k$ coordinate, and the result is summed over the index $i$. The terms involving the constraint reactions $\boldsymbol{\Phi}_i$ vanish, and it becomes straightforward to identify the generalized forces $Q_k$ on the right-hand side, and—after some manipulations—the Lagrange binomials (14.17) on the left.

### 14.4.1 Lagrangian Determinism

The free motion equations (14.22) are of order two in time. We will show in this section how to write them in normal form, i.e. formally solved for the second derivatives $\ddot{q}_k$. Cauchy's theorem will then guarantee the existence of a unique solution to the initial-value problem with given position and velocity at any initial instant, provided the forces (hence the Lagrangian components) depend in a sufficiently regular manner upon position and velocity.

Lagrange's equations are therefore *deterministic*: knowing the velocity distribution at some instant $t_0$ warrants, theoretically at least, the complete characterization of the motion at successive instants. Lagrangian determinism does not guarantee the solution of the initial-value problem can be extended to *all* times after $t_0$: even in presence of extremely regular Lagrangian components there might be solutions that develop singularities in finite time. Proving the Lagrangian determinism avoids, for the large class of holonomic systems with ideal bilateral constraints, the need to postulate the Principle of mechanical determinism of Sect. 8.2.

Before we put (14.22) in normal form, let us recall how the kinetic energy depends upon the Lagrangian velocities (see Sect. 6.3.2). Given a holonomic system with generalized coordinates $\{q_1, \dots, q_N\}$, its kinetic energy can be written

$$T = \frac{1}{2} \sum_{j,k=1}^{N} a_{jk}\, \dot{q}_j\, \dot{q}_k + \sum_{k=1}^{N} b_k\, \dot{q}_k + c, \tag{14.25}$$

where the coefficients $a_{jk}$, $b_k$, $c$ are functions of the generalized coordinates and possibly time, not of the Lagrangian velocities:

$$\begin{aligned} a_{jk}(q_1, \dots, q_N; t) &= \sum_{i=1}^{n} m_i \frac{\partial P_i}{\partial q_j} \cdot \frac{\partial P_i}{\partial q_k}, \\ b_k(q_1, \dots, q_N; t) &= \sum_{i=1}^{n} m_i \frac{\partial P_i}{\partial q_k} \cdot \frac{\partial P_i}{\partial t}, \\ c(q_1, \dots, q_N; t) &= \frac{1}{2} \sum_{i=1}^{n} m_i \frac{\partial P_i}{\partial t} \cdot \frac{\partial P_i}{\partial t}. \end{aligned}$$

In terms of the mass matrix $\mathsf{A}$, of the vectors $\mathsf{q} = \{q_1, \dots, q_N\}$, $\dot{\mathsf{q}} = \{\dot{q}_1, \dots, \dot{q}_N\}$, $\mathsf{b} = \{b_1, \dots, b_N\}$ and of the scalar $c$,

$$T(\mathsf{q}, \dot{\mathsf{q}}, t) = \frac{1}{2} \dot{\mathsf{q}} \cdot \mathsf{A}(\mathsf{q}, t)\dot{\mathsf{q}} + \mathsf{b}(\mathsf{q}, t) \cdot \dot{\mathsf{q}} + c(\mathsf{q}, t).$$

**Theorem 14.9** (Determinism) *Consider a holonomic system with ideal bilateral constraints and generalized coordinates* $\mathsf{q} = \{q_1, \dots, q_N\}$. *The initial-value problem*

$$\begin{cases} \dfrac{d}{dt}\left(\dfrac{\partial T}{\partial \dot{q}_k}\right) - \dfrac{\partial T}{\partial q_k} = Q_k(\mathsf{q}, \dot{\mathsf{q}}, t) & \quad \textit{for} \;\; k = 1, \dots, N \\ q_k(t_0) = q_{k0}, \qquad \dot{q}_k(t_0) = \dot{q}_{k0} & \end{cases}$$

*has a unique solution on some interval* $t \in [t_0, t_1]$, *with* $t_1 > t_0$, *if the Lagrangian components are Lipschitz functions (Definition 8.2) of the generalized coordinates and their time derivatives.*

***Proof*** From (14.25), the $N$ Lagrange equations are comprised by the vectorial equation

$$\mathsf{A}\ddot{\mathsf{q}} = \mathsf{F}\big(\mathsf{q}, \dot{\mathsf{q}}, t\big) + \mathsf{Q}, \tag{14.26}$$

where $\mathsf{A}$ is the mass matrix, $\ddot{\mathsf{q}} = \{\ddot{q}_1, \ldots, \ddot{q}_N\}$, $\mathsf{Q} = \{Q_1, \ldots, Q_N\}$ and $\mathsf{F} = \{F_1, \ldots, F_N\}$ with

$$F_k(\mathsf{q}, \dot{\mathsf{q}}, t) = \sum_{h,j=1}^{N} \left( \frac{1}{2} \frac{\partial a_{hj}}{\partial q_k} - \frac{\partial a_{kj}}{\partial q_h} \right) \dot{q}_h \dot{q}_j + \sum_{j=1}^{N} \left( \frac{\partial b_j}{\partial q_k} - \frac{\partial b_k}{\partial q_j} - \frac{\partial a_{jk}}{\partial t} \right) \dot{q}_j + \frac{\partial c}{\partial q_k} - \frac{\partial b_k}{\partial t} .$$

For our argument the detailed structure of the terms in $\mathsf{F}$ is immaterial. What counts is that they depend on time, on the generalized coordinates and their time derivatives, but not on higher-order time derivatives.

Lagrange's equations in form (14.26) can be solved for the $\ddot{\mathsf{q}}$ since the mass matrix Eq. 6.32 is invertible: $\ddot{\mathsf{q}} = \mathsf{A}^{-1} \left( \mathsf{F}(\mathsf{q}, \dot{\mathsf{q}}, t) + \mathsf{Q} \right)$, to which Cauchy's Theorem applies as long as the active forces are regular enough. □

### 14.4.2 *The Lagrangian*

In Sect. 7.7 (in particular sections (7.48)) we found that when all active forces are conservative their Lagrangian components are derivatives in the generalized coordinates $\mathsf{q} = (q_1, \ldots, q_N)$ of the system's potential $U(\mathsf{q}, t)$:

$$Q_k = \frac{\partial U}{\partial q_k} . \tag{14.27}$$

Recall that conservative forces are positional, so $Q_k$ and the potential $U(\mathsf{q}, t)$ itself may depend on generalized coordinates and possibly time in presence of moving constraints, but certainly *not* on the $\dot{q}_k$.

Substituting (14.27) in (14.22) we may recast Lagrange's equations as

$$\frac{d}{dt} \left( \frac{\partial T}{\partial \dot{q}_k} \right) - \frac{\partial (T + U)}{\partial q_k} = 0.$$

Introduce now the *Lagrangian* function

$$\mathcal{L}(\mathsf{q}, \dot{\mathsf{q}}, t) = T(\mathsf{q}, \dot{\mathsf{q}}, t) + U(\mathsf{q}, t).$$

As the potential *does not* depend on the $\dot{q}_k$, the derivatives of the Lagrangian with respect to the $\dot{q}_k$ coincide with the respective derivatives of the kinetic energy:

$$\frac{\partial \mathcal{L}}{\partial \dot{q}_k} = \frac{\partial (T + U)}{\partial \dot{q}_k} = \frac{\partial T}{\partial \dot{q}_k} .$$

By this identity it is possible to rewrite the Lagrange equations in the conservative case as:

$$\frac{d}{dt}\left(\frac{\partial \mathcal{L}}{\partial \dot{q}_k}\right) - \frac{\partial \mathcal{L}}{\partial q_k} = 0\,. \tag{14.28}$$

**Remark 14.10** In presence of conservative active forces (of the potential $U$) and also non-conservative forces, it is still possible to define the Lagrangian $\mathcal{L} = T + U$, and Lagrange's equations read

$$\frac{d}{dt}\left(\frac{\partial \mathcal{L}}{\partial \dot{q}_k}\right) - \frac{\partial \mathcal{L}}{\partial q_k} = Q_k^{(\text{n.c.})},$$

where $Q_k^{(\text{n.c.})}$ denotes the Lagrangian component of the sole non-conservative active forces. □

**Example 14.11** Let us use the Lagrange equations to find the motion of the system in Fig. 14.4. On a vertical plane, the inextensible string $ABDEGP$, of negligible mass, is fixed at $A$ and carries at $P$ a particle of mass $\mu$. The portions $BD$ and $EG$ rest on two homogeneous pulleys (disks of radii $r$, $R$ and masses $m$, $M$), the second of which is constrained to rotate around its fixed center $F$. The string's contact with the latter disk is rough, so the string cannot slip. We want to compare two situations: when the contact with the first disk (centered at $C$) prohibits slipping, and when this contact is smooth. At the initial instant the system is at rest.

Our system lends itself to a few interesting kinematical considerations. Let us introduce the following coordinates, many of which will be shown to be related. Call $y_C$, $y_P$ the ordinates (oriented downwards) of the points $C$, $P$ and $\theta$, $\phi$ the (counter-clockwise) rotation angles of the disks at $C$, $F$. The choice of orientation for the $y$-axis implies the $z$-axis enters the plane in the figure, and that the disk's angular velocities are $-\dot{\theta}\mathbf{k}$ and $-\dot{\phi}\mathbf{k}$.

Let us see what it means the string is inextensible. First, the length is constant, i.e.

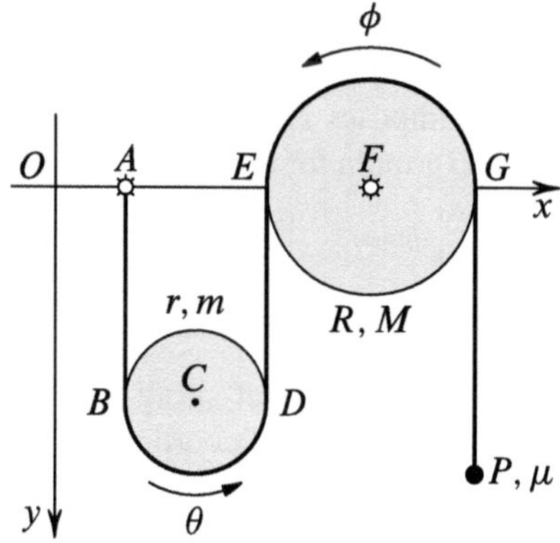


Fig. 14.4 A string fixed at $A$, wrapped around two pulleys and subject to a weight at P

$$\overline{AB} + \overset{\frown}{BD} + \overline{DE} + \overset{\frown}{EG} + \overline{GP} \equiv \text{constant}$$
$$\implies \quad y_B + \underbrace{\text{constant}}_{r\pi} + y_B + \underbrace{\text{constant}}_{R\pi} + y_P \equiv \underbrace{\text{constant}}_{L_{\text{filo}}}$$

so differentiating in time gives

$$2\dot{y}_B + \dot{y}_P = 0. \tag{14.29}$$

Consider the points at the ends of the straight portions that move parallel to itself, such as $AB$, $DE$ and $GP$ in the example. The velocities of said points are parallel to the string, and must necessarily be equal. If not, the points would either move closer (curling up the string) or away from one another (stretching it). Consequently

$$\mathbf{v}_B = \mathbf{v}_A (= \mathbf{0}), \qquad \mathbf{v}_E = \mathbf{v}_D, \qquad \mathbf{v}_G = \mathbf{v}_P. \tag{14.30}$$

Due to the notation of Sect. 6.2.3 it is important to stress that the position we label $P$ is occupied by the same point at all times (the endpoint, where the particle of mass $\mu$ is). On the contrary the geometric point $B$ is occupied in turn by different particles. Therefore

$$\mathbf{v}_P = \dot{P} = \dot{y}_P \mathbf{j}, \qquad \text{ma} \qquad \mathbf{v}_B \neq \dot{B} = \dot{y}_B \mathbf{j}.$$

Recall in fact that to compute $\dot{P}$ and $\dot{B}$ we just have to find the point's coordinates (with respect to a fixed origin) and differentiate them in time. Further considerations gives us the velocity. For example, $\mathbf{v}_P$ follows from the previous reasoning: the particle at position $P$ does not change, so $\mathbf{v}_P = \dot{P}$. As for $B$, constraint (14.30) gives its velocity, since it sets it equal to the velocity of $A$, which is zero. Note finally that because $AC = r\mathbf{i} + y_B\mathbf{j}$, and what we argued for $P$ holds for $C$ too, we have $\mathbf{v}_C = \dot{C} = \dot{y}_B\mathbf{j} (\neq \mathbf{v}_B)$.

The possible relationship between $y_B$, $y_P$ and $\theta$, $\phi$ depends on the kind of contact between string and disk. If smooth, it does not add any further kinematical relation. But if the constraint is rough and forbids the string from slipping, the velocity of any point of the string touching the disk will coincide with the velocity of the corresponding disk point. Consider in particular the point $G$ of Fig. 14.4. Constraint (14.30) imposes $\mathbf{v}_G^{(\text{filo})} = \mathbf{v}_P = \dot{y}_P\mathbf{j}$. On the other hand, applying (2.22) (the fundamental formula for rigid kinematics) to the disk at $F$ gives

$$\mathbf{v}_G^{(\text{disco})} = \underbrace{\mathbf{v}_F}_{\mathbf{0}} + \underbrace{\boldsymbol{\omega}}_{-\dot{\phi}\mathbf{k}} \times \underbrace{FG}_{R\mathbf{i}} = -R\dot{\phi}\mathbf{j} \quad \implies \quad R\dot{\phi} = -\dot{y}_P = 2\dot{y}_B.$$

Whenever the contact string-disk at $C$ is smooth, the system has two degrees of freedom (generalized coordinates $y_B$, $\theta$). If this contact is rough like the other one, we invite the reader to infer the further kinematical relation $r\dot{\theta} = -\dot{y}_B$ by arguing on $B$ as we did for $G$. Hence, the system has one degree of freedom.

All our constraints are ideal (in particular, the rough contact string-disk is equivalent to the disk's pure rolling on the string). Therefore the free motion equations arise from Lagrange's equations. As

$$\begin{aligned} T &= \left(\frac{1}{2}mv_C^2 + \frac{1}{4}mr^2\dot{\theta}^2\right) + \left(\frac{1}{4}MR^2\dot{\phi}^2\right) + \frac{1}{2}\mu v_P^2 \\ &= \begin{cases} \left(\frac{1}{2}m + M + 2\mu\right)\dot{y}_B^2 + \frac{1}{4}mr^2\dot{\theta}^2 & \text{(2gdl)} \\ \left(\frac{3}{4}m + M + 2\mu\right)\dot{y}_B^2 & \text{(1gdl)} \end{cases} \\ U &= mgy_C + Mgy_F + \mu g y_P = (m - 2\mu)gy_B + \text{constant}, \end{aligned}$$

we define the Lagrangian $\mathcal{L} = T + U$ and eventually find the free motion equations (14.28).

(2dof) When the contact is smooth,

$$\ddot{y}_B = \frac{(m - 2\mu)g}{m + 2M + 4\mu} \qquad \text{and} \qquad \ddot{\theta} = 0.$$

Therefore the angular velocity of the disk at $C$ stays constant, hence zero as at the initial instant. This disk then translates with velocity $\mathbf{v}_C = \dot{y}_B\mathbf{j}$. The motion relative to the generalized coordinates $y_B$ is uniformly accelerated, and depends on the sign of $m - 2\mu$. If for instance the mass of the disk at $C$ ($m > 2\mu$) is dominant, this disk descends and $P$ consequently goes up because of (14.29).

(1dof) If the contact prevents the string from slipping, then

$$\ddot{y}_B = \frac{(m - 2\mu)g}{\frac{3}{2}m + 2M + 4\mu}. \tag{14.31}$$

In this case both disks rotate, while the motion of $C$ and $P$ depends on the sign of $m - 2\mu$. We emphasize how now the motion of the two points is slower than the previous one (notice the factor $\frac{3}{2}$ in (14.31)). The reason is that part of the system's energy now goes to making the disk at $C$ rotate. □

## 14.5 Lagrangian First Integrals

Lagrange's equations are a system of $N$ coupled ordinary differential equations of order two, for which one cannot generally find the solution by analytical means. Nonetheless, there are cases in which a simple analysis of the Lagrangian's structure allows to detect the presence of *first integrals of motion*, i.e. functions of the generalized coordinates, their time derivatives and possibly time, that remain constant along the motion (see Sect. 11.2).

In this section we will study two types of first integrals that descend directly from the Lagrangian: conjugate momenta and the Hamiltonian.

### 14.5.1 First Integrals of Conjugate Momenta

When the Lagrangian does not depend explicitly on one of the generalized coordinates $q_k$, the corresponding Lagrange equations can be integrated to produce in a straightforward manner a first integral:

$$\frac{d}{dt}\left(\frac{\partial \mathcal{L}}{\partial \dot{q}_k}\right) = 0 \qquad \Longrightarrow \qquad \frac{\partial \mathcal{L}}{\partial \dot{q}_k} = \text{constant}$$

The conserved quantity

$$p_k = \frac{\partial \mathcal{L}}{\partial \dot{q}_k}$$

is called *conjugate momentum*, and the missing generalized coordinate in the Lagrangian is a *cyclic* coordinate.

**Remark 14.12** Conjugate momenta depend only on the kinetic energy, since the potential does not depend on the velocities. Moreover, (14.25) allows to prove the conjugate momenta depend on the Lagrangian velocities at most linearly:

$$p_k = \frac{\partial \mathcal{L}}{\partial \dot{q}_k} = \frac{\partial T}{\partial \dot{q}_k} = \sum_{j=1}^{N} a_{jk}(\mathsf{q}, t)\,\dot{q}_j + b_k(\mathsf{q}, t). \qquad \square$$

**Example 14.13** (*Integral of the linear momentum*) Consider a free particle $P$ of mass $m$, subject to a potential not depending on one Cartesian coordinate of the point: $U(P) = U(y, z)$. The missing coordinate ($x$ in our case) is cyclic. The associated conjugate momentum is the component of the linear momentum along that direction:

$$\mathcal{L} = \frac{1}{2}\,m\,\left(\dot{x}^2 + \dot{y}^2 + \dot{z}^2\right) + U(y, z) \quad \Longrightarrow \quad p_x = \frac{\partial \mathcal{L}}{\partial \dot{x}} = m\,\dot{x} = \text{constant}.$$

The first integral could have been found via the equation of motion of the point, since the $x$-component of the force acting on it is zero:

$$F_x = \frac{\partial U}{\partial x} = 0 \quad \Longrightarrow \quad m\,\ddot{x} = 0 \quad \Longrightarrow \quad m\,\dot{x} = \text{constant}. \qquad \square$$

**Example 14.14** (*Integrals of angular momentum*) Consider the motion of a rigid body whose position is described by the coordinates of the center of mass with respect to a fixed point $OG = (x_G, y_G, z_G)$ and by the Euler angles $\{\theta, \psi, \phi\}$ the comoving central principal axes of inertia $\{\mathbf{e}_{G1}, \mathbf{e}_{G2}, \mathbf{e}_{G3}\}$ determine with respect to a fixed basis $\{\mathbf{i}_1, \mathbf{i}_2, \mathbf{i}_3\}$. We have seen (see (3.14)) that the angular velocity reads, in term of the Euler angles,

$$\omega = (\dot{\theta}\cos\phi + \dot{\psi}\sin\theta\sin\phi)\mathbf{e}_{G1} - (\dot{\theta}\sin\phi - \dot{\psi}\sin\theta\cos\phi)\mathbf{e}_{G2} + (\dot{\phi} + \dot{\psi}\cos\theta)\,\mathbf{e}_{G3}. \tag{14.32}$$

Therefore the kinetic energy is

$$T = \frac{1}{2}m\left(\dot{x}_G^2 + \dot{y}_G^2 + \dot{z}_G^2\right) + \frac{1}{2}\Big(I_{G1}\big(\dot{\theta}\cos\phi + \dot{\psi}\sin\theta\sin\phi\big)^2 + I_{G2}\big(\dot{\theta}\sin\phi - \dot{\psi}\sin\theta\cos\phi\big)^2 + I_{G3}\big(\dot{\phi} + \dot{\psi}\cos\theta\big)^2\Big), \tag{14.33}$$

where $m$ is the mass of the rigid body, and $I_{G1}, I_{G2}, I_{G3}$ are the principal central moments of inertia. Expression (14.33) shows that when the principal moments of inertia are distinct, the precession angle $\psi$ can be a cyclic coordinate, while the angle of proper rotation can be cyclic if $I_{G1} = I_{G2}$. Finally, the nutation angle is never cyclic. Note that we say these angles *may be* cyclic coordinates, since we are only looking at the kinetic energy, and a cyclic coordinate cannot appear in the potential either.

The conjugate momenta associated with the precession and proper rotation angles have simple mechanical interpretations. Using (3.13) for the angular velocity with respect to the fixed basis, it is easy to show the conjugate momentum associated with precession coincides with the component of the barycentric angular momentum along the fixed axis $\mathbf{i}_3$:

$$p_\psi = \frac{\partial T}{\partial \dot{\psi}} = \mathbf{K}_G \cdot \mathbf{i}_3.$$

Similarly, by (14.32) one can show the conjugate momentum associated with the proper rotation coincides with the component of the barycentric angular momentum along the comoving axis $\mathbf{e}_3$:

$$p_\phi = \frac{\partial T}{\partial \dot{\phi}} = \mathbf{K}_G \cdot \mathbf{e}_3.$$

□

### 14.5.2 Hamiltonian

**Definition 14.15** (*Hamiltonian*) We call Hamiltonian of a holonomic system with generalized coordinates $\{q_1, \ldots, q_N\}$ the function

$$\mathcal{H}(\mathsf{q}, \dot{\mathsf{q}}, t) = \sum_{k=1}^{N} \dot{q}_k \frac{\partial \mathcal{L}}{\partial \dot{q}_k} - \mathcal{L}. \tag{14.34}$$

**Proposition 14.16** *The time derivative of* $\mathcal{H}$ *is related to the fact the Lagrangian depends explicitly on time:*

$$\frac{d\mathcal{H}}{dt} = -\frac{\partial \mathcal{L}}{\partial t}\,. \tag{14.35}$$

***Proof*** Differentiate in time (14.34), recalling that the Lagrangian's derivative is computed with the chain rule:

$$\begin{aligned}\frac{d\mathcal{H}}{dt} &= \underbrace{\sum_{k=1}^{N} \ddot{q}_k \frac{\partial \mathcal{L}}{\partial \dot{q}_k}} + \sum_{k=1}^{N} \dot{q}_k \frac{d}{dt}\left(\frac{\partial \mathcal{L}}{\partial \dot{q}_k}\right) - \left(\sum_{k=1}^{N} \frac{\partial \mathcal{L}}{\partial q_k}\dot{q}_k + \underbrace{\sum_{k=1}^{N} \frac{\partial \mathcal{L}}{\partial \dot{q}_k}\ddot{q}_k} + \frac{\partial \mathcal{L}}{\partial t}\right)\\ &= \sum_{k=1}^{N} \dot{q}_k \left[\frac{d}{dt}\left(\frac{\partial \mathcal{L}}{\partial \dot{q}_k}\right) - \frac{\partial \mathcal{L}}{\partial q_k}\right] - \frac{\partial \mathcal{L}}{\partial t} = -\frac{\partial \mathcal{L}}{\partial t}\,.\end{aligned}$$

In the first row the round brackets contain terms coming from the time derivative of the Lagrangian. The underbraced terms are opposite and cancel out. In the second row we gathered up the term $\dot{q}_k$: the factor in square brackets is identically zero by Lagrange's equation. □

Immediate consequence of (14.35) is that the Hamiltonian is conserved every time the Lagrangian does not depend explicitly on time. This occurs, for example, when the constraints are fixed and the active forces do not depend explicitly on time.

**Theorem 14.17** (Energy integral) *When the constraints are fixed, and the active forces are conservative, the Hamiltonian function coincides with the system's mechanical energy* $E = T - U$.

Before we prove the equality of Hamiltonian and mechanical energy let us remember the definition of homogeneous function and the theorem of Euler about them.

**Definition 14.18** A function $f : \mathbb{R}^m \to \mathbb{R}$ is *homogeneous of degree* $\alpha$ when

$$f(\lambda x_1, \lambda x_2, \ldots, \lambda x_m) = \lambda^\alpha f(x_1, \ldots, x_m), \qquad \text{for any } \lambda \in \mathbb{R}^+.$$

**Theorem 14.19** (Euler's homogeneous function theorem) *Let* $f : \mathbb{R}^m \to \mathbb{R}$ *be a differentiable homogeneous map of degree* $\alpha$. *Then:*

$$\sum_{i=1}^{m} x_i \frac{\partial f}{\partial x_i} = \alpha f. \tag{14.36}$$

***Proof* of the energy integral theorem** We will apply Euler's theorem to study the Hamiltonian. Expression (14.25) shows the kinetic energy of any holonomic system is the sum of three terms, each of which is homogeneous in the Lagrangian velocities $\dot{\mathsf{q}} = \{\dot{q}_1, \ldots, \dot{q}_N\}$. The first term, which we will call $T_2$, is quadratic, i.e.

homogeneous of degree 2. The second, $T_1$, is linear, so homogeneous of degree 1. The last, $T_0$, does not depend on the $\dot{q}_k$, whence it is homogeneous of degree 0.

Now let us examine the definition of Hamiltonian (14.34), in particular the combination of derivatives of $\mathcal{L}$ in the first summand. Since the potential $U$ does not depend on the velocities, first of all we can replace in this summand the Lagrangian by the kinetic energy:

$$\begin{aligned}\sum_{k=1}^{N} \dot{q}_k \frac{\partial \mathcal{L}}{\partial \dot{q}_k} &= \sum_{k=1}^{N} \dot{q}_k \frac{\partial (T+U)}{\partial \dot{q}_k} = \sum_{k=1}^{N} \dot{q}_k \frac{\partial T}{\partial \dot{q}_k} \\ &= \sum_{k=1}^{N} \dot{q}_k \frac{\partial T_2}{\partial \dot{q}_k} + \sum_{k=1}^{N} \dot{q}_k \frac{\partial T_1}{\partial \dot{q}_k} + \sum_{k=1}^{N} \dot{q}_k \frac{\partial T_0}{\partial \dot{q}_k}\,. \end{aligned} \tag{14.37}$$

In row two we have replaced the kinetic energy with the combination $T_2 + T_1 + T_0$ in order to separate the terms by their degree of homogeneity.

Each sum on the second row of (14.37) has the same structure as the sum of products in (14.36), with the $\dot{q}_k$ replacing the $x_i$. The sums then follow from Euler's theorem, taking care of applying to each summand the coefficient $\alpha$ equal to its homogeneous degree. Hence

$$\sum_{k=1}^{N} \dot{q}_k \frac{\partial \mathcal{L}}{\partial \dot{q}_k} = \cdots = \sum_{k=1}^{N} \dot{q}_k \frac{\partial T_2}{\partial \dot{q}_k} + \sum_{k=1}^{N} \dot{q}_k \frac{\partial T_1}{\partial \dot{q}_k} + \sum_{k=1}^{N} \dot{q}_k \frac{\partial T_0}{\partial \dot{q}_k} = 2T_2 + T_1\,.$$

Back to the full Hamiltonian, we obtain

$$\mathcal{H} = \sum_{k=1}^{N} \dot{q}_k \frac{\partial \mathcal{L}}{\partial \dot{q}_k} - \mathcal{L} = (2T_2 + T_1) - (T_2 + T_1 + T_0 + U) = T_2 - T_0 - U\,.$$

This holds for any type of constraint. When the constraints are fixed, both $T_1$ and $T_0$ vanish so

$$T = T_2 \quad \Longrightarrow \quad \mathcal{H} = T - U = E\,. \qquad \square$$

By a simple example we shall verify how in some cases, under conservative active forces and *moving* ideal constraints, the Lagrangian does not depend explicitly on time, $\partial_t \mathcal{L} = 0$. If so, $\mathcal{H}$ is an integral of motion, though now it *does not* coincide with the system's mechanical energy.

**Example 14.20** Let $P$ be a point of mass $m$ constrained to a smooth guide that rotates uniformly on a horizontal plane with constant angular velocity $\omega$ around the origin. A spring with constant $k$ and length $s$ connects $P$ to the origin, as per Fig. 14.5. In the inertial frame, the kinetic energy of the point, subject to a *moving* ideal bilateral constraint, is

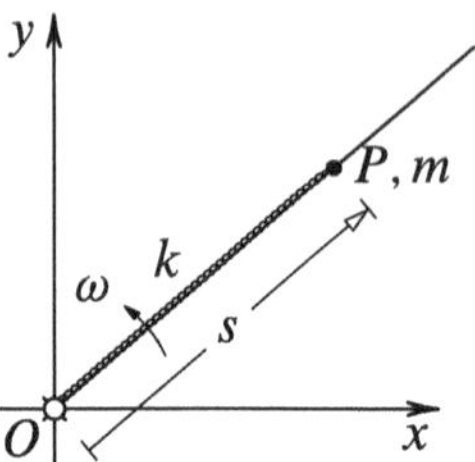


**Fig. 14.5** A particle $P$ constrained to a rotating guide

$$T = \frac{1}{2}m(\dot{s}^2 + s^2\omega^2)$$

while the potential reduces to the spring's $U = -ks^2/2$. Hence

$$\mathcal{L}(s, \dot{s}) = \frac{1}{2}m(\dot{s}^2 + s^2\omega^2) - ks^2/2$$

and obviously, $\partial_t \mathcal{L} = 0$. Therefore $\mathcal{H}$, given by

$$\begin{aligned}\mathcal{H} &= \frac{\partial \mathcal{L}}{\partial \dot{s}}\dot{s} - \mathcal{L} = m\dot{s}^2 - \frac{1}{2}m(\dot{s}^2 + s^2\omega^2) + ks^2/2 \\ &= \frac{1}{2}m\dot{s}^2 - \frac{1}{2}ms^2\omega^2 + ks^2/2\,,\end{aligned}$$

remains constant and is an integral of motion. However, this quantity *is not* the mechanical energy $E = T - U$, which by the way is *not* constant since the constraint reaction of the moving guide has non-zero effective power.

A careful inspection reveals that $\mathcal{H}$ corresponds instead to the mechanical energy $E_{\mathrm{rel}} = T_{\mathrm{rel}} - U_{\mathrm{rel}}$ measured in the frame rotating with the guide. The quantity $m\dot{s}^2/2$ is the relative kinetic energy, while $ms^2\omega^2/2$ and $-ks^2/2$ are the potentials of the centrifugal and elastic force respectively. Indeed, viewed in the *rotating* frame, the system has fixed ideal bilateral constraints with conservative active force, so it satisfies the theorem of conservation of the energy (Theorem 11.20). □

The importance and usefulness of Lagrange's equations can be assessed from two points of view at least. First, they essentially provide an "automatic" method (once the Lagrangian is written properly) to deduce the free motion equations, without the need to find the constraint reactions, since it does not require that we decompose the system into its components. This property is shared by the Principle of virtual work, to which the Lagrange equations have a tight theoretical relationship.

Now, though, we wish to highlight a second aspect: the presence of first integrals of motion, such as integrals of the conjugate momenta, is sometimes revealed by the Lagrangian, even when the conserved quantity is not evident by direct inspection of the system.

There certainly are many situations where there is a first integral associated with a conjugate momentum, a consequence of the Lagrangian's non-dependence on a specific generalized coordinate. Often, however, one proves this conserved conjugate momentum is trivially understandable as the part of a mechanical quantity we already knew was constant, following a preliminary look at the system. In the ensuing example there is an integral of motion that, albeit easily deducible from the Lagrangian, cannot be guessed that easily in advance.

**Example 14.21** Figure 14.6 shows a homogeneous disk of radius $r$ and mass $m$ in a vertical plane, constrained to roll without slipping along a fixed horizontal guide. A rod $AB$ of length $3r$ and mass $2m$ is pinned at the center $A$. The disk's rotation angle is $\theta$, while $\phi$ is the angle between the rod and the vertical direction.

The system is initially at rest with $\phi = \pi/3$, and we ask whether we can find the velocity distribution when the rod becomes vertical, with $\phi = 0$. The pair $(\theta, \phi)$ forms a system of generalized coordinates and in function of them and their derivatives we shall express the main mechanical quantities. We know that a problem of this type is in general solvable quite easily once we find two first integrals.

As the constraints are perfect and fixed, and the active forces (the weights) conservative, we can already say that we will have an energy first integral: $T - U =$ const. Regarding the second first integral, we may think of $Q_x =$ const.. This would be *wrong*, as shown in Fig. 14.6 right, because there is a horizontal component of the guide's constraint reaction acting on the disk (caused by pure rolling). Hence it is *not true* that $R_x^{\text{ext}} = 0$. Although a priori we cannot see if there is a second integral of motion, let us in any case write down the Lagrangian.

The velocity of $G$ is found from its coordinates

$$\begin{aligned} x_G &= r\theta + 3r/2 \sin\phi + \text{const.} \\ y_G &= r - 3r/2 \cos\phi \end{aligned} \quad \Rightarrow \quad \begin{aligned} \dot{x}_G &= r\dot{\theta} + 3r/2\dot{\phi}\cos\phi \\ \dot{y}_G &= 3r/2\dot{\phi}\sin\phi \end{aligned}$$

so by König's theorem the kinetic energy of the rod equals

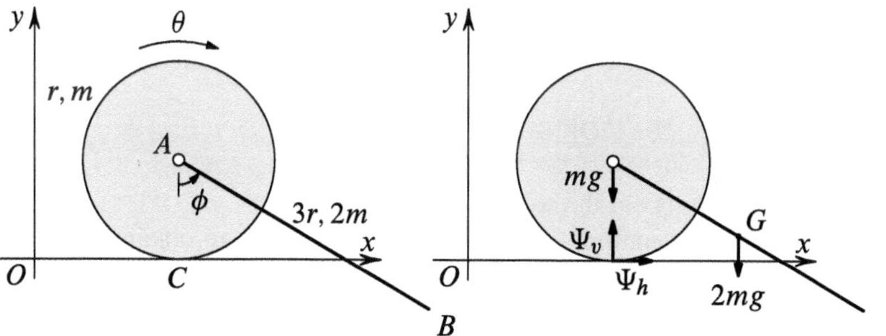


**Fig. 14.6** Lagrange equations and conservation of a conjugate momentum

$$\begin{aligned} T^{\text{rod}} &= \frac{1}{2}2m(\dot{x}_G^2+\dot{y}_G^2)+\frac{1}{2}I_{z,G}\dot{\phi}^2 = \frac{1}{2}2m(\dot{x}_G^2+\dot{y}_G^2)+\frac{1}{2}\frac{2m(3r)^2}{12}\dot{\phi}^2 \\ &= \frac{7}{4}mr^2\dot{\theta}^2+3mr^2\dot{\phi}^2+3mr^2\dot{\phi}\dot{\theta}\cos\phi. \end{aligned}$$

The disk's kinetic energy is

$$T^{\text{disk}} = \frac{1}{2}I_{z,C}\dot{\theta}^2 = \frac{1}{2}\frac{3mr^2}{2}\dot{\theta}^2 = \frac{3}{4}mr^2\dot{\theta}^2$$

with weight potential

$$U = -2mgy_G + \text{const.} = 3mgr\cos\phi + \text{const.}$$

Overall, the Lagrangian $\mathcal{L} = T + U$ is

$$\mathcal{L} = \frac{7}{4}mr^2\dot{\theta}^2+3mr^2\dot{\phi}^2+3mr^2\dot{\phi}\dot{\theta}\cos\phi+3mgr\cos\phi.$$

Note immediately that the function $\mathcal{L}$ *does not* depend on time explicitly (though we could have anticipated this), and this is coherent with the conservation of the energy. But more interestingly it is *independent* of the generalized coordinate $\theta$. We therefore have integrals of the conjugate momenta:

$$\frac{\partial\mathcal{L}}{\partial\theta}=0 \quad\Rightarrow\quad \frac{d}{dt}\frac{\partial\mathcal{L}}{\partial\dot{\theta}}=0 \quad\Leftrightarrow\quad \frac{\partial\mathcal{L}}{\partial\dot{\theta}}=\text{const.}$$

The derivative of $\mathcal{L}$ with respect to $\dot{\theta}$ is

$$\frac{7}{2}mr^2\dot{\theta}+3mr^2\dot{\phi}\cos\phi=0 \tag{14.38}$$

where the constant on the right has been set to zero since initially $\dot{\phi}=\dot{\theta}=0$. Note

$$Q_x^{\text{tot}} = 3mr\dot{\theta}+3mr\dot{\phi}\cos\phi,$$

so as anticipated $Q_x$ is *not* the conserved quantity.

In (14.38) we can multiply by two and divide by $mr^2$ to obtain a relationship involving $\dot{\theta}$ and $\dot{\phi}$:

$$7\dot{\theta}+6\dot{\phi}\cos\phi=0\,,$$

that for $\phi_f = 0$ becomes

$$7\dot{\theta}_f+6\dot{\phi}_f=0 \quad\Leftrightarrow\quad \dot{\phi}_f=-\frac{7}{6}\dot{\theta}_f\,. \tag{14.39}$$

Moreover,

$$7\dot{\theta} + 6\dot{\phi}\cos\phi = 0 \quad \Leftrightarrow \quad \frac{d}{dt}[7\theta + 6\sin\phi] = 0$$

so

$$7\theta + 6\sin\phi = \text{const.}$$

Inserting the initial and final values of $\theta$ and $\phi$ gives

$$7\theta_0 + 6\sin\phi_0 = 7\theta_f + 6\sin\phi_f \quad \Leftrightarrow \quad 7(\theta_f - \theta_0) = 3\sqrt{3}$$

from which we deduce that the rotation angle $\theta$ varies, from the first to the second configuration, by

$$\Delta\theta = \frac{3\sqrt{3}}{7}.$$

The total mechanical energy is

$$T - U = \frac{7}{4}mr^2\dot{\theta}^2 + 3mr^2\dot{\phi}^2 + 3mr^2\dot{\phi}\dot{\theta}\cos\phi - 3mgr\cos\phi$$

and the relation $T_f - U_f = T_0 - U_0$ reads

$$\frac{7}{4}mr^2\dot{\theta}_f^2 + 3mr^2\dot{\phi}_f^2 + 3mr^2\dot{\phi}_f\dot{\theta}_f - 3mgr = -\frac{3}{2}mgr$$

(we know $\phi_0 = \pi/3$, $\phi_f = 0$). Substituting here the value of $\dot{\phi}_f$ from (14.39), a few calculations produce

$$\dot{\theta}_f^2 = \frac{9g}{14r}.$$

The conjugate momentum associated with the generalized coordinate $\theta$ is therefore the integral of motion whose existence we could not have somehow guessed, since it does not coincide with one of the usual mechanical quantities. Remarkably, the Lagrangian has signalled straightaway the presence of this noteworthy property of the system, which allowed us to find $\dot{\phi}_f$ and $\dot{\theta}_f$.

Can we make explicit sense, a posteriori, of the surprising integral of motion just found? The answer is yes, and in a not too complicated way. Indicate by $\Phi_h$ the rightward-pointing horizontal component of the force the rod exerts on the disk's center $A$. The second balance equation of Dynamics for the disk, with pole $C$, is

$$\frac{d}{dt}[K_C^z] = \frac{d}{dt}[I_{z,C}\dot{\theta}] = \frac{d}{dt}\left[\frac{3mr^2}{2}\dot{\theta}\right] = r\Phi_h.$$

On the other hand the first balance equation for the rod along the horizontal direction says

$$\frac{d}{dt}Q_x^{\text{rod}} = \frac{d}{dt}2m[r\dot{\theta} + 3r/2\dot{\phi}\cos\phi] = -\Phi_h$$

where the minus sign on the right is due to the action-reaction principle. Comparing the two equations gives

$$r\frac{d}{dt}Q_x^{\text{rod}} + \frac{d}{dt}[K_C^z] = 0$$

and so

$$\frac{d}{dt}\left[rQ_x^{\text{rod}} + K_C^z\right] = 0.$$

The conserved quantity is therefore

$$rQ_x^{\text{rod}} + K_C^z,$$

which indeed is (14.38). □

## 14.6 State Space

In Chap. 4, devoted to constraints, we saw how the configuration of a system is prescribed uniquely by the generalized coordinates $\mathsf{q} = (q_1, q_2, \ldots, q_N)$, which vary in a domain of $\mathbb{R}^N$. As the velocity of a generic point is expressible though (4.21) it is immediate to see that if at some instant we know both the $\mathsf{q} = (q_1, q_2, \ldots, q_N)$ and the derivatives $\dot{\mathsf{q}} = (\dot{q}_1, \dot{q}_2, \ldots, \dot{q}_N)$ we can find the *configuration* and the *velocity distribution* of the system. For this reason it seems natural to think of a $2N$-dimensional space (twice the number of generalized coordinates), called *state space*, whose coordinates are the $\mathsf{q}$ and the $\dot{\mathsf{q}}$, so that each point corresponds to a *state* of the system, meaning a *configuration* jointly with a *velocity distribution*.

Obviously we can draw the state space only for systems with one generalized coordinate, for then the dimension is two, with coordinates $(q, \dot{q})$. The ensuing considerations will nevertheless hold more generally.

In the sequel, for simplicity we will suppose systems have fixed ideal constraints and time-independent active forces, so that the associated system of second-order equations, arising from Lagrange's equations, will not exhibit time explicitly.

Given a state $(\mathsf{q}_0, \dot{\mathsf{q}}_0)$, Theorem 14.9 tells there is a unique motion $\mathsf{q}(t)$, so one pair of functions $(\mathsf{q}(t), \dot{\mathsf{q}}(t))$ such that $(\mathsf{q}(0), \dot{\mathsf{q}}(0)) = (\mathsf{q}_0, \dot{\mathsf{q}}_0)$. This pair $(\mathsf{q}(t), \dot{\mathsf{q}}(t))$ traces out a curve in the space of states called an *orbit*.

As the equations of motion do not depend explicitly on time, if $(\tilde{\mathsf{q}}(t), \dot{\tilde{\mathsf{q}}}(t))$ is a different solution passing through state $(\mathsf{q}_0, \dot{\mathsf{q}}_0)$ at the generic time $\tau$, then $(\tilde{\mathsf{q}}(t), \dot{\tilde{\mathsf{q}}}(t)) = (\mathsf{q}(t-\tau), \dot{\mathsf{q}}(t-\tau))$. That is, through *every* point in state space passes *a unique* orbit.

Figure 14.7 shows a state space with an orbit passing through a generic state (the dot). The points $(\mathsf{q}, 0)$ (on the "horizontal" subspace) correspond to *rest states*,

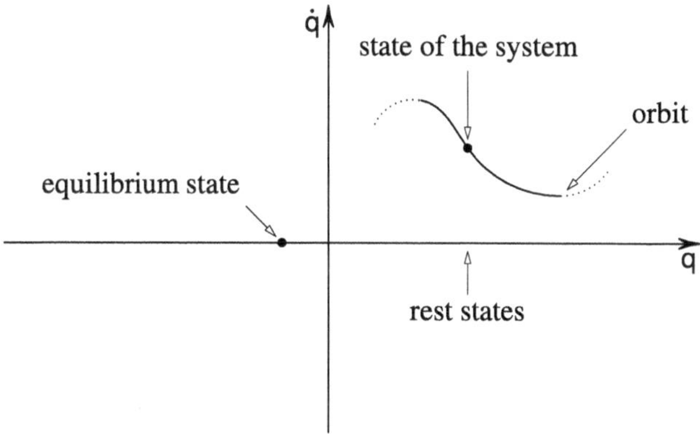


**Fig. 14.7** The state space

for if the $\dot{\mathsf{q}}$ vanish the system's velocity distribution is zero. Among rest states are *equilibrium states*, whose peculiarity is to initiate orbits that reduce to a point: each "motion" starting at an equilibrium state $(\mathsf{q}_0, \mathsf{0})$ stays in that state: $(\mathsf{q}(t), \dot{\mathsf{q}}(t)) = (\mathsf{q}_0, \mathsf{0})$ at any instant.

The role of integrals of motion (if present) is of great importance. Integrals of motion $I(\mathsf{q}, \dot{\mathsf{q}})$ have the property that

$$I(\mathsf{q}(t), \dot{\mathsf{q}}(t)) = I_0$$

where $I_0$ is the value attained by $I(\mathsf{q}, \dot{\mathsf{q}})$ at a generic instant. Hence an integral of motion implicitly defines in state space a hypersurface (of dimension $2N - 1$) on which the orbit lies. Orbits belong on the level surfaces of the integral of motion $I$: on *each orbit* the value of the integral of motion $I$ is *constant*.

**Remark 14.22** Only in two dimensions (systems with one generalized coordinate $q$), as suggested by Fig. 14.7, states in the upper half-plane correspond to $\dot{q} > 0$. An orbit in that region has coordinate $q$ increasing with time, whence the orbit moves to the right. Similarly, orbits in the lower half-plane are oriented leftwards ($q$ decreases). This observation is more difficult to extend to systems with several generalized coordinates. □

**Example 14.23** It is instructive to construct an explicit state space, to witness first-hand how powerful a tool it is to visualize, and especially carry out, the qualitative analysis of the system's possible motions. Consider a rod under its weight and pinned at one end to a fixed point, as in Fig. 14.8. The state space is two-dimensional, with coordinates $(q, \dot{q}) = (\theta, \dot{\theta})$.

A first observation is that the system displays a natural periodic nature with respect to the generalized coordinates $\theta$: the state $(\theta, \dot{\theta})$ coincides with $(\theta + 2\pi, \dot{\theta})$, so it suffices to restrict the state space to the strip $-\pi \leq \theta \leq \pi$, as in Fig. 14.9. This suggests that the best way to describe states is to use a round cylinder of infinite

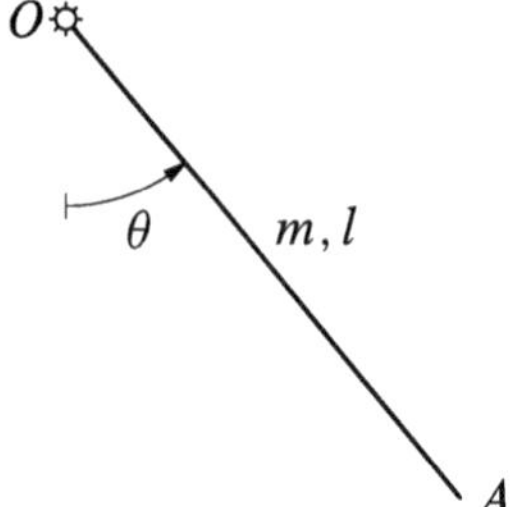


**Fig. 14.8** A homogeneous rod with one end fixed at $O$

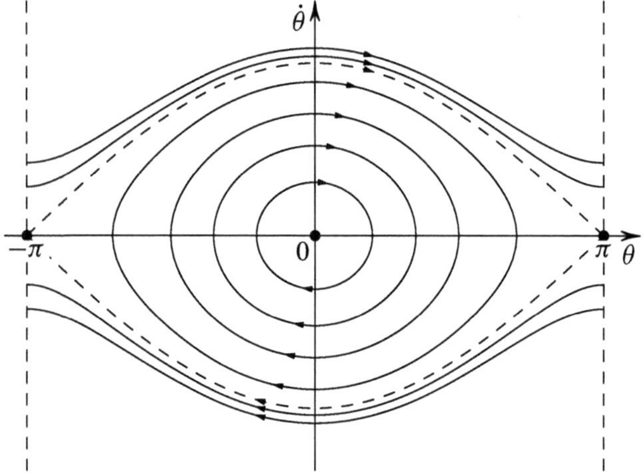


**Fig. 14.9** The orbits for the motions of the rod in Fig. 14.8

height, where the angle on the base circle corresponds to the generalized coordinate $\theta$, while the height of a generic point is associated with $\dot{\theta}$. This fact holds more generally, beyond the present circumstances: the geometry of the appropriate spaces of configurations and states might be that of a hypersurface (a differentiable manifold, to use a technical name) and not just a flat space. All this, which we could have addressed in Chap. 4, leads to advanced topics that we shall not tackle in our treatise. We shall merely remember this patent periodicity when looking at Fig. 14.9, without further comment.

It is an easy matter to compute $T$, $V$ and the mechanical energy $E = T + V$ by

$$E(\theta, \dot{\theta}) = \frac{1}{6} m l^2 \dot{\theta}^2 - mgl/2 \cos\theta \tag{14.40}$$

and then write its conservation as

$$\frac{1}{6} m l^2 \dot{\theta}^2 - mgl/2 \cos\theta = E_0$$

where

$$E_0 = \frac{1}{6} m l^2 \dot\theta_0^2 - mgl/2 \cos\theta_0$$

and $(\theta_0, \dot\theta_0)$ is the initial state. States travel along orbits that we may describe as the (oriented) level curves of the function $E(\theta, \dot\theta)$, defined by (14.40), on the plane $(\theta, \dot\theta)$.

The value of the mechanical energy is necessarily bigger than or equal to $-mgl/2$, and there exist two equilibrium states: $(\theta, \dot\theta) = (0, 0)$ (the rod hangs vertically), and $(\theta, \dot\theta) = (\pm\pi, 0)$ (the rod is standing vertically). The mechanical energy of these states is respectively $E = -mgl/2$ (minimum value) and $E = mgl/2$, and the orbits originating at these states reduce to points.

Easy manipulations transform the conservation of energy into

$$\dot\theta^2 = \frac{3g}{l} \cos\theta + \frac{6E_0}{ml^2}$$

and we deduce (in this simple instance) the explicit expression for the orbits, represented in detail in Fig. 14.9. A more comprehensive study (we omit the details) allows to conclude the following:

- for each value of $\theta$ there are two opposite values of $\dot\theta$ corresponding to symmetric states with respect to the $\theta$-axis.
- Not all orbits meet a rest state. In Fig. 14.9 the orbits with mechanical energy larger than $E_0 = mgl/2$ do not pass through rest states (the derivative $\dot\theta$ never vanishes) and are represented in the upper and lower part of Fig. 14.9, together with their orientation. These orbits describe motions during which the rod makes complete rotations around $O$ with nowhere vanishing angular velocity, due to the sufficiently large energy.
- The orbits with energy strictly between $-mgl/2$ and $mgl/2$ are closed paths circling around the equilibrium state $(\theta, \dot\theta) = (0, 0)$, and correspond to oscillations of the rod between a maximum angle $\bar\theta > 0$ and a minimum angle $-\bar\theta < 0$, at which $\dot\theta$ is zero (the motions are harmonic).
- There are two trajectories called *separatrices* (dashed in the figure) that seem to start and end at the equilibrium states $(\pm\pi, 0)$, thus separating the differently behaved orbits of the previous cases. We might wrongly think that in this case uniqueness fails. A careful analysis reveals that the *time* required for the rod to reach configuration $\theta = \pm\pi$ is infinite, so the separatrices describe motions that tend asymptotically to the upper equilibrium configuration, without ever reaching it.

Another remark regarding the equilibrium state $(\theta, \dot\theta) = (0, 0)$ (the origin of the plane in Fig. 14.9): the orbits starting nearby this point stay close to it. Put differently, for any arbitrarily small "rectangle" $B$ centered at $(0, 0)$ we can find another rectangle such that each orbit born in the latter will be forever confined inside $B$. This property expresses the idea of *stability* for the configuration corresponding to the equilibrium state $(0, 0)$.

The same cannot be said for the equilibrium states $(\pm\pi, 0)$. Looking at Fig. 14.9, irrespective of how close a trajectory passes to this state, it will eventually move away from it. This characterizes the *instability* of the corresponding equilibrium configuration.

The property that distinguishes clearly the two equilibrium configurations is that the former (stable) is a *minimum* for the potential energy while the other (unstable) is a *maximum*.

As we will see later, this observation is not accidental and can be elaborated further to produce a number of theorems. □

## 14.7 Stability of Equilibria

**Definition 14.24** (*Lyapunov stability*) Consider a system of $n$ particles. An equilibrium configuration $\mathcal{C}^\circ = \left(P_1^\circ, \ldots, P_n^\circ\right)$ is said to be *Lyapunov stable* if for any $\varepsilon, \varepsilon' > 0$ there exist $\delta, \delta' > 0$ (depending on $\varepsilon, \varepsilon'$) such that

$$\left.\begin{array}{ll} \left|P_i^\circ P_i(t_0)\right| < \delta & \forall i \\ \left|\mathbf{v}_i(t_0)\right| < \delta' & \forall i \end{array}\right\} \Longrightarrow \left\{\begin{array}{ll} \left|P_i^\circ P_i(t)\right| < \varepsilon & \forall i\,,\ \forall t \geq t_0 \\ \left|\mathbf{v}_i(t)\right| < \varepsilon' & \forall i,\ \forall t \geq t_0. \end{array}\right. \tag{14.41}$$

Definition (14.41) thus identifies as *stable* those equilibrium configurations that allow for the following procedure. Choose arbitrarily a maximum distance $\varepsilon$ from the equilibrium configuration and a maximum speed $\varepsilon'$ for the particles. Given $\varepsilon$ and $\varepsilon'$, we want to find a $\delta$ and a $\delta'$ (non-zero) satisfying the following property: *every* motion starting at a distance less than $\delta$ from $\mathcal{C}^\circ$ and with speed less than $\delta'$ will remain within $\varepsilon$ from the equilibrium, and its speed will always be bounded by $\varepsilon'$. It is particularly important to say that Lyapunov stability guarantees that not only the positions are bounded, but the velocities are as well.

### *14.7.1 Dirichlet-Lagrange Stability Theorem*

We shall devote this section to the proof and examination of a fundamental result regarding the Lyapunov stability of equilibrium configurations corresponding to local maxima of the potential $U$, or equivalently, local minima of the potential energy $V = -U$. Because it is more natural for the presentation and discussion, we will formulate the result using the potential energy $V$.

**Theorem 14.25** (Dirichlet-Lagrange) *Consider a conservative holonomic system with fixed constraints, whose potential energy $V = -U$ is a continuous map of the positions. Let $\mathfrak{q}^\circ$ be an equilibrium configuration. If $\mathfrak{q}^\circ$ is an isolated local minimum*

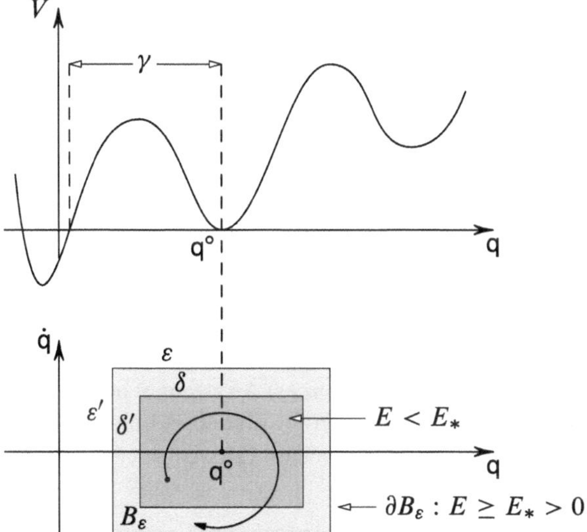


**Fig. 14.10** The top picture shows an example of a potential energy with an isolated local minimum at $\mathsf{q}^\circ$. The bottom picture shows the procedure for the proof of the theorem. Given $\varepsilon$, $\varepsilon'$ we define $E_* > 0$ to be the minimum value of the mechanical energy on the boundary of the light region. Next, the dark region is characterized by the property that at all its points the mechanical energy is strictly less than $E_*$. Thus any motion starting in the darker region does not have sufficient energy to escape from the lighter region

*of $V$ (i.e. an isolated local maximum of the potential energy $U = -V$) then $q^\circ$ is Lyapunov stable.*

***Proof*** We start by noting that it suffices to add a constant to the potential energy to be able to assume it is zero at the equilibrium configuration: $V(\mathsf{q}^\circ) = 0$.

If $\mathcal{C}^\circ$ is an isolated local minimum of the potential energy, there exists a region in the space of configurations around $\mathcal{C}^\circ$ where the potential energy is positive. In formulas, there exists $\gamma > 0$ such that $V(\mathsf{q}) > 0$ if $0 < |\mathsf{q} - \mathsf{q}^\circ| \leq \gamma$. On the other hand the kinetic energy $T$ is larger than zero for each velocity distribution where some velocity vector is non-zero.

Putting all that together, the mechanical energy $E = T + V$ will attain absolute minimum (equal to 0) at the velocity distribution of the equilibrium $\mathcal{A}^\circ = \{\mathsf{q}^\circ, \mathbf{0}\}$, and will be positive for any other velocity distribution (either because the kinetic energy is positive, or the potential energy is positive, or both).

Given now $\varepsilon, \varepsilon' > 0$, suppose $\varepsilon \leq \gamma$.[1] Let $B_\varepsilon$ be the set, in Fig. 14.10, formed by the velocity distributions such that $|\mathsf{q} - \mathsf{q}^\circ| \leq \varepsilon$ and $|\dot{\mathsf{q}}| \leq \varepsilon'$. The boundary of $B_\varepsilon$, denoted $\partial B_\varepsilon$, is the set of velocity distributions for which the distance to the

[1] If $\varepsilon > \gamma$, it suffices to follow the steps and replace $\varepsilon$ with $\gamma$ everywhere: the result will produce a motion confined inside the region of width $\gamma$, hence a fortiori in the region of width $\varepsilon$.

equilibrium configuration equals $\varepsilon$, and/or the speeds equal $\varepsilon'$. In Fig. 14.10, $\partial B_\varepsilon$ is the boundary of the light region, a closed and bounded set (hence compact).

The mechanical energy is the difference of continuous functions and hence is continuous, so the Weierstrass theorem ensures it has a maximum value $E^*$ and a minimum value $E_*$ on such set. In particular, the minimum $E_*$ will be strictly positive since $E$ vanishes only for the velocity distribution of the equilibrium $\mathcal{A}^\circ$. As $E(\mathcal{A}^\circ) = 0$, and $E$ is continuous, there exists a rectangular neighbourhood of $\mathcal{A}^\circ$ (of edges $\delta$, $\delta'$, dark in Fig. 14.10) whose velocity distributions will have energy strictly less than $E_*$. None of the motions starting in this neighbourhood will ever leave the rectangle of edges $\varepsilon, \varepsilon'$, because to do so the energy (an integral of motion) would have to increase. □

**Definition 14.26** (*Active dissipative forces*) A system of active forces is *dissipative* when its effective power is always non-positive. In a holonomic system, a system of active forces of Lagrangian components $\mathsf{Q}(\mathsf{q}, \dot{\mathsf{q}}) = \{Q_1, \ldots, Q_N\}$ is therefore dissipative when (see (7.45))

$$\Pi = \mathsf{Q} \cdot \dot{\mathsf{q}} = \sum_{k=1}^{N} Q_k(\mathsf{q}, \dot{\mathsf{q}})\, \dot{q}_k \le 0 \qquad \forall \dot{\mathsf{q}}. \tag{14.42}$$

The system of active forces is *completely dissipative* if we have equality in (14.42) if *and only if* all velocities are zero.

Put equivalently, dissipative active forces cannot increase the mechanical energy, but they can leave it invariant. Completely dissipative active forces, on the other hand, lower the mechanical energy the instant some part of the systems starts moving.

The Dirichlet-Lagrange theorem is still valid (with the same proof as above) in case the holonomic system also has non-conservative forces, provided they are dissipative, i.e. if they do not increase the energy. If the non-conservative active forces are completely dissipative, one can prove that the motion not only stays forever close to the equilibrium configuration, but it converges to it as $t \to \infty$ (in this case we speak of *asymptotic stability*).

**Remark 14.27** Examining carefully the proof of the Dirichlet-Lagrange theorem it becomes clear that we can weaken the assumption of having an *isolated* local maximum for the equilibrium configuration to be stable. The same argument applies if $\mathsf{q}^\circ$ is a limit point of local maxima, as long as the following condition holds: *for each $\varepsilon > 0$ there exists a closed surface $S_\varepsilon$, surrounding $\mathsf{q}^\circ$ and whose points have distance to $\mathsf{q}^\circ$ not exceeding $\varepsilon$, such that $U\big|_{S_\varepsilon} < U(\mathsf{q}^\circ)$.* □

**Example 14.28** Consider a point that satisfies the previous condition and is not an isolated local maximum (the example is ad hoc, but not terribly so). Take the function $g(x) = -x^2 \sin^2(1/x)$ for any $x \neq 0$, and $g(0) = 0$. Clearly $g(x) \le 0$ for any $x$, so the origin $x = 0$ is an absolute maximum point, whence local, for $g$. It is not isolated because the sequence $x_k = 1/(k\pi)$, $k \in \mathbb{N}$ (all zeroes of $g(x)$) accumulates at $x = 0$. Now, for any $\varepsilon > 0$ we can surround the origin with a surface $S_\varepsilon$ (for a one-variable function it will reduce to two points, one to the left and one to the right of $x = 0$, with distance less than $\varepsilon$) on which the function is strictly negative. We can seek such points among those of the form $x'_k = 1/(\frac{\pi}{2} + k\pi)$: these cluster at $x = 0$ and at them $g$ is negative. □

We finally emphasize that the Dirichlet-Lagrange theorem proves that in conservative systems, the notion of *dynamical* stability (Lyapunov stability) is coherent with *statical* stability (Sect. 9.10): both definitions produce the same stable configurations. The differences between the two concepts manifest themselves mainly in systems subject to velocity-dependent forces.

### *14.7.2 Instability Criteria*

The Dirichlet-Lagrange theorem labels as *stable* the isolated local maxima of the potentials of holonomic conservative systems. It says nothing about the stability or instability of minima, or of critical points that are neither maxima nor minima. This issue, referred to as the *inversion of the Dirichlet-Lagrange stability theorem*, is extremely complicated. In this section we will review a number of instability criteria that ensure the Lyapunov instability of many equilibrium configurations that do not satisfy the hypotheses of Dirichlet's theorem. All things said, even accounting for all criteria, there still remain the equilibrium configurations whose stability cannot be decided a priori, and must be examined case by case.

**Theorem 14.29** (Lyapunov's instability criterion) *Let $q^\circ$ be an equilibrium configuration of a holonomic system with fixed constraints and conservative active forces whose potential $U$ is at least of class $C^2$ around the equilibrium configuration. If the potential's Hessian at $q^\circ$ has one positive eigenvalue at least, the equilibrium configuration $q^\circ$ is Lyapunov unstable.*

We will not prove the criterion, nor the ensuing results. The proof of the above can be deduced by the analysis of the motion nearby the equilibrium position that we will carry out in Sect. 14.9. At that stage we will come back to this point.

Lyapunov's instability criterion applies to potentials admitting a Hessian. Depending on the signs of the latter's eigenvalues, we have the following possibilities.

- All eigenvalues are negative. Then we are certain the equilibrium point is an isolated local maximum of the potential, and the Dirichlet-Lagrange theorem guarantees stability.
- There is at least one positive eigenvalue. Lyapunov's criterion guarantees instability.
- One eigenvalue vanishes (at least one, possibly all), and the remaining ones are negative.

When the holonomic system has one or two degrees of freedom, Lyapunov's criterion can be improved.

- In holonomic conservative systems with one degree of freedom it is possible to prove (see Sect. 14.8) that any equilibrium position $q^\circ$ satisfying the following is unstable.
  *There exists a right and/or left neighborhood of $q^\circ$ where the potential is never less than $U(q^\circ)$.*
  In other words, while the Dirichlet-Lagrange theorem guarantees the stability of isolated maxima of the potential, the above property guarantees the instability of minima, of horizontal inflections, and also of *locally flat* maxima, those near which the potential is constant.
- For systems with two degrees of freedom, Painlevé demonstrated the instability of the equilibrium configurations that are not maxima of the potential (assuming the potential is analytic[2] near an equilibrium), even when one eigenvalue of the Hessian is negative and the other one is null.

Several criteria are known that address some of the cases the Dirichlet-Lagrange theorem and Lyapunov's instability criterion leave open. We list two below.

**Theorem 14.30** (Chetaev instability criterion) *Let $q^\circ$ be an equilibrium configuration of a holonomic system with fixed constraints and conservative active forces with potential $U$. If the potential function is* homogeneous *of degree $\lambda > 1$ in $(q - q^\circ)$, and $q^\circ$ is not a maximum of $U$, then $q^\circ$ is Lyapunov unstable.*

Let us recall Definition 14.18 of homogeneous function. Chetaev's criterion only applies to homogeneous potentials. But for those, it answers the stability question even in cases where Lyapunov's criterion is not applicable.

**Theorem 14.31** (Hagedorn-Taliaferro instability criterion) *Let $q^\circ$ be an equilibrium configuration of a holonomic system with fixed constraints and conservative active forces with potential $U$. If the potential is at least of class $C^1$ and $q^\circ$ is a local minimum of $U$, then $q^\circ$ is Lyapunov unstable.*

This criterion only applies to equilibrium configurations whose potential has a local minimum (isolated or not). At the same time, it does not ask much as regards $U$, nor does it require the Hessian to have positive eigenvalues.

## 14.8 Stability of Systems with One Degree of Freedom

We discovered in Sect. 10.4 several qualitative properties of the motions of conservative systems with one degree of freedom. Here we will complete that analysis by studying the Lyapunov stability of the equilibrium configurations of those systems.

Consider a holonomic system, with fixed constraints and a unique degree of freedom, subject to a conservative active force. Let $q$ denote the generalized coordinate.

[2] A real-valued function is *analytic* on a domain if it admits Taylor series at all points, and it coincides with its series on some non-trivial neighbourhood.

The theorem on the potential's critical points (see Sect. 9.9.3) guarantees the equilibrium configurations are the stationary points of the potential $U(q)$. If $q^\circ$ is an equilibrium configuration, we have

$$U'(q^\circ) = 0. \tag{14.43}$$

The point $q^\circ$ may correspond to a maximum, a minimum or an inflection (with horizontal tangent) of the potential. We will prove in the sequel that the first case corresponds to a Lyapunov stable equilibrium configuration, while the minima and inflections characterize unstable equilibrium positions. In holonomic systems with one degree of freedom, establishing the stability of an equilibrium position is not that same as characterizing the critical point of the position representing it. As we saw in the previous section, and we will corroborate in the next, the situation is more complicated for systems with several degrees of freedom.

To study the motions occurring near $q^\circ$ we change variable

$$q(t) = q^\circ + \varepsilon\, \eta(t). \tag{14.44}$$

The variable $\eta$ will replace $q$ as generalized coordinate, while the small parameter $\varepsilon$ will steer the qualitative analysis. In the sequel we will find the equation of motion, and after the change (14.44) we will neglect higher-order terms in $\varepsilon$, thus obtaining an equation with a simple explicit solution.

### *14.8.1 Linearized Equation of Motion*

In a holonomic system with fixed constraints, expression (14.25) of the kinetic energy simplifies, and $T$ depends quadratically on $\dot{q}$:

$$T(q, \dot{q}) = \frac{1}{2}\, a(q)\, \dot{q}^2. \tag{14.45}$$

The Lagrangian is then $\mathcal{L}(q, \dot{q}) = \frac{1}{2}\, a(q)\, \dot{q}^2 + U(q)$, and the Lagrange equation reads

$$a(q)\, \ddot{q} + \frac{1}{2}\, a'(q)\, \dot{q}^2 - U'(q) = 0. \tag{14.46}$$

Now let us implement the change (14.44) in (14.46). Setting $\dot{q} = \varepsilon\dot{\eta}$, and obviously $\ddot{q} = \varepsilon\ddot{\eta}$, we obtain

$$a\big(q^\circ + \varepsilon\eta\big)\, \varepsilon\, \ddot{\eta} + \frac{1}{2}\, a'\big(q^\circ + \varepsilon\eta\big)\, \varepsilon^2 \dot{\eta}^2 - U'\big(q^\circ + \varepsilon\eta\big) = 0. \tag{14.47}$$

To comprehend better what the leading term in $\varepsilon$ in the equation of motion (14.47) is, we use Taylor expansions with small $\varepsilon$ for each summand:

$$
\begin{aligned}
a\big(q^\circ + \varepsilon\eta\big)\,\varepsilon\,\ddot\eta &= \varepsilon\,a(q^\circ)\,\ddot\eta \quad + \varepsilon^2\,a'(q^\circ)\,\eta\,\ddot\eta \quad + o(\varepsilon^2) \\
\tfrac{1}{2}\,a'\big(q^\circ + \varepsilon\eta\big)\,\varepsilon^2\dot\eta^2 &= \frac{\varepsilon^2}{2}\,a'(q^\circ)\,\dot\eta^2 \quad + o(\varepsilon^2) \\
U'\big(q^\circ + \varepsilon\eta\big) &= U'(q^\circ) + \varepsilon\,U''(q^\circ)\,\eta + \frac{\varepsilon^2}{2}\,U'''(q^\circ)\,\eta^2 + o(\varepsilon^2).
\end{aligned}
\tag{14.48}
$$

The term $O(1)$ in the potential's derivative is zero, since $U'(q^\circ) = 0$ (see (14.43)). The leading term in (14.47) is therefore linear, both in $\varepsilon$ and in $\eta$. The linearized equation of motion is then

$$
\ddot\eta(t) - \frac{U''(q^\circ)}{a(q^\circ)}\,\eta(t) = 0. \tag{14.49}
$$

### 14.8.2 Frequency of Small Oscillations

The nature of the solutions to (14.49) depends on the signs of its constant coefficients. As $a(q^\circ) > 0$ due to the positive-definiteness of the kinetic energy (see Sect. 6.3.3), the approximated motion is characterized by the sign of the second derivative of the potential, computed at the equilibrium position $q^\circ$. More precisely (see Sect. A.5 as well):

$U''(q^\circ) < 0 \implies$ approximately harmonic motion with $\omega = \sqrt{-\dfrac{U''(q^\circ)}{a(q^\circ)}}$ :
$\eta(t) = \eta(0)\cos\omega t + \frac{\dot\eta(0)}{\omega}\sin\omega t$;

$U''(q^\circ) = 0 \implies$ approximately linear motion:
$\eta(t) = \eta(0) + \dot\eta(0)\,t$

$U''(q^\circ) > 0 \implies$ approximately hyperbolic motion:
$\eta(t) = \eta(0)\cosh\omega t + \frac{\dot\eta(0)}{\omega}\sinh\omega t$

In the first case $\omega$ is called *frequency of small oscillations.*

### 14.8.3 Stability of Equilibria

The Dirichlet-Lagrange stability theorem guarantees $q^\circ$ is stable if $U''(q^\circ) < 0$, since the equilibrium configuration corresponds to an isolated maximum of the potential. If on the contrary $U''(q^\circ) > 0$, it is the Lyapunov instability criterion Theorem 14.29 that ensures the approximately exponential motion detects an unstable equilibrium position.

We still know nothing about the stability of $q^\circ$ if $U''(q^\circ) = 0$. In fact, in that case the linear approximation in $\varepsilon$ cannot be relied on, since expansions (14.48) were stopped to an order such that the approximate equation does not retain any information on the potential (whence on the active forces). It should not be a surprise then that equation prescribes a uniform motion, exactly what we obtain in absence of external forces. A trustworthy prediction would require we put in the approximate equation all higher-order terms in $\varepsilon$ up to the first non-zero Taylor term in the potential. A glance at (14.48), though, is enough to convince ourselves that the approximate equation of motion is all but easily solvable explicitly.

In each case, the Dirichlet-Lagrange stability theorem declares as stable also the configurations with $U''(q^\circ) = 0$ whose higher derivatives show $q^\circ$ is an isolated maximum (even if in this case the approximate motion will not have a characterizing frequency of small oscillations). In all other cases the equilibrium position is Lyapunov unstable, as we will show in the sequel.

**Proposition 14.32** *Consider an equilibrium configuration $q^\circ$ of a holonomic system, subject to ideal and fixed constraints and conservative active forces (with potential $U$). If there is a right and/or left neighbourhood of $q^\circ$ where the potential is not less than $U(q^\circ)$, then $q^\circ$ is Lyapunov unstable.*

***Proof*** Consider an equilibrium configuration $q^\circ$ with the stated property. Without loss of generality we may assume that the neighbourhood of $q^\circ$ where the potential is at least $U(q^\circ)$ is a right neighbourhood. In other words, suppose there exists $q_1 > q^\circ$ such that

$$U(q) \geq U(q^\circ) \qquad \forall q \in [q^\circ, q_1].$$

Let us construct explicitly a motion starting from $q^\circ$ with arbitrarily small speed, that will nevertheless reach $q_1$ in finite time: set $q(0) = q^\circ$ and $\dot{q}(0) = \dot{q}_0 > 0$. From these initial conditions we compute the mechanical energy (a first integral of motion):

$$E = \frac{1}{2} a(q)\, \dot{q}^2 - U(q)\bigg|_{t=0} = \frac{1}{2} a(q^\circ)\, \dot{q}_0^2 - U(q^\circ).$$

At each instant

$$\frac{1}{2} a\big(q(t)\big)\, \dot{q}^2(t) - U\big(q(t)\big) = \frac{1}{2} a(q^\circ)\, \dot{q}_0^2 - U(q^\circ),$$

i.e.

$$\dot{q}^2(t) = \frac{a(q^\circ)}{a\big(q(t)\big)}\, \dot{q}_0^2 + \frac{2\Big(U\big(q(t)\big) - U(q^\circ)\Big)}{a\big(q(t)\big)}. \tag{14.50}$$

Under the hypotheses on the potential, the right-hand-side quantity in (14.50) does not vanish whenever the system stays in the interval $[q^\circ, q_1]$. Hence the velocity, initially positive, will stay positive at least until we exit that interval. Not only that. By letting

$$a_{\mathrm{M}} = \max_{q \in [q^\circ, q_1]} a(q) > 0,$$

we have the estimate

$$\dot{q}^2(t) = \frac{a(q^\circ)}{a\big(q(t)\big)}\dot{q}_0^2 + \frac{2\Big(U\big(q(t)\big) - U(q^\circ)\Big)}{a\big(q(t)\big)} \geq \frac{a(q^\circ)}{a_M}\dot{q}_0^2 \;\Rightarrow\; \dot{q}(t) \geq \sqrt{\frac{a(q^\circ)}{a_M}}\,\dot{q}_0.$$

By the above, the system will reach configuration $q_1$ at time

$$t_1 \leq \sqrt{\frac{a_M}{a(q^\circ)}}\,\frac{q_1 - q^\circ}{\dot{q}_0}.$$

Therefore it is impossible to confine the system to any neighborhood of width less than $(q_1 - q^\circ)$, which implies the equilibrium configuration is unstable. □

**Example 14.33** Let us analyze the equilibrium positions of two rods $OA$ and $AB$, both homogeneous of mass $m$ and length $2l$, a system we discussed in Example 9.11.2 (see Fig. 9.27). The potential, found in (9.52), is

$$U(\theta) = -8kl^2\left(\cos^2\theta + 2\lambda\sin\theta\right), \tag{14.51}$$

where $k$ is the elastic constant between $O$ and $B$, and $\lambda = mg/(8kl)$. The equilibrium configurations satisfy condition (14.51), i.e.

$$U'(\theta) = 16kl^2\cos\theta\,(\sin\theta - \lambda) = 0 \quad\Longrightarrow\quad \begin{cases} \theta_1^* = \frac{\pi}{2} \\ \theta_2^* = \arcsin\lambda \quad (\text{if } \lambda \leq 1) \\ \theta_3^* = -\frac{\pi}{2}\,. \end{cases}$$

To establish the stability of the above we look at the second derivative

$$U''(\theta) = 16kl^2(1 + \lambda\sin\theta - 2\sin^2\theta)$$
$$\Longrightarrow \begin{cases} U''(\theta_1^*) = -16kl^2(1-\lambda) \\ U''(\theta_2^*) = 16kl^2(1-\lambda^2) \quad (\text{if } \lambda \leq 1) \\ U''(\theta_3^*) = -16kl^2(1+\lambda). \end{cases}$$

We conclude the following.

($\lambda < 1$) The values $\theta_1^*$, $\theta_3^*$ correspond a stable configurations; the value $\theta_2^*$ shows an unstable equilibrium.

($\lambda = 1$) The configurations $\theta_1^*$ and $\theta_2^*$ coincide. As $U''(\theta_1^*) = 0$, their study requires we examine higher derivatives. After a few calculations we find $U'''(\theta_1^*) = 0$ and $U^{(iv)}(\theta_1^*) = 48kl^2 > 0$, so these configurations are unstable equilibria. In contrast, $\theta_3^*$ is a stable equilibrium.

($\lambda > 1$) Position $\theta_1^*$ is unstable, while $\theta_3^*$ is always stable; in this regime there are no further equilibrium configurations.

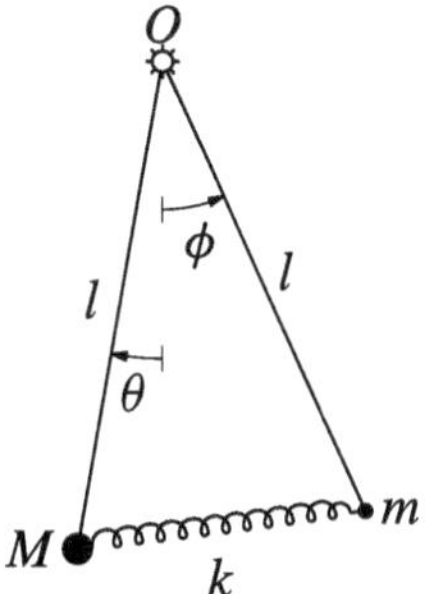


**Fig. 14.11** Two pendulums linked by a spring of elastic constant $k$. The ratio $m/M$ is assumed very small

To sum up: the configuration with $\theta = \pi/2$ (the rods are standing vertically) is stable for $\lambda < 1$ and unstable for $\lambda \geq 1$; the configuration $\theta = -\pi/2$ (the rods are hanging vertically) is always stable; the configuration with $\theta = \arcsin\lambda$ ($\lambda \leq 1$) (intermediate positions) is stable for $\lambda > 1$ and unstable for $\lambda \leq 1$.

In correspondence to the stable equilibrium positions we can compute the frequency of small oscillations. For that we need to find the kinetic energy. Calling $G_2$ the center of mass of $AB$ (see Fig. 9.27) we have $\dot{O}G_2 = 3l\cos\theta\mathbf{i} + l\sin\theta\mathbf{j}$, so $\mathbf{v}_{G_2} = l\dot{\theta}(-3\sin\theta\mathbf{i} + \cos\theta\mathbf{j})$. Therefore

$$T = \left(\frac{2}{3}ml^2\dot{\theta}^2\right) + \left(\frac{2}{3}ml^2\dot{\theta}^2 + 4ml^2\dot{\theta}^2\sin^2\theta\right) = \frac{4}{3}\left(1 + 3\sin^2\theta\right)ml^2\dot{\theta}^2.$$

The kinetic energy has the format of (14.45), with mass matrix reduced to the coefficient

$$a(\theta) = \frac{8}{3}\left(1 + 3\sin^2\theta\right)ml^2.$$

The small oscillations around the stable equilibrium positions have frequencies

$$\omega_1 = \sqrt{-\frac{U''(\theta_1^*)}{a(\theta_1^*)}} = \sqrt{\frac{6k(1-\lambda)}{4m}}, \quad \omega_3 = \sqrt{-\frac{U''(\theta_3^*)}{a(\theta_3^*)}} = \sqrt{\frac{6k(1+\lambda)}{4m}}. \qquad \square$$

### 14.8.4 Small Oscillations of Two Coupled Pendulums

In the context of small oscillations, which—as seen—practically reduce to harmonic motions, it is interesting to show how a link of elastic type between two bodies of very different mass can produce a time-dependent forcing term (see the discussion in Sect. 10.3.1). We present a simple example to clarify the matter.

Figure 14.11 shows two particles of mass $M$ and $m$, representing pendulums of length $l$ hanging from $O$, joined by a spring of elastic constant $k$. Later we will suppose the ratio $m/M$ is very small.

It is straightforward to find the kinetic energy and the weights' potential, using as generalized coordinates the angles $\theta$ and $\phi$ depicted:

$$T = \frac{1}{2}Ml^2\dot{\theta}^2 + \frac{1}{2}ml^2\dot{\phi}^2 \qquad U = Mgl\cos\theta + mgl\cos\phi$$

The spring's length squared is

$$\begin{aligned} s^2 &= \left[(l\sin\phi + l\sin\theta)^2 + (l\cos\phi - l\cos\theta)^2\right] \\ &= 2l^2\left[-\cos(\phi+\theta)\right] + \text{const.} \end{aligned}$$

The spring's potential is therefore

$$U = -\frac{1}{2}ks^2 = kl^2\cos(\phi+\theta) + \text{const.}$$

and so from the Lagrangian

$$\mathcal{L} = \frac{1}{2}Ml^2\dot{\theta}^2 + \frac{1}{2}ml^2\dot{\phi}^2 + Mgl\cos\theta + mgl\cos\phi + kl^2\cos(\phi+\theta)$$

we obtain the equations of motion

$$\begin{cases} \ddot{\theta} + \dfrac{g}{l}\sin\theta + \dfrac{k}{M}\sin(\phi+\theta) = 0 \\ \ddot{\phi} + \dfrac{g}{l}\sin\phi + \dfrac{k}{m}\sin(\phi+\theta) = 0\,. \end{cases}$$

In a regime of small oscillations we may approximate $\sin x \sim x$ and rewrite the equations as

$$\begin{cases} \ddot{\theta} + \dfrac{g}{l}\theta + \dfrac{k}{m}\dfrac{m}{M}(\phi+\theta) = 0 \\ \ddot{\phi} + \dfrac{g}{l}\phi + \dfrac{k}{m}(\phi+\theta) = 0\,. \end{cases} \tag{14.52}$$

If now $m/M$ is infinitesimal, the first equation reduces to

$$\ddot{\theta} + \frac{g}{l}\theta = 0$$

and its solution is the generic harmonic motion

$$\theta(t) = r\sin(\sqrt{g/l}\,t + \delta)\,.$$

Substituting in the other equation of (14.52) one sees how the mass $M$ becomes the source of a periodic *forcing term* for the harmonic motion of the mass $m$:

$$\ddot{\phi} + \frac{mg + kl}{ml}\phi = \psi(t)$$

where $\psi(t) = -k\theta(t)/m$.

## 14.9 Normal Modes for Systems with Several Degrees of Freedom

Let us extend the technique developed in the previous section to systems with any number of degrees of freedom. We stick to systems with fixed holonomic constraints, subject to conservative active forces whose potentials are at least $C^2$. Call $\mathsf{q}^\circ = (q_1^\circ, \ldots, q_N^\circ)$ the generalized coordinates of the equilibrium configuration. To study motions in proximity of $\mathsf{q}^\circ$, using a variable change we re-introduce the generalized coordinates $\boldsymbol{\eta} = \{\eta_1, \ldots, \eta_N\}$:

$$q_k(t) = q_k^\circ + \varepsilon\,\eta_k(t)\,, \qquad k = 1, \ldots, N. \tag{14.53}$$

The small adimensional parameter $\varepsilon$ will control the approximation of the equations of motion near the equilibrium configuration.

### 14.9.1 Linearization of the Equations of Motion

We will find the Lagrangian in function of the new coordinates $\boldsymbol{\eta}$ and their time derivatives. Later the Lagrangian will be used to write the equations of motion. At last, we shall employ the small parameter $\varepsilon$ introduced in (14.53) to linearize the equations of motion around the equilibrium configuration.

- The kinetic energy follows from (14.25), recalling that in presence of fixed constraints both the linear term in $\dot{\mathsf{q}}$ and the term independent of the Lagrangian velocities vanish:

$$T = \frac{1}{2}\sum_{i,j=1}^{N} a_{ij}(\mathsf{q})\,\dot{q}_i\,\dot{q}_j = \frac{\varepsilon^2}{2}\sum_{i,j=1}^{N} a_{ij}(\mathsf{q}^\circ + \varepsilon\,\boldsymbol{\eta})\,\dot{\eta}_i\,\dot{\eta}_j = \frac{\varepsilon^2}{2}\,\dot{\boldsymbol{\eta}}\cdot\mathsf{A}(\mathsf{q}^\circ + \varepsilon\,\boldsymbol{\eta})\dot{\boldsymbol{\eta}},$$

  where $\mathsf{A} = \{a_{ij} : i, j = 1, \ldots, N\}$ is the mass matrix (6.32). As for the potential, we just need to make the variable change (14.53): $U(\mathsf{q}) = U(\mathsf{q}^\circ + \varepsilon\,\boldsymbol{\eta})$. The Lagrangian equals

$$\mathcal{L}(\boldsymbol{\eta}, \dot{\boldsymbol{\eta}}) = \frac{\varepsilon^2}{2}\,\dot{\boldsymbol{\eta}}\cdot\mathsf{A}(\mathsf{q}^\circ + \varepsilon\,\boldsymbol{\eta})\dot{\boldsymbol{\eta}} + U(\mathsf{q}^\circ + \varepsilon\,\boldsymbol{\eta}).$$

Lagrange's equation for the $k$-th generalized coordinate ($k = 1, \dots, N$) then is

$$\varepsilon^2 \frac{d}{dt}\left(\sum_{j=1}^{N} a_{jk}\,\dot{\eta}_j\right) = \frac{\varepsilon^2}{2}\sum_{i,j=1}^{N}\left[\frac{\partial}{\partial \eta_k} a_{ij}(\mathsf{q}^\circ + \varepsilon\,\mathsf{\eta})\right]\dot{\eta}_i\,\dot{\eta}_j + \frac{\partial}{\partial \eta_k} U(\mathsf{q}^\circ + \varepsilon\,\mathsf{\eta}). \tag{14.54}$$

- All terms of (14.54) are at least quadratic in the small parameter $\varepsilon$ appearing in (14.53). This property, evident in the left side and in the first summand on the right, is also true for the last term, which contains the potential's derivatives. In fact if we expand $U$ in $\varepsilon$,

$$\begin{aligned} U(\mathsf{q}^\circ + \varepsilon\,\mathsf{\eta}) &= U(\mathsf{q}^\circ) + \varepsilon \sum_{j=1}^{N} \left.\frac{\partial U}{\partial q_j}\right|_{\mathsf{q}=\mathsf{q}^\circ} \eta_j \\ &\qquad + \frac{\varepsilon^2}{2}\sum_{i,j=1}^{N} \left.\frac{\partial^2 U}{\partial q_i \partial q_j}\right|_{\mathsf{q}=\mathsf{q}^\circ} \eta_i \eta_j + o(\varepsilon^2) \\ &= U(\mathsf{q}^\circ) + \frac{\varepsilon^2}{2}\sum_{i,j=1}^{N} b^\circ_{ij}\,\eta_i\eta_j + o(\varepsilon^2) \\ &= U(\mathsf{q}^\circ) + \frac{\varepsilon^2}{2}\mathsf{\eta}\cdot\mathsf{B}^\circ\mathsf{\eta} + o(\varepsilon^2). \end{aligned} \tag{14.55}$$

In passing from row one to two in (14.55) we used the fact that the equilibrium configuration $\mathsf{q}^\circ$ is a critical point of the potential. We have also introduced the matrix $\mathsf{B}^\circ$, which is nothing but the potential's Hessian matrix at the equilibrium configuration:

$$b^\circ_{ij} = \left.\frac{\partial^2 U}{\partial q_i \partial q_j}\right|_{\mathsf{q}=\mathsf{q}^\circ}.$$

Using (14.55) to compute the derivatives in (14.54) we obtain

$$\frac{\partial}{\partial \eta_k} U(\mathsf{q}^\circ + \varepsilon\,\mathsf{\eta}) = \varepsilon^2 \sum_{j=1}^{N} b^\circ_{kj}\,\eta_j + o(\varepsilon^2),$$

thus confirming that the last summand in (14.54) is (at least) or order two in $\varepsilon$.
- The previous remark clarifies that the lowest term to which it makes sense to approximate Lagrange's equations (14.54) is precisely $\varepsilon^2$. Let us then use this order, and look at the Taylor expansion of the mass matrix. For reason that will become clear in a moment, we stop the expansion at order one in $\varepsilon$:

$$a_{ij}(\mathsf{q}^\circ + \varepsilon\,\eta) = a_{ij}(\mathsf{q}^\circ) + \varepsilon \sum_{k=1}^{N} \left.\frac{\partial a_{ij}}{\partial q_k}\right|_{\mathsf{q}=\mathsf{q}^\circ} \eta_k + o(\varepsilon),$$

and then

$$\frac{\partial}{\partial \eta_k} a_{ij}(\mathsf{q}^\circ + \varepsilon\,\eta) = \varepsilon \left.\frac{\partial a_{ij}}{\partial q_k}\right|_{\mathsf{q}=\mathsf{q}^\circ} + o(\varepsilon).$$

The partial derivatives of the components of the mass matrix are already of magnitude $\varepsilon$, so that the first summand on the right in (14.54) has order three in $\varepsilon$ at least, and can be neglected for the current analysis near the equilibrium. A somewhat similar argument holds for the left side of (14.54). There, $\varepsilon^2$ multiplies the time derivative, so any dependence on $\varepsilon$ in the time derivative can be ignored:

$$\begin{aligned}\varepsilon^2 \frac{d}{dt}\left(\sum_{j=1}^{N} a_{jk}(\mathsf{q}^\circ + \varepsilon\,\eta)\,\dot\eta_j\right) &= \varepsilon^2 \frac{d}{dt}\left(\sum_{j=1}^{N} a_{jk}(\mathsf{q}^\circ)\,\dot\eta_j + O(\varepsilon)\right)\\ &= \varepsilon^2 \sum_{j=1}^{N} a_{jk}(\mathsf{q}^\circ)\,\ddot\eta_j + o(\varepsilon^2).\end{aligned}$$

To justify the last passage, notice that the entries of the mass matrix at the equilibrium position are constant, so they commute with the derivative.

- Gathering up the previous remarks, we approximate the equations of motion to second order in $\varepsilon$:

$$\varepsilon^2 \sum_{j=1}^{N} A^\circ_{jk}\,\ddot\eta_j = \varepsilon^2 \sum_{j=1}^{N} b^\circ_{jk}\,\eta_j + o(\varepsilon^2) \qquad k = 1, \ldots, N,$$

where we set $A^\circ_{jk} = a_{jk}(\mathsf{q}^\circ)$. Define the matrix $\mathsf{A}^\circ = \{A^\circ_{jk} : j, k = 1, \ldots, N\}$ and neglect all higher-order terms in $\varepsilon$. The approximated equations of motion near the equilibrium configuration $\mathsf{q}^\circ$ read in compact form:

$$\mathsf{A}^\circ\,\ddot\eta = \mathsf{B}^\circ\,\eta, \tag{14.56}$$

where we recall that the matrices $\mathsf{A}^\circ$, $\mathsf{B}^\circ$ are the mass matrix and the potential's Hessian, *both evaluated at the equilibrium configuration* $\mathsf{q}^\circ$. They are therefore constant matrices, meaning they do not depend on $\eta$.

### 14.9.2 Discussion of the Linearized Motion

The equations of motion (14.56) form a system of homogeneous linear differential equations of order two with constant coefficients in the unknown $\eta(t)$. Let us solve them and finish the study of motions for conservative holonomic systems with fixed constraints, near the equilibrium configurations. According to the classical theory of linear differential equations with constant coefficients, we seek a solution of the form

$$\eta(t) = \eta_0 \, e^{\lambda t}, \tag{14.57}$$

where $\eta_0 = \{\eta_{0,1}, \dots, \eta_{0,N}\}$ is a (non-zero) vector to be determines and $\lambda \in \mathbb{C}$. Substituting and differentiating twice, it is easy to see (14.57) solves system (14.56) if and only if

$$\lambda^2 \, \mathsf{A}^\circ \, \eta_0 = \mathsf{B}^\circ \eta_0 \quad \Longleftrightarrow \quad \left(\mathsf{B}^\circ - \lambda^2 \, \mathsf{A}^\circ\right) \eta_0 = 0, \tag{14.58}$$

i.e. when $\lambda^2$ is a root of the characteristic equation

$$\det\left(\mathsf{B}^\circ - \lambda^2 \, \mathsf{A}^\circ\right) = 0. \tag{14.59}$$

We call *eigenvalues of* $\mathsf{B}^\circ$ *relative to* $\mathsf{A}^\circ$ the roots $\lambda^2$ of the characteristic polynomial (14.59), and *eigenvectors of* $\mathsf{B}^\circ$ *relative to* $\mathsf{A}^\circ$*, associated with* $\lambda^2$, the non-zero vectors $\eta_0$ solving (14.58).

It can be proved that the eigenvectors of $\mathsf{B}^\circ$ relative to $\mathsf{A}^\circ$ are all real, and that the corresponding eigenvectors form a basis, in general not orthogonal for the standard dot product. We rely on the Appendix (see Sect. A.4) for the proof, in order to not pause our discussion. At present we will focus on the consequences.

### Normal Oscillatory Modes. Eigenfrequencies

We move to the analysis of the nature of the solutions to (14.56) associated with each relative eigenvalue. First of all we look at the case where a root $\lambda^2$ of (14.59) is negative. Let $\eta_0$ be the associated eigenvector, i.e. a non-identically zero solution of (14.58). Both $\eta_+(t) = \eta_0 \exp(i\sqrt{-\lambda^2}\, t)$ and $\eta_-(t) = \eta_0 \exp(-i\sqrt{-\lambda^2}\, t)$ solve system (14.56). As is customary for linear differential equations with constant coefficients, we consider as fundamental solutions the real and imaginary parts of the solutions thus found. This gives

$$\eta(t) = \eta_0 \left(C_1 \cos \omega t + C_2 \sin \omega t\right), \qquad \text{where} \quad \omega = \sqrt{-\lambda^2} \qquad (C_1, C_2 \in \mathbb{R}).$$

Hence *each root* $\lambda^2 < 0$ *of the characteristic polynomial (*14.59*) has an associated solution of the approximate equations of motion (*14.56*), in which all parts of the system oscillate with the same frequency. Such a motion is called* oscillatory normal mode, *and the frequency* $\omega = \sqrt{-\lambda^2}$ *is the* eigenfrequency.

### Linear Normal Modes

Consider the motion associated with each root $\lambda^2 = 0$ of the characteristic polynomial (14.59). Calling again $\eta_0$ an eigenvector associated with that eigenvalue, the fundamental solutions are linear in time:

$$\eta(t) = \eta_0 \left(C_1 + C_2 t\right) \qquad (C_1, C_2 \in \mathbb{R}),$$

and give rise to a *linear normal mode*.

### Hyperbolic Normal Modes

At last, when (14.59) has a root $\lambda^2 > 0$ with eigenvector $\eta_0$, the fundamental solutions are exponential-like:

$$\eta(t) = \eta_0 \left(C_1 \, \mathrm{e}^{\lambda t} + C_2 \, \mathrm{e}^{-\lambda t}\right) \qquad (C_1, C_2 \in \mathbb{R}).$$

As these can be written using hyperbolic functions (hyperbolic sine and cosine), the motion goes by the name of *hyperbolic normal mode*.

In presence of linear and/or hyperbolic normal modes, the approximate equations (14.56) possess solutions where the $\eta$ are unbounded as $t$ grows. As $\eta$ represents the displacement from the equilibrium configuration (see (14.53)), clearly for sufficiently large times these solutions describe a motion that no longer stays close to $\mathsf{q}^\circ$. Therefore the crucial assumption leading to (14.56) fails. We will return to this point shortly, but we anticipate that these normal modes should be interpreted more as indications of instability of the equilibrium configuration rather than precise descriptors of the approximate motion.

### Relative Contribution of Normal Modes

Equations (14.56) are linear. Their general integral is then a linear combination of normal modes; the latter are oscillatory, linear or hyperbolic according to the signs of the roots of (14.59). We will show how, given initial conditions, we can compute the relative weight of each normal mode in the linear combination that gives the approximate motion. In particular, it will follow how one should choose the initial conditions to be sure that certain modes (e.g. the unstable ones) do not affect the approximate motion.

Write the general solution of (14.56) as:

$$\eta(t) = \sum_{k=1}^{N_1} \eta_k^{(s)} \left(A_k \cos\omega_k t + \frac{B_k}{\omega_k} \sin\omega_k t\right)$$
$$+ \sum_{k=N_1+1}^{N_2} \eta_k^{(0)} \left(C_k + D_k\, t\right) \tag{14.60}$$
$$+ \sum_{k=N_2+1}^{N} \eta_k^{(i)} \left(E_k \cosh\lambda_k t + \frac{F_k}{\lambda_k} \sinh\lambda_k t\right),$$

corresponding to the first $N_1$ relative eigenvectors being negative, the next $N_2 - N_1$ zero, and the remaining $N - N_2$ positive. The $2N$ constants to find in (14.60) follow from the initial values:

$$\eta(t_0) = \sum_{k=1}^{N_1} \eta_k^{(s)} A_k + \sum_{k=N_1+1}^{N_2} \eta_k^{(0)} C_k + \sum_{k=N_2+1}^{N} \eta_k^{(i)} E_k \tag{14.61}$$
$$\dot{\eta}(t_0) = \sum_{k=1}^{N_1} \eta_k^{(s)} B_k + \sum_{k=N_1+1}^{N_2} \eta_k^{(0)} D_k + \sum_{k=N_2+1}^{N} \eta_k^{(i)} F_k. \tag{14.62}$$

Let us recall once again the results proved in Sect. A.4. In particular, we are interested in reminding that the $N$ relative eigenvectors can be chosen to form an orthogonal basis for the inner product defined by the mass matrix $\mathsf{A}^\circ$. In other words, defining the inner product

$$(\mathsf{u}\,,\,\mathsf{v}) = \mathsf{u}\cdot\mathsf{A}^\circ\mathsf{v}\,, \tag{14.63}$$

we can choose the eigenvectors so that each distinct pair has (14.63) equal to zero, while (14.63) is 1 if the eigenvector is multiplied with itself.

Formulas (14.61)–(14.62) are just the initial conditions and initial velocities expressed in the above basis of eigenvectors. The $2N$ unknown constants are found through the dot products (14.63):

$$\begin{array}{lll} A_k = \eta(t_0)\cdot\mathsf{A}^\circ\eta_k^{(s)} & C_k = \eta(t_0)\cdot\mathsf{A}^\circ\eta_k^{(0)} & E_k = \eta(t_0)\cdot\mathsf{A}^\circ\eta_k^{(i)}, \\ B_k = \dot{\eta}(t_0)\cdot\mathsf{A}^\circ\eta_k^{(s)} & D_k = \dot{\eta}(t_0)\cdot\mathsf{A}^\circ\eta_k^{(0)} & F_k = \dot{\eta}(t_0)\cdot\mathsf{A}^\circ\eta_k^{(i)}. \end{array}$$

In particular, if $\eta(t_0)$ and $\dot{\eta}(t_0)$ are chosen parallel to a relative eigenvector, the approximate motion will certainly coincide with the normal mode corresponding to that eigenvector. Similarly, if the initial conditions are picked to be orthogonal (with respect to (14.63)) to one or more relative eigenvectors, the contribution of these normal modes will disappear from the approximate motion.

### Approximate Motion

Keeping into account the linear analysis we have just carried out, we can conclude the following.

- If *every* root of the characteristic polynomial (14.59) is negative, the fundamental solutions are oscillatory. The approximate motion is a linear combination of oscillations with eigenfrequencies $\omega_k = \sqrt{-\lambda_k^2}$. In particular, the motion stays bounded and the equilibrium position is stable.
- If one root of the characteristic polynomial is positive, the approximate motion associated with it can move away exponentially from the equilibrium position, which is then unstable. The presence of positive roots does not imply *all* motions will do so. The initial conditions might make the contribution of hyperbolic normal modes in the combination disappear.
- As already said in the previous section, the case of zero roots should be handled with special care. The associated normal mode is linear and will give a uniform (approximate) motion. This should make the system move away from the equilibrium position, which would then automatically be unstable. To understand better the previous hypothetical situation (which will not occur), we must consider a particular case.

  Suppose the potential's Hessian is zero in an equilibrium position. Stopping the expansion of the potential at order two will obviously say nothing about the stability of the equilibrium configuration: this might be a maximum, a minimum, or neither. Any conclusion must necessarily involve discussing higher derivatives of the potential, which in turn forces us to include in the approximation also the terms of similar order coming from the kinetic energy.

  In general, a thorough analysis of approximate motions in presence of zero eigenvalues of $\mathsf{B}^\circ$ requires we study the behaviour of non-linear differential equations, and goes beyond our scope. The linear normal mode can be only interpreted as a qualitative description *of the beginning* of the approximate motion, which might later evolve into an unstable or a stable motion (at any rate, the latter will not be characterized by eigenfrequencies).

  These observations explain how difficult it is to determine the stability of equilibria in the case (already mentioned when discussing Lyapunov's instability criterion) where the potential's Hessian has one or more zero eigenvalues.

We wrap up the study of motions near an equilibrium position by examining the consequences of the following property, which will be proved in the sequel (see again Sect. A.4).

*The matrix* $\mathsf{B}^\circ$ *has as many positive, zero or negative eigenvectors as its positive, zero or negative eigenvectors relative to* $\mathsf{A}^\circ$.

- If the roots of the characteristic polynomial (14.59) are all negative (and all normal modes are then harmonic), the eigenvectors of $\mathsf{B}^\circ$ will be negative, so the equilibrium point will be an isolated local maximum for the potential. Hence we can

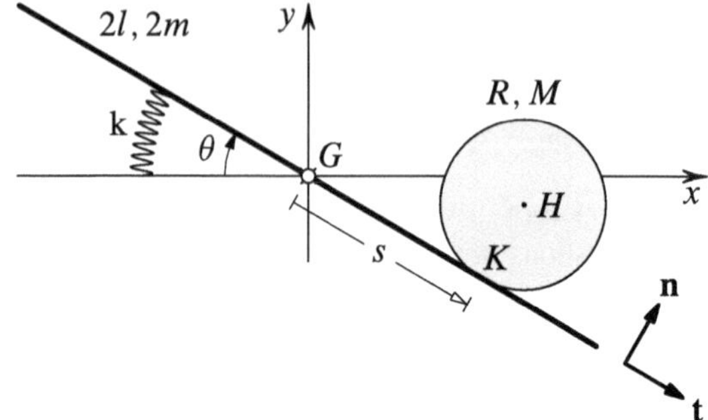


**Fig. 14.12** A disk rolling without slipping on a rod pinned at $G$

complement the Dirichlet-Lagrange theorem (which predicted stability) with the precise description of the type of bounded motions of the system.

- If at least one root of (14.59) is positive (so there is at least one hyperbolic mode), the potential's Hessian $\mathsf{B}^{\circ}$ will have a positive eigenvalue. Our analysis then describes the escape from the equilibrium position, which Lyapunov's instability criterion had established.

We close this section by saying that the stability/instability of an equilibrium position, in conservative holonomic systems with fixed constraints, does not depend on the system's kinetic energy, but only on the potential. It is necessary to examine the kinetic energy as well, in particular the mass matrix, if we seek more precise information (eigenfrequencies, normal modes) about particular motions.

**Example 14.34** Consider the system in Fig. 14.12. In a vertical plane, a homogeneous rod of length $2l$ and mass $2m$ is pinned at its center of mass $G$. A homogeneous disk of radius $R$ and mass $M$ rolls without slipping on the rod. A torsion spring of elastic constant $k$ acts on the rod by pulling it to become horizontal. Let us study the equilibrium positions and the possible harmonic normal modes around them.

The system has two degrees of freedom. If in fact we define the (clockwise) angle $\theta$ between the rod and the horizontal direction, the abscissa $s$ of the contact point $K$ of disk and rod, and the (clockwise) rotation angle $\phi$ of the disk, we can establish a relationship between the coordinates coming from the pure rolling.

To find it we impose that at the contact point $K$ the velocities of rod and disk coincide. The angular velocities of the rigid bodies are $\boldsymbol{\omega}^{(a)} = -\dot{\theta}\mathbf{k}$ and $\boldsymbol{\omega}^{(d)} = -\dot{\phi}\mathbf{k}$. In order to simplify the ensuing computations we define unit vectors $\mathbf{t}$, $\mathbf{n}$, respectively tangent and normal to the rod:

$$\mathbf{t} = \cos\theta\mathbf{i} - \sin\theta\mathbf{j}, \qquad \mathbf{n} = \sin\theta\mathbf{i} + \cos\theta\mathbf{j}. \tag{14.64}$$

They obviously comove with the rod, so the Poisson formulas (2.15) imply $\dot{\mathbf{t}} = \boldsymbol{\omega}^{(a)} \times \mathbf{t} = -\dot{\theta}\mathbf{n}$ and $\dot{\mathbf{n}} = \boldsymbol{\omega}^{(a)} \times \mathbf{n} = \dot{\theta}\mathbf{t}$ (as one can prove directly by differentiating (14.64) in time, by the way).

Calling $H$ the centre of the disk $\big(GH = s\mathbf{t} + R\mathbf{n}\big)$, we obtain $\mathbf{v}_H = \dot{s}\mathbf{t} + s\dot{\mathbf{t}} + R\dot{\mathbf{n}} = (\dot{s} + R\dot{\theta})\mathbf{t} - s\dot{\theta}\mathbf{n}$. Hence

$$\begin{aligned}\mathbf{v}_K^{(\mathrm{d})} &= \mathbf{v}_H + \boldsymbol{\omega}^{(\mathrm{d})} \times HK = \big((\dot{s} + R\dot{\theta})\mathbf{t} - s\dot{\theta}\mathbf{n}\big) - \dot{\phi}\mathbf{k} \times (-R\mathbf{n}) \\ &= (\dot{s} + R\dot{\theta} - R\dot{\phi})\mathbf{t} - s\dot{\theta}\mathbf{n}.\end{aligned} \tag{14.65}$$

On the other hand

$$\mathbf{v}_K^{(\mathrm{a})} = \boldsymbol{\omega}^{(\mathrm{a})} \times GK = -\dot{\theta}\mathbf{k} \times s\mathbf{t} = -s\dot{\theta}\mathbf{n}. \tag{14.66}$$

Comparing (14.65) and (14.66) we obtain the kinematic relation $R\dot{\phi} = \dot{s} + R\dot{\theta}$.

Regarding the equilibrium configurations let us determine the potential of the active forces. Using expression (7.33) for the potential of the spring,

$$U(s,\theta) = -\frac{1}{2}\mathrm{k}\theta^2 - mgy_G - Mgy_H = -\frac{1}{2}\mathrm{k}\theta^2 - Mg(-s\sin\theta + R\cos\theta).$$

As

$$\frac{\partial U}{\partial s} = Mg\sin\theta \quad \text{and} \quad \frac{\partial U}{\partial \theta} = -\mathrm{k}\theta + Mg(s\cos\theta + R\sin\theta),$$

there is a unique equilibrium configuration, given by $s^* = 0, \theta^* = 0$. For its stability we look at the Hessian

$$\begin{aligned} H(s,\theta) &= \begin{pmatrix} 0 & Mg\cos\theta \\ Mg\cos\theta & Mg(R\cos\theta - s\sin\theta) - \mathrm{k}\end{pmatrix} \\ &\implies \mathsf{B}^\circ = H(s^*,\theta^*) = \begin{pmatrix} 0 & Mg \\ Mg & MgR - \mathrm{k}\end{pmatrix}.\end{aligned}$$

As $\det \mathsf{B}^\circ = -M^2g^2 < 0$, we can immediately say the Hessian has a positive and a negative eigenvalue, so the equilibrium position is unstable. For the normal modes (one harmonic and one hyperbolic) we start from the kinetic energy

$$\begin{aligned} T &= \frac{1}{3}ml^2\dot{\theta}^2 + \left(\frac{1}{2}Ms^2\dot{\theta}^2 + \frac{3}{4}M(\dot{s} + R\dot{\theta})^2\right) \\ &= \frac{3}{4}M\dot{s}^2 + \frac{3}{2}MR\dot{s}\dot{\theta} + \left(\frac{1}{3}ml^2 + \frac{1}{2}Ms^2 + \frac{3}{4}MR^2\right)\dot{\theta}^2.\end{aligned}$$

The mass matrix is

$$\begin{aligned} \mathsf{A}(s,\theta) &= \begin{pmatrix} \frac{3}{2}M & \frac{3}{2}MR \\ \frac{3}{2}MR & \frac{2}{3}ml^2 + Ms^2 + \frac{3}{2}MR^2\end{pmatrix} \\ &\implies \mathsf{A}^\circ = \mathsf{A}(s^*,\theta^*) = \begin{pmatrix} \frac{3}{2}M & \frac{3}{2}MR \\ \frac{3}{2}MR & \frac{2}{3}ml^2 + \frac{3}{2}MR^2\end{pmatrix}.\end{aligned}$$

The eigenvectors of $\mathsf{B}^\circ$ relative to $\mathsf{A}^\circ$ are the roots of the characteristic equation (14.59). Define the adimensional (positive) parameters $\alpha = \mathrm{k}/(MgR)$, $\beta = m/M$,

$\gamma = R/l$. Then

$$2\beta l^2(\lambda^2)^2 + 3(\alpha+1)\gamma g l \lambda^2 - 2g^2 = 0$$
$$\implies \lambda^2 = \frac{g}{4\beta l}\left[-3\gamma(\alpha+1) \pm \sqrt{9\gamma^2(\alpha+1)^2 + 16\beta}\right].$$

The negative root $\lambda_-^2$ gives a harmonic mode, with frequency $\omega = \sqrt{-\lambda_-^2}$, while the positive root $\lambda_+^2$ indicates a hyperbolic mode. Let us find the relative eigenvectors, so to identify the initial conditions giving rise to a harmonic motion and those after which the system escapes. In practice we seek a vector $\eta_0 = (s_0, \theta_0)$ such that $\left(\mathsf{B}^\circ - \lambda^2\,\mathsf{A}^\circ\right)\eta_0 = 0$. Such vector is characterized by

$$s_0 = -\gamma l \theta_0 \left(1 - \frac{2g}{3\gamma l \lambda^2}\right). \tag{14.67}$$

In particular, in the harmonic normal mode ($\lambda^2 = \lambda_-^2 < 0$) (14.67) implies $s_0$ and $\theta_0$ have opposite signs, and that their moduli stand in a certain ratio that depends on the system's parameters. The oscillation is therefore characterized by this: when the disk moves away from the origin in some direction, the rod rotates so that the point $K$ lifts above the equilibrium position. □

## 14.10 Dissipation Function

Given a holonomic system with generalized coordinates $\mathsf{q} = (q_1, \ldots, q_N)$, we shall say $\mathcal{D}(\mathsf{q}, \dot{\mathsf{q}})$ is a *(Rayleigh) dissipation function* associated with the system if it contributes to each Lagrangian component (14.14) by the term

$$Q_k^{(\text{diss})} = -\frac{\partial \mathcal{D}}{\partial \dot{q}_k} \qquad \text{for} \quad k = 1, \ldots, N. \tag{14.68}$$

In presence of a dissipation function $\mathcal{D}$ and of conservative forces with potential $U$ the Lagrange equations take the form

$$\frac{d}{dt}\left(\frac{\partial T}{\partial \dot{q}_k}\right) - \frac{\partial T}{\partial q_k} = \frac{\partial U}{\partial q_k} - \frac{\partial \mathcal{D}}{\partial \dot{q}_k}. \tag{14.69}$$

If we define the Lagrangian to be $\mathcal{L} = T + U$, formulas (14.69) read

$$\frac{d}{dt}\left(\frac{\partial \mathcal{L}}{\partial \dot{q}_k}\right) - \frac{\partial \mathcal{L}}{\partial q_k} = -\frac{\partial \mathcal{D}}{\partial \dot{q}_k}.$$

In many applications the dissipation function is a quadratic map in the Lagrangian velocities $\dot{\mathsf{q}} = (\dot{q}_1, \ldots, \dot{q}_N)$:

$$\mathcal{D}(\mathsf{q}, \dot{\mathsf{q}}) = \frac{1}{2} \sum_{j,k=1}^{N} \gamma_{jk}(\mathsf{q})\, \dot{q}_j \dot{q}_k. \tag{14.70}$$

In these cases the Rayleigh function admits a simple mechanical interpretation. Suppose in fact the system admits, apart from the active forces associated with (14.70), only conservative active forces with potential $U$. Then, going through the proof that led us to (14.35) we find

$$\frac{d\mathcal{H}}{dt} = -2\mathcal{D} - \frac{\partial \mathcal{L}}{\partial t}, \tag{14.71}$$

where $\mathcal{H}$ is the system's Hamiltonian (defined in (14.34)), and $\mathcal{L} = T + U$. Indeed

$$\frac{d\mathcal{H}}{dt} = \sum_{k=1}^{N} \dot{q}_k \left[ \frac{d}{dt}\left(\frac{\partial \mathcal{L}}{\partial \dot{q}_k}\right) - \frac{\partial \mathcal{L}}{\partial q_k} \right] - \frac{\partial \mathcal{L}}{\partial t} = - \sum_{k=1}^{N} \dot{q}_k \frac{\partial \mathcal{D}}{\partial \dot{q}_k} - \frac{\partial \mathcal{L}}{\partial t} = -2\mathcal{D} - \frac{\partial \mathcal{L}}{\partial t}.$$

In the last passage we invoked theorem (14.36) for the dissipation function.

If the constraints are fixed, the Lagrangian does not depend on time explicitly and the Hamiltonian is simply the mechanical energy $E$. Property (14.71) implies then that in presence of a quadratic dissipation function,

$$\frac{dE}{dt} = \frac{d}{dt}(T - U) = -2\mathcal{D}.$$

This justifies the name given to $\mathcal{D}$: its value at each instant accounts for the rate of loss of mechanical energy.

**Example 14.35** (*Resistant forces*) Consider a holonomic system with fixed constraints and a system of dissipative forces having the nature of a linear resistance. Suppose, in other words, that on the points of the system (all or perhaps only some) there acts a force opposite to the velocity of the point in question, of modulus proportional to the velocity itself and proportionality factor possibly varying from point to point. Indicate by $\{(P_i, \mathbf{v}_i) : i = 1, \ldots, m\}$ the velocity distribution of the points on which the resistant forces act. Then

$$\mathbf{F}_i^{(\text{diss})} = -\gamma_i\, \mathbf{v}_i\,, \qquad \text{with} \;\; \gamma_i > 0 \;\; \forall i = 1, \ldots, m.$$

It is easy to prove that the dissipation function must be of type (14.70), actually of the form

$$\mathcal{D} = \frac{1}{2} \sum_{i=1}^{m} \gamma_i\, \mathbf{v}_i \cdot \mathbf{v}_i\,. \tag{14.72}$$

In fact, the $k$-th Lagrangian component of the dissipative forces is

$$Q_k^{(\mathrm{diss})} = -\sum_{i=1}^{m} \gamma_i \,\mathbf{v}_i \cdot \frac{\partial P_i}{\partial q_k} = -\sum_{j=1}^{N}\left(\sum_{i=1}^{m} \gamma_i \frac{\partial P_i}{\partial q_j}\cdot\frac{\partial P_i}{\partial q_k}\right)\dot{q}_j, \tag{14.73}$$

where in writing the velocity of $P_i$ in terms of the generalized coordinates we exploited the fact the constraints are fixed (this kills the term with the time derivative). On the other hand, expanding (14.72) we obtain

$$\mathcal{D} = \frac{1}{2}\sum_{i=1}^{m}\gamma_i \,\mathbf{v}_i\cdot\mathbf{v}_i = \frac{1}{2}\sum_{j,k=1}^{N}\left(\sum_{i=1}^{m}\gamma_i \frac{\partial P_i}{\partial q_j}\cdot\frac{\partial P_i}{\partial q_k}\right)\dot{q}_j\,\dot{q}_k = \frac{1}{2}\sum_{j,k=1}^{N}\gamma_{jk}(\mathsf{q})\,\dot{q}_j\,\dot{q}_k,$$

where we have introduced the (positive) coefficients

$$\gamma_{jk}(\mathsf{q}) = \sum_{i=1}^{m}\gamma_i \frac{\partial P_i}{\partial q_j}\cdot\frac{\partial P_i}{\partial q_k}.$$

At this point a simple computation makes us conclude that the dissipation function (14.72) and the Lagrangian components (14.73) satisfy the constitutive relationship (14.68). Note that the matrix $\gamma$, with coefficients $\{\gamma_{jk} : j, k = 1, \ldots, N\}$, formally looks the same as the mass matrix $\mathsf{A}$ used to compute the kinetic energy. The only difference is that the viscosity coefficient for the $i$th point replaces the mass. In particular, $\gamma$ is symmetric and positive semi-definite. It becomes positive definite if the viscous force acts on every point, i.e. if $\gamma_i > 0$ for all $i = 1, \ldots, n$. In the latter situation, and only then, the system of forces generated by the dissipation function is completely dissipative, as per Definition 14.26. □

## 14.11 Nonholonomic Linear Constraints

In the present section we generalize the work we did to obtain Lagrange's equations, and apply it to systems admitting linear nonholonomic constraint, possibly together with other holonomic constraints. Let $\mathsf{q} = (q_1, \ldots, q_N)$ be the generalized coordinates that would describe the system if only holonomic constraints were present. The nonholonomic constraints impose further restrictions on the admissible velocity distributions of any configuration. A nonholonomic constraint is said to be *linear* when it can be expressed as

$$f_0(\mathsf{q}, t) + \sum_{k=1}^{N} f_k(\mathsf{q}, t)\,\dot{q}_k = 0. \tag{14.74}$$

We will also say a nonholonomic constraint is *fixed* when it does not explicitly depend on time and it allows for all Lagrangian velocities to vanish. The linear nonholonomic constraint (14.74) will therefore be fixed if $f_0$ is the zero map and none of the $f_k$ depend on time.

Certain ostensibly nonholonomic constraints may in reality be integrated and turned into holonomic constraints. For linear constraints of the type (14.74), the integration problem consists in determining a function $F(\mathsf{q}, t)$ whose time derivative equals the left-hand side of (14.74), possibly up to some integrating factor. In other words, we seek two functions $F(\mathsf{q}, t)$, $g(\mathsf{q}, t)$ such that

$$g\, f_0 = \frac{\partial F}{\partial t} \quad \text{and} \quad g\, f_k = \frac{\partial F}{\partial q_k}\,.$$

If we do find such maps, we can replace (14.74) with $F(\mathsf{q}, t) = \text{constant}$, and include the constraint among the holonomic ones. In Sects. 4.6.1 and 4.2 we presented two examples of linear nonholonomic constraints, one integrable and the other non-integrable: a disk rolling without slipping on a fixed guide, and ice-skating.

Suppose now that our system has $r < N$ independent constraints of type (14.74):

$$f_0^{(h)}(\mathsf{q}, t) + \sum_{k=1}^{N} f_k^{(h)}(\mathsf{q}, t)\,\dot{q}_k = 0 \qquad \text{for} \quad h = 1, \dots, r. \tag{14.75}$$

We also assume the constraints are linearly independent, i.e. that the matrix of coefficients $\left\{f_k^{(h)}(\mathsf{q}) : h = 1, \dots, r\,,\ k = 1, \dots N\right\}$ has rank $r$ for any $(\mathsf{q}, t)$. Then one can solve (14.75) and express $r$ Lagrangian velocities in function of the remaining $(N - r)$ ones. Alternatively, it is possible (and more useful for the sequel) to introduce $(N - r)$ arbitrary parameters $\boldsymbol{\lambda} = \{\lambda^{(1)}, \dots, \lambda^{(N-r)}\}$ in function of which we express *all* Lagrangian velocities:

$$\dot{q}_k(\mathsf{q}, t, \boldsymbol{\lambda}) = \sum_{h=1}^{N-r} g_k^{(h)}(\mathsf{q}, t)\,\lambda^{(h)} + g_k^{(0)}(\mathsf{q}, t), \tag{14.76}$$

where the coefficients $\{g_k^{(0)}, g_k^{(h)}\}$ come from solving (14.75). We emphasize that in absence of better choices, the *generalized velocities* $\boldsymbol{\lambda}$, introduced in Example 4.2, can be taken to be $(N - r)$ of the Lagrangian velocities, in function of which one writes the other $r$.

In case of fixed constraints the $f_0^{(h)}(\mathsf{q}, t)$ vanish, and then so do the $g_k^{(0)}(\mathsf{q}, t)$. These terms must therefore be excluded when determining the virtual displacements admitted by every constraint acting on the system (holonomic or not). Introducing $(N - r)$ independent parameters $\{\nu^{(1)}, \dots, \nu^{(N-r)}\}$, we will then have

$$\delta q_k = \sum_{h=1}^{N-r} g_k^{(h)}\,\nu^{(h)}. \tag{14.77}$$

The virtual displacements of the generalized coordinates are no longer independent, so in presence of nonholonomic constraints the proof in Sect. 14.7 that $Q_k = \tau_k$ for any $k = 1, \dots, N$ does not work. We have to restart from d'Alembert's equation (14.13), and insert in it the (14.77):

$$\sum_{i=1}^{n} \big(\mathbf{F}_i - m_i\, \mathbf{a}_i\big) \cdot \delta P_i = \cdots = \sum_{k=1}^{N} \big(Q_k - \tau_k\big)\, \delta q_k$$
$$= \sum_{h=1}^{N-r} \left[ \sum_{k=1}^{N} \big(Q_k - \tau_k\big)\, g_k^{(h)} \right] \nu^{(h)} = 0.$$

Let us define the quantity

$$\Psi_h = \sum_{k=1}^{N} \big(Q_k - \tau_k\big)\, g_k^{(h)}, \qquad h = 1, \dots, N - r.$$

Keeping in account that the $\nu^{(h)}$ are arbitrary, and using (14.17), which is always valid, we obtain *Maggi's equations*:

$$\sum_{k=1}^{N} g_k^{(h)} \left[ \frac{d}{dt} \left( \frac{\partial T}{\partial \dot{q}_k} \right) - \frac{\partial T}{\partial q_k} - Q_k \right] = 0 \qquad \forall h = 1, \dots, N - r. \tag{14.78}$$

These are differential equations of order two in the generalized coordinates. They can be viewed as $(N - r)$ equations of first order in the unknowns $(\mathsf{q}, \lambda)$, provided we use (14.76) to express $(\dot{\mathsf{q}}, \ddot{\mathsf{q}})$ in function of $(\lambda, \dot{\lambda})$. Maggi's equations should be solved together with the $N$ first-order equations (14.76) in the variables $(\mathsf{q}, \lambda)$. Altogether, we have $(2N - r)$ equations of order one in the $(2N - r)$ unknowns $(\mathsf{q}, \lambda)$. Of those, (14.76) are already in normal form for $\mathsf{q}$. With arguments similar to those employed in Sect. 14.4.1 it is easy to prove that (14.78) can be put in normal form for the $\lambda$. The initial data for the variables $\lambda$ come from those of $(\mathsf{q}, \dot{\mathsf{q}})$ (which obviously must respect the constraint equations), by inverting (14.76).

# Chapter 15
# Equilibrium of One-Dimensional Continua

The modelling of one-dimensional continuous bodies, such as cables, rods or beams, is traditionally part of Rational Mechanics, for important reasons:

- for such continua, it is possible to introduce the notion of "internal action" rigorously, which is of great importance not only in itself but also as a first step towards the field of Continuum Mechanics, where this idea is extended to a more complex 3-dimensional context.
- Cables play a role in many types of applications, for instance those involving pulleys, and it is important that the concept of "tension" is made rigorous, rather than left to the intuition.
- For flexible beams or elastic strings simple examples of "constitutive relations" are needed, a keystone concept for both Solid and Fluid Mechanics.
- The "vibrating string equation" is discussed in all introductory texts to the field of Partial Differential Equations. It is important that, in the spirit of Rational Mechanics, a correct and rigorous derivation from basic mechanical principles is made available, the more so since this is missing from most such texts.

We therefore offer a brief introduction to the mechanics of *one-dimensional rigid or deformable* bodies, whose geometry allows to describe them, even if approximately, using segments or curves. This hypothesis is certainly reasonable when the configuration can be modelled by a straight segment, or more generally a piece of regular curve. By this we mean that the cross-section is much smaller than the length, hence negligible. Figure 15.1 shows two continua possessing this feature.

Rods, strings, cables and other similar objects can easily be considered one-dimensional continua. Naturally, any schematization will invariably overlook certain aspects that, with hindsight, might be significant. Yet, the situation is not much different from the one in which we consider pointwise or rigid bodies (truth be told, every real body is always deformable to some extent). In any case, these are useful abstractions which should–needless to say–be used in a suitable way when dealing with a concrete situation.

 
P. Biscari et al., *Rational Mechanics*, UNITEXT 177,
https://doi.org/10.1007/978-3-032-07462-1_15

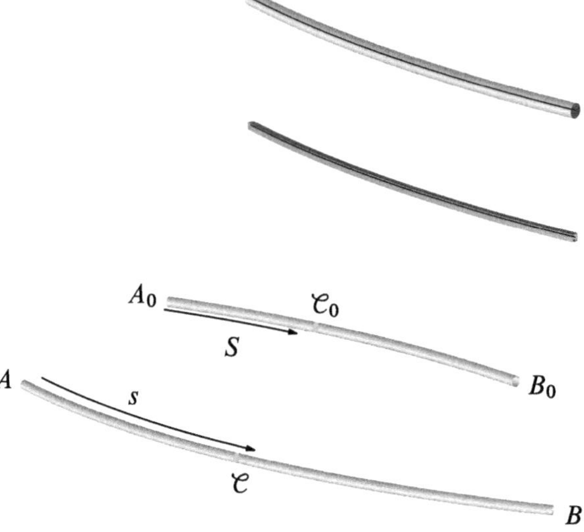


**Fig. 15.1** One-dimensional continua: the first has a round cross-section and the second one a square section. The mechanical description of both reduces to the curve through the section's center, known as the directrix

**Fig. 15.2** A one-dimensional continuuum in the reference configuration $\mathcal{C}_0$ and in the current configuration $\mathcal{C}$. The arclengths $S$ and $s$ correspond to the same point $P$ in both configurations

Consider then a body, with tubular shape and small cross-section, in a certain configuration called *reference configuration*. This is described by a finite and regular, simple (i.e., with no self-intersections) curve in space, which we shall call *directrix* (we are assuming that the section's diameter at every point is much smaller than the directrix' length).

Call $\mathcal{C}_0$ the directrix and $S$ the arclength from $A_0$ to $B_0$. The particles forming the body can be *identified* with the points in space they occupy in the reference configuration.

Under the action of the forces applied to the body we expect the latter to deform, and pass to a *current* configuration $\mathcal{C}$, different than the reference one, whose new arclength we call $s$.

In general the length of a portion of the body *changes* when passing from the reference configuration to the current one. As we see in Fig. 15.2, the arclengths $S$ and $s$ of corresponding points in distinct configurations of the body might be different.

The ratio $ds/dS$ obviously expresses the variation of the length of an infinitesimal piece of the body at a generic point: $ds/dS > 1$ indicates *stretching*, $ds/dS < 1$ a *contraction* of the infinitesimal piece. If in particular $ds/dS = 1$, the length does not change.

A one-dimensional body is said to be *inextensible* if its material structure forces $ds/dS = 1$ always. In this case, obviously, the length is the same in every configuration.

A restriction of this kind should be viewed as a constraint (an inextensibility constraint, here) imposed on the body. This is an *internal constraint*, since the restriction concerns the relative positions of the body's points, rather than their position relative to external objects.

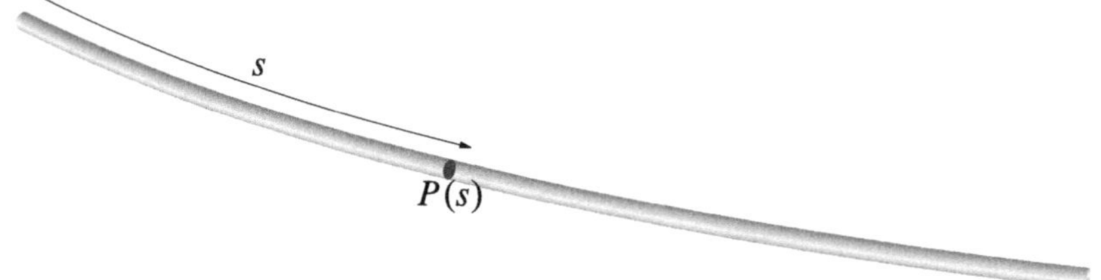


**Fig. 15.3** The point $P(s)$ where an ideal cut of the one-dimensional continuum is located and where the actions exerted by one part on the other one are defined as functions of $s$

In general it is possible to describe the configuration of a body using both a function $P(S)$ of the arclength $S$ in configuration $\mathcal{C}_0$, and a function $P(s)$ of the arclength $s$ in configuration $\mathcal{C}$. We will essentially only treat inextensible bodies, where $s = S$, and briefly mention *elastic strings*, for which that condition does not hold. We finally recall that the *unit tangent vector* $\mathbf{t}$ to the directrix in configuration $\mathcal{C}$ equals the derivative $dP/ds$, as shown in (A.16).

## 15.1 Internal Actions

We will now tackle the problem of describing the interactions of two parts of the body that we imagine are separated by an ideal cut at a generic point $P$. We know that the newly created cross-section is not an actual point but a surface, though with tiny diameter compared to the other dimensions. Therefore, it is reasonable to assume one part acts on the other via a surface force field, characterized by a resultant force, applied at the section's centroid, and by a couple. The action, in other words, is subsumed by the resultant and moment of the forces supposed distributed over the contact surface. The area of such surface is negligible, or more precisely we shall consider it to be negligible in our modelling. Consequently, we assume that the effective distribution of the interaction forces does not affect the surface, apart from the resultant and moment with respect to the point on the curve occupying the centroid, in practice the point on the directrix. This procedure might seem somewhat arbitrary, and it most certainly is, because there exist alternative and more sophisticated choices: the only true guarantee that our hypotheses are adequate lies in the validity of the consequences in the applications and problems of engineering type.

**Postulate** *(Internal actions in a one-dimensional continuum)* Let $P(s)$ be the directrix describing the current configuration of a one-dimensional continuum. The mechanical action enforced by the part that *follows* $P(s)$ (as $s$ increases) on the part *preceding* it is expressed by a force $\mathbf{T}(s)$, called *stress*, and a couple with torque $\mathbf{M}(s)$, called *internal torque*. Both are functions of the point via the arclength $s$.

Under Newton's third law the action exerted by the part *before* $P(s)$ on the part *after* it will obviously be given by a force $-\mathbf{T}(s)$ and a couple with torque $-\mathbf{M}(s)$.

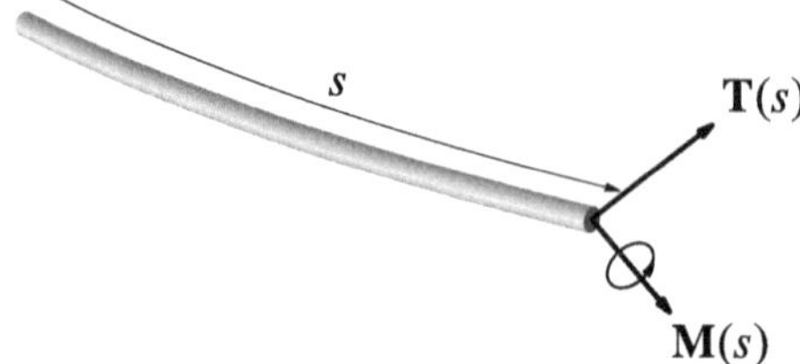


**Fig. 15.4** The action of the part with arclength greater than $s$ on the part with arclength less than $s$: a force $\mathbf{T}(s)$ and a couple with torque $\mathbf{M}(s)$. The oriented circle around $\mathbf{M}(s)$ is to remind that this vector describes a torque and not a force

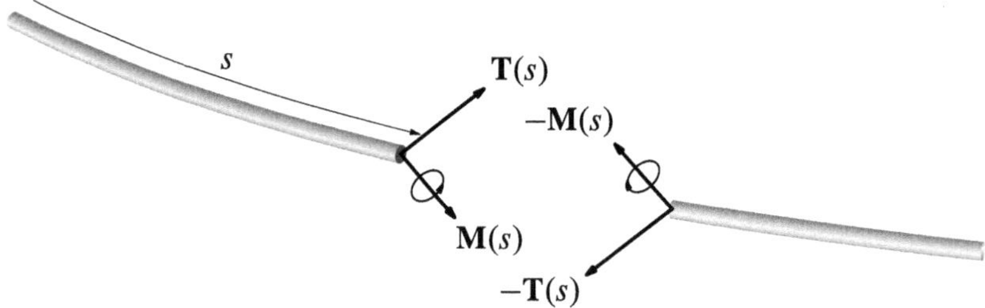


**Fig. 15.5** The internal actions between the parts of the continuum separated at $P(s)$. Notice that $-\mathbf{T}(s)$ and $-\mathbf{M}(s)$ describe the action of the part with arclength *smaller* than $s$ *on* the part with arclength *greater* than $s$

In practice we may think of $\mathbf{T}(s)$ and $\mathbf{M}(s)$ as resultant and moment of a system of forces that one part of the continuum imposes on the other part at $P(s)$.

Figure 15.3 shows the point $P(s)$, where $s$ is the arclength computed starting from one end, at which the two parts are ideally separated, to highlight the interaction of the two parts.

In Fig. 15.4 we see the force $\mathbf{T}(s)$ and the torque of the couple $\mathbf{M}(s)$ that, at the separating surface, the part *following* $P(s)$ exerts *on* the *preceding* part (as $s$ increases). We interpret $\mathbf{T}(s)$ as the resultant of the collection of forces through the surface, and $\mathbf{M}(s)$ as the forces' torque with respect to the section's center.

Finally, Fig. 15.5 shows the action-reaction principle applied to the continuum. Therefore, $-\mathbf{T}(s)$ and $-\mathbf{M}(s)$ describe the total action the part *before* $P(s)$ creates on the part that *follows*.

**Definition 15.1** Let $\mathbf{T}$ and $\mathbf{M}$ denote the stress and torque in a one-dimensional continuum. We call *axial* stress the component of $\mathbf{T}$ perpendicular to the separating cross-section, hence tangent to the curve $P(s)$, and *shear stress* the component on the separating plane, orthogonal to the curve. The analogous components of the torque $\mathbf{M}$ are respectively called *twisting torque* and *bending torque*.

To better grasp these definitions refer to Figs. 15.6 and 15.7.

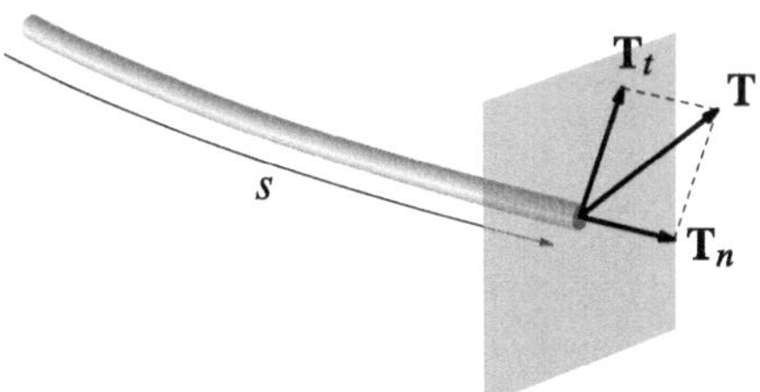


**Fig. 15.6** The decomposition of $\mathbf{T}$ into a part $\mathbf{T}_n$ *normal* to the separating plane, and thus tangent to the directrix, and a *transversal* part $\mathbf{T}_t$ on the plane. The former is known as the *axial stress*, as parallel to the axis of the continuum, and the latter is the *shear* or *transversal* stress, being tangential to the separating plane (the subscripts $n$ and $t$ refer to directions normal and tangential to the separating plane, and not to the generating curve itself)

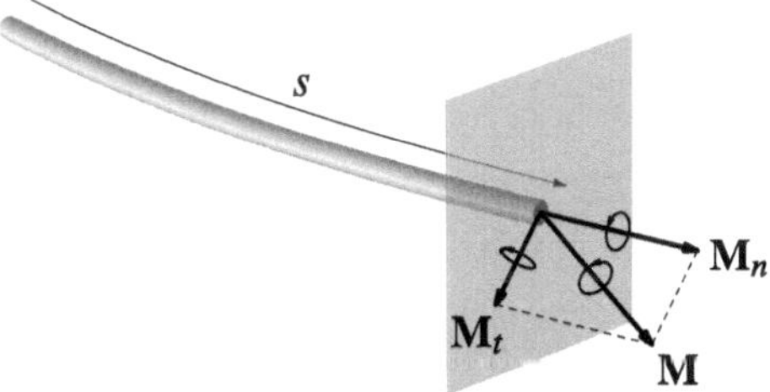


**Fig. 15.7** The decomposition of $\mathbf{M}$ into a part $\mathbf{M}_n$ *normal* to the separating plane, and thus tangent to the continuum–which describes a torsional action and is known as *twisting torque*–and a part $\mathbf{M}_t$ *transversal* to the plane, which tries to bend the continuum and is known as *bending torque* (the subscripts $n$ and $t$ refer to directions normal and tangential to the separating plane, and not to the generating curve itself)

### *15.1.1 Distributed Forces and Actions at the Endpoints*

We must now characterize the forces applied from outside on a one-dimensional continuum. For that, it is quite reasonable to assume there may be a distributed force along the curve describing the body's configuration: it could be the weight for instance, or a constraint reaction due to a surface on which the body lies. In any case the distributed force will be assigned by a force *density* $\mathbf{f}(s)$, of the dimensions of a force per unit of length, whose integral along any portion of curve gives the resultant of the external distributed force acting on that portion. The vector $\mathbf{f}(s)$ hence represents the force per unit of length, measured in the current configuration, acting on the continuum at point $P(s)$.

As shown in Fig. 15.2, it is possible to describe the configuration of the body using as parameter the arclength $S$ in the reference configuration $\mathscr{C}_0$, instead of $s$. If we do so, the external force per unit of length is given by the vector $\mathbf{f}_0(S)$ such that $\mathbf{f}_0(S)dS = \mathbf{f}(s)ds$. Therefore while $\mathbf{f}(s)$ is the external force per unit of length in the deformed configuration, $\mathbf{f}_0(S)$ indicates the same external force per unit of length

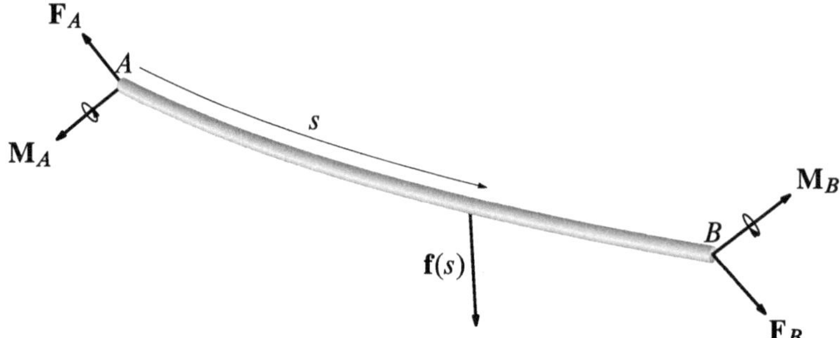


**Fig. 15.8** A one-dimensional continuum subject to forces and torques applied at both ends and to one distributed force

in the reference configuration. Evidently, for inextensible bodies, where $s = S$, the distinction is lost.

Each one of the two ends of the body will furthermore be subject to an external force and moment, in agreement with the description we gave of the internal actions. We will write $\mathbf{F}_A, \mathbf{F}_B$ for the forces applied at the endpoints $A, B$, corresponding to $s = 0, l$. Similarly, we will call $\mathbf{M}_A, \mathbf{M}_B$ the torques of the external couples applied at the ends, as Fig. 15.8 shows.

The model briefly presented above is sophisticated enough to capture a wide class of one-dimensional bodies, which include the systems called *strings* and *canes*. We must point out though that there exist more intricate situations and more complex types of one-dimensional continua, such as beams for example, whose internal forces most definitely require a richer and deeper description from a mathematical standpoint.

## 15.2 Equilibrium of One-Dimensional Continua

The balance equations hold during the motion of every body or system, and are enough to determine the motion of a rigid body, as we saw in Sect. 11.1, Chap. 11 and Theorem 12.3, Chap. 12.

We proved (Theorem 9.12) that a rigid body is in equilibrium if and only if the external forces have zero resultant and moment.

The situation is very different for deformable bodies, even if we concentrate on the problem of equilibrium only. Thinking of an elastic string stretched between the fingers convinces anybody that having zero resultant and moment is not enough to warrant equilibrium, as opposed to the case of rigid bodies.

The equilibrium equations of a deformable continuum arise from postulating that the balance equations must hold not only for the body as a whole, but for *each one of its parts* as well, again in contrast to rigid bodies.

**Postulate** *(Equilibrium of deformable systems)* A deformable body is in equilibrium if and only if the resultant and moment of the forces acting on *every* part of the system vanish.

It is important to stress that the above axiom requires that the system consisting of *all* forces acting on *every* part be balanced. Put equivalently, for any subsystem we must take into account both the external forces acting on it, and the forces originating from the contiguous parts. First of all, then, we should describe the interactions between the various parts of a deformable system, and clarify what kind of action each part imposes on adjacent parts once we assume them separated by imaginary cuts. Such a procedure will allow to write down explicitly the equilibrium balance equations for any deformable subsystem.

For three-dimensional deformable bodies this problem is far from trivial, and leads to Cauchy's theory of internal stresses. As already anticipated, however, the present goal is to focus on the one-dimensional case, which from this point of view is simpler.

## 15.3 Equilibrium Equations

Take a one-dimensional continuum, modelled as we have seen earlier. We shall find a set of differential equations that is equivalent, under fairly general hypotheses, to the equilibrium postulate for deformable systems.

**Theorem 15.2** (Equilibrium of one-dimensional continua) *The balance equations hold on every part of the body if and only if*

$$\frac{d\mathbf{T}}{ds} + \mathbf{f} = \mathbf{0}, \qquad \frac{d\mathbf{M}}{ds} + \mathbf{t} \times \mathbf{T} = \mathbf{0}. \tag{15.1}$$

*The above are known as* equilibrium equations of a continuous body.

***Proof*** Consider the part of a body corresponding to the arclength interval $[s_1, s_2]$ in the current configuration, see Fig. 15.9. On it we have: the stress $\mathbf{T}(s_2)$ (applied at $P(s_2)$) and the torque $\mathbf{M}(s_2)$, representing the action of the part with $s > s_2$ on the system; the stress $-\mathbf{T}(s_1)$ (applied at $P(s_1)$) and the torque $-\mathbf{M}(s_1)$, representing the action of every continuum with $s < s_1$ (note the minus signs, due to the convention adopted above); the system of distributed forces $\{(P(s), \mathbf{f}(s)),\ s \in [s_1, s_2]\}$.

The first equilibrium balance equation for the part $[s_1, s_2]$ gives

$$\mathbf{T}(s_2) - \mathbf{T}(s_1) + \int_{s_1}^{s_2} \mathbf{f}(s)\, ds = \mathbf{0}. \tag{15.2}$$

By the Fundamental theorem of calculus

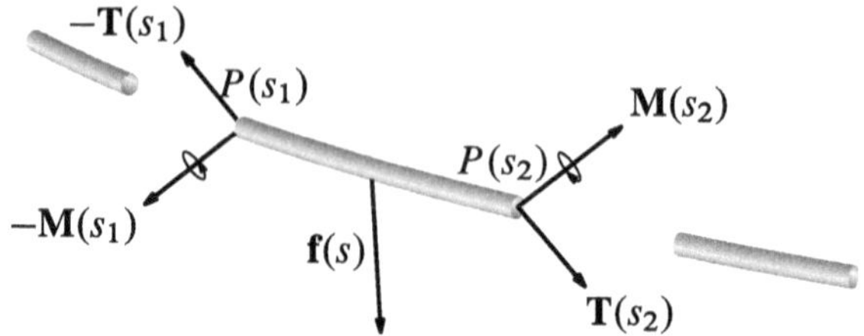

**Fig. 15.9** The forces and moments acting on the part consisting of points $P(s)$ such that $s_1 \le s \le s_2$

$$\mathbf{T}(s_2) - \mathbf{T}(s_1) = \int_{s_1}^{s_2} \frac{d\mathbf{T}}{ds}\, ds,$$

so substituting in (15.2) we obtain

$$\int_{s_1}^{s_2} \left( \frac{d\mathbf{T}}{ds} + \mathbf{f}(s) \right) ds = \mathbf{0}.$$

This integral vanishes for *every* choice of interval if and only if the integrand function is identically zero (if the functions at play are regular enough). Therefore the first equilibrium balance equation holds on any part if and only if

$$\frac{d\mathbf{T}}{ds} + \mathbf{f} = \mathbf{0}. \tag{15.3}$$

Once we choose a pole $O$, the second equilibrium balance equation for $[s_1, s_2]$ tells us

$$\mathbf{M}(s_2) - \mathbf{M}(s_1) + OP(s_2) \times \mathbf{T}(s_2) - OP(s_1) \times \mathbf{T}(s_1) + \int_{s_1}^{s_2} OP(s) \times \mathbf{f}(s)\, ds = \mathbf{0}. \tag{15.4}$$

Now,

$$\mathbf{M}(s_2) - \mathbf{M}(s_1) + OP(s_2) \times \mathbf{T}(s_2) - OP(s_1) \times \mathbf{T}(s_1) = \int_{s_1}^{s_2} \frac{d\mathbf{M}}{ds}\, ds + \int_{s_1}^{s_2} \frac{d}{ds}\big(OP \times \mathbf{T}\big)\, ds,$$

so we rewrite the second balance equation as

$$\int_{s_1}^{s_2} \left( \frac{d\mathbf{M}}{ds} + \frac{d}{ds}\big(OP \times \mathbf{T}\big) + OP \times \mathbf{f} \right) ds = \mathbf{0}.$$

After differentiating in $s$ the cross product on the left, we obtain

$$\int_{s_1}^{s_2} \left( \frac{d\mathbf{M}}{ds} + \frac{dOP}{ds} \times \mathbf{T} + OP \times \frac{d\mathbf{T}}{ds} + OP \times \mathbf{f} \right) ds = \mathbf{0}.$$

The derivative of $OP$ in $s$ equals $\mathbf{t}$, the unit tangent vector to the curve. Keeping into account (15.3), corresponding to the first balance equation, and simplifying the last two terms,

$$\int_{s_1}^{s_2} \left( \frac{d\mathbf{M}}{ds} + \mathbf{t} \times \mathbf{T} \right) ds = \mathbf{0}.$$

Again, this holds for any $[s_1, s_2]$ if and only if

$$\frac{d\mathbf{M}}{ds} + \mathbf{t} \times \mathbf{T} = \mathbf{0},$$

thus ending the proof. □

The differential system (15.1) must obviously be accompanied by the boundary conditions

$$\mathbf{T}(0) = -\mathbf{F}_A, \quad \mathbf{T}(l) = \mathbf{F}_B, \quad \mathbf{M}(0) = -\mathbf{M}_A, \quad \mathbf{M}(l) = \mathbf{M}_B. \tag{15.5}$$

### 15.3.1 Concentrated Forces

In many applications the continuum might be acted on by *concentrated forces* $\{(P_i, \mathbf{F}_i),\ i = 1, \dots, n\}$ as well. In this case Eqs. (15.2) and (15.4) need to be modified by adding

$$\sum_{i^*} \mathbf{F}_i, \quad \sum_{i^*} OP_i \times \mathbf{F}_i,$$

respectively, where the sum runs over indices $i^*$ corresponding to points in $[s_1, s_2]$. In this way it is possible to prove that at every point $P_i$ where there is a concentrated force, the function $\mathbf{T}(s)$ has a jump discontinuity, while $\mathbf{M}(s)$ is continuous but has discontinuous first derivative.

Suppose that at a point $\bar{P}$ with arclength $\bar{s}$ there is a *concentrated* force $\bar{\mathbf{f}}$. We may think that the infinitesimal piece of continuum centered at $\bar{P}$ is subject to a force $\mathbf{T}^+$ caused by the part with $s > \bar{s}$, plus a force $-\mathbf{T}^-$ induced by the part $s < \bar{s}$, where

$$\mathbf{T}^{\pm} = \lim_{s \to \bar{s}^{\pm}} \mathbf{T}(s)$$

(the one-sided limits of $\mathbf{T}(s)$ at $\bar{s}$). Recalling the force $\bar{\mathbf{f}}$ concentrated at $\bar{P}$, from the first equilibrium balance equation we obtain

$$\mathbf{T}^+ - \mathbf{T}^- + \bar{\mathbf{f}} = \mathbf{0}. \tag{15.6}$$

Let $[\![\cdot]\!]$ denote the jump value at a given point in the map's domain. Then (15.6) reads

$$[\![\mathbf{T}]\!] + \bar{\mathbf{f}} = \mathbf{0}.$$

The presence of a component of $\bar{\mathbf{f}}$ *tangent* to the directrix of the continuum makes the *axial stress* discontinuous, while the *normal* part of $\bar{\mathbf{f}}$ produces a discontinuity of the *shear stress*.

Eventually, in the light of the second equation in (15.1), we immediately see that a discontinuity of $\mathbf{T}$ corresponds to a discontinuity of the first derivative of $\mathbf{M}(s)$.

## 15.4 Internal Actions for Planar Systems

The case where the body's directrix lies on the same plane as the system of applied forces is of great applicative importance. Under this hypothesis the torques of the possible couple acting on the endpoints will be perpendicular to the plane, as is always the case for planar forces (cf. Figure 15.10).

Take the dot product of Eq. $(15.1)_1$ with the unit vector $\mathbf{k}$ orthogonal to the plane, to obtain

$$\frac{d\mathbf{T}}{ds} \cdot \mathbf{k} + \mathbf{f} \cdot \mathbf{k} = 0.$$

In our hypothesis $\mathbf{f} \cdot \mathbf{k} = 0$ ($\mathbf{f}$ lies on the plane, too), so if we set $T_z = \mathbf{T} \cdot \mathbf{k}$ then

$$\frac{dT_z}{ds} = 0$$

and $T_z$ will vanish identically because $T_z(0) = -F_{Az} = 0$ (Fig. 15.10).

The stress $\mathbf{T}(s)$ at a generic point will therefore be a vector on the plane, which we can decompose into a part tangent to the directrix (and normal to the plane of the ideal section) and a normal part, still lying on the plane of the directrix.

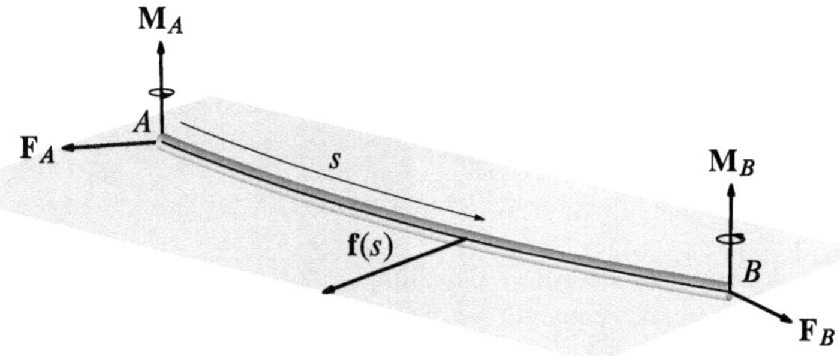


**Fig. 15.10** A one-dimensional continuum with planar generating curve, subject to a planar system of forces

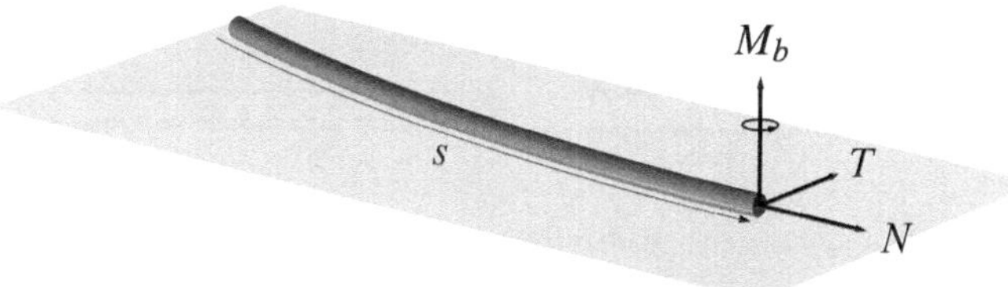


**Fig. 15.11** The internal stresses in a one-dimensional continuum subject to a planar force system reduce to: an axial stress $N$, a shear stress $T$ and a bending torque $M_b$. These three are the stresses' components along the unit vectors

The tangent component is traditionally written $N$, while the component normal to the curve is $T$ (as shown in Fig. 15.11).

Let us now take the dot product of $(15.1)_2$ with the unit vectors $\mathbf{i}$, $\mathbf{j}$ of the plane:

$$\frac{d\mathbf{M}}{ds}\cdot\mathbf{i}+\mathbf{t}\times\mathbf{T}\cdot\mathbf{i}=0 \qquad \frac{d\mathbf{M}}{ds}\cdot\mathbf{j}+\mathbf{t}\times\mathbf{T}\cdot\mathbf{j}=0 \tag{15.7}$$

where, as explained above,

$$\mathbf{t}\times\mathbf{T}\cdot\mathbf{i}=0 \qquad \mathbf{t}\times\mathbf{T}\cdot\mathbf{j}=0$$

since the vectors are coplanar. Putting $M_x=\mathbf{M}\cdot\mathbf{i}$ and $M_y=\mathbf{M}\cdot\mathbf{j}$, from (15.7) we infer

$$\frac{dM_x}{ds}=0 \qquad \frac{dM_y}{ds}=0.$$

But since $M_x(0)=-M_{Ax}=0$ and $M_y(0)=-M_{Ay}=0$, $\mathbf{M}(s)$ is everywhere *orthogonal* to the directrix' plane. The perpendicular component of $\mathbf{M}(s)$ to the plane is called *bending torque* and indicated with $M_b$ (or sometimes $\Gamma$) (Fig. 15.11).

To sum up, the internal actions, or stresses, in a one-dimensional continuum subject to a system of planar forces reduce to: axial stress, shear stress and bending torque.

A straightforward and useful consequence of $(15.1)_2$ comes from noting that the cross product $\mathbf{t}\times\mathbf{T}$ of the tangent vector and the stress vector $\mathbf{T}$ only involves the shear stress $T$ (the axial stress $N$ is parallel to $\mathbf{t}$), and has perpendicular component to the plane $\pm T$. The sign depends on the chosen orientations for $T$ and the plane's normal.

All in all Eq. $(15.1)_2$ reads

$$\frac{dM_b}{ds}\pm T=0. \tag{15.8}$$

We interpret the above as follows: the derivative of the bending torque equals the shear stress up to a sign which depends on the convention adopted for $M_b$ and $T$.

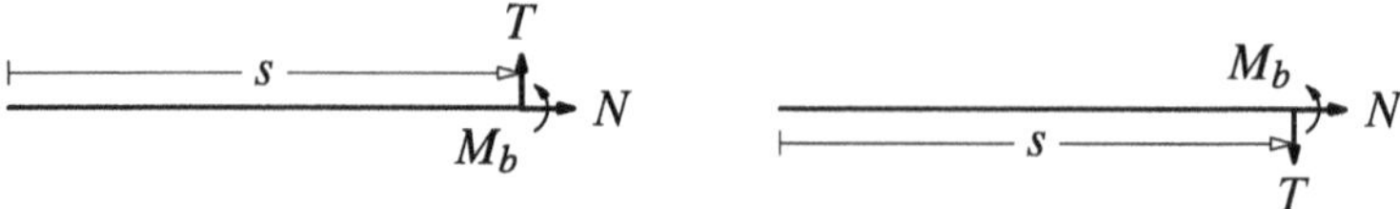


**Fig. 15.12** The internal forces shown with different orientations of the shear stress

For instance, with the choice in Fig. 15.12 left, in (15.8) we have a $+$, while with the choice on the right we have the opposite sign.

**Remark 15.3** In the light of the previous considerations it should be evident that to study the mechanics of one-dimensional continua it suffices to represent the directrix, and ignore the body's cross-section. This is what we will do henceforth. □

## 15.5 Constitutive Relations

The conditions discussed until this point have nothing to do with the actual material which the deformable body is made of. But it is very clear that the equations we need in order to describe, for instance, a bending elastic cane, will depend in some way or another on the physical features of the material. Similarly, we expect that the theory dealing with strings will involve ad-hoc equations, i.e. equations with some essential element that distinguishes them from those of rods and canes.

As already noticed, in fact, system (15.1) plus the boundary conditions (15.5) *is not* enough to determine the unknown functions that naturally crop up when discussing the equilibrium of one-dimensional continua. Suppose for example we have a given force $\mathbf{f}(s)$ together with forces and moments at the endpoints, and we have to find: (1) the body's equilibrium configuration, described by $P(s)$; (2) the internal actions $\mathbf{T}(s)$ and $\mathbf{M}(s)$. A quick computation reveals that the problem involves 9 unknown scalar functions, while there are just 6 scalar differential equations available. So, we will have to introduce further hypotheses to account for the material nature of the body–the so-called *constitutive relations*–or some sort of constraint that will reduce the number of unknowns.

Generally speaking, a constitutive relation links stresses to deformations. For instance, the axial stress $N$ can be taken to be proportional to the stretching of an infinitesimal piece of continuum expressed by the ratio $ds/dS$; the bending torque can be naturally related to the variation of the directrix' curvature. Given the range and great importance of the subject, both theoretical and applicative, it is not surprising that there exist several constitutive relations, some of which very complicated, that allow to characterize many types of one-dimensional continua.

**Remark 15.4** (*Notation and terminology*) The terminology about different types of one-dimensional continua does not seem to be clearly standardized. In order to avoid confusion with other books we spend a few words on this. Our choices are only motivated by simplicity and a desire to avoid using too much notation, which would make this final chapter harder to read.

We call *cable* a one-dimensional body which is inextensible and perfectly flexible, in the sense that it does not resist bending significantly. In a cable the internal moment is always zero. A cable can be heavy or have negligible weight (we say "weightless", more simply). We reserve the expression *elastic string* for a weightless cable which can be (slightly) stretched.

*Rods* are *rigid* one-dimensional continua which, as a consequence, *resist bending* and, thus, exhibit an internal moment in general different from zero.

*Beams* are *flexible* and require some type of constitutive relation involving a measure of their deformation and the internal actions. We shall only consider inextensible beams.

Finally, we reserve the word *cantilever* for a beam which is rigidly constrained at one end to a vertical body, such as a wall, and is free at the other end.

We warn the reader that our terminology is not in agreement with more subtle and varied choices to be found elsewhere, which depend on the cultural context and which we are not going to discuss. □

At present we only wish to briefly present a few examples and applications, among the most classical ones:

- internal actions in planar systems of rigid rods;
- cables, both heavy and weightless;
- elastic strings;
- inextensible beams described by the "Euler model";
- horizontal cantilever subject to its weight;
- weightless beam subject to a fixed vertical load.

## 15.6 Internal Actions in a Planar Isostatic Truss

Systems formed by *rigid* rods, namely one-dimensional continua on which we have imposed the rigidity constraint, are of great importance also form the viewpoint of the applications. In particular, finding the internal actions in planar isostatic systems is a classical subject.

We remind that an *isostatic* system is a collection of points and rigid bodies constrained in a way that does not permit displacements, neither finite nor infinitesimal. One further assumption is that the set of constraints is minimal, meaning that if even one constraint were eliminated the system would admit finite or infinitesimal displacements (see Sect. 4.4, Chap. 4 for the relative discussion).

In planar isostatic systems of rigid rods there are enough equations to compute the internal actions in every rod. Conceptually, if we look at Fig. 15.13, it is evident that if we know all the forces and moments acting on $A$ and along $AP$, we are also able to compute the internal actions using the balance equations for $AP$. In the planar case,

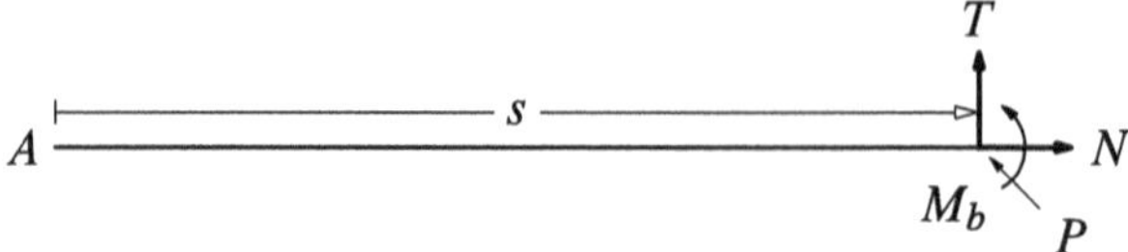


**Fig. 15.13** The internal actions at $P$ can be computed through the equilibrium balance equations for part $AP$, once all other forces acting on $AP$ are known

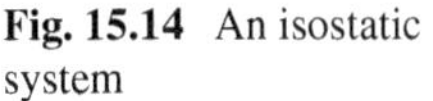
**Fig. 15.14** An isostatic system

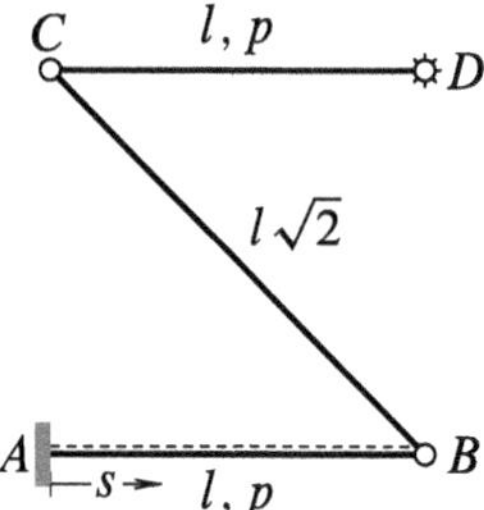


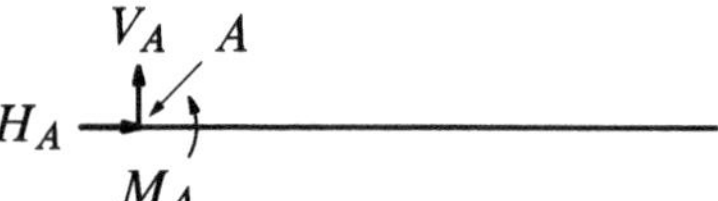


**Fig. 15.15** The actions exerted at $A$ by the fixed-joint constraint

the one of present concern, there are three equations in the three unknowns $N(s)$, $T(s)$ and $M_b(s)$.

**Example 15.5** Referring to Fig. 15.14, we wish to compute the internal actions in a rod $AB$ (dashed line), which is subject to a fixed joint at $A$ and is pinned to a second rod at $B$. The rods' lengths and weights can be read off the picture.

An efficient way to proceed, even if certainly not the only one, is to compute the actions the constraint at $A$ exerts on $AB$. Let us write the moment with respect to $B$ of the forces acting on $AB$, and the moments with respect to $C$ of subsystem $CBA$ and of the whole system $ABCD$. This produces three equations in the unknowns $H_A$, $V_A$, $M_A$ shown in Fig. 15.15 (there exist many conventions as regards the best notation for studying isostatic systems).

Note that $M_A$ is the outward component perpendicular to the plane of the fixed-joint moment at $A$, while $H_A$ and $V_A$ are the horizontal and vertical components of the reactive force due to the constraint.

The three equations are:

$$
\begin{aligned}
M_B^{AB} = 0 \Rightarrow &\qquad l/2p - lV_A + M_A = 0\\
M_C^{ABC} = 0 \Rightarrow &\qquad -l/2p + M_A + lH_A = 0\\
M_D^{ABCD} = 0 \Rightarrow &\qquad l/2p + lH_A + M_A - lV_A + l/2p = 0
\end{aligned}
$$

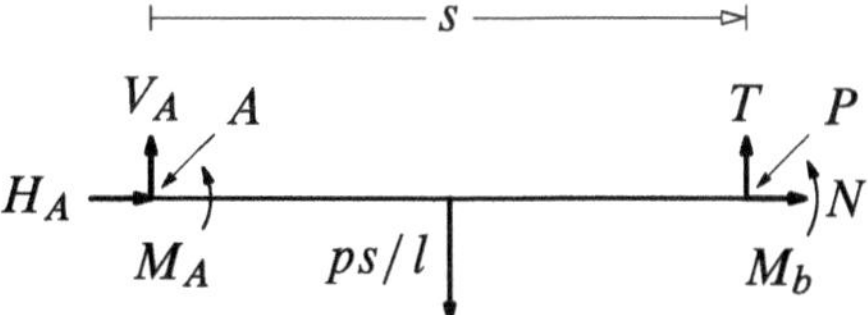


**Fig. 15.16** Forces and moments acting on part $AP$ of rod $AB$

(we chose the moments' positive orientation to be counter-clockwise). Solving the system,

$$H_A = -p/2 \quad V_A = 3p/2 \quad M_A = pl\,.$$

The balance equations for $AP$, in Fig. 15.16, read

$$\begin{aligned} N + H_A = 0 &\Rightarrow & N - p/2 = 0 \\ T + V_A - ps/l = 0 &\Rightarrow & T + 3p/2 - ps/l = 0 \\ M_A + ps^2/2l - V_A s = 0 &\Rightarrow & pl + ps^2/2l - 3ps/2 + M_b = 0 \end{aligned} \tag{15.9}$$

and then

$$N(s) = p/2 \quad T(s) = -3p/2 + ps/l \quad M_b(s) = -pl + 3ps/2 - ps^2/2l\,.$$

The axial stress is in this case constant, and

$$\frac{dM_b}{ds} = 3p/2 - ps/l = -T(s)\,,$$

confirming what (15.8) prescribes.

It is very important to notice that in the balance equations (15.9) for $AP$ we have inserted the part of the weight of $AB$ referring to this portion, i.e. $(p/l)s$. The overall weight should be divided by the rod's length (assumed homogeneous here) to obtain the weight per unit of length $p/l$, and then multiplied by $s$ to find the weight of $AP$, applied at its center of gravity.

Another method to find $T(s)$, $N(s)$, $M_b(s)$, using three independent equations, can be summarized as follows:

$$\begin{aligned} M_B^{PB} = 0 &\Rightarrow & -M_b + T(l-s) + (l-s)^2 p/2l = 0 \\ M_C^{PBC} = 0 &\Rightarrow & -M_b - Ts - Nl - (l+s)p(l-s)/2 = 0 \\ M_D^{PBCD} = 0 &\Rightarrow & -M_b + p(l-s)^2/2l + T(l-s) - Nl + pl/2 = 0 \end{aligned}$$

and we can check the result is the same, even though the above equations are slightly less easy to write (Fig. 15.17). □

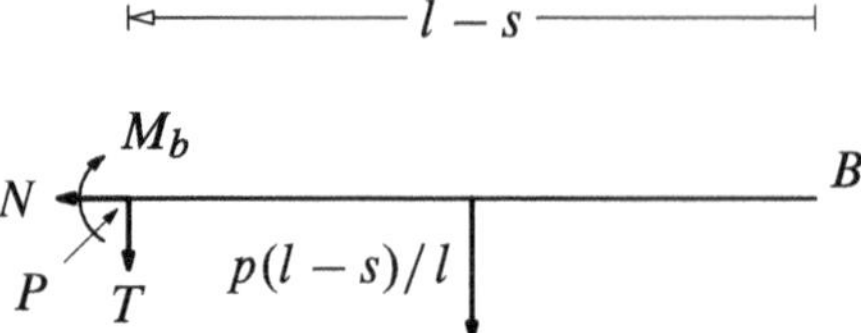


**Fig. 15.17** The internal actions at $P$ on part $PB$ of rod $AB$. The forces which the rod receives from the pin at $B$ are not shown

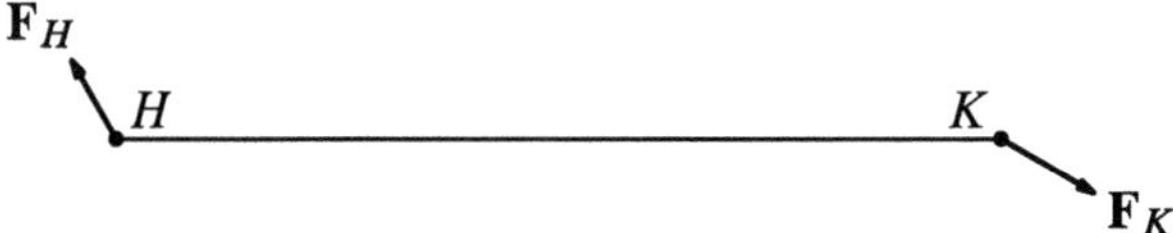


**Fig. 15.18** A rod $HK$ subject only to forces at its ends is known as a *two-force rod*. By letting the resultant of forces and moments be equal to zero we quickly deduce that the forces at the ends must be equal, opposite and directed as the segment joining $H$ with $K$

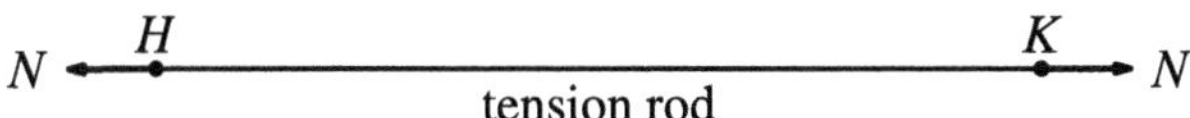


**Fig. 15.19** If the forces acting at the ends are oriented outwards, the element of the truss is known as a *tension rod*

### Two-Force Rods

Another observation regarding the system in Fig. 15.14 is that rod $BC$ has a special feature that is worth mentioning: it is a *two-force rod*. The name refers to rigid rods subject to *only two forces* at the endpoints (*not* moments), and therefore *no load* along the rod (distributed, like the weight, or concentrated).

In this case, referring to Fig. 15.18, by setting to zero the resultant and moment with respect to one end we infer that, to have equilibrium, the two forces must be equal, oppositely oriented and aligned with the line joining the endpoints.

Computing the stresses in a *straight* two-force rod is very easy: it is immediate to see that $T(s) = 0$, $N(s) = \text{constant}$, $M_b(s) = 0$. The only stress is axial, and constant.

Figuring out that a certain rod is of this type is useful, for it immediately reduces the unknowns. For example, in the system of Fig. 15.14 we can view $BC$ (weightless and pinned at the endpoints) as a two-force rod, thus deducing straightaway that the actions it exerts on the pins $B, C$ are parallel to the rod, opposite and of the same magnitude.

Borrowing from technical applications, if it turns out that the action the rod receives at the endpoints tends to stretch it, as in Fig. 15.19, we speak of a "tension rod", and in the opposite case we call it a "compression rod", see Fig. 15.20.

**Fig. 15.20** If the forces acting at the ends are oriented inwards, the element of the truss is known as a *compression rod*

**Fig. 15.21** In a curvilinear two-force rod, the forces applied at the ends are equal and opposite but each one decomposes into an axial and a tangential part. Contrary to what happens in a rectilinear rod, here the internal stress is not purely axial. Notice that at a generic point a bending torque is necessarily present

It is useful to notice that in curvilinear two-force rods the stress *is not* exclusively axial, as if the rod were rectilinear. It has a transverse, shear-like, component as shown in Fig. 15.21 shows.

In summary:

- a *two-force rod* is a rod subject *exclusively* to forces applied at the endpoints. This does not cover rods subject to their weight (a two-force rod necessarily has negligible weight), rods with a fixed end (hence subject to a fixed-joint moment), or rods where other forces or moments, either concentrated or distributed, are applied elsewhere.
- The force a two-force rod exerts or receives at the endpoints has the direction of the segment connecting the endpoints.
- A rectilinear two-force rod only has axial stress (the shear stress and the bending torque are zero), which is constant along the rod.
- A curvilinear two-force rod has internal actions including axial stress, shear stress and bending torque at its generic point.

## 15.7 Cables

The core idea at the heart of the mathematical model of a cable is *perfect flexibility*. We call *cable* a very thin one-dimensional continuum that opposes no resistance to bending. This translates into the hypothesis that the momentum $\mathbf{M}$ is *zero*, both inside and at the endpoints of the body. The equilibrium equations for cables thus reduce to

$$\frac{d\mathbf{T}}{ds} + \mathbf{f} = \mathbf{0}, \qquad \mathbf{t} \times \mathbf{T} = \mathbf{0}. \tag{15.10}$$

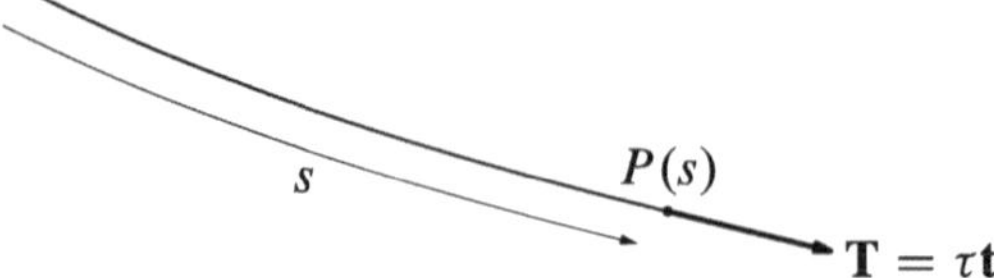


**Fig. 15.22** Tension in a cable. Notice that $\mathbf{T}$ is tangent to the directrix of the body (in this context the notation is different from Fig. 15.4). The scalar quantity $\tau$, assumed always positive, denotes the tension at point $P$ of the cable

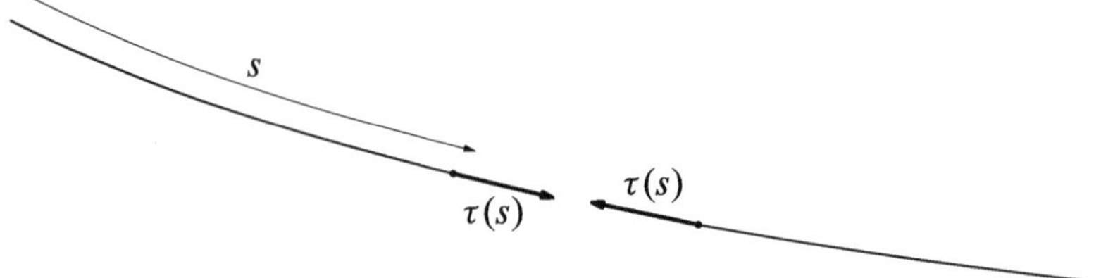


**Fig. 15.23** The tension $\tau$ is never less than zero, because we assume that a cable can only support traction and never compression. The action that a part of the cable exerts on another is thus always directed as in the picture

The second equation is satisfied by imposing that the stress $\mathbf{T}$ and the unit tangent vector $\mathbf{t}$ are parallel. So $\mathbf{T} = \tau\mathbf{t}$, where the scalar quantity $\tau$ is called *tension*. For cables the tension is assumed *positive* (a cable can only be stretched, and does not resist compression), see Figs. 15.22 and 15.23.

**Remark 15.6** *Chains* have the same essential features of cables, if we look at them as one-dimensional continua, and for this reason they share the same mathematical properties. □

Let us now consider a concrete situation: a cable with a known system made of a set of external distributed forces $\{(P(s), \mathbf{f}(s)),\ s \in [0, l]\}$ and two forces $\{(A, \mathbf{F}_A),\ (B, \mathbf{F}_B)\}$ at the endpoints, for which we want to find the possible equilibrium configuration. The unknowns are $P(s)$ and $\tau(s)$, which introduce $3 + 1 = 4$ unknown scalar functions, while we have a vectorial equation

$$\frac{d(\tau\mathbf{t})}{ds} + \mathbf{f} = \mathbf{0}. \tag{15.11}$$

There is only one equation since the second equilibrium equation has already been used to say $\mathbf{T}$ is tangent to the cable. Naively, one would be led to believe (four unknowns and three equations) that we need an extra equation. There are two possibilities: (1) introduce a relationship between the tension and the body's deformation (*elastic* cable); (2) introduce a constraint (*inextensible* cable). In the latter case, which we will discuss initially, the missing equation is the restriction $s = S$ (inextensibility constraint), saying that the arclength function does not change during the deformations and motions of the body.

**Remark 15.7** It is easy but important to observe that for cables not subject to concentrated or distributed actions between the endpoints (hence cables of negligible weight, and not in contact with surfaces or other bodies), (15.10) implies that the tension stays constant and tangent to the cable, which is then in equilibrium when straight. This property justifies in a rigorous way the behaviour of "inextensible and weightless" strings showing up in the problems of earlier chapters about the mechanics of systems of points and rigid bodies. Also note that cables and continua with negligible mass behave in dynamical situations as they do in static ones. □

Equation (15.11) can be profitably decomposed along the Frenet basis of the curve describing the cable's equilibrium configuration.

**Proposition 15.8** (Intrinsic equilibrium equations) *An inextensible cable is in equilibrium if and only if*

$$\frac{d\tau}{ds} + f_t = 0, \qquad c\tau + f_n = 0, \qquad f_b = 0, \tag{15.12}$$

*where* $\tau(s)$ *is the tension at* $P(s)$ *and* $\mathbf{f} = f_t\mathbf{t} + f_n\mathbf{n} + f_b\mathbf{b}$ *is the external force per unit of length acting on the cable.*

***Proof*** Recall that the derivative of the unit tangent vector with respect to the arclength equals $c\mathbf{n}$, where $c$ is the curvature and $\mathbf{n}$ the normal (see (A.19)). Therefore,

$$\frac{d(\tau\mathbf{t})}{ds} = \frac{d\tau}{ds}\mathbf{t} + \tau\frac{d\mathbf{t}}{ds} = \frac{d\tau}{ds}\mathbf{t} + \tau c\mathbf{n}.$$

Decomposing (15.11) along $\{\mathbf{t}, \mathbf{n}, \mathbf{b}\}$ gives (15.12). □

The cable adopts a configuration whereby the binormal component of $\mathbf{f}$ vanishes, while the force's tangent part controls the variation of the tension $\tau$ along the cable.

The first relation in (15.12) also gives us another important condition, valid when $\mathbf{f}$ admits a potential (this is typical when the force is a weight).

**Proposition 15.9** *Let* $\tau$ *be the tension of a cable subject to external, distributed conservative forces, with potential per unit of length* $u$*. Then*

$$\tau + u = constant, \tag{15.13}$$

*meaning that the sum of the tension and the potential per unit of length is the same at all points.*

***Proof*** Suppose the external force $\mathbf{f} = \nabla\,\hat{u}$ comes from a conservative field, where $\hat{u}(P)$ is a function of the points in space. Differentiating $u(s) = \hat{u}(P(s))$ in $s$ gives

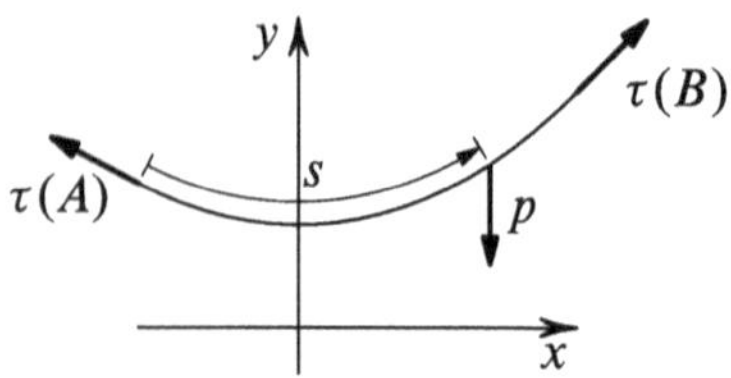


**Fig. 15.24** A homogeneous string or cable with mass

$$\frac{du}{ds} = \frac{\partial \hat{u}}{\partial x}\frac{dx}{ds} + \frac{\partial \hat{u}}{\partial y}\frac{dy}{ds} + \frac{\partial \hat{u}}{\partial z}\frac{dz}{ds} = \nabla \hat{u} \cdot \mathbf{t} = \mathbf{f} \cdot \mathbf{t} = f_t \,.$$

Therefore $(15.12)_1$ turns into

$$\frac{d\tau}{ds} + \frac{du}{ds} = 0,$$

that is to say $\tau + u =$ constant. □

**Remark 15.10** Property (15.13) derives exclusively from the first equilibrium equation in (15.12). That implies that the possible presence of non-conservative (active or reactive) forces directed along the normal or binormal will not affect the property. Also note that (15.13) speaks about the conservation of a certain quantity $(\tau + u)$ along the cable. It should not be confused with a first integral of motion, for which the quantity at stake is constant over time. □

## 15.8 Equilibrium of a Cable Subject to its Weight

Consider a homogeneous inextensible cable under its weight, with weight per unit of length $\mathbf{f} = -p\mathbf{j}$. The unit vector $\mathbf{j}$ is chosen to be opposite to gravity, so $p$ is a positive constant (Fig. 15.24). The internal action $\mathbf{T}$ decomposes as $\mathbf{T} = T_x\mathbf{i} + T_y\mathbf{j}$, where evidently $T_x = \tau \cos\theta$, $T_y = \tau \sin\theta$, with $\theta$ being the angle the unit tangent $\mathbf{t}$, hence the tension, forms with the $x$-axis.

Projecting (15.11) horizontally, the horizontal component of $\mathbf{T} = T_x\mathbf{i} + T_y\mathbf{j}$ is constant (along the cable). Set then $T_x = h$, and write the vertical component of the equilibrium equation as

$$\frac{dT_y}{ds} - p = 0. \tag{15.14}$$

If the cable's shape is given by the graph of the function $y(x)$, so that $OP(x) = x\mathbf{i} + y(x)\mathbf{j}$, we can express the tension's tangency by $T_y = hy'$. Keeping in account that

$$\frac{ds}{dx} = \left|\frac{dP}{ds}\right| = \sqrt{1 + y'^2}\,,$$

and substituting in (15.14), we find

$$\frac{dT_y}{ds} - p = \frac{d(hy')}{dx}\frac{dx}{ds} - p = h\frac{dy'}{dx}\frac{1}{\sqrt{1+y'^2}} - p = 0.$$

Hence $y(x)$ must satisfy

$$y" = \frac{p}{h}\sqrt{1+y'^2}\,.$$

As this second-order differential equation does not contain $y(x)$ explicitly, we set $z(x) = y'(x)$ and write

$$z' = \frac{1}{\alpha}\sqrt{1+z^2}\,, \qquad \text{with } \alpha = h/p\,, \tag{15.15}$$

(the coefficient $\alpha$, dimension-wise a length, is by definition the ratio of the tension's constant horizontal part to the cable's weight).

Equation (15.15) is separable and of order one in $z(x)$. The general integral (see Sect. A.5.1) is

$$z(x) = \sinh\left(\frac{x}{\alpha} + c\right)$$

and $c$ is an arbitrary constant. Integrating once more to find $y(x)$ from $z(x)$ eventually gives

$$y(x) = \alpha\cosh\left(\frac{x}{\alpha} + c\right) + d \qquad \text{where } \alpha = h/p \tag{15.16}$$

where $d$ is a second arbitrary constant. The curve described by a map of this type is called *catenary*.

While the value of $\alpha$ determines the catenary's shape, the constants $c$ and $d$ only affect its position on the plane.

The total length of the cable can be found using the well-known formula

$$l = \int_{x_A}^{x_B} \sqrt{1+y'^2}\,dx\,,$$

where $A, B$ are the endpoints. Therefore, since $y'(x) = \sinh(x/\alpha + c)$, we obtain

$$l = \alpha\left[\sinh\left(\frac{x_B}{\alpha} + c\right) - \sinh\left(\frac{x_A}{\alpha} + c\right)\right].$$

**Remark 15.11** The cable is said to be "tight" when the constant $\alpha$ in (15.16) (a length, dimension-wise) is large compared to the $x$-interval below the cable itself (this is essentially the case when the horizontal component $T_x = h$ is big). Considering for simplicity $c = d = 0$ in Eq. (15.16), and recalling the Taylor expansion for the hyperbolic cosine ($\cosh x = 1 + x^2/2 + o(x^2)$), we deduce that the cable's configuration may be approximated by the parabola $y = \alpha + x^2/\alpha$. □

**Example 15.12** In a vertical plane a homogeneous cable, of weight $p$ per unit of length and length $l$, has one end hooked to a fixed point $A$ and the other end held up

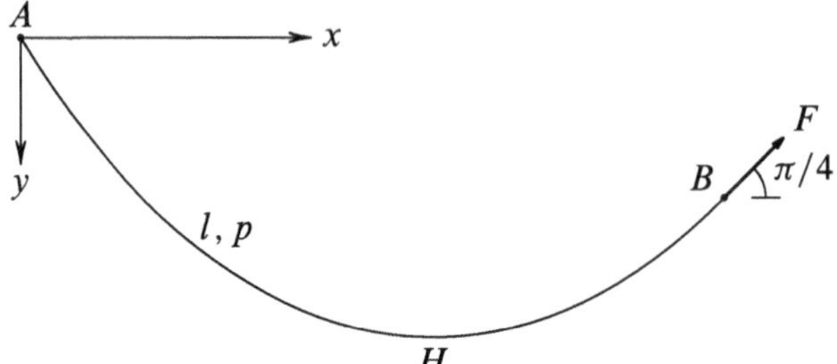


Fig. 15.25 A cable fixed at $A$ and subject to a force $F$ at $B$ at an angle $\pi/4$ with the horizontal

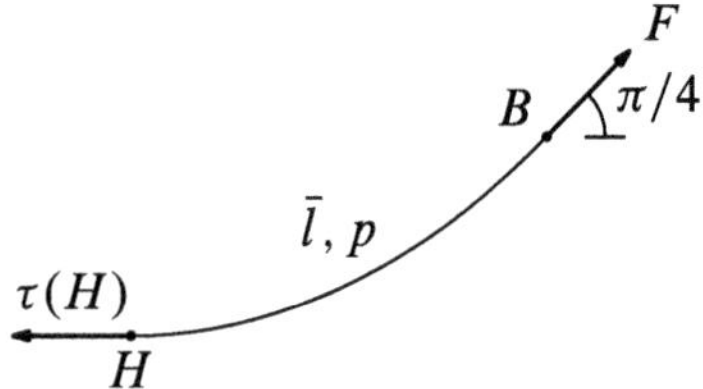


Fig. 15.26 The part of cable between the lowest point $H$ and $B$

by a force $F$ forming an angle $\pi/4$ with the horizontal. We seek the height of the lowest point $H$, as shown in Fig. 15.25.

The cable is subject to a force applied at $A$ with components $\Phi_x\mathbf{i}$ and $\Phi_y\mathbf{j}$, where $\mathbf{i}$, $\mathbf{j}$ are the unit vectors of the coordinate axes in Fig. 15.25. The other forces are the total weight $pl\mathbf{j}$ and the force $F\sqrt{2}/2\mathbf{i} - F\sqrt{2}/2\mathbf{j}$ applied at $B$.

The first balance equation for the forces acting on the cable gives

$$\Phi_x = -F\sqrt{2}/2\,, \qquad \Phi_y = -pl + F\sqrt{2}/2\,, \tag{15.17}$$

The cable's tension at $A$ is equal to the modulus of $\mathbf{\Phi}$, so

$$\tau(A)^2 = f^2 + p^2l^2 - Fpl\sqrt{2}\,.$$

Figure 15.26 shows that the piece $HB$, of length $\bar{l}$, is subject to the (horizontal) tension $\tau(H)$, to its own weight $p\bar{l}$ and to the force $F$ at $B$. Writing the horizontal resultant of the forces we find

$$\tau(H) = F\sqrt{2}/2\,, \qquad \bar{l} = F\sqrt{2}/2p\,.$$

Let us use the relation $\tau + u =$ constant. As the $y$-axis is oriented downwards, this reads $\tau(P) + py(P) =$ constant. With reference to $A$ and $H$ we recast it as

$$\tau(A) + py(A) = \tau(H) + py(H)\,.$$

This allows to conclude

$$y(H) = \frac{\sqrt{F^2 + p^2 l^2 - Fpl\sqrt{2}} - F\sqrt{2}/2}{p}, \tag{15.18}$$

and from that, we recover the height of $H$.

It is interesting to solve the problem by a different approach, involving the catenary's equation (15.16). As the latter refers to an upward $y$-axis (the opposite of the current situation, cf. Figure 15.25), we shall use

$$y = -\alpha \cosh\left(\frac{x}{\alpha} + c\right) + d$$

(the constant $d$ is arbitrary). Imposing $y(0) = 0$ gives $d = \alpha \cosh(c)$. Moreover, recalling that the cable's direction is always parallel to the force at the endpoint, from (15.17) we obtain

$$y'(0) = -\sinh(c) = \frac{F\sqrt{2} - 2pl}{F\sqrt{2}}$$

and so

$$y(x) = -\alpha \cosh\left(\frac{x}{\alpha} + c\right) + \alpha \cosh(c) \tag{15.19}$$

where $\cosh(c) = \sqrt{1 + \sinh^2(c)}$. Consequently

$$\cosh(c) = \sqrt{1 + \left[(F\sqrt{2} - 2pl)/F\sqrt{2}\right]^2}. \tag{15.20}$$

But

$$y'(x) = -\sinh\left(\frac{x}{\alpha} + c\right)$$

and $y'(x_H) = 0$, so $\sinh(x_H/\alpha + c) = 0$ and eventually $\cosh(x_H/\alpha + c) = 1$. Substituting, from (15.19) we end up with

$$y(x_H) = -\alpha + \alpha \cosh(c). \tag{15.21}$$

As the tension's horizontal component has value $F\sqrt{2}/2$ we know that $\alpha = F\sqrt{2}/2p$. Because of (15.20) then, a few computations confirm that (15.21) indeed coincides with (15.18). □

### *15.8.1 Arch Subject to Weight and Axial Stress*

The one-dimensional devices called "arches subject to compression" are able to stay in equilibrium under their own weight without producing shear stresses. In a structure of this kind the constant $h$ in (15.16) is negative, since the axial stress (the only component present), namely the tension $\tau$ for cables, compresses the arch rather than pulling it. For this reason also the constant $\alpha$, defined in (15.15), is *negative*, and (15.16) produces an *upside-down catenary*.

A remarkable example in architecture is Eero Saarinen's famous Gateway Arch in the US. It is the largest monument (a structure with purely celebratory purpose) in the world, and the reader is invited to look it up on the internet.

## 15.9 Equilibrium of a Cable Lying on a Smooth Surface

The study of the equilibrium conditions for a cable lying on a surface, for the time being supposed smooth, has a certain interest, both applicative and theoretical. Assuming the distributed force acting on the cable is the constraint reaction $\phi$, and calling $\mathbf{N}$ the unit normal to the surface (in general this is different from the cable's principal normal $\mathbf{n}$), the equilibrium equation assumes the form

$$\frac{d(\tau\mathbf{t})}{ds} + \phi\mathbf{N} = \mathbf{0},$$

from which

$$\frac{d\tau}{ds}\mathbf{t} + \tau c\,\mathbf{n} + \phi\mathbf{N} = \mathbf{0}.$$

Taking the dot product with the curve's Frenet basis we find

$$\frac{d\tau}{ds} = 0, \qquad \phi\mathbf{N}\cdot\mathbf{b} = 0, \qquad \tau c + \phi\mathbf{N}\cdot\mathbf{n} = 0.$$

From the first equation the tension's modulus remains constant. The second one implies that either the constraint reaction vanishes ($\phi = 0$), or $\mathbf{N}$ is perpendicular to $\mathbf{b}$, and so $\mathbf{N} = \pm\mathbf{n}$. The last equation shows that the first possibility can only occur at points with zero curvature, while in the second case $\phi = \pm\tau c$.

Curves on a surface whose principal normal $\mathbf{n}$ is pointwise parallel to the surface normal $\mathbf{N}$ are called *geodesics*, and enjoy a fundamental property: they minimize lengths among all curves between sufficiently close points.

### Geodesic Cable on a Rough Surface

Consider a tight cable on *rough* surface, so that the distributed force coincides with the constraint reaction $\phi$, and let us neglect any other contribution. Further suppose the cable lies on a *geodesic*, so that the surface normal is parallel to the curve's principal normal at each point. Then (15.12) and the Coulomb-Morin law (8.23) give

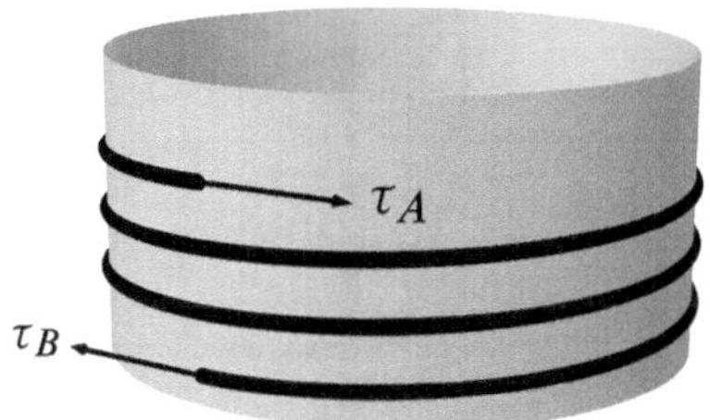


**Fig. 15.27** A cable wrapped around the surface of a cylinder with static friction

$$\frac{d\tau}{ds} + \phi_t = 0, \quad c\tau + \phi_n = 0, \quad \phi_b = 0; \qquad |\phi_T| = \sqrt{\phi_t^2 + \phi_b^2} \le \mu_s |\phi_n|,$$

where $\phi_T$ represents the component of the constraint reaction on the tangent plane to the surface. As $\phi_b = 0$ and $\mathbf{N} = \pm\mathbf{n}$, the tangential part of the constraint reaction is just $\phi_t$, also tangent to the cable. Substituting the first three equations in the last one we obtain

$$\left|\frac{d\tau}{ds}\right| \le \mu_s c\tau \quad \Longrightarrow \quad -\mu_s c\tau \le \frac{d\tau}{ds} \le \mu_s c\tau. \tag{15.22}$$

Assuming the value $\tau(A)$ of the tension at $s = 0$ is known, inequalities (15.22) enable us to determine which values $\tau(B)$ at the other end will allow for equilibrium. To compute these values we rewrite (15.22) as follows:

$$-\mu_s c(s) \le \frac{1}{\tau}\frac{d\tau}{ds} \le \mu_s c(s) \quad \Longrightarrow \quad -\mu_s c(s) \le \frac{d \log \tau(s)}{ds} \le \mu_s c(s). \tag{15.23}$$

Integrating (15.23) between $A$ and $B$ (i.e. from $s = 0$ to $s = l$) we obtain

$$-\mu_s \int_0^l c(s)\, ds \le \underbrace{\int_0^l \frac{d\log\tau(s)}{ds}\, ds}_{\log(\tau(B)/\tau(A))} \le \mu_s \int_0^l c(s)\, ds\,. \tag{15.24}$$

Finally, exponentiating (15.24) we end up with

$$\tau(A)\, \exp\left(-\mu_s \int_0^l c(s)\, ds\right) \le \tau(B) \le \tau(A)\, \exp\left(\mu_s \int_0^l c(s)\, ds\right).$$

The presence of friction consents to counterbalance a force $\mathbf{F}_A = -\tau(A)\mathbf{t}_A$ of even large intensity by using a much smaller force $\mathbf{F}_B = \tau(B)\mathbf{t}_B$. In the particular case where the cable wraps $n$ times around a cylinder of radius $R$ ($c = 1/R, l = 2\pi n R$), we obtain in fact

$$\tau(A)\, \exp\left(-\mu_s 2\pi n\right) \le \tau(B) \le \tau(A)\, \exp\left(\mu_s 2\pi n\right).$$

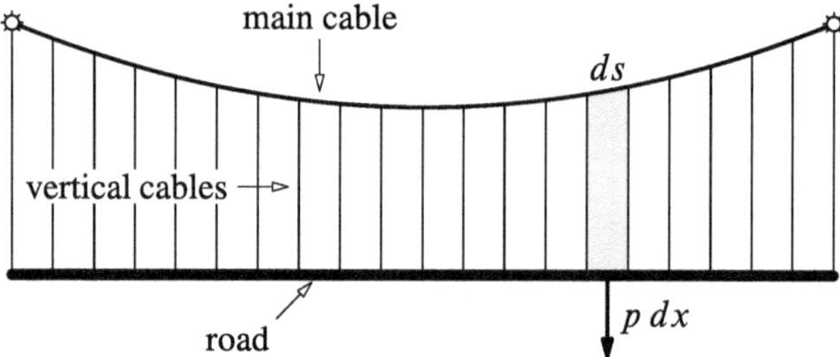


**Fig. 15.28** A suspension bridge: the vertical cables are supposed to be so close to one another that their action on the lower horizontal deck can be reasonably described as a continuously distributed force

If for instance we suppose $\mu_s = 1/2$, and wrap a rough cable three times ($n = 3$, Fig. 15.27), we can balance out $\tau(A)$ by a tension $\tau(B)$ that is 10,000 times smaller:

$$\tau(A)\,\exp(-3\pi) \leq \tau(B) \leq \tau(A)\,\exp(3\pi),$$

where: $\exp(-3\pi) \approx 8.06 \cdot 10^{-5}$, $\exp(3\pi) \approx 12,391.6$.

## 15.10 Suspension Bridges

Consider now the problem of finding the equilibrium configuration of an inextensible cable, of *negligible* weight, that supports a uniformly distributed load along the horizontal direction, as usual denoted by $x$. This model is a reasonable approximation for the cables that hold up suspension bridges through vertical suspender cables, whose weight is negligible compared to the load of the deck below (like a road, see (Fig. 15.28)).

Let $p$ be the (constant) weight per unit of length of the road. We impose that the force every portion of road exerts on the cable above it, say $\mathbf{f}$, is equal to its weight, so that

$$\int_{s_1}^{s_2} \mathbf{f}\,ds = \int_{x_1}^{x_2} (-p\mathbf{j})\,dx,$$

where, obviously, $x$ and $s$ depend on one another in a regular bijective way. Changing variables we immediately obtain

$$\int_{s_1}^{s_2} \mathbf{f}\,ds = \int_{s_1}^{s_2} (-p\mathbf{j})\frac{dx}{ds}\,ds,$$

and so

$$\mathbf{f} = -p\frac{dx}{ds}\mathbf{j}.$$

The force $\mathbf{f}$ is a vector on the $xy$-plane, which can be decomposed into components $f_x$, $f_y$ (horizontal and vertical), $\mathbf{f} = f_x\mathbf{i} + f_y\mathbf{j}$, Clearly the same goes for $\mathbf{T} = T_x\mathbf{i} + T_y\mathbf{j}$. In the case of present concern, in particular,

$$f_x = 0, \qquad f_y = -p\frac{dx}{ds}.$$

The equilibrium equation can be projected along the axes $x$ and $y$ parallel to $\mathbf{i}$ and $\mathbf{j}$:

$$\frac{dT_x}{ds} + f_x = 0, \qquad \frac{dT_y}{ds} + f_y = 0.$$

As $\mathbf{f}$ has no horizontal component ($f_x = 0$), the tension's horizontal part $T_x$ is constant: $T_x = h$ ($h$ constant).

Substituting the geometrical relationship $T_y = hy'$ into the vertical component of the equilibrium equation finally shows that

$$h\frac{dy'}{ds} - p\frac{dx}{ds} = 0$$

so

$$h\frac{dy'}{dx}\frac{dx}{ds} - p\frac{dx}{ds} = 0.$$

This reads $y'' = p/h$, which can be integrated to give

$$y(x) = \frac{p}{2h}x^2 + cx + d,$$

with $c, d$ arbitrary constants. We therefore are in presence of a *parabolic arch*.

## 15.11 Vibrating String Equation

The bulk of this chapter is limited to a discussion of the *statics* of one-dimensional continuous bodies. Here, we make an exception: with the help of d'Alembert's principle (see Sect. 14.1 in Chap. 14), we introduce some ideas related to the *dynamics* of an elastic string. Our aim is to provide a rigorous justification for the equation which describes its motion, under appropriate conditions.

*Elastic strings* are defined by means of a relationship between the tension's modulus $\tau$ and

$$\delta = \frac{ds}{dS},$$

measuring the length's variation (stretching or compression) of an infinitesimal piece of cable at the position corresponding to the value of $S$ in the reference configuration (the non-deformed one).

More precisely, suppose there is a regular map $\hat{\tau}$ such that

$$\tau(S) = \hat{\tau}(\delta(S)) \tag{15.25}$$

The function $\hat{\tau}$ assigns to each $\delta$ at a point of abscissa $S$ a value for the tension $\tau$. The assumption's underpinning idea is naturally that the stress is determined uniquely by the local deformation status, as measured by $\delta$ (ratio of current length to reference length of an infinitesimal piece of cable). We could certainly envision a more complicated dependence between tension and local deformation and introduce viscous elastic cables or the like, which will not be addressed here.

The only application of the constitutive model of an elastic cable that we shall discuss is the one leading up to the well-known "wave equation", also known as "vibrating string equation". We will show how one can *rigorously* derive this equation within the theory of elastic strings, where the necessary assumptions are clearly laid out together with the role of the model's material and constitutive constants.

### *15.11.1 Vibrating Elastic String*

Consider an elastic string, assumed straight in the reference configuration, that coincides with the interval $[0, L]$ on the $x$-axis of the Cartesian plane. We further suppose that in this configuration the string is in equilibrium and tight, with tension $\tau_0$ (later we will prove this must be constant). The arclength $S$ then coincides with the abscissa $x$ of the generic point in this reference configuration.

The constitutive relation of type (15.25) that we select is the simplest, and given by the linear dependence between $\tau$ and $\delta$

$$\tau = \tau_0\, \delta, \tag{15.26}$$

where $\tau_0$ is precisely the value of the tension in the reference configuration, when $\delta = ds/dS = 1$.

We intend to study the spontaneous oscillations induced on the string starting from a given initial state when the force per unit of length $\mathbf{f}$ is zero (this implies the weight is negligible).

For this we need to use the equations governing the *dynamics* of strings, which we did not discuss earlier so to keep the treatise simpler and more agile. We shall avoid to present full details and limit ourselves to observe that if we impose the first and second balance equation of *Dynamics* to every portion of string, in analogy to what we saw earlier, especially in Sects. 15.3 and 15.7, without major complications we arrive at

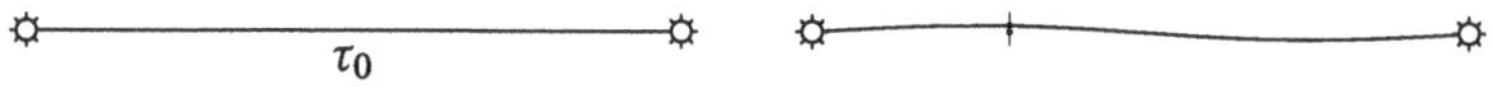

**Fig. 15.29** An elastic string in the reference configuration, with tension $\tau_0$ (left); during motion every particle has fixed abscissa and only moves vertically (right)

$$\frac{d\mathbf{T}}{ds} + \mathbf{f} = \rho\mathbf{a}, \qquad \mathbf{t} \times \mathbf{T} = \mathbf{0}. \tag{15.27}$$

where $\rho(s)$ and $\mathbf{a}(s)$ are the density and acceleration at the point corresponding to arclength $s$ (note that (15.27) extend (15.10) to the dynamics of strings).

We can also view (15.27) as a consequence of the d'Alembert principle applied to (15.10) (see Chap. 14).

As in the static case, the second equation in (15.27) immediately implies $\mathbf{T} = \tau\mathbf{t}$, where the tension's modulus $\tau$ depends on $s$ and time in general, and $\mathbf{t}$ is the unit tangent vector to the string's current configuration.

For the problem we wish to address it is convenient to assign the position of a generic point on the string in function of $S$ in the reference configuration, as well as its density and acceleration, rather than with respect to the variable $s$ in the deformed configuration (recall that at present $s \neq S$). Therefore we recast (15.27) as

$$\frac{d\mathbf{T}}{dS}\frac{dS}{ds} + \mathbf{f} = \rho\mathbf{a}\,,$$

that is

$$\frac{d\mathbf{T}}{dS} + \mathbf{f}_0 = \rho\frac{ds}{dS}\mathbf{a}\,, \tag{15.28}$$

where $\mathbf{f}_0 = \mathbf{f}\,ds/dS$ is the external force per unit of length in the reference configuration.

The mass of an infinitesimal portion of string is $dm = \rho_0 dS$ or $dm = \rho ds$, where $\rho_0$ and $\rho$ are the densities in the two configurations, so $\rho_0 = \rho\,ds/dS$. Now (15.28) can be written

$$\frac{d\mathbf{T}}{dS} + \mathbf{f}_0 = \rho_0\mathbf{a}\,.$$

The assumption that $\mathbf{f}_0 = \mathbf{0}$ (no distributed forces along the string) reduces the above to

$$\frac{d\mathbf{T}}{dS} = \rho_0\mathbf{a}\,. \tag{15.29}$$

Immediately, the reference configuration (where $\mathbf{t} = \mathbf{i}$ is the unit vector of the $x$-axis, with $\mathbf{T} = \tau\mathbf{i}$) is an equilibrium (putting $\mathbf{a} = \mathbf{0}$) if and only if

$$\frac{\partial\tau}{\partial S}\mathbf{i} = \mathbf{0}$$

namely if the tension in this configuration is *constant*, equal to $\tau_0$, independent of $S$.

To recover the equation governing the motion of the string we add to the constitutive relation (15.26) the further hypotheses that *the motion of each particle is vertical*, along a unit vector **j** orthogonal to **i**, as Fig. 15.29 shows, and also that the density $\rho_0$ in the reference configuration is *constant*, independent of $S$.

The position of a point with abscissa $S$ is then

$$P(S,t) = (S,\, f(S,t)) \tag{15.30}$$

where $f(S,t)$ prescribes the vertical position at time $t$, while the horizontal coordinate does not change. In view of (15.30) the acceleration equals

$$\mathbf{a}(S,t) = \frac{\partial^2 f(S,t)}{\partial t^2}\mathbf{j}.$$

As the **i**-component of the acceleration is *zero*, projecting (15.29) onto the axes we obtain

$$\frac{\partial T_x}{\partial S} = 0, \quad \frac{\partial T_y}{\partial S} = \rho_0 \frac{\partial^2 f(S,t)}{\partial t^2}. \tag{15.31}$$

The immediate consequence of the first equation is that the tension's horizontal component is constant along the string, even if we still cannot exclude that it may depend on time.

For a curve parametrized by (15.30), adapting (A.15) leads us to

$$\delta = \frac{ds}{dS} = \sqrt{1+(f')^2}, \quad \text{where } f' := \frac{\partial f}{\partial S},$$

so that the constitutive assumption $\tau = \tau_0\delta$ can now take the form

$$\tau = \tau_0\sqrt{1+(f')^2}. \tag{15.32}$$

As $\mathbf{T} = \tau\mathbf{t}$ is pointwise *tangent* to the string, the geometric meaning of the derivative allows us to see that its components $T_x$ and $T_y$ satisfy

$$T_y = T_x f', \tag{15.33}$$

and since

$$\tau = \sqrt{T_x^2 + T_y^2}$$

we conclude

$$\tau = T_x\sqrt{1+(f')^2}. \tag{15.34}$$

Finally, comparing (15.34) and (15.32) we find $T_x = \tau_0$, so the horizontal tension is independent of both $S$ and time.

From (15.33) we deduce

$$T_y = \tau_0 f',$$

and substituting in the second equation of (15.31) and swapping the two sides,

$$\frac{\partial^2 f(S,t)}{\partial t^2} = \frac{\tau_0}{\rho_0}\frac{\partial^2 f(S,t)}{\partial S^2}.$$

Put $c^2 := \tau_0/\rho_0$, and, as customary, use $x$ instead of $S$ and $y(x,t)$ instead of $f(S,t)$. Then we have

$$\frac{\partial^2 y}{\partial t^2} = c^2\frac{\partial^2 y}{\partial x^2}.$$

This is the so-called *wave equation*, a 'must' in every book and lecture course on *partial differential equations*.

Relying on the rigorous argument presented, it is possible to interpret the coefficient $c^2$ neatly, and therefore tell how the solution depends both on the density and on the tension in the reference configuration.

To make our discussion more concrete let us imagine that, under non-specified forces, the body in question leaves the reference configuration and is put at rest in a given initial configuration, described by $y = g(x)$ $(0 \le x \le L)$. The body is then released and starts to move by inertia, assuming over time configurations given by the solution $y(x,t)$ to the linear partial differential equation

$$\frac{\partial^2 y}{\partial t^2} = c^2\frac{\partial^2 y}{\partial x^2}, \quad 0 \le x \le L, \quad t \ge 0,$$

with initial conditions

$$y(x,0) = g(x), \quad \frac{\partial y}{\partial t}(x,0) = 0, \quad 0 \le x \le L$$

and boundary conditions

$$y(0,t) = y(L,t) = 0.$$

The latter constrain the endpoints to not move. There is an obvious compatibility condition whereby $g(x)$, giving the initial configuration, must also satisfy the restrictions $g(0) = g(L) = 0$.

## 15.12 Euler Model for Beams

There are several models available for describing inextensible beams. We shall examine the simplest instance, the so-called *Euler model*, and apply it to study a beam, of negligible weight, bolted to a wall and subject to a given fixed load at the other end. The model is also known as a *cantilever*. As we will see, the problem is not that trivial and will make room for a number of interesting observations.

What distinguishes beams from cables is essentially the resistance to bending. It will therefore be necessary to introduce a constitutive relation to describe this fact properly. Let us see how one proceeds, at least in the simplest situation.

Suppose a beam of length $L$ is *straight* in a reference configuration, in which the internal stresses are assumed *zero*. Moveover, we make the following working hypotheses.

- The beam is inextensible, so $s = S$.
- The deformations occur on a fixed plane, spanned by the orthonormal vectors $\mathbf{i},\ \mathbf{j}$.
- The applied forces lie on the plane, the torques are perpendicular.
- The modulus of the bending torque is proportional to the curvature.

The first three requests aim at simplifying the problem geometrically. The last one instead is a *constitutive equation*, based on experimental evidence. Let us try to translate it into a formula. The curvature is defined by (A.17):

$$c = \left|\frac{d\mathbf{t}}{ds}\right|.$$

As shown in Fig. 15.30, on a plane curve the unit tangent vector $\mathbf{t}$ is determined by the angle $\theta(s)$ it forms with the unit vector $\mathbf{i}$ at the point of abscissa $s$ ($s = S$): $\mathbf{t} = \cos\theta\mathbf{i} + \sin\theta\mathbf{j}$, from which

$$c = \left|\frac{d\mathbf{t}}{ds}\right| = \left|(-\sin\theta\mathbf{i} + \cos\theta\mathbf{j})\theta'\right| = \left|\frac{d\theta}{ds}\right|,$$

where the dash denotes the derivative of $\theta$ with respect to $s$.

The torque $\mathbf{M}(s)$ at $s$ along the beam is orthogonal to the plane,

$$\mathbf{M}(s) = M(s)\mathbf{k},$$

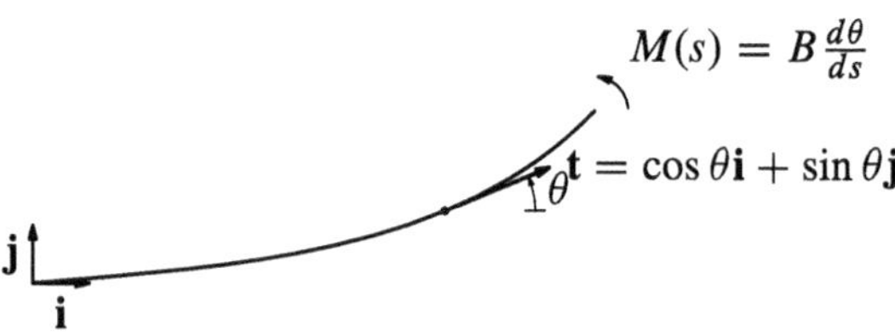


**Fig. 15.30** The bending torque in the Euler model

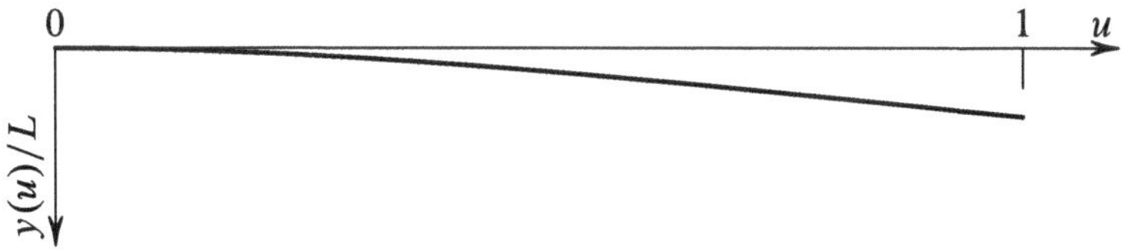


**Fig. 15.31** A horizontal beam bolted at one end and subject to its own weight

where $\mathbf{k}$ forms with $\mathbf{i}$ and $\mathbf{j}$ a right-handed orthonormal basis. The constitutive hypothesis requests that the modulus of $\mathbf{M}(s)$ is proportional to the curvature $c$, so

$$|M| = B\left|\frac{d\theta}{ds}\right| \quad \Longrightarrow \quad M = \pm B\frac{d\theta}{ds}$$

where $B$ is a characteristic constant of the beam's material and of its cross-section. The curvature $c(s)$ has the dimensions of the inverse of a length ($\theta$ is a number), while the torque $\mathbf{M}(s)$ is a force times a length, so that $B$ has the dimensions of a force times a length squared.

Simple physical considerations induce us to believe that the behaviour of a flexible beam is properly described assuming the above sign to be

$$M = B\frac{d\theta}{ds} \tag{15.35}$$

as in Fig. 15.30.

Let us make a list of the equations at our disposal:

$$\frac{d\mathbf{T}}{ds} + \mathbf{f} = \mathbf{0}, \quad \frac{dM}{ds}\mathbf{k} + \mathbf{t} \times \mathbf{T} = \mathbf{0}, \quad M = B\frac{d\theta}{ds}.$$

To these we will add the appropriate boundary conditions for the case at hand. There is a large showcase of mechanical phenomena of some applicative interest that have been studied using the above model.

### 15.12.1 Horizontal Cantilever Subject to its Weight

An inextensible beam with one end rigidly connected to a fixed vertical wall is also known as a "cantilever", in the engineering literature. Consider a beam $AB$ of length $L$ and weight per unit of length $p$, placed horizontally in the reference configuration, with end $A$ bolted to fixed vertical wall (Fig. 15.31).

Take the reference frame with origin at $A$ and axes with unit vectors $\mathbf{i}$, horizontal and pointing to $B$, and $\mathbf{j}$ vertical and pointing downwards. In this frame the force per unit of length acting on the beam is

$$\mathbf{f} = -p\mathbf{j}.$$

Let us also assume to concentrate a load $\mathbf{q} = -q\mathbf{j}$ at the free end $B$. The equilibrium equation $(15.1)_1$, projected horizontally and vertically, says that

$$T_x = \text{constant}, \quad \frac{dT_y}{ds} - p = 0.$$

As, at $B$, we have $\mathbf{T}(L) = \mathbf{q}$, and since $\mathbf{q}$ is vertical, at each point $T_x = 0$. Integrating the second equation is immediate and gives $T_y(s) = ps + d$, where $d$ is a constant we determine from the boundary condition $T_y(L) = -q$, so eventually

$$T_y(s) = p(s - L) - q. \tag{15.36}$$

Let us write now the second equilibrium equation as

$$\frac{d\mathbf{M}}{ds} + \mathbf{t} \times \mathbf{T} = \mathbf{0}, \tag{15.37}$$

and compute first the cross product $\mathbf{t} \times \mathbf{T}$. As $\mathbf{t} = \cos\theta\mathbf{i} + \sin\theta\mathbf{j}$ and $\mathbf{T} = T_x\mathbf{i} + T_y\mathbf{j}$ we have

$$\mathbf{t} \times \mathbf{T} = (T_y \cos\theta - T_x \sin\theta)\mathbf{k},$$

where $\mathbf{k} = \mathbf{i} \times \mathbf{j}$ is the unit vector perpendicular to the plane of the system. Inserting $T_x = 0$ and the constitutive relation $\mathbf{M} = B\theta'\mathbf{k}$ into (15.37), we obtain

$$B\theta'' + T_y(s)\cos\theta = 0, \tag{15.38}$$

where we will replace $T_y(s)$ with (15.36). The resulting second-order differential equation in the unknown $\theta(s)$ should be accompanied by the boundary conditions

$$\theta(0) = 0, \quad \theta'(L) = 0, \tag{15.39}$$

corresponding to the fixed-point constraint at $A$ ($\theta(0) = 0$) and to the torque vanishing at $B$ ($M(L) = B\theta'(L) = 0$).

We obtain a remarkable and significant simplification in the problem by considering very "small" deformations of the straight reference configuration. More precisely, suppose that the angle $\theta$–measuring the pointwise bending–is so small that

$$\sin\theta \approx \theta, \quad \cos\theta \approx 1. \tag{15.40}$$

The differential equation (15.38) becomes

$$\theta'' = \frac{p}{B}(L - s) + \frac{q}{B},$$

and with conditions (15.39) it can be integrated to

$$\theta(s) = \frac{p}{6B}[(L-s)^3 - L^3] + \frac{q}{2B}[(s-L)^2 - L^2].$$

As we are actually interested in the function that describes the deformed configuration, we integrate

$$\frac{dx}{ds} = \cos\theta(s), \quad \frac{dy}{ds} = \sin\theta(s),$$

and write

$$x(s) = \int_0^s \cos\theta(s^*)\, ds^*, \quad y(s) = \int_0^s \sin\theta(s^*)\, ds^*.$$

Exploiting (15.40) we then obtain

$$x(s) = s, \quad y(s) = \int_0^s \frac{p}{6B}[(L-s^*)^3 - L^3] + \frac{q}{2B}[(s^*-L)^2 - L^2]\, ds^*,$$

and easy calculations produce

$$y(u) = Lu^2\left\lfloor -\alpha(u^2 - 4u + 6) + \beta(u-3)\right\rfloor,$$

where we have introduced the *dimensionless* quantities

$$\alpha = \frac{pL^3}{24B}, \quad \beta = \frac{qL^2}{6B}, \quad u = \frac{x}{L}.$$

Notice that from the values of $L$, $q$, $B$, $p$ we can immediately find the function that maps any value of the dimensionless coordinate $u$ (fraction of the beam) to the vertical displacement of the generic point. More precisely,

$$u^2\left[-\alpha(u^2 - 4u + 6) + \beta(u-3)\right]$$

is the (dimensionless) ratio between the generic particle's vertical displacement and the beam's length $L$. It can be instructive to plot this function using reasonable values of $\alpha$ and $\beta$. Putting $\alpha = 1/100$ and $\beta = 2/100$, for instance, produces the graph in Fig. 15.31.

## 15.13 Weightless Beam Subject to a Prescribed Load

Consider a beam bolted perpendicularly to a vertical wall, along the unit vector $\mathbf{j}$. In the reference configuration the beam occupies the interval $[0, L]$ on the $x$-axis, with origin at the fixed point, see Fig. 15.32 (note that the joint is an external constraint for

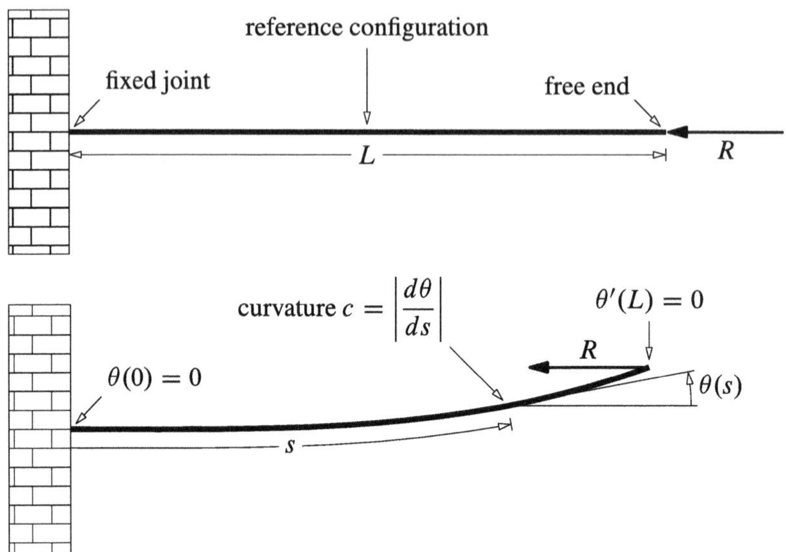


**Fig. 15.32** A weightless beam with a fixed joint at the left end

the beam). Now assume the *weight is negligible*, and that no other force acts along the beam, so $\mathbf{f} = \mathbf{0}$. But we do assume that at the free end $B$ there is a force

$$\mathbf{F} = -R\mathbf{i}.$$

We are therefore studying a bolted beam subject to a buckling force that tends to compress it. This is, of course, a very basic model for describing the behavior of an elastic column with negligible weight, rigidly fixed to the ground, which is supporting a vertical load.

The first equilibrium equation reduces to

$$\frac{d\mathbf{T}}{ds} = \mathbf{0},$$

which trivially means $\mathbf{T}$ is constant, hence pointwise equal to its value

$$\mathbf{T}(s) = -R\mathbf{i}.$$

at the free end. Substituting the cross product

$$\mathbf{t} \times \mathbf{T} = (\cos\theta\mathbf{i} + \sin\theta\mathbf{j}) \times (-R\mathbf{i}) = R\sin\theta\mathbf{k}$$

in the second equilibrium equation, and using the constitutive relation (15.35), easily leads to a single equation

$$\frac{d^2\theta}{ds^2} + \frac{R}{B}\sin\theta = 0.$$

Let us now deal with the boundary conditions. At the fixed end ($s = 0$) we know nothing about the forces and torque that the wall applies, which are unknown. What we can do, due to the constraint, is impose $\theta(0) = 0$. At the free end, on the other hand, the torque is zero, so by the constitutive relation the derivative of $\theta$ in $s$ vanishes: $\theta'(L) = 0$.

To study the system mathematically it is better to write it using dimensionless quantities. Define the dimensionless quantity

$$u = \frac{s}{L}.$$

The second derivative of $\theta$ in $s$ can be rewritten

$$\frac{d^2\theta}{ds^2} = \frac{d^2\theta}{du^2}\frac{1}{L^2},$$

so that the equilibrium equation reads

$$\theta'' + \alpha\sin\theta = 0, \quad \text{where } \theta'' = \frac{d^2\theta}{du^2}, \quad \alpha = \frac{RL^2}{B},$$

where $\alpha$ is a positive *dimensionless* quantity (a number). Overall, we end up with the *boundary problem*

$$\theta'' + \alpha\sin\theta = 0, \quad (0 \le u \le 1), \quad \theta(0) = 0, \quad \theta'(1) = 0.$$

Observe that there is the trivial solution, $\theta(u) = 0$ for any value of $u$: the beam stays in equilibrium and maintains a straight shape. But because the problem *is not* a Cauchy problem (the two conditions are computed at different points, $u = 0$ and $u = 1$), we cannot conclude the solution is *unique*. Let us delve a bit deeper into this.

### 15.13.1 Critical Load

The differential equation

$$\theta'' + \alpha\sin\theta = 0, \quad 0 \le u \le 1,$$

admits a first integral, since is arises from differentiating in $u$ the equation

$$\frac{1}{2}\left(\frac{d\theta}{du}\right)^2 - \alpha\cos\theta = c\,.$$

Above, $c$ is constant and determined by the boundary condition $\theta'(1) = 0$, so

$$c = -\alpha \cos\theta_f, \quad \text{with } \theta_f = \theta(1),$$

where $\theta_f$ is the (unknown) value of $\theta$ at the free end where the load $R$ is applied.

If $\theta(u)$ solves the boundary value problem

$$\theta'' + \alpha \sin\theta = 0, \quad 0 \le u \le 1, \quad \theta(0) = 0, \quad \theta'(1) = 0,$$

then certainly $-\theta(u)$ will solve it too. Without loss of generality we will examine only solutions satisfying $\theta'(0) > 0$ (if $\theta'(0) = 0$, in addition to $\theta(0) = 0$, the only solution would be the trivial one due to the existence and uniqueness theorem).

From the first integral we have

$$\left(\frac{d\theta}{du}\right)^2 = 2\alpha(\cos\theta - \cos\theta_f) \tag{15.41}$$

and so the derivative vanishes only when $\cos\theta = \cos\theta_f$, and in general $\cos\theta > \cos\theta_f$. Suppose now for simplicity that $\theta'(u) > 0$ from $u = 0$ to $u = 1$ (excluded). Later we will see how to address the general case.

Equation (15.41) is separable and we integrate it between 0 and $u$, choosing the positive root since we have assumed $\theta'(u) \ge 0$:

$$\int_0^{\theta(u)} \frac{d\theta}{\sqrt{\cos\theta - \cos\theta_f}} = \int_0^u \sqrt{2\alpha}\, dx = u\sqrt{2\alpha}.$$

In this way we reduced the solution to the differential equation to computing a definite integral.

The first boundary condition holds ($\theta(0) = 0$). For the second condition ($\theta'(1) = 0$), we must check $\theta_f$ is attained for $u = 1$. To that end we must put

$$\int_0^{\theta_f} \frac{d\theta}{\sqrt{\cos\theta - \cos\theta_f}} = \sqrt{2\alpha}. \tag{15.42}$$

As $\theta_f = \pi$ would give a divergent integral, necessarily $0 \le \theta \le \theta_f < \pi$. We change variable so to write in a simpler form this integral that depends on the parameter $\theta_f$.

For $\theta_f \in [0, \pi)$, define the constant $k \in [0, 1)$ by the 1-1 correspondence

$$k = \sin(\theta_f/2)$$

Similarly, for $\theta \in [0, \theta_f)$, let the variable $\phi \in [0, \pi/2)$ be given via the implicit regular bijection

$$\sin\phi = \sin(\theta/2)/k. \tag{15.43}$$

Differentiating the latter we find

$$2k\cos\phi\, d\phi = \cos(\theta/2)\, d\theta$$

whence

$$d\theta = \frac{2k\cos\phi}{\sqrt{1-k^2\sin^2\phi}} d\phi.$$

Note

$$\cos\theta = 1 - 2\sin^2(\theta/2) = 1 - 2k^2\sin^2\phi, \quad \cos\theta_f = 1 - 2\sin^2(\theta_f/2) = 1 - 2k^2,$$

and so

$$\sqrt{\cos\theta - \cos\theta_f} = \sqrt{2}k\sqrt{1-\sin^2\phi} = \sqrt{2}k\cos\phi.$$

Now change variable from $\theta$ to $\phi$ in the integral, so that

$$\begin{aligned}\int_0^{\theta_f} \frac{d\theta}{\sqrt{\cos\theta - \cos\theta_f}} &= \int_0^{\pi/2} \frac{2k\cos\phi}{\sqrt{2}k\cos\phi\sqrt{1-k^2\sin^2\phi}} d\phi \\ &= \sqrt{2}\int_0^{\pi/2} \frac{d\phi}{\sqrt{1-k^2\sin^2\phi}}.\end{aligned}$$

Relation (15.42) is recast as

$$I(k) := \int_0^{\pi/2} \frac{d\phi}{\sqrt{1-k^2\sin^2\phi}} = \sqrt{\alpha}$$

expressing the necessary and sufficient condition under which the boundary problem admits a non-trivial solution. The function $I(k)$, whose value is the above integral and depends on the parameter $k$, must equal $\sqrt{\alpha}$. The values of $k$ $(0 \le k < 1)$ for which that happens, if any, correspond to $\sin(\theta_f/2)$ where $\theta_f$ is the bending angle at the free end.

A few simple calculations show that

1. $I(0) = \pi/2$;
2. $\lim_{k\to 1^-} I(k) = +\infty$;
3. $I(k)$ is an increasing map over the interval $[0, 1)$.

$I(k)$ is a special type of *inverse Jacobi function*, and its graph can be plotted by a computer, resulting in Fig. 15.33. From that observation we deduce that

- if $\sqrt{\alpha} < \pi/2$ the problem admits only the trivial solution, since there exists no $k$ such that $I(k) = \sqrt{\alpha}$;
- if $\sqrt{\alpha} \ge \pi/2$ the problem admits a non-trivial solution.

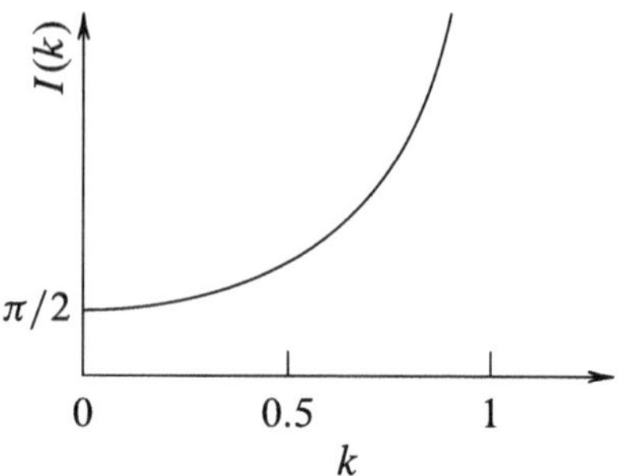


**Fig. 15.33** Graph of the function $I(k)$

We know that $\alpha = RL^2/B$, so the condition for the existence of non-trivial solutions becomes

$$R \geq \frac{\pi^2 B}{4L^2}.$$

The load $R$ must therefore have modulus larger than

$$R_c = \frac{\pi^2 B}{4L^2},$$

called *critical load*. To sum up: below the critical load the problem admits only the trivial solution (a line) and the beam does not bend. Above the critical load the beam can deform, and the bending, measured by the number $\theta_f$, increases with the load.

The mathematical model in use does not allow us to decide which, among several solutions, is the most plausible one physically. Getting out of this rut would require introducing energy considerations, which in turn that would quickly lead us to address the *stability* or *instability* of the solutions found. This topic falls outside of our (narrow) scope.

We end the section with a brief discussion of the (more general) case where, before reaching the value $\theta_f$ at $u = 1$, the function $\theta(u)$ has zero derivative at $n$ points inside $(0, 1)$. From the first integral we know that $\theta' = 0$ only when $\theta = \pm\theta_f$. Now, on intervals where $\theta' > 0$ we must use the $+$ sign for the square root that gives $\theta'$ in terms of $\theta$, and the $-$ sign where $\theta' < 0$. This allows us to write the relationship that determines $\theta_f$ as

$$\int_0^{|\theta_f|} \frac{d\theta}{\sqrt{\cos\theta - \cos\theta_f}} - \int_{|\theta_f|}^{0} \frac{d\theta}{\sqrt{\cos\theta - \cos\theta_f}} - \int_0^{-|\theta_f|} \frac{d\theta}{\sqrt{\cos\theta - \cos\theta_f}} + \int_0^{|\theta_f|} \frac{d\theta}{\sqrt{\cos\theta - \cos\theta_f}} + \cdots = \sqrt{2\alpha}. \tag{15.44}$$

The sum contains $2n + 1$ integrals, where $n$ is the number of points between 0 and 1 where the first derivative of $\theta(u)$ vanishes. Recalling that the modulus of $\theta'$ equals the curvature, and that the points where this derivative changes sign are the inflections, we conclude that (15.44) is a necessary and sufficient condition for the problem to admit at least one non-trivial solution with $n$ inflections. On the other hand, since $\cos\theta$ is an even function,

$$\int_0^{|\theta_f|} \frac{d\theta}{\sqrt{\cos\theta - \cos\theta_f}} = -\int_{|\theta_f|}^{0} \frac{d\theta}{\sqrt{\cos\theta - \cos\theta_f}}$$
$$= -\int_0^{-|\theta_f|} \frac{d\theta}{\sqrt{\cos\theta - \cos\theta_f}} = \ldots$$

and therefore (15.44) becomes

$$(2n+1)\int_0^{|\theta_f|} \frac{d\theta}{\sqrt{\cos\theta - \cos\theta_f}} = \sqrt{2\alpha}\,. \tag{15.45}$$

Putting again $k = \sin(|\theta_f|/2)$ and using the previous variable change (15.43), we obtain

$$\int_0^{|\theta_f|} \frac{d\theta}{\sqrt{\cos\theta - \cos\theta_f}} = \sqrt{2}\int_0^{\pi/2} \frac{d\phi}{\sqrt{1 - k^2\sin^2\phi}} = \sqrt{2}I(k) \tag{15.46}$$

Comparing (15.45) and (15.46) we eventually see that the condition for the existence of a non-trivial solution with $n$ inflections is

$$(2n+1)I(k) = \sqrt{\alpha},$$

possible only if

$$\sqrt{\alpha} \geq (2n+1)\pi/2.$$

Obviously, for constant $L$ and $B$, a bigger load is needed to produce solutions with more inflections. Here, too, we cannot know which solutions the system prefers from a physical perspective.

### 15.13.2 *Linearized Model*

The discussion is greatly simplified if we adopt the reasonable assumption of studying only small bending. More precisely, suppose the angle $\theta$ stays small for any value of $u$, so that the approximation

$$\sin\theta \approx \theta$$

holds. The boundary problem thus transforms into

$$\theta'' + \alpha\theta = 0, \quad \theta(0) = 0, \quad \theta'(1) = 0,$$

and this can be solved explicitly because now the differential equation is linear and with constant coefficients. The general integral is

$$\theta(u) = A\sin(\sqrt{\alpha}u) + B\cos(\sqrt{\alpha}u),$$

and $\theta(0) = 0$ implies $B = 0$. Condition $\theta'(1) = 0$ forces $A = 0$ (trivial solution) or

$$\sqrt{\alpha} = (2n+1)\pi/2.$$

Corresponding to each of these values of $\alpha$ we have non-trivial solutions

$$\theta(u) = A\sin\left((2n+1)u\pi/2\right),$$

with arbitrary $A$. The conclusion greatly resembles what we would find with an in depth analysis, since the values of the critical loads coincide. That here the modelling is more simplistic might lead us to believe we can have non-trivial solutions only for values of $\alpha$ *equal* to those of a critical load, which is patently not the case. The linearization process, however, if accompanied by a little physical common sense, produces useful results with a mild mathematical effort.

# Appendix A Reminder of Vectors, Curves and Differential Equations

## A.1 Points and Vectors

### Points

Let $O$ denote a point in three-dimensional Euclidean space $\mathcal{E}$ where the system's motion takes place. Using a basis $\{\mathbf{e}_1, \mathbf{e}_2, \mathbf{e}_3\}$, we identify the position of any point $P \in \mathcal{E}$ with its *Cartesian coordinates* $(x_1, x_2, x_3)$ as follows:

$$OP = x_1\,\mathbf{e}_1 + x_2\,\mathbf{e}_2 + x_3\,\mathbf{e}_3. \tag{A.1}$$

The *origin* $O$ and the basis $\{\mathbf{e}_1, \mathbf{e}_2, \mathbf{e}_3\}$ used to describe the position of points in the system form a *reference frame*. Note that in (A.1), as in the entire book, vectors are written in boldface. The only exceptions to this rule are the vectors joining two points, such as $OP$ in (A.1), which is the vector *starting at $O$ and ending at $P$*.

A basis is called *orthonormal* when the vectors that form it are pairwise perpendicular and have length 1. In general, we will call *unit vector* any vector of length one. Throughout the text we will exclusively use orthonormal bases, even when not said so explicitly.

### Dot Product

Given two vectors $\mathbf{a}$ and $\mathbf{b}$, we define their *dot product* to be

$$\mathbf{a} \cdot \mathbf{b} = |\mathbf{a}|\;|\mathbf{b}|\;\cos\alpha, \tag{A.2}$$

where $|\mathbf{a}|$, $|\mathbf{b}|$ are the lengths (or moduli) of the vectors $\mathbf{a}$, $\mathbf{b}$, and $\alpha \in [0, \pi]$ is the angle between the two vectors' directions, oriented coherently.

Formula (A.2) clearly defines a symmetric operation, since $\mathbf{a} \cdot \mathbf{b} = \mathbf{b} \cdot \mathbf{a}$. It is moreover possible to show that the dot product is linear, i.e.



P. Biscari etal., *Rational Mechanics*, UNITEXT 177,
https://doi.org/10.1007/978-3-032-07462-1

$$\big(\lambda \mathbf{a} + \mu \mathbf{b}\big) \cdot \mathbf{c} = \lambda\, \mathbf{a} \cdot \mathbf{c} + \mu\, \mathbf{b} \cdot \mathbf{c}. \tag{A.3}$$

From definition (A.2) we deduce that the dot product of any two unit vectors in an orthonormal basis is 0 if the vectors are different and 1 if we dot-multiply one by itself

$$\mathbf{e}_i \cdot \mathbf{e}_j = \delta_{ij} = \begin{cases} 1 & \text{if } i = j \\ 0 & \text{if } i \neq j, \end{cases} \tag{A.4}$$

where we use the function $\delta_{ij}$ called *Kronecker delta*. Now, using linearity (A.3) one proves that when we decompose vectors along an orthonormal basis the dot product becomes a linear combination of the components:

$$\begin{cases} \mathbf{a} = a_1\, \mathbf{e}_1 + a_2\, \mathbf{e}_2 + a_3\, \mathbf{e}_3 \\ \mathbf{b} = b_1\, \mathbf{e}_1 + b_2\, \mathbf{e}_2 + b_3\, \mathbf{e}_3 \end{cases} \quad \Longrightarrow \quad \mathbf{a} \cdot \mathbf{b} = \sum_{i=1}^{3} a_i b_i\,.$$

Let $\mathbf{e}$ be any unit vector. We call *component along* $\mathbf{e}$ *of a vector* $\mathbf{v}$ the dot product

$$v_e = \mathbf{v} \cdot \mathbf{e}\,,$$

while we say $\mathbf{v}_e = (\mathbf{v} \cdot \mathbf{e})\mathbf{e} = v_e\mathbf{e}$ is the (vector) component of $\mathbf{v}$ along $\mathbf{e}$.

Similarly, given any point $P$ and a straight line $r$ through $O$ and parallel to $\mathbf{e}$, one builds (see Fig. A.1) the projection of $P$ on $r$ as the point $P_r$ such that:

$$OP_r = \big(OP \cdot \mathbf{e}\big)\, \mathbf{e},$$

which coincides with the vector component of $OP$ along $\mathbf{e}$.

Pythagoras' theorem provides a simple means to express the distance of point $P$ to $r$ in terms of two dot products. Indeed, calling $\alpha$ the angle between $OP$ and the unit vector $\mathbf{e}$ we have:

$$\begin{aligned} d(P, r) &= |PP_r| = |OP|\, \sin\alpha = \sqrt{|OP|^2 - |OP_r|^2} \\ &= \sqrt{OP \cdot OP - \big(OP \cdot \mathbf{e}\big)^2}\,. \end{aligned} \tag{A.5}$$

Let now $\pi$ denote the plane orthogonal to the unit vector $\mathbf{e}$ and passing through $O$. Given any point $P$, we build (see Fig. A.1) the projection of $P$ onto $\pi$ to be the point $P_\pi$ such that:

$$OP_\pi = OP - \big(OP \cdot \mathbf{e}\big)\, \mathbf{e}.$$

Again, Pythagoras' theorem allows us to write the distance of $P$ to the plane $\pi$. Let in fact $\alpha$ be the angle between $OP$ and $\mathbf{e}$ (note that $\alpha$ is complementary to the angle formed by $OP$ and $\pi$). Then:

$$d(P, \pi) = |PP_\pi| = |OP|\, |\cos\alpha| = \big|OP \cdot \mathbf{e}\big|.$$

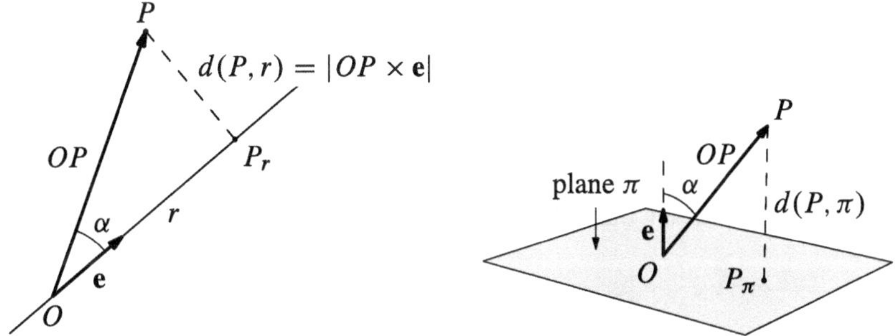


**Fig. A.1** Orthogonal projection of a point $P$ on a straight line $r$ and on a plane $\pi$

The dot product $\left(OP \cdot \mathbf{e}\right)$ is positive if $P$ belongs in the half-space defined by $\pi$ and containing $\mathbf{e}$; it is zero if $P$ lies on $\pi$, and it is negative if $P$ is in the other half-space.

## Cross Product

Given two vectors $\mathbf{a}$ and $\mathbf{b}$ we define their *cross product* $\mathbf{a} \times \mathbf{b}$ to be the vector with the following properties:

- its modulus is
$$\left|\mathbf{a} \times \mathbf{b}\right| = |\mathbf{a}|\ |\mathbf{b}|\ \sin\alpha, \tag{A.6}$$
where $\alpha$ is the angle between the directions of $\mathbf{a}$ and $\mathbf{b}$;
- its direction is orthogonal to both $\mathbf{a}$ and $\mathbf{b}$ (hence orthogonal to the plane they define);
- its orientation is fixed using the *right-hand rule*: if we align the four right fingers (pointing finger to little finger) with the first vector $\mathbf{a}$, and then close the fist towards the vector $\mathbf{b}$, the thumb will give the orientation of the cross product $\mathbf{a} \times \mathbf{b}$. An equivalent definition would be to align the right hand's thumb, pointing finger and middle finger perpendicularly, as if we were counting 1 to 3: if the thumb represents $\mathbf{a}$ and the pointing finger $\mathbf{b}$, the middle finger gives $\mathbf{a} \times \mathbf{b}$.

Definition (A.6) shows how the cross product vanishes if and only if: the first vector is zero, and/or the second vector is zero, and/or the two are parallel (with the same or opposite orientations). Note the difference with the dot product, which vanishes if and only if either factor is zero or the vectors are orthogonal.

The right-hand rule explained in the definition shows that the cross product is skew-symmetric:
$$\mathbf{a} \times \mathbf{b} = -\mathbf{b} \times \mathbf{a}.$$

Furthermore, just like the dot product, also the cross product defines a linear operation:

$$(\lambda \mathbf{a} + \mu \mathbf{b}) \times \mathbf{c} = \lambda\, \mathbf{a} \times \mathbf{c} + \mu\, \mathbf{b} \times \mathbf{c}.$$

The cross product divides in two groups the sets of unit vectors we can use to define an orthonormal frame. These groups are distinguished by the result of $\mathbf{e}_1 \times \mathbf{e}_2$. By definition, that cross product must be parallel to $\mathbf{e}_3$, so there only are two possibilities

- $\mathbf{e}_1 \times \mathbf{e}_2 = +\mathbf{e}_3$; we say that $\{\mathbf{e}_1, \mathbf{e}_2, \mathbf{e}_3\}$ is a *right-handed basis* (or *positively oriented*)
- $\mathbf{e}_1 \times \mathbf{e}_2 = -\mathbf{e}_3$; we say $\{\mathbf{e}_1, \mathbf{e}_2, \mathbf{e}_3\}$ is a *left-handed basis* (or *negatively oriented*).

These names refer to the property that if in a left-handed (respectively, right-handed) frame we stand as the unit vector $\mathbf{e}_3$ and observe the plane $\{\mathbf{e}_1, \mathbf{e}_2\}$, to move the first unit vector onto the second one we must rotate counter-clockwise (clockwise).

Given any basis (left- or right-handed) we can produce a basis of the other type simply flipping the sign of one of the vectors, or swapping two of them. Instead, a *cyclic permutation* of the vectors, where each vector takes the place of the one after it (respectively, before it) in the basis, and the last one (first one) moves to the first place (last), does not change the overall orientation. Applying the definition (or cyclic permutations) it is easy to see that in a right-handed basis

$$\begin{aligned} &\mathbf{e}_1 \times \mathbf{e}_2 = +\mathbf{e}_3, \qquad \mathbf{e}_3 \times \mathbf{e}_1 = +\mathbf{e}_2, \quad \text{and} \quad \mathbf{e}_2 \times \mathbf{e}_3 = +\mathbf{e}_1, \\ &\mathbf{e}_2 \times \mathbf{e}_1 = -\mathbf{e}_3, \qquad \mathbf{e}_1 \times \mathbf{e}_3 = -\mathbf{e}_2, \quad \text{and} \quad \mathbf{e}_3 \times \mathbf{e}_2 = -\mathbf{e}_1. \end{aligned}$$

Just as we assumed that bases are orthonormal, in the whole book we shall suppose all bases right-handed.

Using the linearity and the above formulas for the cross product of elements of a basis we can find the expression for the cross product of arbitrary vectors in terms of their components. In this way

$$\begin{cases} \mathbf{a} = a_1\, \mathbf{e}_1 + a_2\, \mathbf{e}_2 + a_3\, \mathbf{e}_3 \\ \mathbf{b} = b_1\, \mathbf{e}_1 + b_2\, \mathbf{e}_2 + b_3\, \mathbf{e}_3 \end{cases} \implies \begin{aligned} \mathbf{a} \times \mathbf{b} = &(a_2 b_3 - a_3 b_2)\, \mathbf{e}_1 \\ &+ (a_3 b_1 - a_1 b_3)\, \mathbf{e}_2 \\ &+ (a_1 b_2 - a_2 b_1)\, \mathbf{e}_3. \end{aligned}$$

The second and third component of the cross product also arise from the first by cyclic permutations of the type $1 \to 2 \to 3 \to 1$.

Here are other properties of the cross product.

- Given two vectors $\mathbf{a}$ and $\mathbf{b}$, we have

$$|\mathbf{a} \times \mathbf{b}|^2 + (\mathbf{a} \cdot \mathbf{b})^2 = |\mathbf{a}|^2 |\mathbf{b}|^2 = (\mathbf{a} \cdot \mathbf{a})\, (\mathbf{b} \cdot \mathbf{b}).$$

- Using the cross product we can write the distance of a point $P$ to a line $r$ and to a plane $\pi$ (both passing through $O$ and parallel (the line) or orthogonal (the plane) to the unit vector $\mathbf{e}$):

$$d(P, r) = |OP| \sin\alpha = |\, OP \times \mathbf{e} \,| \tag{A.7}$$

$$d(P, \pi) = |OP|\,|\cos\alpha| = \sqrt{|OP|^2 - |OP \times \mathbf{e}|^2}\,.$$

- Given three vectors **a**, **b** and **c**, we have the identities

$$\mathbf{a} \times (\mathbf{b} \times \mathbf{c}) = (\mathbf{a} \cdot \mathbf{c})\mathbf{b} - (\mathbf{a} \cdot \mathbf{b})\mathbf{c} \tag{A.8}$$

$$(\mathbf{a} \times \mathbf{b}) \times \mathbf{c} = (\mathbf{a} \cdot \mathbf{c})\mathbf{b} - (\mathbf{c} \cdot \mathbf{b})\mathbf{a}. \tag{A.9}$$

  In particular, the result of taking the cross product twice depends on the order of the operations (as such, it is not associative).

### Mixed Product

Given three vectors **a**, **b** and **c**, we call mixed product:

$$\mathbf{a} \cdot \mathbf{b} \times \mathbf{c} = \mathbf{a} \cdot (\mathbf{b} \times \mathbf{c})$$

Swapping any two factors results in a change of sign:

$$\mathbf{a} \cdot \mathbf{b} \times \mathbf{c} = -\mathbf{b} \cdot \mathbf{a} \times \mathbf{c} = \mathbf{c} \cdot \mathbf{a} \times \mathbf{b} = -\mathbf{c} \cdot \mathbf{b} \times \mathbf{a} = \cdots$$

As a consequence of the previous property, any cyclic permutation of the vectors in a mixed product, exactly as the swap of the operations $\cdot$ and $\times$, does not change the result:

$$\mathbf{a} \times \mathbf{b} \cdot \mathbf{c} = \mathbf{a} \cdot \mathbf{b} \times \mathbf{c} = \mathbf{c} \cdot \mathbf{a} \times \mathbf{b} = \mathbf{b} \cdot \mathbf{c} \times \mathbf{a}\,. \tag{A.10}$$

The mixed product is zero if and only if the three vectors are coplanar. Its absolute value equals the volume of the parallelepiped with edges **a**, **b** and **c**.

### Solving the Vectorial Linear Equation a × v = b

We wish to examine the following linear equation in the vectorial unknown **v**:

$$\mathbf{a} \times \mathbf{v} = \mathbf{b}, \tag{A.11}$$

where **a** and **b** are given vectors (with $\mathbf{a} \neq \mathbf{0}$).

- As the cross product is orthogonal to either factor, a *necessary* condition for (A.11) to make sense and admit solutions is that the given **a** and **b** are orthogonal to one another. If $\mathbf{a} \cdot \mathbf{b} \neq 0$, then (A.11) has no solution.
- Let us take the cross product of (A.11) with **a**, to obtain:

$$\big(\mathbf{a} \times \mathbf{v}\big) \times \mathbf{a} = \mathbf{b} \times \mathbf{a}\,,$$

or equivalently

$$\mathbf{v} = \frac{\mathbf{a} \cdot \mathbf{v}}{a^2}\mathbf{a} + \frac{\mathbf{b} \times \mathbf{a}}{a^2}. \tag{A.12}$$

Equation (A.12) has a simple interpretation: the vectors $\mathbf{v}$ solving (A.11) have an arbitrary component parallel to $\mathbf{a}$ (indeed this component vanishes in the cross product) and a given component orthogonal to $\mathbf{a}$, equal to the second summand in (A.12). The most general solution to (A.11) then reads:

$$\mathbf{v} = \lambda\,\mathbf{a} + \frac{\mathbf{b} \times \mathbf{a}}{a^2}, \qquad \text{with} \quad \lambda \in \mathbb{R}.$$

## A.2 Curves

A *curve* is a function $P : I \to \mathcal{E}$ that associates a point $P(t) \in \mathcal{E}$ with every value of a real parameter $t \in I \subseteq \mathbb{R}$. We shall call *regular* any curve that is differentiable at least twice with continuous second derivative. In the sequel we will assume all curves are regular. For simplicity we will also suppose the first derivative $\dot{P}(t)$ is never zero (a curve with $\dot{P}$ identically zero would simply describe a point).

We will call *image* $S_P$ of the curve $P$ the geometric locus of points visited by the curve:

$$S_P = \big\{ Q \in \mathcal{E} \quad \text{such that} \quad P(t) = Q \quad \text{for some } t \in I \big\}.$$

Clearly the same locus can be the image of a myriad different curves, just like a car can drive on the same road in infinitely many ways.

### Straight Lines

The simplest curves are straight lines, or portions of lines. For these the law prescribing how the point $P(t)$ changes is affine in the parameter $t$:

$$OP(t) = OP_0 + \mathbf{v}\,t, \qquad t \in I \subseteq \mathbb{R}. \tag{A.13}$$

- Depending on the type of interval $I \subseteq \mathbb{R}$ in which the parameter $t$ lives, the image of (A.13) is the segment joining $P(t_1)$ to $P(t_2)$ if $I = [t_1, t_2]$; it is a half-line with initial point $P(t_1)$ if $I = [t_1, +\infty)$ or $I = (-\infty, t_1]$; it is a straight line if $I = \mathbb{R}$.
- The curve in (A.13) belongs to the line passing through $P_0 = P(0)$ and parallel to the vector $\mathbf{v}$. Given two points $P' = P(t')$ and $P'' = P(t'')$, in fact:

$$P'P'' = \mathbf{v}\,(t'' - t') \quad \Longrightarrow \quad P'P'' \times \mathbf{v} = \mathbf{0}.$$

- A curve whose image is the segment from $Q$ to $R$ is:

$$OP(t) = OQ + QR\,t\,, \qquad t \in [0, 1]. \tag{A.14}$$

  The line through $Q$, $R$ can be found by extending (A.14) to $t \in \mathbb{R}$.

### Tangent Line

Let $P$ be a regular curve. We call *tangent line to* $P$, at the point $P_0 = P(t_0)$, the line obtained as limit of the secant lines through $P(t_0)$ and $P(t_0 + \varepsilon)$ as $\varepsilon \to 0$. A simple Taylor expansion of $P(t_0 + \varepsilon)$ in $\varepsilon$ to order one shows that such a line $R(t)$ is:

$$OR(t) = OP_0 + (t - t_0)\,\dot{P}(t_0).$$

It clearly passes through $P_0$ (for $t = t_0$). Moreover, is parallel to the *unit tangent vector to* $P(t)$ *at* $P_0$

$$\mathbf{t}(t_0) = \frac{\dot{P}(t_0)}{\left|\dot{P}(t_0)\right|}.$$

In the previous expressions, as in all the following ones, we use a dot to denote the derivative with respect to the parameter $t$.

### Arclength

Let $P$ be a regular curve. We call *arclength* $s$ from $P_1 = P(t_1)$ the length of the portion of its image between $P_1$ and $P(t)$:

$$s(t) = \int_{t_1}^{t} \left|\dot{P}(u)\right| du.$$

- The Fundamental theorem of calculus implies that

$$\frac{ds}{dt} = \left|\dot{P}(t)\right| > 0 \qquad \text{for any } t, \tag{A.15}$$

  provided the first derivative never vanishes. If so, the function $s(t)$ is invertible.
- Let $P : [t_1, t_2] \to \mathcal{E}$ be a curve whose first derivative is never zero. Call $t = \hat{t}(s)$ the inverse function of arclength, and $L = s(t_2)$ the length of the image of $P$. We define a new curve $\hat{P}(s)$, taking values in the interval $s \in [0, L]$, by:

$$[0, L] \ni s \to \hat{P}(s) = P\big(\hat{t}(s)\big) \in \mathcal{E}.$$

The curve $\hat{P}$ has the same image as $P$, but it travels along it uniformly. In fact, if we first use the chain rule and then differentiate the inverse map, we find that:

$$\frac{d\hat{P}}{ds} = \dot{P}\big(\hat{t}(s)\big)\,\frac{d\hat{t}}{ds} = \frac{\dot{P}\big(\hat{t}(s)\big)}{\big|\dot{P}\big(\hat{t}(s)\big)\big|} = \mathbf{t}\big(\hat{t}(s)\big) \tag{A.16}$$

so in particular $|d\hat{\mathbf{P}}/ds| = 1$. The parametrization $\hat{P}(s)$ then enjoys the remarkable property of depending only on the image of $P$, not on how we travel along it.

### Curvature and Curvature Radius

We now wish to introduce a quantitative measure of how straight the image of a curve is. For a line the unit tangent vector does not change, irrespective of the point at which we compute it. So it is natural to associate the measure we seek with the rate of variation of the unit tangent vector along the curve itself. More precisely, we define the *curvature* to be:

$$c = \left|\frac{d\mathbf{t}}{ds}\right|. \tag{A.17}$$

- The definition of $c$ via the derivative of the unit tangent vector with respect to the arclength (instead of the original parameter $t$) guarantees that all curves with that same image will measure the same curvature at the point.
- In terms of the parametrization $P(t)$,

$$c = \frac{|\dot{P} \times \ddot{P}|}{|\dot{P}|^3} \in [0, +\infty). \tag{A.18}$$

Formula (A.18) is extremely practical to compute a curve's curvature directly, since Definition (A.17) is not viable before we find the arclength $s(t)$ and make the inverse function $\hat{t}(s)$ explicit. For completeness we write below the steps that build the proof of (A.18):

$$c = \left|\frac{d\mathbf{t}}{ds}\right| = \left|\frac{d\mathbf{t}}{dt}\frac{d\hat{t}}{ds}\right| = \frac{1}{|\dot{P}|}\left|\frac{d}{dt}\frac{\dot{P}}{|\dot{P}|}\right| = \frac{1}{|\dot{P}|}\left|\frac{\ddot{P}}{|\dot{P}|} - \frac{\ddot{P}\cdot\dot{P}}{|\dot{P}|^3}\dot{P}\right|$$
$$= \frac{1}{|\dot{P}|^2}\left|\ddot{P} - (\ddot{P}\cdot\mathbf{t})\,\mathbf{t}\right| = \frac{|\mathbf{t}\times\ddot{P}|}{|\dot{P}|^2} = \frac{|\dot{P}\times\ddot{P}|}{|\dot{P}|^3}$$

- The curvature has the dimension of the inverse of a length. Its reciprocal

$$\rho = \frac{1}{c} = \frac{|\dot{P}|^3}{|\dot{P} \times \ddot{P}|} \in (0, +\infty].$$

then has the dimension of a length, and is called *curvature radius*. The name's origin will become clear shortly.

## Unit Normal

Consider a point on a regular curve $P$ where the curvature is non-zero. Knowing that the unit tangent vector changes direction, we call *(principal) unit normal* (or simply *normal*) the unit vector giving the direction towards which $\mathbf{t}$ is turning:

$$\mathbf{n} = \frac{\dfrac{d\mathbf{t}}{ds}}{\left|\dfrac{d\mathbf{t}}{ds}\right|} = \frac{1}{c}\frac{d\mathbf{t}}{ds}. \tag{A.19}$$

- As shown in Fig. A.2, the unit normal points by construction towards the image's concavity.
- The unit normal is independent of the image's orientation, whereas the unit tangent vector changes sign when we travel along the image in the opposite direction.
- The unit normal is always orthogonal to the tangent line. This is a special case of the following lemma.

**Lemma A.1** *Let* $\mathbf{f}$ *be a differentiable vectorial function, depending on the real variable* $t$*. The vector* $\mathbf{f}(t)$ *has constant modulus as* $t$ *varies if and only if it is always orthogonal to its derivative:*

$$|\mathbf{f}|^2 = constant \qquad \Longleftrightarrow \qquad \mathbf{f} \cdot \dot{\mathbf{f}} = 0 \ \ \forall t. \tag{A.20}$$

***Proof*** Equivalence (A.20) is proved simply by noting that

$$\frac{d}{dt}|\mathbf{f}|^2 = \frac{d}{dt}\mathbf{f} \cdot \mathbf{f} = 2\mathbf{f} \cdot \dot{\mathbf{f}}.$$

The change in the modulus of $\mathbf{f}(t)$ thus depends on the dot product of $\mathbf{f}$ and $\dot{\mathbf{f}}$. In particular, the dot product is zero if and only if $|\mathbf{f}(t)|$ is constant. □

A straightforward consequence of the lemma is the orthogonality of the unit tangent vector, whose modulus is constant (equal 1) and the unit normal, which is parallel to the unit tangent vector's derivative.

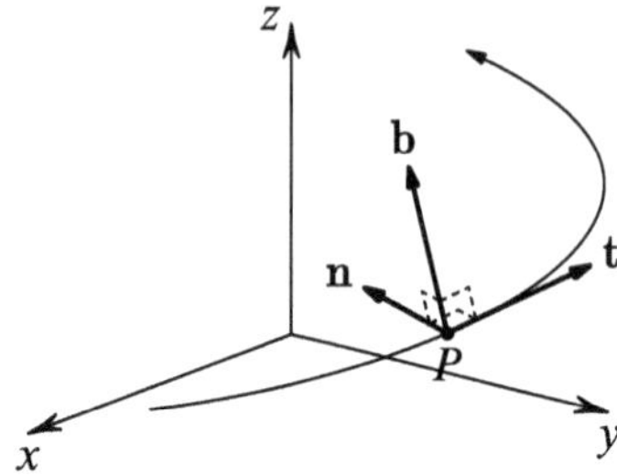


**Fig. A.2** The Frenet frame $\{\mathbf{t}, \mathbf{n}, \mathbf{b}\}$

### Binormal Vector

Consider again a point on a regular curve $P$ with non-zero curvature. Since the unit tangent and normal are orthogonal we introduce the *binormal vector*:

$$\mathbf{b} = \mathbf{t} \times \mathbf{n}\,. \tag{A.21}$$

The orthonormal basis $\{\mathbf{t}, \mathbf{n}, \mathbf{b}\}$ thus built is called *Frenet frame* at $P$, and is shown in Fig. A.2.

In order to calculate unit normals without having to resort to the arclength we can use the following:

$$\mathbf{b} = \frac{\dot{P} \times \ddot{P}}{|\dot{P} \times \ddot{P}|} \qquad \text{and} \qquad \mathbf{n} = \mathbf{b} \times \mathbf{t}\,. \tag{A.22}$$

The second relation in (A.22) follows from (A.21) by a cyclic permutation. The latter is allowed in the expression relating a unit vector in an orthonormal basis to the cross product of the other two. Below we write the main steps to prove the first formula in (A.22). In the proof we will use the scalars $\lambda$, $\lambda'$, $\lambda''$, $\lambda'''$ to comprise all terms that do not appear explicitly in the cross product that gives the direction of $\mathbf{b}$. These scalars contribute to making the binormal a unit vector, but this normalization is already trivially warranted by the denominator of $(\text{A.22})_1$:

$$\begin{aligned}\mathbf{b} = \mathbf{t} \times \mathbf{n} = \lambda\,\mathbf{t} \times \frac{d\mathbf{t}}{ds} = \lambda'\,\mathbf{t} \times \frac{d\mathbf{t}}{dt} = \lambda''\,\dot{P} \times \frac{d}{dt}\left(\frac{\dot{P}}{|\dot{P}|}\right)\\ = \lambda''\,\dot{P} \times \left(\frac{\ddot{P}}{|\dot{P}|} + \dot{P}\,\frac{d}{dt}\frac{1}{|\dot{P}|}\right) = \lambda'''\,\dot{P} \times \ddot{P}.\end{aligned}$$

By definition the binormal vector is orthogonal to both the tangent vector and the unit normal. On plane curves these two directions always lie on the curve's plane. Therefore the binormal is constrained to be perpendicular to the plane.

But note that at certain points (those with zero curvature, to be precise) it is not possible to define the unit normal, and hence it also is impossible to define the binormal direction. So it may happen that a planar curve has constant binormal vector along some section where the curvature is non-zero, does not admit **n** nor **b** at an inflection point where the curvature vanishes, and goes back to having constant binormal (but oppositely oriented) on a successive section with non-zero curvature.

### Osculating Circle

Let $P$ be a regular curve. The *osculating circle* to $P$, at point $P_0 = P(t_0)$, is the limit of the circles through $P(t_0)$, $P(t_0+\varepsilon)$, $P(t_0+\varepsilon')$ as $\varepsilon \neq \varepsilon'$ tend to 0. It can be proved that the osculating circle's radius is equal to the curvature radius $\rho$, and its center has distance $\rho$ from $P_0$, in the direction to which the unit normal $\mathbf{n}$ points.

## A.3 Linear Transformations, Matrices

We gather in this section a number of important properties of linear maps and linear algebra.

By definition a *linear map* (or *tensor*) $\mathbf{A}$ is a function that transforms elements $\mathbf{u}$ in a vector space $U$ into vectors $\mathbf{v}$ of another vector space $V$ (which may or not coincide with the former), so that to satisfy linearity:

$$\mathbf{A}\big(\lambda_1\mathbf{u}_1 + \lambda_2\mathbf{u}_2\big) = \lambda_1\mathbf{A}(\mathbf{u}_1) + \lambda_2\mathbf{A}(\mathbf{u}_2), \qquad \text{for any } \lambda_1, \lambda_2 \in \mathbb{R}.$$

We shall try to omit the parentheses whenever possible and just write $\mathbf{Au}$ to mean $\mathbf{A}(\mathbf{u})$.

Let $\mathbf{u}$, $\mathbf{v}$ be vectors of dimension $n$, and $\mathbf{A}, \mathbf{B} : \mathbb{R}^n \to \mathbb{R}^n$ linear maps from $\mathbb{R}^n$ to itself. If $\{\mathbf{e}_1, \ldots, \mathbf{e_n}\}$ is an orthonormal basis of $\mathbb{R}^n$, we denote by $\{u_i, i = 1, \ldots, n\}$ and $\{a_{ij}, i, j = 1, \ldots, n\}$ the respective components of the vector $\mathbf{u}$ and the elements of the matrix representing $\mathbf{A}$, in the given basis.

### Components and Elements of a Matrix

The $i$-th component of the vector $\mathbf{u}$ is $u_i = \mathbf{e}_i \cdot \mathbf{u}$. Similarly, the element $a_{ij}$ of the matrix representing the linear map $\mathbf{A}$ in the same basis is

$$a_{ij} = \mathbf{e}_i \cdot \mathbf{A}\mathbf{e}_j\,. \tag{A.23}$$

### Image of a Vector

Applying the linear map $\mathbf{A}$ to the vector $\mathbf{u}$ we obtain the vector $\mathbf{v} = \mathbf{Au}$ of components

$$v_i = \mathbf{e}_i \cdot \mathbf{v} = \mathbf{e}_i \cdot \mathbf{Au} = \mathbf{e}_i \cdot \mathbf{A}\Big(\sum_{j=1}^{n} u_j\, \mathbf{e}_j\Big)$$
$$= \sum_{j=1}^{n} \big(\mathbf{e}_i \cdot \mathbf{Ae}_j\big)\, u_j = \sum_{j=1}^{n} a_{ij} u_j = a_{ij}\, u_j .$$

In the last passage above we have used Einstein's summation convention and suppressed the symbol of sum over the repeated index $j$. Henceforth we shall use Einstein's notation if no ambiguity arises.

### Transpose Map

We call transpose of $\mathbf{A}$, written $\mathbf{A}^T$, the map satisfying

$$\mathbf{u} \cdot \mathbf{Av} = \mathbf{A}^T \mathbf{u} \cdot \mathbf{v} \qquad \forall\, \mathbf{u}, \mathbf{v} \in \mathbb{R}^n . \tag{A.24}$$

It is easy to show from (A.23) that the transpose exists, is unique, and its matrix elements are:

$$\big(a^T\big)_{ij} = a_{ji} .$$

The transpose of the product of two linear maps does not equal to product of the transposed maps. In fact:

$$\mathbf{u} \cdot \big(\mathbf{AB}\big)\mathbf{v} = \mathbf{u} \cdot \mathbf{A}\big(\mathbf{Bv}\big) = \mathbf{A}^T \mathbf{u} \cdot \mathbf{Bv} = \mathbf{B}^T \mathbf{A}^T \mathbf{u} \cdot \mathbf{v} ,$$

from which

$$\big(\mathbf{AB}\big)^T = \mathbf{B}^T \mathbf{A}^T .$$

A linear map is called *symmetric* if it coincides with its transpose, and *skew-symmetric* if it equals minus the transpose:

$$\begin{aligned} &\mathbf{S} \text{ symmetric} && \Longleftrightarrow && \mathbf{S} = \mathbf{S}^T \\ &\mathbf{W} \text{ skew-symmetric} && \Longleftrightarrow && \mathbf{W} = -\mathbf{W}^T . \end{aligned} \tag{A.25}$$

Any linear map $\mathbf{A}$ can be written uniquely as sum of a symmetric and a skew-symmetric linear map:

$$\mathbf{A} = \mathbf{S} + \mathbf{W} , \qquad \text{with} \;\; \mathbf{S} = \tfrac{1}{2}\big(\mathbf{A} + \mathbf{A}^T\big) \;\; \text{and} \;\; \mathbf{W} = \tfrac{1}{2}\big(\mathbf{A} - \mathbf{A}^T\big) .$$

### Product of Linear Maps

Given linear maps **A**, **B**, the elements of the matrix of the composite map (the product) **AB** are $(AB)_{ij} = a_{ik}\, b_{kj}$. We point out that the composition of linear maps, and therefore also the product of their matrices, are not commutative: $\mathbf{AB} \neq \mathbf{BA}$.

### Tensor Product

Starting from any vectors **a**, **b** one can define their *tensor product* $\mathbf{a} \otimes \mathbf{b}$. This is the linear map that acts on a generic vector **u** by transforming it as follows:

$$\left(\mathbf{a} \otimes \mathbf{b}\right)\mathbf{u} = \left(\mathbf{b} \cdot \mathbf{u}\right)\mathbf{a}.$$

From the definition it is possible to deduce directly the elements of the matrix of the tensor product in any basis

$$\left(\mathbf{a} \otimes \mathbf{b}\right)_{ij} = \mathbf{e}_i \cdot \left(\mathbf{a} \otimes \mathbf{b}\right)\mathbf{e}_j = \mathbf{e}_i \left(\mathbf{b} \cdot \mathbf{e}_j\right)\mathbf{a} = a_i b_j. \tag{A.26}$$

The tensor product is not symmetric. From (A.26) in fact, we see that swapping the factors in a tensor product is the same as taking the transpose:

$$\left(\mathbf{a} \otimes \mathbf{b}\right)^T = \mathbf{b} \otimes \mathbf{a}.$$

Working out the definition one can also find the following expression for the multiplication of tensor products:

$$\left(\mathbf{a} \otimes \mathbf{b}\right)\left(\mathbf{c} \otimes \mathbf{d}\right) = \left(\mathbf{b} \cdot \mathbf{c}\right)\mathbf{a} \otimes \mathbf{d}.$$

### Inverse Transformation

Let **I** be the identity map, namely the transformation such that $\mathbf{Iu} = \mathbf{u}$ for any **u**, whose matrix elements are $I_{ij} = \delta_{ij}$, where $\delta_{ij}$ is the Kronecker delta of (A.4). We call inverse transformation of **A**, written $\mathbf{A}^{-1}$, the mapping satisfying $\mathbf{AA}^{-1} = \mathbf{A}^{-1}\mathbf{A} = \mathbf{I}$. As we will see in the sequel, not all linear maps admit an inverse. The inverse of the product of two transformations (if it exists) does not equal the product of the inverses. In fact:

$$\left(\mathbf{AB}\right)\left(\mathbf{AB}\right)^{-1} = \mathbf{I} \quad \Longrightarrow \quad \mathbf{B}\left(\mathbf{AB}\right)^{-1} = \mathbf{A}^{-1} \quad \Longrightarrow \quad \left(\mathbf{AB}\right)^{-1} = \mathbf{B}^{-1}\mathbf{A}^{-1}.$$

### Trace, Determinant

Any combination of the matrix elements of a linear map that do not depend on the basis chosen to represent the matrix is called a *scalar invariant*. One proves that a linear map with values in $\mathbb{R}^n$ admits $n$ functionally independent scalar invariants. The two most used in our treatise (and the only ones for two-dimensional mappings) are the trace and the determinant.

- The trace of a linear map is the sum of the diagonal elements of any matrix representing it:

$$\operatorname{tr}\mathbf{A} = \sum_{i=1}^{n} a_{ii} = \mathbf{e}_i \cdot \mathbf{A}\mathbf{e}_i \,.$$

  The trace is a linear operator:

$$\operatorname{tr}\big(\lambda\,\mathbf{A} + \mu\,\mathbf{B}\big) = \lambda \operatorname{tr}\mathbf{A} + \mu \operatorname{tr}\mathbf{B} \qquad \text{for} \quad \lambda, \mu \in \mathbb{R}.$$

- The determinant of a linear map $\mathbf{A} : \mathbb{R}^n \to \mathbb{R}^n$ is defined recursively using Laplace's rule.
  - If $n = 1$, then $\det\mathbf{A} = a_{11} = \mathbf{e}_1 \cdot \mathbf{A}\mathbf{e}_1$. (In this case the determinant equals the trace because in one dimension there can only be one invariant.)
  - If $n > 1$, for any $i, j = 1, \ldots, n$ we introduce the matrix obtained from $\mathbf{A}$ by erasing the $i$-th row and the $j$-th column. Calling $m_{ij}$ the determinant of the resulting matrix, we set

$$\det\mathbf{A} = \sum_{j=1}^{n} (-1)^{i+j} a_{ij} m_{ij} \qquad \big(\text{for any } i\big).$$

  This definition allows to express the determinant of an $n \times n$ matrix as linear combination of $n$ determinants of order $(n-1)$.

  In particular, if $n = 2$ we have $\det\mathbf{A} = a_{11}a_{22} - a_{12}a_{21}$, while for $n = 3$:

$$\begin{aligned}\det\mathbf{A} = {} & a_{11}a_{22}a_{33} + a_{21}a_{32}a_{13} + a_{12}a_{23}a_{31} \\ & - a_{13}a_{22}a_{31} - a_{12}a_{21}a_{33} - a_{23}a_{32}a_{11} \,.\end{aligned}$$

  The determinant satisfies the following properties:

$$\begin{aligned}&\det\big(\mathbf{AB}\big) = \det\big(\mathbf{BA}\big) = \big(\det\mathbf{A}\big)\big(\det\mathbf{B}\big) \\ &\det\big(\mathbf{A}^T\big) = \det\mathbf{A}\,, \qquad \det\big(\mathbf{A}^{-1}\big) = \big(\det\mathbf{A}\big)^{-1}.\end{aligned}$$

  Further, in case $n = 3$ we have

$$\mathbf{u} \times \mathbf{v} = \det \begin{pmatrix} \mathbf{e}_1 & \mathbf{e}_2 & \mathbf{e}_3 \\ u_x & u_y & u_z \\ v_x & v_y & v_z \end{pmatrix}, \qquad \mathbf{u} \cdot \mathbf{v} \times \mathbf{w} = \det \begin{pmatrix} u_x & u_y & u_z \\ v_x & v_y & v_z \\ w_x & w_y & w_z \end{pmatrix}$$

$$\mathbf{Au} \cdot \mathbf{Av} \times \mathbf{Aw} = \big(\det \mathbf{A}\big)\, \mathbf{u} \cdot \mathbf{v} \times \mathbf{w}\,.$$

For completeness we mention another invariant (independent of the trace and the determinant) that arises for three-dimensional linear maps. It is a type of *principal invariant* and is defined by

$$II_A = \tfrac{1}{2}\left((\mathrm{tr}\,\mathbf{A})^2 - \mathrm{tr}\left(\mathbf{A}^2\right)\right) = a_{11}a_{22} + a_{11}a_{33} + a_{22}a_{33} - a_{12}a_{21} - a_{13}a_{31} - a_{23}a_{32}.$$

## Invertibility

One can prove that a linear map $\mathbf{A}$ is invertible if and only if its determinant is non-zero. Under this hypothesis, the elements of the inverse matrix are built as follows. Let again $m_{ij}$ denote the determinant of the matrix obtained suppressing the $i$-th row and $j$-th column from the matrix of $\mathbf{A}$. Then:

$$\left(a^{-1}\right)_{ij} = \frac{(-1)^{i+j} m_{ji}}{\det \mathbf{A}}\,.$$

## Orthogonal Group

A linear map $\mathbf{Q}$ is called *orthogonal* if it preserves the dot product, in other words if

$$\mathbf{Qu} \cdot \mathbf{Qv} = \mathbf{u} \cdot \mathbf{v} \qquad \forall \mathbf{u}, \mathbf{v} \in \mathbb{R}^n. \tag{A.27}$$

In particular, an orthogonal transformation preserves the modulus of all vectors and the angles between vectors. When applied to an orthonormal basis, therefore, an orthogonal map produces a new orthonormal basis. It can be proved that the converse holds too: the transformation sending an orthonormal basis to an orthonormal basis is necessarily orthogonal.

The composite of two orthogonal mappings is orthogonal. In fact, if $\mathbf{Q}$ and $\mathbf{R}$ are orthogonal,

$$(\mathbf{RQ})\mathbf{u} \cdot (\mathbf{RQ})\mathbf{v} = \mathbf{R}(\mathbf{Qu}) \cdot \mathbf{R}(\mathbf{Qv}) = \mathbf{Qu} \cdot \mathbf{Qv} = \mathbf{u} \cdot \mathbf{v} \quad \forall \mathbf{u}, \mathbf{v} \in \mathbb{R}^n.$$

Equation (A.27) implies that orthogonal maps are characterized by the property

$$\mathbf{QQ}^T = \mathbf{Q}^T\mathbf{Q} = \mathbf{I} \qquad \Longleftrightarrow \qquad \mathbf{Q}^{-1} = \mathbf{Q}^T. \tag{A.28}$$

In particular, all orthogonal maps are invertible, and both their transposes and their inverses are orthogonal. We call *orthogonal group* the set of all orthogonal maps.

The determinant of an orthogonal transformation can only take the values $+1$ and $-1$. In fact, (A.28) implies

$$\det\big(\mathbf{Q}\mathbf{Q}^T\big) = \big(\det\mathbf{Q}\big)\,\det\big(\mathbf{Q}^T\big) = \big(\det\mathbf{Q}\big)^2 = \det\mathbf{I} = 1 \quad\Longrightarrow\quad \det\mathbf{Q} = \pm 1.$$

In particular, we call *rotations* the orthogonal maps with determinant $+1$.

### Changes of Bases

Let $\{\mathbf{e}_1, \ldots, \mathbf{e_n}\}$ and $\{\mathbf{f}_1, \ldots, \mathbf{f_n}\}$ be orthonormal bases, and $\mathbf{Q}$ the orthogonal transformation that maps one to the other:

$$\mathbf{e}_i = \mathbf{Q}\mathbf{f}_i\,, \quad \mathbf{f}_i = \mathbf{Q}^{-1}\mathbf{e}_i = \mathbf{Q}^T\mathbf{e}_i\,, \qquad \text{per } i = 1, \ldots, n.$$

Let respectively $\{u_i, i = 1, \ldots, n\}$ and $\{u'_i, i = 1, \ldots, n\}$ denote the components of a vector $\mathbf{u}$ in the two bases. These components are related by the following equation:

$$u'_i = \mathbf{f}_i \cdot \mathbf{u} = \mathbf{Q}^T\mathbf{e}_i \cdot \mathbf{u} = \mathbf{e}_i \cdot \mathbf{Q}\mathbf{u} = (Qu)_i = q_{ij}u_j.$$

Similarly, calling $\{a_{ij}, i, j = 1, \ldots, n\}$ and $\{a'_{ij}, i, j = 1, \ldots, n\}$ the elements of the matrix representing $\mathbf{A}$ in the two bases, we have:

$$a'_{ij} = \mathbf{f}_i \cdot \mathbf{A}\mathbf{f}_j = \mathbf{Q}^T\mathbf{e}_i \cdot \mathbf{A}\mathbf{Q}^T\mathbf{e}_j = \mathbf{e}_i \cdot \mathbf{Q}\mathbf{A}\mathbf{Q}^T\mathbf{e}_j = \big(QAQ^T\big)_{ij} = q_{ik}a_{kl}q_{jl}. \tag{A.29}$$

### Eigenvalues, Eigenvectors

A scalar $\lambda$ is called an eigenvalue of the linear map $\mathbf{A}$, and a vector $\mathbf{u} \neq \mathbf{0}$ is the eigenvector of $\mathbf{A}$ associated with $\lambda$, when

$$\mathbf{A}\mathbf{u} = \lambda\,\mathbf{u}\,.$$

One can prove that the eigenvalues are precisely the roots of the *characteristic polynomial*:

$$\det\big(\mathbf{A} - \lambda\,\mathbf{I}\big) = 0.$$

### Diagonalizability

A linear map is called *diagonalizable* if there exists an orthonormal basis $\{\mathbf{e}_1, \ldots, \mathbf{e_n}\}$ in which its matrix is diagonal, meaning

$$\mathbf{e}_i \cdot \mathbf{A}\mathbf{e}_j = 0 \quad \text{if } i \neq j \qquad \Longrightarrow \qquad a_{ij} = \delta_{ij}\lambda_i \,. \tag{A.30}$$

If a map is diagonalizable, the basis that diagonalizes it is made of its eigenvectors, and the diagonal elements are the eigenvalues. In fact, from (A.30) it follows that

$$\mathbf{A}\mathbf{e}_i = \lambda_i \, \mathbf{e}_i \,. \tag{A.31}$$

Using formula (A.29) for basis change, we note that if a map is diagonalizable it must be symmetric. In fact, in any basis obtained from the former under the orthogonal transformation $\mathbf{Q}$,

$$a'_{ij} = q_{ik} a_{kN} q_{jN} = q_{ik} \delta_{kN} \lambda_k q_{jN} = q_{ik} \lambda_k q_{jk} = a'_{ji} .$$

One proves that the symmetry is not only necessary, but also a sufficient condition for a linear map's diagonalizability: *the diagonalisable maps are precisely the symmetric ones*.

### Skew-Symmetric Transformations

Let $\mathbf{W} = -\mathbf{W}^T$. The image $\mathbf{W}\mathbf{v}$ of an arbitrary vector $\mathbf{v}$ is orthogonal to the vector itself. In fact

$$\mathbf{W}\mathbf{v} \cdot \mathbf{v} = \mathbf{v} \cdot \mathbf{W}^T \mathbf{v} = -\mathbf{v} \cdot \mathbf{W}\mathbf{v} \quad \Longrightarrow \quad \mathbf{W}\mathbf{v} \cdot \mathbf{v} = 0 \quad \forall \mathbf{v} \in \mathbb{R}^n .$$

Moreover $\det \mathbf{W} = 0$ if the dimension $n$ of the space on which $\mathbf{W}$ acts is odd, since

$$\det \mathbf{W} = \det \mathbf{W}^T = \det \big( -\mathbf{W} \big) = (-1)^n \det \mathbf{W} .$$

Consider at last the case $n = 3$ more carefully. By explicit computation it is possible to show that there exists a unique vector $\boldsymbol{\omega}$, called *axis* of $\mathbf{W}$, such that

$$\mathbf{W}\mathbf{u} = \boldsymbol{\omega} \times \mathbf{u} \qquad \forall \mathbf{u} \in \mathbb{R}^3 . \tag{A.32}$$

The components of $\boldsymbol{\omega}$ in any basis are related to the matrix elements of $\mathbf{W}$ as follows:

$$\omega_1 = w_{32} = -w_{23}, \qquad \omega_2 = w_{13} = -w_{31}, \qquad \omega_3 = w_{21} = -w_{12} .$$

Let us stress how the second and third relations above are easy consequences of the first under cyclic permutation of the subscripts $1 \to 2 \to 3 \to 1$.

## A.4 Simultaneous Diagonalization of Symmetric Matrices

We restate and prove the property used in Sect. 14.9.2 regarding the roots of the characteristic polynomial (14.59).

**Theorem A.2** (Simultaneous diagonalization) *Let* $\mathbf{A}$ *be a symmetric and positive-definite* $N \times N$ *matrix, i.e. such that* $\mathbf{A} = \mathbf{A}^T$ *and* $\mathbf{v} \cdot \mathbf{A}\mathbf{v} > 0$ *for any* $\mathbf{v} \neq 0$. *Let* $\mathbf{B}$ *be a symmetric* $N \times N$ *matrix. There exists a basis of vectors* $\{\mathbf{v}_1, \ldots, \mathbf{v}_N\}$, *orthonormal for the inner product*

$$(\mathbf{u}\,,\,\mathbf{v}) = \mathbf{u} \cdot \mathbf{A}\mathbf{v}\,,$$

*in which* $\mathbf{A}$ *is the identity matrix and* $\mathbf{B}$ *is diagonal. Furthermore, the roots of the characteristic polynomial*

$$\det\left(\mathbf{B} - \mu\,\mathbf{A}\right) = 0 \tag{A.33}$$

*are all real (and are the diagonal elements in the above representation of* $\mathbf{B}$*).*

***Proof*** Call $\mathbf{u} \cdot \mathbf{v}$ the standard dot product, and define the following inner product associated with $\mathbf{A}$:

$$(\mathbf{u}\,,\,\mathbf{v}) = \mathbf{u} \cdot \mathbf{A}\mathbf{v}. \tag{A.34}$$

It is easy to check that the bilinear operation in (A.34) enjoys all of the properties that define the dot product, due to the symmetry of $\mathbf{A}$ and it being positive definite.

The matrix $\mathbf{C} = \mathbf{A}^{-1}\mathbf{B}$ is symmetric for the dot product (A.34). In fact

$$(\mathbf{u}\,,\,\mathbf{C}\mathbf{v}) = \mathbf{u} \cdot \mathbf{A}\left(\mathbf{A}^{-1}\mathbf{B}\right)\mathbf{v} = \mathbf{B}\mathbf{u} \cdot \mathbf{v} = \mathbf{A}\left(\mathbf{A}^{-1}\mathbf{B}\right)\mathbf{u} \cdot \mathbf{v} = (\mathbf{C}\mathbf{u}\,,\,\mathbf{v}).$$

By the Spectral theorem there exists a basis $\{\mathbf{v}_1, \ldots, \mathbf{v}_N\}$ of eigenvectors of $\mathbf{C}$ that is orthonormal for (A.34):

$$\mathbf{C}\mathbf{v}_k = \mu_k \mathbf{v}_k \qquad \forall\, k = 1, \ldots, N. \tag{A.35}$$

Moreover, $(\mathbf{v}_j, \mathbf{v}_k) = \delta_{jk}$, where $\delta_{jk}$ is the Kronecker delta, and the eigenvalues $\{\mu_k\,,\, k = 1, \ldots, N\}$ are real.

In the previous basis $\mathbf{A}$ and $\mathbf{B}$ are represented by the identity matrix and the diagonal matrix made of the eigenvalues of $\mathbf{C}$. In fact, calling $a_{jk}$ and $b_{jk}$ the matrix elements, we have

$$\begin{aligned} a_{jk} &= \mathbf{v}_j \cdot \mathbf{A}\mathbf{v}_k = (\mathbf{v}_j\,,\,\mathbf{v}_k) = \delta_{jk}, \\ b_{jk} &= \mathbf{v}_j \cdot \mathbf{B}\mathbf{v}_k = \mathbf{v}_j \cdot \mathbf{A}(\mathbf{C}\mathbf{v}_k) = \mathbf{v}_j \cdot \mathbf{A}(\mu_k \mathbf{v}_k) = \mu_k\, \delta_{jk}. \end{aligned}$$

If we substitute the definition of $\mathbf{C}$ in (A.35) we immediately discover that its eigenvalues are precisely the roots of the characteristic polynomial (A.33):

$$\mathbf{C}\mathbf{v}_k = \mathbf{A}^{-1}\mathbf{B}\mathbf{v}_k = \mu_k \mathbf{v}_k \quad \Rightarrow \quad \mathbf{B}\mathbf{v}_k = \mu_k \, \mathbf{A}\mathbf{v}_k \quad \Rightarrow \quad \det(\mathbf{B} - \mu_k \, \mathbf{A}) = 0\,,$$

thus ending the proof. □

Now we move on to the proof of a result that was used in Sect. 14.9.2.

**Theorem A.3** (Signature of a symmetric matrix) *Let* $\mathbf{A}$ *be a symmetric, positive-definite* $N \times N$ *matrix, i.e. such that* $\mathbf{A} = \mathbf{A}^T$ *and* $\mathbf{v} \cdot \mathbf{A}\mathbf{v} > 0$ *for any* $\mathbf{v} \neq 0$*. Let* $\mathbf{B}$ *be a symmetric* $N \times N$ *matrix. The number of positive, zero and negative roots of the characteristic polynomial*

$$\det(\mathbf{B} - \mu \, \mathbf{A}) = 0$$

*coincides with the number of positive, zero and negative eigenvectors of* $\mathbf{B}$.

***Proof*** We introduce again the standard dot product $\mathbf{u} \cdot \mathbf{v}$ and the inner product $(\mathbf{u}\,,\mathbf{v}) = \mathbf{u} \cdot \mathbf{A}\mathbf{v}$ associated with $\mathbf{A}$.

Let $\{\mathbf{e}_1, \ldots, \mathbf{e}_N\}$ be an orthonormal basis of eigenvectors of $\mathbf{B}$, chosen so that the corresponding eigenvalues $\{\nu_1, \ldots, \nu_N\}$ are ordered decreasingly. In particular, the first $r$ eigenvectors will be positive, the next $s$ will be zero and the remaining $N - r - s$ negative. Let $V_+$, $V_0$ and $V_-$ be the subspaces spanned by the first $r$ eigenvectors, the $s$ null ones, and the negative ones, respectively.

The matrix $\mathbf{B}$ is positive definite on $V_+$, zero on $V_0$ and negative definite on $V_-$, irrespective of the inner product adopted to compute its sign. In fact, given

$$\mathbf{v} = \sum_{i=1}^{r} v_i \, \mathbf{e}_i \in V_+ \qquad (\mathbf{v} \neq 0)$$

we have

$$\begin{aligned}(\mathbf{B}\mathbf{v}\,,\mathbf{v}) = \mathbf{B}\mathbf{v} \cdot \mathbf{A}\mathbf{v} &= \sum_{i,j=1}^{r} v_i v_j \mathbf{B}\mathbf{e}_i \cdot \mathbf{A}\mathbf{e}_j = \sum_{i,j=1}^{r} \nu_i v_i v_j \mathbf{e}_i \cdot \mathbf{A}\mathbf{e}_j \\ &\geq \nu_r \sum_{i,j=1}^{r} v_i v_j \mathbf{e}_i \cdot \mathbf{A}\mathbf{e}_j = \nu_r \mathbf{v} \cdot \mathbf{A}\mathbf{v} > 0.\end{aligned}$$

In the above proof we used that $\nu_r$ is the smallest positive eigenvalue, and $\mathbf{A}$ is positive definite. Analogously one proves that $\mathbf{B}\mathbf{v} = \mathbf{0}$ if $\mathbf{v} \in V_0$, and finally $(\mathbf{B}\mathbf{v}\,,\mathbf{v}) < 0$ for $\mathbf{v} \in V_- \setminus \{\mathbf{0}\}$.

Changing inner product, then, does not alter the subspaces $V_+$, $V_0$ and $V_-$ on which $\mathbf{B}$ has a clear sign. Hence also the number of positive, zero, negative eigenvalues of $\mathbf{B}$ relative to the inner product's matrix $\mathbf{A}$ is invariant. These numbers, which equal the dimensions of the eigenspaces $V_+$, $V_0$, $V_-$, define the *signature* of the symmetric matrix $\mathbf{B}$. □

## A.5 Ordinary Differential Equations: Basics

The equations of motion are differential equations of second order. To simplify their solution in this section we will summarise some results about them, together with the techniques for solving some of the most common types of differential equations of mechanical interest.

**Definition A.4** A differential equation is a scalar equation whose unknown is a function of one variable $y(t)$ that involves one or several derivatives of $y(t)$. The order $n \in \mathbb{N}$ of a differential equation is the order of the highest derivative appearing in the equation.

The differential equations we shall deal with here (and in the book) are usually called *ordinary*. There also are *partial differential equations*, that involve unknown functions in several variables. As we will only concern ourselves with the former, we shall always suppress the word *ordinary*.

**Definition A.5** A *differential problem* is a differential equation together with a number of *initial conditions* or *boundary conditions*, namely constraints that fix the value of the unknown function or its derivatives at one or more points of its domain.

A differential problem may admit any number of solutions (finite or infinite). The basic result that ensures the existence and uniqueness of solutions to a differential problem is Cauchy's Theorem 8.1.

Let us move to the study of special types of equations and differential problems that can be solved. For each one we will provide the form of the possible solutions, without proof. Interested readers may find the demonstrations in Analysis textbooks.

### A.5.1 *Separable Differential Equations*

**Definition A.6** A differential equation of order one is called *separable* (or *with separable variables*) if it can be put in the form $\dot{y}(t) = a(t)\, b(y)$.

**Proposition A.7** *The differential problem*

$$\dot{y}(t) = a(t)\, b(y), \qquad y(t_0) = y_0$$

*admits a unique solution in an open interval containing* $t = t_0$ *if the function* $a$ *is continuous and* $b$ *is of class* $C^1$ *(that is, differentiable with continuous first derivative).*

*Moreover, if* $b(y_0) = 0$ *the solution is constant:* $y(t) \equiv y_0$ *for any* $t$. *If instead* $b(y_0) \neq 0$, *the solution satisfies*

$$\int_{y_0}^{y(t)} \frac{du}{b(u)} = \int_{t_0}^{t} a(\tau)\, d\tau.$$

Separable differential equations are of interest in Mechanics when we study systems with forces that depend on the velocity but not on the position. In such a situation the equation $F(v) = ma$ becomes separable in the unknown $v(t)$, as we show below.

**Example A.8** Let us determine the motion of a particle of mass $m$ constrained to move along the $x$-axis under the viscous force $F(v) = -\gamma(t)v$, with increasing viscosity coefficient $\gamma$: $\gamma(t) = \eta t$.

Let us solve the differential problem

$$\dot{v}(t) = -\eta t\, v, \qquad v(0) = v_0.$$

We have:

$$\begin{aligned} a(t) &= -\eta t &&\Longrightarrow \quad A(t) = -\int a(t)\,dt = -\frac{1}{2}\eta t^2 \\ b(v) &= v &&\Longrightarrow \quad B(v) = \int \frac{dv}{v} = \log|v|. \end{aligned}$$

Assuming $v_0 > 0$, the motion of the point satisfies $v(t) = v_0 \mathrm{e}^{-\eta t^2/2}$. □

### Differential Equations Integrable by Quadratures

A second-order differential equation of type

$$\ddot{y}(t) = f(y(t)) \tag{A.36}$$

can be reduced to a separable equation. It describes the motion of a system with one freedom degree, subject to a positional force $f(y)$. By multiplying (A.36) by $\dot{y}$ we can find the first integral:

$$\dot{y}\ddot{y} = \dot{y}\, f(y) \quad \Longrightarrow \quad \frac{d}{dt}\left[\frac{1}{2}\dot{y}^2 - U(y)\right] = 0 \quad \Longrightarrow \quad \frac{1}{2}\dot{y}^2 - U(y) = E, \tag{A.37}$$

where $U(y)$ is a primitive of $f(y)$ and $E$ is an integration constant that can be determined using the initial conditions. Equation (A.37) is now a separable differential equation of order one. We can in fact set

$$\dot{y} = \pm\sqrt{2(E + U(y))}\,,$$

where the sign of $\dot{y}$ is chosen in accordance to the sign of $\dot{y}(t_0)$. Taking for instance the plus sign we obtain

$$\int_{y_0}^{y(t)} \frac{du}{\sqrt{2(E + U(u))}} = t - t_0,$$

which gives the solution $y(t)$ in implicit form.

## A.5.2 Linear Differential Equations

**Definition A.9** A differential equation of order $n$ is called *linear* if it has the following form

$$a_n(t)y^{(n)}(t) + a_{n-1}(t)y^{(n-1)}(t) + \cdots + a_1(t)\dot{y}(t) + a_0(t)y(t) = f(t).$$

A linear differential equation is called *homogeneous* if the source term $f(t)$ is zero. It is said to have *constant coefficients* if the coefficients $\{a_k,\ k = 0, 1, \ldots, n\}$ do not depend on time.

**Proposition A.10** *The solutions to a homogeneous differential equation of order $n$ form a vector space of dimension $n$. In other words, they coincide with all linear combinations of $n$* fundamental solutions $\{y_1(t), \ldots, y_n(t)\}$.

*The solutions to a non-homogeneous linear differential equation of order $n$ can be written as sums of an arbitrary* particular solution $y_\mathrm{p}(t)$ *plus the solutions to the associated homogeneous equation (the one obtained putting the source $f(t)$ equal zero).*

Hence the problem of solving a linear differential equation can be split in two phases: computing $n$ fundamental solutions for the associated homogeneous equation, and finding a particular solution.

### Linear Differential Equations of Order One

Consider the differential problem

$$\dot{y}(t) + a(t)\, y(t) = f(t), \qquad y(t_0) = y_0. \tag{A.38}$$

The associated homogeneous equation is separable. A fundamental solution is then $y_1(t) = \mathrm{e}^{-A(t)}$, where $A(t)$ is a primitive of $a(t)$.

A particular solution to the equation in (A.38) is

$$y_\mathrm{p}(t) = F(t)\mathrm{e}^{-A(t)}, \qquad \text{with} \quad F(t) = \int f(t)\, \mathrm{e}^{A(t)} dt.$$

Eventually, the general solution is $y(t) = y_\mathrm{p}(t) + C_1 y_1(t)$, where $C_1$ is found by imposing the initial condition in (A.38).

**Example A.11** (*First-order linear differential equations with constant coefficients*) In the special case $a(t) \equiv a_0$ we have $A(t) = a_0 t$, and

$$y(t) = \big(F(t) + C_1\big)\, \mathrm{e}^{-a_0 t}, \qquad \text{with} \quad F(t) = \int f(t)\, \mathrm{e}^{a_0 t} dt. \qquad \square$$

## Second-Order Linear Differential Equations with Constant Coefficients

Consider now the differential equation

$$a\ddot{y}(t) + b\dot{y}(t) + cy(t) = f(t). \tag{A.39}$$

Equations of motion of this kind represent the motion of a point (of mass $a$ in (A.39)) under the following forces: a viscous force $F_r = -b\dot{y}$ of coefficient $b$, an elastic force $F_{el} = -cy$ with elastic constant $c$, and an external force $f(t)$, possibly depending on time.

The structure of the fundamental solutions depends on the roots $\lambda_1, \lambda_2$ of the associated characteristic equation

$$a\lambda^2 + b\lambda + c = 0. \tag{A.40}$$

Let us emphasize that in the previous example the coefficients $a, b, c$ are non-negative, which implies the roots' real parts in (A.40) will be non-positive.

- If $\lambda_1 \neq \lambda_2 \in \mathbb{R}$ (that is, $b^2 - 4ac > 0$) the fundamental solutions are

$$y_1(t) = \mathrm{e}^{\lambda_1 t} \qquad \text{and} \qquad y_2(t) = \mathrm{e}^{\lambda_2 t}.$$

- If $\lambda_1 = \lambda_2 = \lambda \in \mathbb{R}$ (that is, $b^2 - 4ac = 0$) the fundamental solutions are

$$y_1(t) = \mathrm{e}^{\lambda t} \qquad \text{and} \qquad y_2(t) = t\,\mathrm{e}^{\lambda t}.$$

- If $\lambda_{1,2} = \alpha \pm \mathrm{i}\beta \in \mathbb{C}$ (that is, $b^2 - 4ac < 0$) the fundamental solutions are

$$y_1(t) = \mathrm{e}^{\alpha t} \cos \beta t \qquad \text{and} \qquad y_2(t) = \mathrm{e}^{\alpha t} \sin \beta t.$$

From the above discussion of the sign we deduce that in the aforementioned applications the fundamental solutions will be bounded for any $t$. If furthermore the roots' real part is strictly negative (which happens in case $b > 0$), the fundamental solutions will decay to zero as time passes. For this reason these solutions are often called *transient*.

Seeking a particular solution can be more challenging, in general. We consider now a special forcing term $f(t)$, which subsumes a wide class of concrete applications. This is the case $f(t) = A\cos\omega t$, where $A$ is the forcing term's *amplitude*, and $\omega$ its *frequency*. The particular solution's structure depends on $\omega$ and on the roots of the characteristic polynomial (A.40). More precisely, we have to check whether one of the fundamental solutions is $\cos\omega t$.

- If $\lambda_{1,2}^2 \neq -\omega^2$ the fundamental solutions either do not oscillate, or they oscillate with a different frequency from the forcing frequency. Then the particular solution has the form

$$y_p(t) = \frac{A}{a(\omega^2 + \lambda_1^2)(\omega^2 + \lambda_2^2)} \left[ \left( \lambda_1 \lambda_2 - \omega^2 \right) \cos \omega t - (\lambda_1 + \lambda_2)\omega \sin \omega t \right]. \tag{A.41}$$

Especially interesting is the case where $\lambda_{1,2} = \pm\omega_o$: this is the situation where the system, if there were no forcing term, would oscillate with its own *eigenfrequency* $\omega_o$. Then (A.41) simplifies to

$$y_p(t) = \frac{A \cos \omega t}{a(\omega_o^2 - \omega^2)}. \tag{A.42}$$

The structure fo the particular solution shows how the amplitude of the oscillation cause by the forcing term grows without bounds as $\omega \to \omega_o$, that is, when the forcing frequency tends to the system's eigenfrequency.

- If $\lambda_{1,2}^2 = -\omega^2$, the particular solution (A.42) does not exist, since $\omega_o = \omega$ and its denominator is zero. In this case the particular solution becomes

$$y_p(t) = \frac{A}{4a\omega^2} \left[ \cos \omega t + 2\omega t \sin \omega t \right]. \tag{A.43}$$

Solution (A.43) grows without bound as time passes. This phenomenon is known as resonance and may induce devastating effects on mechanical systems (see Sect. 10.3.1).

## A.6 Differential Forms

We will recall very briefly a few properties of differential forms. Consider a general one-form

$$\Psi_1 (x_1, \ldots, x_n)\, dx_1 + \Psi_2 (x_1, \ldots, x_n)\, dx_2 + \cdots + \Psi_n (x_1, \ldots, x_n)\, dx_n. \tag{A.44}$$

Given a curve $\gamma(t) = \big(x_1(t), \ldots, x_n(t)\big)$, with $t \in [t_0, t_1]$, the integral of (A.44) along $\gamma$ is defined to be

$$\int_{t_1}^{t_2} \Big[ \Psi_1 (x_1(t), \ldots, x_n(t))\, \dot{x}_1 + \cdots + \Psi_n (x_1(t), \ldots, x_n(t))\, \dot{x}_n(t) \Big] dt.$$

A differential 1-form is said to be exact (or integrable) if it equals the differential of some function $f(x_1, \ldots, x_n)$:

$$df = \Psi_1 dx_1 + \cdots + \Psi_n dx_n.$$

A necessary condition for that to happen is that the coefficients $\{\Psi_i,\ i = 1, \ldots, n\}$ (assumed at least $C^1$) coincide with the partial derivatives of $f$:

$$\Psi_i = \frac{\partial f}{\partial x_i} \qquad \forall i = 1, \ldots, n. \tag{A.45}$$

Since equations (A.45) should hold, the Schwarz theorem furnishes the necessary *compatibility condition* for (A.44) to be an exact 1-form:

$$\frac{\partial \Psi_i}{\partial x_j} = \frac{\partial \Psi_j}{\partial x_i} \qquad \forall i \neq j = 1, \ldots, n. \tag{A.46}$$

Condition (A.46) becomes sufficient as well if the domain $\mathcal{D}$ is *simply connected.*

When a differential 1-form is exact the computation of its integrals is greatly simplified, because

$$\int_{t_1}^{t_2} \sum_{i=1}^{n} \Psi_i \, \dot{x}_i(t) \, dt = \int_{t_1}^{t_2} \sum_{i=1}^{n} \frac{\partial f}{\partial x_i} \dot{x}_i(t) \, dt = \int_{t_1}^{t_2} \frac{df}{dt} \, dt = f(t_2) - f(t_1). \tag{A.47}$$

# Bibliography

The bibliography accompanying a text on Rational Mechanics typically includes, first and foremost, references to appropriate workbooks. It may also list textbooks that the authors deem noteworthy for their comprehensiveness and potential to offer students a broader perspective on some or most of the topics covered.

Here, we take a different approach and limit our bibliography primarily to books and authors who contributed to the flourishing of Rational Mechanics in Italy, as this is the cultural environment to which we belong and in which our book was written. This section is therefore intended as a modest tribute to the Italian school of Mathematical Physics and its interpretation of Classical Mechanics.

There is broad consensus that Rational Mechanics, in its classical Italian form, reached a defining moment in 1923 with the publication of *Lezioni di Meccanica Razionale*[1] by Levi-Civita and Amaldi (1923). The international renown of Levi-Civita, celebrated for his significant contributions to modern Mathematical Physics, combined with Amaldi's conceptual clarity, made this three-volume work a fundamental reference for teachers of Rational Mechanics, a status it has retained to this day. This work was so much needed that a compendium, a shorter version of it, was quickly published in 1928 (Levi-Civita and Amaldi 1928), for the students' convenience. As stated in the preface, this second work gathers "as much of our Lectures on Rational Mechanics as roughly corresponds to the standard syllabus of an institutional course at a university or polytechnic."

In Italy, Rational Mechanics—a subject for a while also taught by Nobel laureate Enrico Fermi—has a rich tradition dating back to the late 19th century, historically intertwined with—and to some extent indebted to—the French school, particularly the work of Appell (1896). Thus, the *Lezioni di Meccanica Razionale* did not appear in a vacuum since, well before, Gian Antonio Maggi—renowned for his studies on nonholonomic systems—had authored a comprehensive series of volumes that can be regarded as an early and significant contribution to the field (Antonio Maggi 1896). Interestingly, the bibliography of Levi-Civita and Amaldi (1923) cites as

[1] *Lectures on Rational Mechanics.*



P. Biscari etal., *Rational Mechanics*, UNITEXT 177,
https://doi.org/10.1007/978-3-032-07462-1

forerunners also the works of Burali-Forti and Boggio (1921), Marcolongo (1905), and Burgatti (1919). Notably, some of these authors were associated with the so-called *Scuola Vettorialistica Italiana*,[2] which insisted on using vector calculus and, in many respects, anticipated the absolute and intrinsic notation later championed by Clifford Truesdell and his school.

The development of Rational Mechanics in Italy was significantly shaped by a major reorganization of higher education in 1923. As a result of this reform, the curricula of Engineering Schools adopted a structure consisting of a two-year preparatory phase within the Faculty of Science, followed by a three-year program dedicated to engineering training. This arrangement inevitably pushed Rational Mechanics to retain a stronger connection to practical applications, serving as a bridge to Applied Mechanics and other core engineering disciplines, rather than veering mostly into the domain of abstract mathematics.

Italian literature on Rational Mechanics remained rich and vibrant for many years after the Second World War. Numerous textbooks were published in this period, some continuing the legacy of the great pioneers and others introducing significant conceptual and methodological innovations.

We refrain from providing an exhaustive list of references and instead highlight four books that, for different reasons, have had a significant impact on the development of this discipline in Italy.

We recall the work of Signorini (1941), a direct student of Levi-Civita, which contains elements of notable originality, as well as that of Grioli (1951), whose contributions significantly shaped the teaching of Rational Mechanics, owing to the prominence of the author's personality and his long-standing influence within the Italian scientific community.

Given the historical importance of the Milan Engineering School, it is worth mentioning Bruno Finzi's *Meccanica Razionale*[3] (Finzi 1946), which served as the reference textbook for generations of students—from the post-war period through the 1990s and beyond and, like its predecessors, was deeply rooted in Levi-Civita s lectures. In Italy, it was not until the late 1970s and early 1980s that the first significant efforts to modernize textbooks on Rational Mechanics began to emerge. Perhaps the earliest of these was the 1976 textbook by Cercignani (1976), a world-renowned expert in kinetic theory and the Boltzmann equation. Cercignani s book adopts a modern approach and covers all classical subjects, from the fundamentals of mechanics to important applications, up to an introductory treatment of continuum mechanics, and several advanced topics typically found in texts on analytical mechanics.

[2] *Italian Vectorial School.*

[3] Rational Mechanics.

## References

Antonio Signorini. *Meccanica razionale con elementi di statica grafica.* D.U.S.A., Roma, 1941.

Bruno Finzi. *Meccanica Razionale – Volumi I e II.* Zanichelli, Bologna, 1946.

Carlo Cercignani. *Spazio, tempo, movimento: introduzione alla meccanica razionale.* Zanichelli, Bologna, 1976.

Cesare Burali-Forti and Tommaso Boggio. *Meccanica razionale.* S. Lattes & C., Torino-Genova, 1921.

Gian Antonio Maggi. *Principii della Teoria Matematica del Movimento dei Corpi: Corso di Meccanica Razionale.* Ulrico Hoepli, Milano, 1896.

Giuseppe Grioli. *Lezioni di Meccanica Razionale.* Commissionaria Libreria Universitaria, Padova, 1951.

Paul Appell. *Traité de mécanique rationnelle.* Gauthier-Villars, Paris, 1896.

Pietro Burgatti. *Lezioni di Meccanica Razionale.* Zanichelli, Bologna, 1919.

Roberto Marcolongo. *Meccanica razionale – Volumi I e II.* Ulrico Hoepli, Milano, 1905.

Tullio Levi-Civita and Ugo Amaldi. *Compendio di Meccanica Razionale.* Zanichelli, Bologna, 1928.

Tullio Levi-Civita and Ugo Amaldi. *Lezioni di meccanica razionale.* Zanichelli, Bologna, 1923. A new edition, featuring historical comments and a critical introduction by Italian scholars, was published in 2013 by Compomat, Configni (RI), Italy. ISBN 978-88-95706-33-7.

# Index




P. Biscari etal., *Rational Mechanics*, UNITEXT 177,
https://doi.org/10.1007/978-3-032-07462-1

GPSR Compliance

*The European Union's (EU) General Product Safety Regulation (GPSR) is a set of rules that requires consumer products to be safe and our obligations to ensure this.*

*If you have any concerns about our products, you can contact us on ProductSafety@springernature.com*

In case Publisher is established outside the EU, the EU authorized representative is:

Springer Nature Customer Service Center GmbH
Europaplatz 3
69115 Heidelberg, Germany

**Batch number: 10399426**

Printed by Printforce, the Netherlands